COMPUTATIONAL FLUID DYNAMICS

by

PATRICK J. ROACHE

Consultant
Albuquerque, New Mexico 87108

Revised Printing

hermosa publishers
P.O. Box 8172 Albuquerque, N.M. 87108

International Standard Book Number 0-913478-05-9
Library of Congress Catalog Card Number 72-89970

Additional copies may be obtained from the publishers at $17.50 each, plus 50¢ handling charge per order.

PREFACE

At the time of this writing, it is obvious that the general area of computer simulation of physical processes, and the particular area of computational fluid dynamics, is rapidly expanding. One need only glance through the titles in any of the scientific abstracting indexes to see a disproportionate number of doctoral dissertations in computational fluid dynamics. Everyone with a computer is computing.

Unfortunately, research progress has been hindered by a widely scattered literature source. Scarcely a month goes by without someone rediscovering upwind differencing or expounding with the intensity of a religious convert on the aesthetic satisfaction of the asymptotic unsteady-flow approach to steady-flow problems. But these ideas have been around for a long time. It is hoped that this book will contribute to starting next year's crop of researchers off from approximately the same point, so that research may proceed.

Several excellent texts, covering numerical solutions of partial differential equations, are available, notably those of Forsythe and Wasow (1960), Richtmyer (1957), Richtmyer and Morton (1967), Ames (1965, 1969), and Mitchell (1969).* The present book differs from these in the material covered and in the approach used.

As for the material covered, the reader is advised that this is not a mathematics book (see Forsythe, 1970; "Pitfalls in Computation, or Why a Math Book Isn't Enough"). The basic interior point-differencing methods are presented in a direct and hopefully intelligible manner. Also, the importance of numerical boundary conditions is discussed. This topic has heretofore received no attention in texts and little in research papers, although there is a growing awareness of its dominant importance. This book also treats the equally important and neglected topics of special finite-difference mesh systems, specialized forms of the differential equations, the problems of initial conditions and convergence criteria, plotting methods and other information-processing techniques, and even some specific suggestions on programming practices. In short, this text addresses the messy problems associated with actually obtaining numerical solutions to the fluid dynamics problems, rather than nice mathematical propositions associated with merely related problems.

As for the approach used, the reader is again advised that this is not a mathematics book. Even the mathematics books admit the necessity of physical intuition, heuristic reasoning, and numerical experimentation in this area, although they do not often use the intuitive, heuristic, or experimental approaches. Some of the mathematical research is certainly valuable, but our interest is the primary interest of the engineers, physicists, and chemists; i.e., our first interest is in the physical phenomena, and in the mathematics only as it relates to the physics. This difference in approach is not merely of subjective value, but often has led to entirely different problem formulations, especially in regard to boundary conditions. Generally, the physical simulation approach has been more successful.

In this regard, it is interesting to note that most practitioners of computational fluid dynamics are converted theoreticians who still regard themselves as such. The present author's background has been largely experimental. It is hoped that my biases will combine with those of previous authors to produce something new. For it is this author's contention that computational fluid dynamics is a separate discipline, distinct from experimental fluid dynamics and from theoretical fluid dynamics, with its own techniques, its own difficulties, and its own utility.

---

*A list of references, arranged alphabetically by the first author and by the year, will be found at the end of this book. For additional texts and a concise evaluation of each, see the literature compilation by Price (1966).

ACKNOWLEDGMENTS

When you work on a book, off and on, for four years, you acquire an impressive indebtedness. As all those authors claim, it really does become impossible to acknowledge everyone who has contributed to the book. But I must acknowledge several people in particular.

Most of the creative (and pleasant) part of this work was done at the University of Kentucky. My good friend, Dr. Charles Knapp, suggested this project to me in 1968, and in 1970 he arranged for my visit to the University. He and Ann arranged housing for my family and provided moral support and friendship to us during our stay. More than any other person except my wife, Dr. Knapp is responsible for the completion of this work.

My stay at Kentucky was made possible by the Chairman of the Mechanical Engineering Department, Dr. Roger Eichhorn. My sincere thanks go to him, to the students in my course, and to those many kind people who gave furniture, dishes, and friendship, particularly Dr. Cliff Cremers, Dr. John Lienhard, Dr. Shiva Singh, Dr. Frank Saggendorf, and Delores Black. My stay at Kentucky was also made possible by my management at Sandia Laboratories, who allowed my leave of absence and who have patiently encouraged the completion of this work; my thanks to Dr. Fred Blottner, Dr. Ken Touryan, and Alan Pope.

In 1967, when this subject was in its infancy, I had the rare good luck to be taught a course in it at Notre Dame by Dr. Steve Piacsek; he has remained a source of enlightenment and inspiration. Also, I would never have become involved in this fascinating area had it not been for the guidance of Dr. T.J. Mueller.

Professor William Oberkampf of the University of Texas at Austin taught a one-semester course from the first three chapters of the manuscript. He found an embarrassing number of manuscript errors and gave many valuable suggestions for improving the textual clarity. I only regret that the remainder of the book has not benefited from the thorough scrutiny of him and his students. Also, Dr. C.W. Hirt kindly reviewed those sections based on his work and other work from LASL.

Eva Marie Franks typed most of the difficult manuscript in both the rough-draft and final forms. Bettye Hollingsworth also provided some excellent typing. Rosemary Teasdale edited the manuscript, and Ruth Barth prepared the line drawings. The many people who generously provided figure material are usually acknowledged in the figure caption, but I should especially mention Dr. Francis Harlow, Dr. C.W. Hirt, and Dr. David Thoman.

Many friends (and critics) have provided suggestions for the corrections and revisions in this revised printing (March 1976). My sincere thanks go to all, with special mention of Prof. W.L. Oberkampf, Prof. A.J. Chorin, Dr. E.D. Martin, Dr. R.S. Hirsh, Dr. L. Bertram, Prof. R.A. Dalrymple, Dr. U. Schumann, Prof. J.F. Thompson, Prof. U. Ghia and and Prof. K. Ghia.

My wife, Catharine, and our children have been very good sports about this whole thing. Needless to say, it would have been impossible without their support.

I gratefully acknowledge those many friends, colleagues, and acquaintances who encouraged this undertaking. Their support meant a lot.

Finally, my sincere thanks go to Tommy Potter, who found Chapter VI.

P.J.R.

This book is dedicated to Mary and to her University, Notre Dame du Lac.

TABLE OF CONTENTS

(continued)

(continued)

CHAPTER I

INTRODUCTION

"Fluid mechanics is especially rich in nonlinearities," (Ames, 1965), as every student of the subject well knows. It is also "rich" in mixed hyperbolic and elliptic partial differential equations, in mathematical singularities of various orders, and in problems with boundary conditions at infinity. In previous years, fluid dynamics has provided a major impetus to the development of partial differential equation theory, complex variable theory, vector and tensor analysis, and nonlinear methods. It is not surprising that, today, fluid dynamics is benefiting from and greatly contributing to the current developments in finite-difference numerical analysis.

However, this book is not concerned with all branches of numerical analysis as applied to fluid dynamics problems. We do not treat the interesting two-point boundary-value problem of ordinary differential equations, which are so fundamental to the calculation of similarity solutions in boundary-layer theory, nor even the important method of characteristics. Rather, we are concerned with that new, and yet emerging, discipline which is perhaps best described by the expression "numerical simulation" of fluid dynamics. The label "computational fluid dynamics" is now coming into use, to vaguely distinguish it from the broader area of "numerical fluid dynamics."

I-A. The Realm of Computational Fluid Dynamics

In earlier days, fluid dynamics, like other physical sciences, was divided into theoretical and experimental branches. The question is, where does computational fluid dynamics stand in relation to these older branches? The answer is that it is separate from each, although it has aspects of both, and that it supplements rather than replaces them.

Computational fluid dynamics is certainly not pure theoretical analysis -- if anything, it is closer to the experimental branch. The existing mathematical theory for numerical solutions of nonlinear partial differential equations is still inadequate. There are no rigorous stability analyses, error estimates, or convergence proofs. Some progress has been made in the questions of existence and uniqueness, but not enough to provide unequivocal answers to specific problems of interest. In computational fluid dynamics, it is still necessary to rely heavily on rigorous mathematical analysis of simpler, linearized, more or less related problems, and on heuristic reasoning, physical intuition, wind tunnel experience, and trial-and-error procedures.

The applied mathematician Biot once made some remarks (Biot, 1956) about applied mathematics in general which seem especially applicable today to computational fluid dynamics. After quoting H. Bateman, who characterized the applied mathematician as a "mathematician without mathematical conscience," Biot went on to discuss the relationship between the applied mathematician and the rigorous science of mathematics. "One could understand the feelings of the artist who undoubtedly benefits from the scientific study of colors but who would be constantly reminded of proceeding with rigorous conformity to the dictates of physics and psychology." The newcomer to computational fluid dynamics is forewarned: In this field, there is at least as much artistry as science.

The numerical simulation of fluid dynamics is then closer to experimental than to theoretical fluid dynamics. The performance of each particular calculation on a computer closely resembles the performance of a physical experiment, in that the analyst "turns on" the equations and waits to see what happens, just as the physical experimenter does. Actual discovery of physical phenomena is possible; thus, Campbell and Mueller (1968) discovered the phenomenon of subsonic ramp-induced separation in a numerical experiment before they verified it in the wind tunnel. But the numerical experimenter has advantages. He has complete control over fluid properties such as density, viscosity, etc. His experimental probe does not disturb the flow. He can run a truly two-dimensional experiment, something virtually impossible in the laboratory. He has enormous flexibility in the choice of flow parameters, e.g., he can arbitrarily select an initial boundary-layer thickness and velocity profile, independent of the Reynolds number per foot and the Mach number, which would be impossible in a wind tunnel. Perhaps most importantly, the numerical experimenter can do what neither the theoretician nor the physical experimenter can do -- he can test the sensitivity of phenomena to independent theoretical approximations, such as constant viscosity coefficient, neglect of buoyancy forces, unit Prandtl number, the boundary-layer approximations, etc. (One is reminded about the old gag of the novice wind tunnel experimentalist who ordered a railroad tank car full of inviscid, nonconducting, ideal gas.) He can also test the adequacy of the basic constitutive equations, e.g., for a new non-Newtonian fluid model.

But in no sense can numerical experimentation ever replace physical experimentation or theoretical analysis. For one obvious reason, the constitutive continuum equations can never hope to be called exact. For another, the numerical experimenter does not work with the continuum equations. It does not matter that the analogy becomes exact in some limiting case of vanishing mesh size, since we never attain this limit. The act of discretization changes not only the quantitative accuracy, but often changes the qualitative behavior of the equations. Thus, certain forms of the discrete analogs will introduce a kind of viscous behavior, even though the numerical experimenter intended to be using inviscid equations. Another very important limitation is the inability of the numerical experiment to properly account for turbulence or, more generally, any physical phenomenon such as turbulence, slip lines, corner eddies, etc., which are on too small a scale to be accurately resolved in the finite-difference mesh, but which may influence the larger features of the flow. An example of this would be the effect of turbulence in a boundary layer on the separation point. Also, there are examples of physical flows which appear to be two-dimensional but in practice are not, e.g., the reattachment line following "planar" flow separation, and planar flow over a cavity. In such cases, the apparent advantage of true two-dimensionality of the numerical experiment could be deceptive.

Finally, it should be noted that the numerical experiment is as limited as the physical experiment, in that it only gives discrete data for a particular parametric combination. It does not provide any functional relationships, beyond those obtained by dimensional analysis of the governing equations. It is therefore no substitute for even the simplest of theories.

Computational fluid dynamics is, then, a separate discipline, distinct from and supplementing both experimental and theoretical fluid dynamics, with its own techniques, its own difficulties, and its own realm of utility, offering new perspectives in the study of physical processes.

I-B. Historical Outline of Computational Fluid Dynamics

In 1910, L. F. Richardson presented to the Royal Society a 50-page paper which must be considered the cornerstone of modern numerical analysis of partial differential equations. Sheppard had done some previous groundwork in finite-difference operators, but Richardson's contributions overshadow the previous work. He treated the iterative solution of Laplace's equation, the biharmonic equation, and others. He distinguished between steady-state problems "according as the integral can or cannot be stepped out from a part of the boundary," i.e., between hyperbolic and elliptic problems, in modern terminology. He carefully treated numerical boundary conditions, including those at a sharp corner and those at infinity. He obtained error estimates and gave an accurate method of extrapolating answers toward the zero grid space limit, and further suggested checking the accuracy of numerical methods with exact solutions of simple geometries such as a cylinder. Finally, he was the first to actually apply these methods to a large-scale practical problem, that of determining stresses in a masonry dam.*

In Richardson's iteration method for elliptic equations, each point in the mesh in turn at the n-th iteration is made to satisfy the finite-difference equation, which involves "old"

---

*Richardson also presented what, in modern vocabulary, must be called a "cost-effectiveness" study of the method, using human computers.

"So far I have paid piece rates for the operation (Laplacian) of about $\frac{n}{18}$ pence per coordinate point, n being the number of digits. The chief trouble to the computers has been the intermixture of plus and minus signs. As to the rate of working, one of the quickest boys averaged 2,000 operations (Laplacian) per week, for numbers of three digits, those done wrong being discounted." (Richardson, 1910, p. 325)

We may all be thankful that social conditions have changed since 1910. Many a computational fluid dynamicist would end up in the poorhouse if he were paid a certain fee per calculation, with "those done wrong being discounted."

values from the (n-1)-th iteration at neighboring points. In 1918, Liebmann showed how to greatly improve the convergence rate merely by using all "new" values as soon as they are available. In this "continuous substitution" scheme, each n-th iteration involves some neighboring point values at the old (n-1)-th iteration and some at the new n-th iteration. In each Liebmann iteration cycle, the most resistant errors are reduced as much as in two cycles of the Richardson procedure (Frankel, 1950).

This comparison exemplifies the tone of numerical analysis of partial differential equations. Seemingly minor modifications of finite-difference forms, iterative schemes, or boundary conditions can result in large improvements. The converse is also true -- plausible and apparently accurate techniques can result in numerical catastrophe. The classic historical example is Richardson's explicit method for the parabolic heat conduction equation. It involves accurate, centered finite-difference approximations for both space and time derivatives; it has been shown (O'Brien et al., 1950) to be unstable for all time steps.*

The emphasis in pre-electronic-computer days was on elliptic equations, or the so-called "jury problem." An early rigorous mathematical treatment of convergence and error bounds for iterative solutions of elliptic equations by Liebmann's method was given by Phillips and Wiener (1923). In 1928, the classic paper of Courant, Friedrichs, and Lewy appeared. The authors were primarily interested in using finite-difference formulations as a tool for pure mathematics (Lax, 1967). By first discretizing the continuum equations, then proving convergence of the discretized system to the continuum system, and finally using algebraic methods to establish the existence of finite-difference solutions, they proved existence and uniqueness theorems for continuum elliptic, parabolic, and hyperbolic systems.** Their work became the guide for practical finite difference solutions in later years.

The first numerical solution of the partial differential equations for a viscous fluid-dynamics problem was given by Thom in 1933. The most sophisticated version of what is still essentially Liebmann's method was presented in 1938 by Shortley and Weller. They developed block relaxation, the trial function method, error relaxation, methods for refining mesh spacings, and error extrapolation. They were also the first to precisely identify and analyze convergence rates.

Southwell (1946) developed a more efficient relaxation method for solving elliptic equations. In this residual relaxation*** method, each point of the mesh is not calculated in turn. Rather, the mesh is scanned for the larger "residuals", and new values are calculated at those points. (In the steady-state heat conduction equation, a "residual" is proportional to the rate of accumulation of energy in the finite-difference cell; hence, a steady-state solution is approached when all residuals approach zero.) An account of sophisticated variants of Southwell's residual relaxation method, including rules for over- and underrelaxation (in which the residual is not set exactly equal to zero), choosing the mesh point to be relaxed, and block relaxation, was given by Fox (1948).

Allen and Southwell (1955) applied Southwell's residual relaxation method by hand calculations in solving the viscous incompressible flow over a cylinder. This was an innovative work in numerical fluid dynamics, in several respects. The authors used a conformal transformation to represent the circular boundary in a regular rectangular mesh. They achieved a computationally stable solution at a Reynolds number of 1000, which is above the physically stable limit. In their hand calculations, they also gained "a distinct impression of instability" at a Reynolds number of 100, and linked this with a physical tendency toward

---

*The instability did not manifest itself in Richardson's sample calculations only because of the small number of time-step calculations performed.

**Three companion articles discussing the significance of the 1928 Courant-Friedrichs-Lewy paper, plus an English translation of the paper, appear in the March 1967 IBM Journal (Lax, 1967; Parter, 1967; Widlund, 1967).

***Older papers reserve the term "relaxation" only for Southwell's residual relaxation method. We use the description "residual relaxation" to distinguish clearly from iterative methods such as Liebmann's, which today are also known as relaxation methods.

instability, thus foreshadowing the modern philosophy of numerical simulation. Their work must also stand as a model of resourcefulness in obtaining research funding, being supported by a grant made to London's Imperial College in 1945 by the Worshipful Company of Clothworkers!

Southwell's method is not so readily adaptable to electronic computers. The human computer can scan a matrix for the largest residual much faster than he can perform the arithmetic. For the electronic computer, the gain of scanning speed over arithmetic speed is not nearly as large, and it becomes more efficient to simply relax every mesh point residual in turn to zero, which is identically Liebmann's method.

Electronic computers thus gave motivation to the further development of a Liebmann-type procedure which might utilize the advantage of the overrelaxation concept of Southwell's residual relaxation. In 1950, Frankel (and independently Young, 1954) presented a method which he called the "extrapolated Liebmann method," which has since come to be known as the "successive overrelaxation" (Young, 1954), or "optimum overrelaxation" method. Frankel also noted the analogy between iterative solutions of elliptic equations and the time-step solution of parabolic equations, which was to have important consequences.

The development of electronic computers also focused attention on parabolic problems, since time-history solutions now became feasible. In Richtmyer's first book (1957), which contributed greatly to the development of one-dimensional fluid dynamics, he presented more than 10 numerical schemes. The first of the implicit methods which, for multidimensional problems, require iterative solutions at each time step, was the Crank-Nicolson method published in 1947. This method is still one of the most popular, and is the basis for the most widely used non-similar boundary-layer calculation method (Blottner, 1970).

Of undeterminable origin is the idea of using an asymptotic time solution of the unsteady flow equations to obtain a steady-state solution. It is doubtful that anyone could have seriously considered it before the age of electronic computers.

Much of the pioneering work in computational fluid dynamics was done at the Los Alamos Scientific Laboratory. It was at Los Alamos that J. von Neumann, during World War II, developed his criterion for stability of parabolic finite-difference equations and presented a method of analyzing a linearized system. His brief open-literature description of the method (Charney et al., 1950) did not appear until 1950.* This important paper also presented the first large-scale meteorological calculation of the nonlinear vorticity equations. The authors explained the advantage of stability of the vorticity equations over the primitive equations and presented heuristic arguments for their treatment of the mathematically incomplete problem of upstream and downstream boundary conditions for a nonsteady problem.

In the mid-fifties, Peaceman and Rachford (1955) and Douglas and Rachford (1956) presented efficient methods for solving an implicit form of the parabolic equation which allowed for arbitrarily large time steps. Known as the "alternating direction-implicit" (or simply ADI) methods, they were also applied to the solution of elliptic problems, the transfer being based on the analogy of Frankel (1950) of the time step of a parabolic equation and the iteration number for an elliptic equation. ADI methods are probably the most popular methods used for incompressible flow problems using the vorticity transport equation.

In 1953, DuFort and Frankel presented their "leapfrog" method for parabolic equations, which, like ADI methods, allows for arbitrarily large time steps (in the absence of advection terms) and has the advantage of being fully explicit. This method was used by Harlow and Fromm (1963) in their well-known numerical solutions of the time-dependent vortex street.

Their *Scientific American* article (Harlow and Fromm, 1965) especially served to excite widespread interest among the U.S. scientific community in the potentialities of computational fluid dynamics. At almost the same time, a similar article by Macagno appeared in the French magazine *La Houille Blanche* (Macagno, 1965). In both these articles, the concept of numerical simulation, or computer experiment, was clearly stated for the first time. With the appearance of these two articles, we can date the emergence of computational fluid dynamics as a separate discipline.

All the time-dependent solutions mentioned so far had encountered an upper Reynolds-number limit to computational stability. (Fundamentally, the limit is on the Reynolds number of the

---

*A more complete elucidation was given in 1950 by O'Brien, Hyman, and Kaplan.

finite-difference cell spacing.) In 1966, Thoman and Szewczyk achieved apparently unlimited computational stability by the use of upwind differencing of advection terms and by careful attention to boundary conditions. Their calculations of flow over a cylinder extended to a Reynolds number of one million, and they were even able to "spin" the cylinder and obtain Magnus lift, with no computational instability developing. The agreement of their calculations with experimental values, in spite of the fact that the method has only first-order accuracy, forced a reappraisal of the importance of the formal truncation order in partial differential equations. In this regard, the work of Cheng (1968), which clarified the dominant effect of numerical boundary conditions, is of considerable importance.

The direct (noniterative) Fourier method for solving the elliptic Poisson equation had been known for some time (e.g., see Forsythe and Wasow, 1960) but had not been used in a fluid dynamics problem. Hockney (1965) used a related (but faster) method to solve large Poisson problems very efficiently. Since his paper, direct methods for the Poisson equation have been used more extensively.

The methods described so far have been applicable to subsonic incompressible fluid flow problems. Supersonic flow problems differ from the subsonic in several important aspects, most importantly in that shock waves (discontinuities in the solutions) may develop in supersonic flow.

The fundamental paper on the numerical treatment of hyperbolic equations was published by Courant, Friedrichs, and Lewy in 1928. The "characteristic" properties are discussed and the well-known method of characteristics is outlined. They also presented and explained the famous Courant-Friedrichs-Lewy necessary stability condition; that, in a numerical grid which does not follow the characteristic directions, the finite-difference domain of dependence must at least include the continuum domain of dependence. This CFL stability requirement (which, stated in modern vocabulary, simply requires that the "Courant number" be less than one) holds in both Lagrangian and in Eulerian systems.

Lagrangian methods which "follow the particles" were developed to a high sophistication by the Los Alamos group (Fromm, 1961). Eulerian methods are generally preferable for two-dimensional problems, but the problem of shock resolution is aggravated in the Eulerian system. If the mesh spacing is not smaller than the shock thickness, oscillations develop which destroy accuracy. These oscillations in a finite mesh are physically meaningful (Richtmyer, 1957). The ordered kinetic energy of the velocity defect across the shock is being converted to internal energy by random collisions of molecules, the computational molecules being the finite-difference cells.

The most common treatment of shock waves in an Eulerian mesh involves smearing out the shock over several cells by the introduction of an artificial dissipation, explicit or implicit, which does not destroy the solution away from the shock waves. In 1950, von Neumann and Richtmyer proposed their artificial dissipation scheme in which the "viscosity coefficient" is proportional to the square of the velocity gradient. Ludford, Polacheck, and Seeger (1953) simply used large values of physical viscosity in the viscous equations in a Lagrangian mesh, but unrealistically high values of viscosity are required in this method.

Instead of using explicit viscosities, the implicit dissipation of the finite differences may be enough to smear out the shock. This is used in the well-known Particle-in-Cell, or PIC, method of Evans, Harlow, and others at Los Alamos*, in Lax's method (Lax, 1954) and others.

The 1954 paper by Lax is far less important for the numerical scheme presented therein than for the form of the differential equations used -- the conservation form. Lax showed that, by recasting the usual equations which have velocities, density, and temperature as dependent variables, a set of equations which have momenta, density, and stagnation specific internal energy can be derived. This new set of equations shows the nature of the physical conservation laws involved and allows overall integral properties to be maintained identically in the finite-difference system. This system of equations is now in common use in calculations of shock development, regardless of the finite-difference methods used, since the exact planar shock speed will be produced by any stable method (see Longley, 1960, and Gary, 1964).

Shock smearing is also accomplished by implicit dissipation terms in several other methods. The Lax-Wendroff method (1960), or two-step versions of it such as Richtmyer's (1963), is now

*See Evans and Harlow (1957, 1958, 1959), Harlow et al., (1959), Evans et al., (1962), and Harlow (1963).

widely used. The PIC technique (see previous footnote) and the EIC (explosive-in-cell) variation by Mader (1964) attain smearing by the use of a finite number of computational particles. This concept also permits delineation of fluid interfaces (Harlow and Welch, 1965 and 1966; Daly, 1967). The PIC method, like the earlier Courant-Isaacson-Rees method (1952), uses one-sided differences for the first spatial derivatives, thus introducing a kind of numerical viscosity (see Chapter III); but these methods do retain the true characteristic sense of the differential equations. Although all these methods contain implicit shock-smearing dissipation terms, each does require additional explicit artificial dissipation terms to achieve stability for particular conditions.

As an alternate to smearing the shock over several computational cells, one may actually maintain the discontinuity. Moretti and Abbett (1966B) and Moretti and Bleich (1967) calculated the inviscid supersonic flows using shock fitting, and this approach has become very popular in the early 1970's.

For an excellent broad treatment of nonlinear numerical methods, the recent books by Ames (1965, 1969) are recommended. The books by Richtmyer (1957) and Richtmyer and Morton (1967) are recommended for the mathematical aspects of numerical treatment of parabolic and hyperbolic systems, including shock-wave problems and neutron diffusion. The rigorous mathematical text of Forsythe and Wasow (1960) is recommended for elliptic equations. A forthcoming Academic Press book by Moretti will give details of the shock-fitting method.

I-C. Existence and Uniqueness of Solutions

The mathematical problems of existence and uniqueness of solutions to the partial differential equations (PDE) of fluid flow are far from settled, for both the continuum equations (PDE) and for their analogous finite-difference equations (FDE). Ladyzhenskaya (1963) has devoted her monograph to these problems for steady viscous incompressible flow. A readable précis of her work is given by Ames (1965). Based on comparison of the incompressible Navier-Stokes equations with other problems, Ames (page 480) conjectures that a unique steady solution exists only below an undefined Reynolds-number limit, that several solutions exist for a range of Re above this limit, and that no solutions exist above a second undefined Reynolds number. (However, he also rightly questions the validity of the steady-state Navier-Stokes equations themselves above a certain Re, since turbulence then occurs.) Such problems may be aggravated in the finite-difference analog when boundary conditions are vague.

For the compressible-flow problem with completely hyperbolic equations (supersonic flow), the existence for the inviscid limit is easily shown. Foy (1964) has also shown that a continuous viscous solution exists between any two states which can be connected by a sufficiently weak shock. For more general states and for mixed-flow problems, it appears that nothing really helpful is known.*

Existence of solutions is somewhat less a problem if we compute the unsteady equations, the approach which has proved most generally successful for the complete viscous flow equations. Since we have some confidence in the time-dependent Navier-Stokes equations, we are inclined to believe that a numerical solution which proceeds from a physically reasonable initial condition has validity. If no steady-state solution exists, we may detect that fact with the time-dependent finite-difference simulation. It does occur, however, that a continuum flow which is unstable to small disturbances will remain stable in a computational simulation. This can arise in both large-scale instability such as vortex shedding, and in small-scale shear-layer turbulence. Further, any approximation made to the complete Navier-Stokes equations (e.g., a Bousinesq linearization) also degrades confidence in the existence of solutions. This is especially true if one is experimenting with unproved constitutive equations, or equations of state. Godunov and Semendyayev (1962) say that the numerical solution to gas-dynamics problems can be made nonunique by a certain class of equations of state.

The question of uniqueness of an attained numerical solution is even more worrisome, simply because there are many examples, physical and purely mathematical, of nonuniqueness of steady-state solutions. The most obvious examples of physically nonunique flow situations are provided by the bistable fluid flip-flop devices, and the bistable vortex filament orientation caused by flow over a spherical cavity (Snedeker and Donaldson, 1966). These involve a

*Current research papers which treat these questions may be found in the journal, *Archive for Rational Mechanics and Analysis*, edited by C. Truesdell and J. Serrin, and published by Springer-Verlag.

kind of mirror symmetry choice. A more significant nonunique flow is exemplified by the stall hysterisis of an airfoil -- with identical boundary conditions, the airfoil exhibits completely different flow patterns at angles of attack near stall, depending on whether that angle of attack was approached from the low (unstalled) side or the high (stalled) side. McGlaughlin and Greber(1967) give other examples where flow separation is nonunique. Piacsek (1968) has calculated nonunique patterns of steady-state natural convection vortex cells, which probably have physical counterparts. The present author has calculated examples of what resembles supersonic diffuser stall (see Chapter III, Section C-8). This is a particular example of the computational nonuniqueness arising from numerical treatment of boundary conditions, although the physical counterpart does exist.

Ames (1965) gives an example of a quasi-linear elliptic differential equation which does not possess uniqueness. Another simple mathematical example of nonuniqueness is provided by classical oblique shock-wave theory. For supersonic inviscid flow over a wedge, three solutions to Thompson's cubic equation (Anon., 1953) exist. One solution causes a decrease in entropy and is discarded.* Of the remaining two, the "weaker" solution is known to correspond to physical flow over a wedge, whereas the "stronger" solution applies to detached shock problems.

In all these examples, the pertinent question is this: Towards *which* solution, if any, would a numerical scheme converge? No answer is available. We must depend on physical experience, i.e., experiment and intuition, to check the reasonableness of the solutions. More rigorous criteria must await better mathematical groundwork, which appears far in the future.

I-D. Preliminary Remarks on Consistency, Convergence, and Stability of Solutions

The mathematical foundations for the questions of convergence and stability of numerical schemes are well-developed only for linear systems. The results from linear theory are used as guidelines to nonlinear problems, the justification depending on numerical experiments.

The essence of a finite-difference numerical scheme is the replacement of derivatives such as $\partial f/\partial x$ in the differential equations by ratios of finite differences such as $\Delta f/\Delta x$ in the finite-difference equation. A *convergent* finite-difference scheme is defined mathematically as one in which all values of the finite-difference solution approach the continuum differential-equation solution as the finite-difference mesh size approaches zero.

This concept is more subtle than it appears at first glance. It is not simply a restatement of Newton's limit definition of a derivative; the limit behavior here is the limit of the whole solution of the differential equation, not merely the individual terms (differentials) of the equation. This latter property is termed the "consistency" condition (Lax and Richtmyer, 1956). For example, a finite-difference analog to a differential equation may have *consistent* finite differences, but be unstable and therefore *not convergent*. Further, the problem of practical convergence *criteria* has been neglected.

Stability has been defined by O'Brien, Hyman, and Kaplan (1950) and by Eddy (1949) in terms of the growth or decay of roundoff errors. Lax and Richtmyer (1956) define stability more generally, by requiring a bounded extent to which any component of the initial data can be amplified in the numerical procedure.

The equivalence theorem of Lax is of fundamental importance. It states that, for a consistent finite-difference scheme, stability is a necessary and sufficient condition for convergence, for a linear system of equations.

Von Neumann's stability criterion (Charney et al., 1950; O'Brien et al., 1950) is that the largest eigenvalue of the amplification matrix of the iteration scheme be less than unity minus terms of the order of the truncation error. Lax and Richtmyer (1956) demonstrate that this is sufficient for stability for a linear system with constant coefficients and that, if the amplification matrix satisfies any of three sets of properties, it is also sufficient for convergence. These and other stability questions are discussed in Chapter III, Section A, and in Richtmyer and Morton (1967).

At this stage, we point out that neither the linearized analysis nor even the stability definitions are completely satisfactory. Phillips (1959) has given an example of what he

---

*Discarding this solution of a rarefaction shock can be automatically satisfied in a finite-difference scheme (see Lax, 1954).

termed nonlinear stability; it occurs because of nonconstant coefficients (Lilly, 1965). Thommen (1966) has shown that the two-step Lax-Wendroff, or Lax-Wendroff-Richtmyer (Richtmyer, 1963) method develops a nonlinear instability near a stagnation point. These examples show that linearized analysis and constant-coefficient analysis can fail to predict instabilities. Even more fundamentally, the definition of stability appears inadequate. Lilly (1965) shows that the midpoint leapfrog rule applied to a model equation results in oscillations, not bounding the correct solution. The equation used corresponds to the infinite Reynolds-number limit of the incompressible vorticity equation. The present author has shown that the oscillation also persists at decreased amplitude, even at low Reynolds number, when a steady-state solution is approached. This oscillation is something we would like to call a numerical "instability," yet the results are "stable" by the commonly used definitions of error growth or boundedness. Also, since the results do not oscillate about the correct solution, we cannot say with certainty that the correct solution will be approached as Δx, $\Delta t \to 0$. Yet we do know that, as Re is decreased, the correct solution is approached, so that, at a low but nonzero Re, we may get "close enough" to the solution for practical purposes. Thus, the results may be nonconvergent in the mathematical sense, but convergent in a practical sense.

Further, none of the analyses to date have accounted for the effect of such mathematically untidy boundary conditions as the various schemes used at downstream boundaries. Eddy (1949) and a few more recent authors have considered the stability effect of gradient boundary conditions. The destabilizing effect of boundary treatments is very frequently of first importance.

It is clear, then, that elegant mathematical analyses and definitions of stability for numerical schemes should not be treated as an end in themselves, but only as rational supports and guidelines for the numerical experimentation. It is this approach that will be emphasized in this book.

CHAPTER II

INCOMPRESSIBLE FLOW EQUATIONS IN RECTANGULAR COORDINATES

In this chapter, the equations used in solving two-dimensional incompressible flow problems in rectangular coordinates will be presented. Following the statement of the "primitive" equations, the vorticity-stream function equations will be derived. The "conservation" form of the vorticity transport equation will then be presented, although its significance will not become clear until the next chapter, and different normalizing systems will be discussed. Finally, two one-dimensional model equations for vorticity transport, the Burgers equation and the linearized one-dimensional advection-diffusion equation, will be presented.

II-A. Primitive Equations

The fundamental equations for two-dimensional incompressible flow of a Newtonian fluid with no body forces and constant properties are the two momentum equations (Navier-Stokes) and the continuity equation (see, e.g., Lamb, 1945, or Schlichting, 1968).

$$\frac{\partial \bar{u}}{\partial \bar{t}} + \bar{u}\frac{\partial \bar{u}}{\partial \bar{x}} + \bar{v}\frac{\partial \bar{u}}{\partial \bar{y}} = -\frac{1}{\bar{\rho}}\frac{\partial \bar{P}}{\partial \bar{x}} + \bar{\nu}\left(\frac{\partial^2 \bar{u}}{\partial \bar{x}^2} + \frac{\partial^2 \bar{u}}{\partial \bar{y}^2}\right) \qquad (2\text{-}1)$$

$$\frac{\partial \bar{v}}{\partial \bar{t}} + \bar{u}\frac{\partial \bar{v}}{\partial \bar{x}} + \bar{v}\frac{\partial \bar{v}}{\partial \bar{y}} = -\frac{1}{\bar{\rho}}\frac{\partial \bar{P}}{\partial \bar{y}} + \bar{\nu}\left(\frac{\partial^2 \bar{v}}{\partial \bar{x}^2} + \frac{\partial^2 \bar{v}}{\partial \bar{y}^2}\right) \qquad (2\text{-}2)$$

$$\frac{\partial \bar{u}}{\partial \bar{x}} + \frac{\partial \bar{v}}{\partial \bar{y}} = 0 \qquad (2\text{-}3)$$

The overbars represent dimensional quantities. These equations are written in the primitive variables of velocity components ,u and v, and pressure, P, and the fluid properties of mass density, ρ, and kinematic viscosity, ν. They are based on the following physical laws: equations (2-1) and (2-2) are the vector components of Newton's second law of motion $\vec{F} = m\vec{a}$, with viscous forces related to rate of strain through a linear Newtonian shear stress law; equation (2-3) is a statement of conservation mass. The equations are written in an Eulerian frame of reference, i.e., a space-fixed reference through which the fluid flows. (The alternate Lagrangian description, in which the reference frame moves with the fluid, will not be used in this book, although some references will be given in Chapter VI.) Although it is possible to obtain numerical solutions from these equations (see Section III-G), most successful numerical solutions have utilized the vorticity-stream function approach. The advantages and disadvantages of the vorticity-stream function approach will be discussed in a later section; for now, we note the pedagogical advantage that only *one* transport equation need be treated.

II-B. Stream Function and Vorticity Transport Equations for Planar Flows

The pressure is eliminated from equations (2-1) and (2-2) by cross-differentiating equation (2-1) with respect to y and equation (2-2) with respect to x. Defining the vorticity as*

$$\bar{\zeta} = \frac{\partial \bar{u}}{\partial \bar{y}} - \frac{\partial \bar{v}}{\partial \bar{x}} \qquad (2\text{-}4)$$

one obtains the parabolic vorticity transport equation

$$\frac{\partial \bar{\zeta}}{\partial \bar{t}} = -\bar{u}\frac{\partial \bar{\zeta}}{\partial \bar{x}} - \bar{v}\frac{\partial \bar{\zeta}}{\partial \bar{y}} + \bar{\nu}\left(\frac{\partial^2 \bar{\zeta}}{\partial \bar{x}^2} + \frac{\partial^2 \bar{\zeta}}{\partial \bar{y}^2}\right) \equiv -\vec{\bar{V}}\cdot(\bar{\nabla}\bar{\zeta}) + \bar{\nu}\bar{\nabla}^2\bar{\zeta} \qquad (2\text{-}5)$$

---

*In three dimensions, vorticity is customarily defined as $\nabla \times \vec{V}$, which, when reduced to two dimensions, gives the negative of the present definition.

or, using the substantive derivative form,

$$\frac{D\bar{\zeta}}{D\bar{t}} = \bar{\nu}\bar{\nabla}^2\bar{\zeta} \qquad (2\text{-}6)$$

Defining the stream function $\bar{\psi}$ by

$$\frac{\partial\bar{\psi}}{\partial\bar{y}} = \bar{u} \quad , \quad \frac{\partial\bar{\psi}}{\partial\bar{x}} = -\bar{v} \qquad (2\text{-}7)$$

equation (2-4) is recast as an elliptic Poisson equation,

$$\bar{\nabla}^2\bar{\psi} = \bar{\zeta} \qquad (2\text{-}8)$$

The vorticity transport equation (2-5) consists of the unsteady term $\frac{\partial\bar{\zeta}}{\partial\bar{t}}$, the advective (or convective*) terms $\bar{u}\frac{\partial\bar{\zeta}}{\partial\bar{x}}$ and $\bar{v}\frac{\partial\bar{\zeta}}{\partial\bar{y}}$, and the viscous diffusion term, $\bar{\nu}\bar{\nabla}^2\bar{\zeta}$. The equation is nonlinear in the advective terms, since $\bar{u}$ and $\bar{v}$ are functions of the dependent variable $\bar{\zeta}$ via equations (2-7) and (2-8). The equation is parabolic in time, i.e., it poses a "marching" or initial-value problem, wherein the solution is stepped out from some initial condition. The stream-function equation (2-8) is elliptic, i.e., it poses a "jury" or boundary-value problem, which usually has been solved by iterative methods. In many problems of practical concern, we are not interested in the time-dependent behavior, but only in the "steady state" solution. In such a case, we could set $\frac{\partial\bar{\zeta}}{\partial\bar{t}} = 0$ on the left-hand side of (2-5) and thus eliminate one independent variable, time. Certainly this approach is universally used in analytical studies. It therefore usually comes as a surprise to those unexposed to computational fluid dynamics to find that most (though not all) of the successful numerical studies of even steady-state fluid dynamics problems are based on the time-dependent equations, the steady-state solution being obtained, if it exists, as the asymptotic time limit of the unsteady equations.

It is also noteworthy that the vorticity transport equation (2-5) serves as a model for many other transfer processes, and that the techniques presented in the next chapter are often applicable to a wide variety of transfer processes, including the compressible flow equations to be covered in Chapter IV.** The compressible flow equations are much more complicated than the vorticity equation, but the relation is close enough so that a study of the simpler vorticity equation is unquestionably beneficial to the study of the compressible equations.

Mathematics texts usually are content with classifying (linear) PDE's in the categories of parabolic, elliptic, or hyperbolic. This classification does not distinguish between the parabolic vorticity transport equation (2-5) and the diffusion equation $\partial\zeta/\partial t = \alpha\partial^2\zeta/\partial x^2$. As we shall see, the presence of the first-order derivative (advection term) in equation (2-5) makes it qualitatively different from the diffusion equation, and the numerical treatment of the advection term is important. Unfortunately, the best numerical method may be different for the two terms.

---

*The terms are practically synonomous, "convective" implying that the vorticity is carried along with the flow, and "advective" implying that vorticity is carried forward by the flow. The former is more popular with engineers, the latter with meteorologists, who reserve the term "convection" for vertical motions of the atmosphere.

**For an exceptionally coherent development of general transfer processes, see Fulford and Pei (1969). The generality of these concepts is well illustrated by the fact that compressible flow equations can be used to model freeway traffic problems.

II-C. Conservation Form

The continuity equation (2-3) is

$$\frac{\partial \bar{u}}{\partial \bar{x}} + \frac{\partial \bar{v}}{\partial \bar{y}} = 0$$

or, in terms of the total velocity vector $\vec{\bar{V}}$,

$$\bar{\nabla} \cdot \vec{\bar{V}} = 0 \qquad (2\text{-}9)$$

Consider $\bar{\nabla} \cdot (\vec{\bar{V}}\bar{\zeta})$. By vector algebra,

$$\bar{\nabla} \cdot (\vec{\bar{V}}\bar{\zeta}) = \vec{\bar{V}} \cdot (\bar{\nabla}\bar{\zeta}) + \bar{\zeta}(\bar{\nabla} \cdot \vec{\bar{V}}) = \vec{\bar{V}} \cdot (\bar{\nabla}\bar{\zeta})$$

Thus, in equation (2-5), we may replace $\vec{\bar{V}} \cdot (\bar{\nabla}\bar{\zeta})$ with $\nabla \cdot (\vec{\bar{V}}\bar{\zeta})$ to obtain the conservation form of the vorticity transport equation.

$$\frac{\partial \bar{\zeta}}{\partial \bar{t}} = - \bar{\nabla} \cdot (\vec{\bar{V}}\bar{\zeta}) + \bar{\nu}\bar{\nabla}^2\bar{\zeta} = - \frac{\partial (\bar{u}\bar{\zeta})}{\partial \bar{x}} - \frac{\partial (\bar{v}\bar{\zeta})}{\partial \bar{y}} + \bar{\nu}\left(\frac{\partial^2 \bar{\zeta}}{\partial \bar{x}^2} + \frac{\partial^2 \bar{\zeta}}{\partial \bar{y}^2}\right) \qquad (2\text{-}10)$$

The meaning and beneficial effects of using this "conservation" or "divergence" form will be discussed in Section III-A-3.

II-D. Normalizing Systems

The normalizing system used throughout this book is based on the advective time scale $\bar{L}/\bar{U}_0$, where $\bar{L}$ is a characteristic length and $\bar{U}_0$ is a characteristic velocity of the problem; for example, if $\bar{L}$ = chord length of an airfoil and $\bar{U}_0$ = free-stream speed, then $\bar{L}/\bar{U}_0$ is the time that a free-stream particle takes to pass over the airfoil. We define dimensionless quantities by

$$u = \bar{u}/\bar{U}_0 \quad , \quad v = \bar{v}/\bar{U}_0$$

$$x = \bar{x}/\bar{L} \quad , \quad y = \bar{y}/\bar{L} \qquad (2\text{-}11)$$

$$\zeta = \bar{\zeta}/(\bar{U}_0/\bar{L})$$

$$t = \bar{t}/(\bar{L}/\bar{U}_0)$$

Then equations (2-10) and (2-8) become

$$\frac{\partial \zeta}{\partial t} = - \nabla \cdot (\vec{V}\zeta) + \frac{1}{Re} \nabla^2 \zeta \qquad (2\text{-}12)$$

$$\nabla^2 \psi = \zeta \qquad (2\text{-}13)$$

where Re is the dimensionless Reynolds number,

$$Re = \bar{U}_0\bar{L}/\bar{\nu} \qquad (2\text{-}14)$$

Thus, for any one set of boundary conditions, the flow is characterized by a single dimensionless parameter, the Reynolds number.

For high Reynolds-number flows (Re >> 1), the advective term in equation (2-12) dominates the viscous diffusion term, and $(\bar{L}/\bar{U}_0)$ is the time interval which actually characterizes the flow. Thus a dimensionless time $t = \bar{t}/(\bar{L}/\bar{U}_0) >> 1$ would be a relevant criterion for the attainment of steady state, for example. But, for low Reynolds-number flows (Re << 1), a diffusion time constant better characterizes the flow. Defining a dimensionless time by

$$t' = \bar{t}(\bar{\nu}/\bar{L}^2) \tag{2-15}$$

and other dimensionless quantities as in equation (2-11), we obtain the same Poisson equation (2-13) for the stream function, and the vorticity transport equation becomes

$$\frac{\partial\zeta}{\partial t'} = - \text{Re}\ \nabla\cdot(\vec{V}\zeta) + \nabla^2\zeta \tag{2-16}$$

The term $(\bar{L}^2/\bar{\nu})$ obviously has dimensions, or "units" of time. Its physical significance as a time scale in diffusion-dominated problems can be appreciated by noting that equation (2-12) becomes singular in the limit Re → 0, but equation (2-16) is well behaved, with the advection terms dropping out. Likewise, equation (2-12) is well behaved in the limit of Re → ∞ , with the diffusion term dropping out.* For flows with Re >> 1 though finite, or Re << 1 though nonzero, the use of the appropriate time constant will also minimize roundoff errors, which may be important on some computers. The use of equation (2-12), based on the advective time scale, also has the advantage that the particle position vector (or Lagrangian coordinate) $\vec{\bar{r}} = \vec{\bar{r}}_0 + \int \bar{\vec{V}}\, d\bar{t}$ retains the same form in the normalized equations, $\vec{r} = \vec{r}_0 + \int \vec{V}\, dt$.

Other time scales may be appropriate for other flow equations. For example, W. P. Crowley (1968) gives four characteristic times associated with diffusion, advection, mean vorticity gradients, and buoyancy in a problem of natural convection flow stability. However, equation (2-12), based on the advective time scale, will suffice for our purposes.

Exercise: Normalize the primitive momentum equations (2-1) and (2-2), with reference to the advective time constant, velocity, and distance references as in equation (2-11). Refer the pressure to twice the "dynamic pressure," i.e., define $P = \bar{P}/\bar{\rho}\bar{U}_0^2$.

Exercise: Write the primitive momentum equation (2-1) in conservation form. That is, recast the advective terms $\bar{V}\cdot\bar{\nabla}\bar{u}$ into a divergence form, $\vec{\nabla}\cdot(\ \ )$.

II-E. One-Dimensional Model Transport Equations

The vorticity transport equation, in either its non-conservation form or its conservation form (2-12), is parabolic in time, contains two independent space coordinates, and is coupled to the elliptic Poisson stream-function equation (2-13) through the nonlinear advective terms. No analysis of the stability properties of the analogous finite-difference equations has ever been performed which accounts for all these features. But many aspects of the behavior of this equation can be studied, and the essential features of many finite-difference schemes can be illustrated, using either of two decoupled one-dimensional model transport equations.

The first model transport equation is the linearized one-dimensional advection-diffusion equation (Allen, 1968; W. P. Crowley, 1968A) written in either a conservation form,

$$\frac{\partial\zeta}{\partial t} = - \frac{\partial(u\zeta)}{\partial x} + \alpha\frac{\partial^2\zeta}{\partial x^2} \tag{2-17}$$

---

*This limit will not give correct numerical solutions to inviscid or potential flow problems unless the boundary conditions are also appropriate to inviscid flow.

or a non-conservation form,

$$\frac{\partial \zeta}{\partial t} = - u \frac{\partial \zeta}{\partial x} + \alpha \frac{\partial^2 \zeta}{\partial x^2} \tag{2-18}$$

In these equations, ζ now models vorticity or any other advected and diffused flow property,* α is a generalized diffusion coefficient corresponding to 1/Re in the vorticity transport equation, and u is a linearized advection speed. Unless otherwise specified, u is a constant in x, although equation (2-17) may also be used to study the stability effects of spatially varying u(x).

The second model transport equation is the Burgers equation (Burgers, 1948; Hopf, 1950; Rodin, 1970).

$$\frac{\partial u}{\partial t} = - u \frac{\partial u}{\partial x} + \alpha \frac{\partial^2 u}{\partial x^2} \tag{2-19}$$

wherein u is regarded as a generalized velocity. This equation retains the nonlinear aspect of the vorticity transport equation and the Navier-Stokes equations. Because of this nonlinearity, it has been used as a prototype equation for both turbulence and for shock-wave studies (see Section IV-G), as well as for finite-difference studies (Richtmyer, 1963; Allen and Cheng, 1970; B. K. Crowley, 1967; Freudiger et al., 1967; Greenspan, 1967; Allen, 1968; Cheng, 1968; W. P. Crowley, 1968A; Kofoed-Hansen, 1968). It also has an equivalent conservation form,

$$\frac{\partial u}{\partial t} = - \frac{\partial}{\partial x} \frac{u^2}{2} + \alpha \frac{\partial^2 u}{\partial x^2} \tag{2-20}$$

Since several analytic solutions are available for the Burgers equation, it is well-suited for demonstrating the advantages of the conservation form in the finite-difference formulation.

The reader is forewarned that, although the study of one-dimensional model equations is both convenient and illuminating, the limitations are significant. Some aspects of computational fluid dynamics are *essentially* dimensional, with qualitative differences between 1-, 2- and 3-dimensional problems. These aspects will be discussed in the following chapters.

---

*The 1-D equations are not vorticity transport equations, since vorticity does not exist in 1-D, but the equations model some aspects of the multi-dimensional equations. Physically, these equations govern the advection and diffusion of one colored liquid in another (say ink in water). Lelevier referred to equation (2-18) as the "color equation" in the 1950's. (W. P. Crowley, 1968A).

CHAPTER III

BASIC COMPUTATIONAL METHODS FOR INCOMPRESSIBLE FLOW

In this chapter, the basic computational methods for solving incompressible flow problems in rectangular meshes with regular boundaries are presented. Many of the topics covered here will also be applicable to the compressible flow problems as well, since the same model equations are applicable to the transport equations in both systems. Also, the conservative property and other properties defined herein will later apply to the compressible flow problems, as will the discussion on the dominant importance of boundary conditions and the discussions on stability and convergence.

Before we get lost in the details, subproblems, variations and peripheral questions, it will probably be worthwhile to outline the overall procedure for solving a complete fluid dynamics problem. For concreteness, we will describe the computation cycle for only the simplest time-dependent approach.

First, the continuum region of interest is overlayed with a finite-difference "mesh" or "net". At the mesh intersections ("node points" or "mesh points"), the finite-difference solution will be defined.

The solution starts with the establishment of initial values of ψ, ζ, u and v everywhere, at time $t = 0$. These initial conditions may correspond to some real initial situation for a transient problem of interest, or to a rough guess for the steady-state solution if only the steady state is of interest.

Then the computational *cycle* begins as some FDE (finite difference equation) analog of the continuum equation for vorticity transport, (2-12), is used to calculate an approximation to $\partial\zeta/\partial t$ at all interior points in the computational field. The new values of ζ are calculated at a new time level, increased by an increment Δt, by "marching" the vorticity transport equation forward in time. For example, (new ζ) = (old ζ) + $\Delta t \times \partial\zeta/\partial t$. The next step in the computational cycle is to solve an FDE analog of the Poisson equation (2-13) for new values of stream function ψ, using the new interior values of ζ in the "source" term of equation (2-13). Significantly, this Poisson solution for new ψ does not depend on the boundary values for new ζ, which are not yet known. Typically, this solution for new ψ is itself iterative, which means that the ψ iteration is "nested" within the overall computational cycle. At this point, new velocity components can be evaluated by finite-difference analogs of (2-14). The last step in the computational cycle is to calculate new values of ζ on the boundaries of the computational region. Typically, these new boundary values of ζ will depend on the new values of ψ and ζ (already calculated) at interior points near the boundary points. Then the computational cycle is repeated until the desired time is reached, or until some convergence criterion for a steady state is satisfied.

Schematically, the procedure is illustrated on the facing page. Some aspects of this procedure will change with the details of the subproblems, but we now have a framework established.

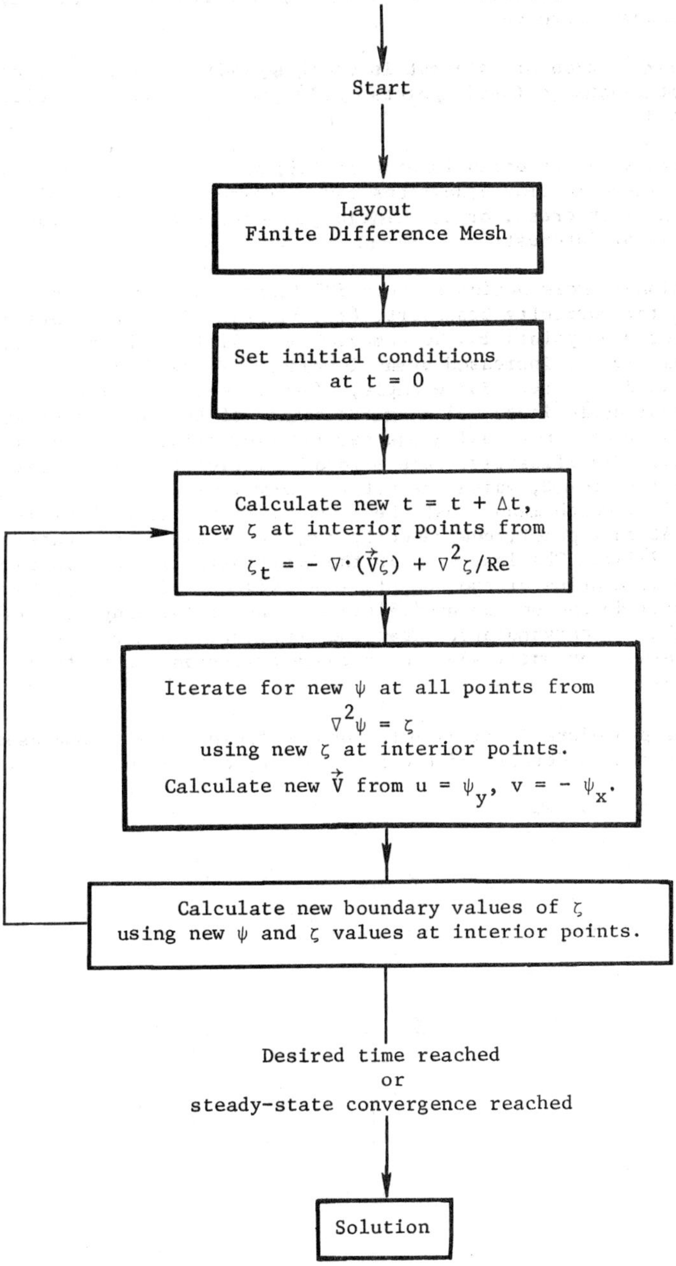
Start
Layout
Finite Difference Mesh
Set initial conditions
at t = 0
Calculate new t = t + Δt,
new ζ at interior points from
$\zeta_t = -\nabla\cdot(\vec{V}\zeta) + \nabla^2\zeta/Re$
Iterate for new ψ at all points from
$\nabla^2\psi = \zeta$
using new ζ at interior points.
Calculate new $\vec{V}$ from $u = \psi_y$, $v = -\psi_x$.
Calculate new boundary values of ζ
using new ψ and ζ values at interior points.
Desired time reached
or
steady-state convergence reached
Solution

III-A. METHODS FOR SOLVING THE VORTICITY TRANSPORT EQUATION

It is natural to separate the discussion of the parabolic vorticity transport equation from the discussion of the elliptic Poisson equation, since the methods are clearly distinguishable. It is well to note at the outset, however, that in the actual solution of a numerical fluid dynamics problem, there is a feedback between the coupled problems. For example, since these equations are solved cyclically, increases in the allowable time steps of the vorticity transport equation will be offset somewhat by increases in the number of iterations required in an iterative solution of the Poisson equation. Also, maltreatment of a boundary condition in one equation can cause a drifting in the other.

More importantly, it is clearly artificial to separate the methods for solving the interior points from the evaluation of the boundary conditions, since these two types of calculations must be made compatible, but we must start somewhere.

The final choice of what method to use on the vorticity transport equation will depend on many factors.* The choice will not always be clear-cut, and the reader is advised that a hard-and-fast recommendation on the best method will not be forthcoming.

Further, the purpose in this section is not to present an assortment of methods in cookbook form. Rather, the purpose is to define classes of methods, to study the behavior of such classes, to present techniques of analysis of the methods, and generally to get the student *thinking* about the methods rather than just blithely *programming* the methods.

III-A-1. Some Basic Finite Difference Forms

a. Taylor Series Expansions

The basic finite-difference forms for partial derivatives can be derived from partial Taylor series expansions. The rectangular mesh used is shown in Figure 3-1. Subscripts (i,j) refer to x,y indices, whereas superscripts (n) refer to time levels. The mesh spacing in the i,j directions are Δx and Δy. (Until Chapter VI, we consider for simplicity that the mesh spacings Δx and Δy are each constant.) The variable f is any function.**

The uncentered first derivative forms for $\partial f/\partial x$ can be derived as follows. We assume continuity of derivatives, and expand forward in a Taylor series. The superscript is omitted for clarity.

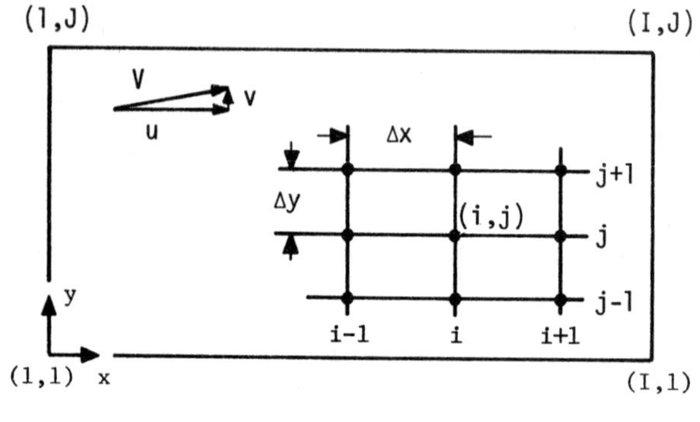

Figure 3-1. Rectangular mesh geometry.

---

*Such as the boundary conditions, the geometry of the problem, the type of solution desired (e.g., steady-state or transient), whether or not pressure and temperature solutions are required in the transient calculation, the range of parameters involved (particularly Reynolds number), and the time available to develop a computer program.

**We use the same notation f for the continuum function $f(x,y,t)$ and the discrete function $f(i,j,n)$.

$$f_{i+1,j} = f_{ij} + \left.\frac{\partial f}{\partial x}\right|_{ij} (x_{i+1,j} - x_{ij}) + \frac{1}{2} \left.\frac{\partial^2 f}{\partial x^2}\right|_{ij} (x_{i+1,j} - x_{ij})^2 + \ldots$$

$$= f_{ij} + \left.\frac{\partial f}{\partial x}\right|_{ij} \Delta x + \frac{1}{2} \left.\frac{\partial^2 f}{\partial x^2}\right|_{ij} \Delta x^2 + \text{HOT} \tag{3-1}$$

where HOT is an abbreviation for higher-order terms.

Solving for $\frac{\partial f}{\partial x}$ gives

$$\left.\frac{\partial f}{\partial x}\right|_{ij} = \frac{f_{i+1,j} - f_{ij}}{\Delta x} - \frac{1}{2} \left.\frac{\partial^2 f}{\partial x^2}\right|_{ij} \Delta x + \text{HOT}$$

$$\left.\frac{\partial f}{\partial x}\right|_{ij} = \frac{f_{i+1,j} - f_{ij}}{\Delta x} + 0(\Delta x) \tag{3-2}$$

where $0(\Delta x)$ is read "terms of order Δx", and refers to additional terms with factors of Δx, Δx^2, Δx^3, etc.

We denote any finite-difference analog of $\frac{\partial f}{\partial x}$ by the notation $\frac{\delta f}{\delta x}$. Then the forward difference approximation $\frac{\delta f}{\delta x}$ is

$$\left.\frac{\delta f}{\delta x}\right|_{ij} = \frac{f_{i+1,j} - f_{ij}}{\Delta x} \tag{3-3}$$

and has a truncation error of order Δx; that is, a "first order" accuracy.

By expanding backwards to $f_{i-1,j}$, we obtain another finite-difference analog $\frac{\delta f}{\delta x}$, this one being the backward-difference approximation,

$$\left.\frac{\delta f}{\delta x}\right|_{ij} = \frac{f_{ij} - f_{i-1,j}}{\Delta x} \tag{3-4}$$

which is also first-order accurate. The centered difference approximation $\frac{\delta f}{\delta x}$ is obtained by subtracting the forward and backward expansions.

$$f_{i+1,j} = f_{ij} + \left.\frac{\partial f}{\partial x}\right|_{ij} \Delta x + \frac{1}{2} \left.\frac{\partial^2 f}{\partial x^2}\right|_{ij} \Delta x^2 + \frac{1}{6} \left.\frac{\partial^3 f}{\partial x^3}\right|_{ij} \Delta x^3 + \frac{1}{24} \left.\frac{\partial^4 f}{\partial x^4}\right|_{ij} \Delta x^4 + 0(\Delta x^5) \tag{3-5}$$

$$f_{i-1,j} = f_{ij} - \left.\frac{\partial f}{\partial x}\right|_{ij} \Delta x + \frac{1}{2} \left.\frac{\partial^2 f}{\partial x^2}\right|_{ij} \Delta x^2 - \frac{1}{6} \left.\frac{\partial^3 f}{\partial x^3}\right|_{ij} \Delta x^3 + \frac{1}{24} \left.\frac{\partial^4 f}{\partial x^4}\right|_{ij} \Delta x^4 + 0(\Delta x^5) \tag{3-6}$$

Subtracting (3-6) from (3-5) gives

$$f_{i+1,j} - f_{i-1,j} = 2 \left.\frac{\partial f}{\partial x}\right|_{ij} \Delta x + \frac{1}{3} \left.\frac{\partial^3 f}{\partial x^3}\right|_{ij} \Delta x^3 + \text{HOT}$$

Solving for $\frac{\partial f}{\partial x}$ gives

$$\left.\frac{\partial f}{\partial x}\right|_{ij} = \frac{f_{i+1,j} - f_{i-1,j}}{2\Delta x} - \frac{1}{6}\left.\frac{\partial^3 f}{\partial x^3}\right|_{ij} \Delta x^2 + \text{HOT} = \frac{f_{i+1,j} - f_{i-1,j}}{2\Delta x} + 0(\Delta x^2) \qquad (3\text{-}7)$$

The centered difference form of $\frac{\delta f}{\delta x}$ is then

$$\left.\frac{\delta f}{\delta x}\right|_{ij} = \frac{f_{i+1,j} - f_{i-1,j}}{2\Delta x} \qquad (3\text{-}8)$$

which has a truncation error of order Δx^2; that is, a "second order" accuracy. Analogous expressions follow immediately for y- and t-derivatives. For example, the centered difference analog of $\frac{\partial f}{\partial t}$ is

$$\left.\frac{\delta f}{\delta t}\right|_{ij}^{n} = \frac{f_{ij}^{n+1} - f_{ij}^{n-1}}{2\Delta t} \qquad (3\text{-}9)$$

We derive a centered difference analog of $\frac{\partial^2 f}{\partial x^2}$ by adding the expansions (3-5) and (3-6).

$$f_{i+1,j} + f_{i-1,j} = 2f_{ij} + \left.\frac{\partial^2 f}{\partial x^2}\right|_{ij} \Delta x^2 + \frac{1}{12}\left.\frac{\partial^4 f}{\partial x^4}\right|_{ij} \Delta x^4 + \text{HOT} \qquad (3\text{-}10)$$

Solving for $\frac{\partial^2 f}{\partial x^2}$ gives

$$\left.\frac{\partial^2 f}{\partial x^2}\right|_{ij} = \frac{f_{i+1,j} + f_{i-1,j} - 2f_{ij}}{\Delta x^2} + 0(\Delta x^2) \qquad (3\text{-}11)$$

The centered difference form of $\frac{\delta^2 f}{\delta x^2}$ is then

$$\left.\frac{\delta^2 f}{\delta x^2}\right|_{ij} = \frac{f_{i+1,j} + f_{i-1,j} - 2f_{ij}}{\Delta x^2} \qquad (3\text{-}12)$$

which is second-order accurate.

Exercise: Derive equation (3-12) by applying equation (3-8) to $g = \frac{\partial f}{\partial x}$. Use half-spaces $\Delta x/2$, so that equation (3-8) becomes

$$\left.\frac{\delta g}{\delta x}\right|_{ij} = \frac{g_{i+1/2,j} - g_{i-1/2,j}}{\Delta x} \qquad (3\text{-}13)$$

Exercise: Derive the following expression for $\frac{\delta^2 f}{\delta x \delta y}$, the analog of the cross-derivative term $\frac{\partial^2 f}{\partial x \partial y}$.

$$\frac{\delta^2 f}{\delta x \delta y} = \frac{f_{i+1,j+1} - f_{i+1,j-1} - f_{i-1,j+1} + f_{i-1,j-1}}{4\Delta x \Delta y} \qquad (3\text{-}14)$$

Derive it first by applying equation (3-8) for $\frac{\delta f}{\delta x}$ to $g = \frac{\delta f}{\delta y}$, and second by a double Taylor series expansion in x and y. Note that a casual carrying of the order of truncation terms in the first method would indicate only first-order accuracy. But, by carefully carrying high-order terms in the second method, one can demonstrate second-order accuracy, with truncation error $E = 0(\Delta x^2 + \Delta y^2)$.

Note that $\frac{\delta^2 f}{\delta x \delta y}$ in equation (3-14) obeys the continuum operative rule, $\frac{\partial^2 f}{\partial x \partial y} = \frac{\partial^2 f}{\partial y \partial x}$. It is always desirable, all other things being equal, for our finite-difference expressions to model such qualitative behavior of the continuum equations. Many other such examples will be noted later.

Combinations of such finite-difference expressions for partial *derivatives* alone may be put together to form finite-difference expressions for the partial differential *equations*. For example, the Laplace equation $\nabla^2 f \equiv \frac{\partial^2 f}{\partial x^2} + \frac{\partial^2 f}{\partial y^2} = 0$ becomes

$$\frac{\delta^2 f}{\delta x^2} + \frac{\delta^2 f}{\delta y^2} = \frac{f_{i+1,j} + f_{i-1,j} - 2f_{ij}}{\Delta x^2} + \frac{f_{i,j+1} + f_{i,j-1} - 2f_{ij}}{\Delta y^2}$$

or

$$f_{i+1,j} + f_{i-1,j} + \beta^2(f_{i,j+1} + f_{i,j-1}) - 2(1 + \beta^2)f_{ij} = 0 \tag{3-15}$$

where β is the mesh aspect ratio, $\beta = \Delta x/\Delta y$. This is commonly referred to as the "5-point analog." For $\beta = 1$, we obtain the well-known equation

$$f_{ij} = \frac{1}{4}(f_{i+1,j} + f_{i-1,j} + f_{i,j+1} + f_{i,j-1}) \tag{3-16}$$

which says that the value f_{ij} is the average of f at the four neighboring points. These forms are indicated schematically in Figure 3-2.

Figure 3-2. Schematic representation of the 5-point Laplace equation. $\beta = \Delta x/\Delta y$.

A simple expression for the linear model equation (2-18), using second-order-accurate expressions for both space and time derivatives would be

$$\frac{\zeta_i^{n+1} - \zeta_i^{n-1}}{2\Delta t} = -\frac{u\zeta_{i+1}^n - u\zeta_{i-1}^n}{2\Delta x} + \alpha\frac{\zeta_{i+1}^n + \zeta_{i-1}^n - 2\zeta_i^n}{\Delta x^2} \tag{3-17}$$

from which one could solve explicitly for ζ_i^{n+1} in terms of previous values. This method is, in fact, unacceptable. For all $\alpha > 0$ and for all time steps $\Delta t > 0$, the method is numerically *unstable*, meaning that chaotic solutions with no relation to the continuum solution will be generated. This behavior emphasizes the difference between an accurate finite-difference analog for a *derivative*, and accuracy for a *differential equation*.

If, instead of all centered differences, we use forward differences in time, we obtain an analog of the linear model equation (2-18) which is second-order accurate in space, and only first order in time.

$$\frac{\zeta_i^{n+1} - \zeta_i^n}{\Delta t} = - \frac{u\zeta_{i+1}^n - u\zeta_{i-1}^n}{2\Delta x} + \alpha \frac{\zeta_{i+1}^n + \zeta_{i-1}^n - 2\zeta_i^n}{\Delta x^2} \tag{3-18}$$

We will refer to this method, using forward-time and centered-space differences, as the FTCS method.

We will show later that this expression is *stable*, at least for some conditions, depending on Δt, u, α, and Δx. But before we cover stability analyses, we consider a few other aspects of finite-difference equations.

b. Basic Finite-Difference Forms; Polynomial Fitting

Another method of obtaining finite-difference expressions is to fit an analytical function with free parameters to the mesh-point values, and then to analytically differentiate the function. This is a common method of obtaining derivatives from experimental data. Ideally, the functional form of the fit would be determined by an approximate analytical solution. More commonly, polynomials are used. We demonstrate the method with a parabolic fit.

Given data f at (i-1), (i), and (i+1), we assume a parabolic form

$$f(x) = a + bx + cx^2 \tag{3-19}$$

For convenience, the origin $x = 0$ is translated to location (i). Then equation (3-19) written at (i-1), (i), and (i+1) gives respectively

$$\begin{aligned} f_{i-1} &= a - b\Delta x + c\Delta x^2 \\ f_i &= a \\ f_{i+1} &= a + b\Delta x + c\Delta x^2 \end{aligned} \tag{3-20}$$

Adding the f_{i-1} and f_{i+1} equations gives

$$c = \frac{f_{i+1} + f_{i-1} - 2f_i}{2\Delta x^2} \tag{3-21}$$

Solving for b gives

$$b = \frac{f_{i+1} - f_{i-1}}{2\Delta x} \tag{3-22}$$

Differentiating equation (3-19) and evaluating at (i) gives

$$\left.\frac{\delta f}{\delta x}\right|_i = \left[b + 2cx\right]_{x=0} = b \tag{3-23}$$

and

$$\left.\frac{\delta^2 f}{\delta x^2}\right|_i = 2c \tag{3-24}$$

These two equations, with equations (3-21) and (3-22), give identically the same results as those of the centered second-order Taylor series expansions, equations (3-8) and (3-12), respectively. If a first-order polynomial is assumed, $f = a + bx$, then the forward or backward difference for $\frac{\delta f}{\delta x}$ is obtained, depending on whether f_i and f_{i+1} are used to fit a and b, or f_i and f_{i-1}. Obviously, no expression can be obtained for $\frac{\delta^2 f}{\delta x^2}$ by fitting f with with a straight line. But, if a straight line is used to fit the first-derivative analogs $\left.\frac{\delta f}{\delta x}\right|_{i+1/2}$ and $\left.\frac{\delta f}{\delta x}\right|_{i-1/2}$ which are respectively evaluated by forward and backward differences, the same centered difference expression (3-12) for $\frac{\delta^2 f}{\delta x^2}$ will result.

Higher order expressions result from higher order polynomials. Beyond the second-order polynomial, the expressions obtained are not identical to those from higher order Taylor series expansions. The truncation error of any expression must still be checked by way of a Taylor-series expansion, in any case. The polynomial fit method has not generally been used in computational fluid dynamics, with the one exception of evaluating derivatives near a boundary. (This aspect will be covered in Section III-C-2.) For now, we note the following shortcoming of higher order polynomial fits which is well known to data analysts. As the order of fit increases, the fit becomes sensitive to "noise," or small amplitude, more-or-less randomly distributed errors in the data. Thus, a sixth-order polynomial passed through seven points, which are algebraically on a straight line, will yield a straight-line fit, as shown in Figure 3-3a. But with the addition of noise, the polynomial coefficients adjust to the noisy data. The analytic evaluation of derivatives at i can then give absurd answers,* as demonstrated in Figure 3-3b.

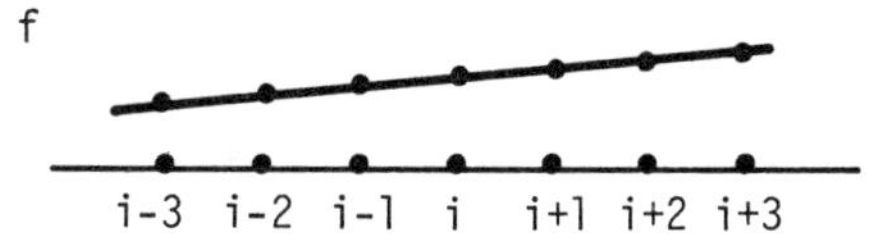

a. Algebraically perfect data.

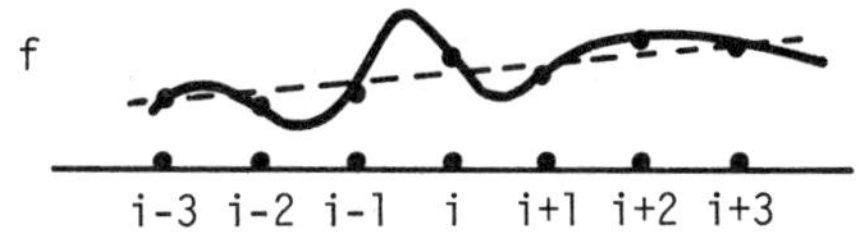

b. Noisy data.

Figure 3-3. Sixth-order polynomial fit.

A quadratic fit cannot indicate the presence of an inflection point in the data, i.e., a point where $\frac{\partial^3 f}{\partial x^3} = 0$. For this reason, a cubic polynomial fit can be justified in data analysis. (The spline function technique is often used, guaranteeing continuity of derivatives from one node point to the next.) For our purposes, the equations governing the physics display no dependence on an inflection point or on a third derivative, so we need not concern ourselves with the resolution of one.

c. Basic Finite-Difference Forms: Integral Method

In the integral method, we approximately satisfy the governing equations in an integral sense, rather than a differential sense. For this derivation, it is clearer to use subscripts x and superscripts t, rather than i and n, to indicate spatial position and time level. We write the linear model equation in conservation form as

$$\frac{\partial \zeta}{\partial t} = -\frac{\partial(u\zeta)}{\partial x} + \alpha \frac{\partial^2 \zeta}{\partial x^2} \qquad (3\text{-}25)$$

*The sensitivity of polynomial fits can be reduced by using data from 3N or 4N data points to fit an N-th order polynomial. The coefficients are then not solved by algebraic elimination, but by a least-squares fit. This time-consuming approach generally has not been used except for some boundary treatments (see Section V-G-6).

We integrate this equation over time from (t) to (t + Δt), and over the spatial region R from (x - Δx/2) to (x + Δx/2), as indicated in Figure 3-4. Since the order of integration,

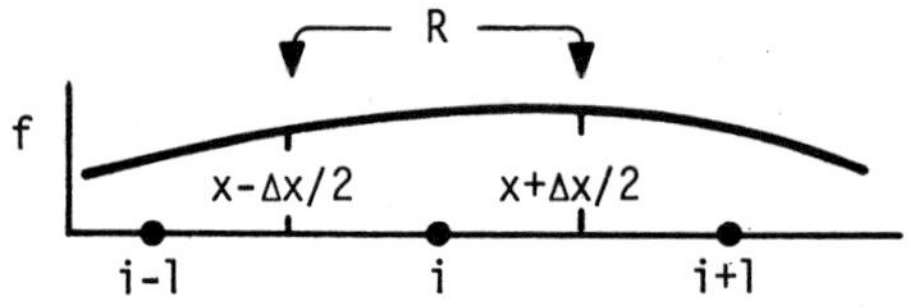

Figure 3-4. Region of integration R for the integral method.

over dt or dx, is immaterial, we choose the order to take advantage of exact differentials.

$$\int_{x-\Delta x/2}^{x+\Delta x/2}\left[\int_t^{t+\Delta t}\frac{\partial\zeta}{\partial t}\,dt\right]dx = -\int_t^{t+\Delta t}\left[\int_{x-\Delta x/2}^{x+\Delta x/2}\frac{\partial(u\zeta)}{\partial x}\,dx\right]dt$$

$$+\alpha\int_t^{t+\Delta t}\left[\int_{x-\Delta x/2}^{x+\Delta x/2}\frac{\partial^2\zeta}{\partial x^2}\,dx\right]dt \qquad (3\text{-}26)$$

Performing the inner integrations,

$$\int_{x-\Delta x/2}^{x+\Delta x/2}\left[\zeta^{t+\Delta t}-\zeta^{t}\right]dx = -\int_t^{t+\Delta t}\left[(u\zeta)_{x+\Delta x/2}-(u\zeta)_{x-\Delta x/2}\right]dt$$

$$+\alpha\int_t^{t+\Delta t}\left[\frac{\partial\zeta}{\partial x}\bigg|_{x+\Delta x/2}-\frac{\partial\zeta}{\partial x}\bigg|_{x-\Delta x/2}\right]dt \qquad (3\text{-}27)$$

The remaining integration is performed numerically. By the mean-value theorem, we can approximate

$$\int_{Z_1}^{Z_1+\Delta Z} f(Z)\,dZ \doteq f(\bar{Z})\cdot\Delta Z$$

where $\bar{Z}\,\varepsilon\left[Z_1, Z_1+\Delta Z\right]$.* Convergence is assured for $\Delta Z \to 0$. If we use a midpoint evaluation for the left member of equation (3-27) and evaluate the integrand values of the right member at the lower integration limit (i.e., Euler integration), we obtain

$$\left[\zeta_x^{t+\Delta t}-\zeta_x^{t}\right]\Delta x = -\left[(u\zeta)^t_{x+\Delta x/2}-(u\zeta)^t_{x-\Delta x/2}\right]\Delta t$$

$$+\alpha\left[\frac{\partial\zeta}{\partial x}\bigg|^t_{x+\Delta x/2}-\frac{\partial\zeta}{\partial x}\bigg|^t_{x-\Delta x/2}\right]\Delta t \qquad (3\text{-}28)$$

The first derivatives can be evaluated by writing

$$\zeta^t_{x+\Delta x} = \zeta^t_x + \int_x^{x+\Delta x}\frac{\partial\zeta}{\partial x}\,dx$$

---

*Throughout this book, we will use this set notation wherein "ε" is read as "is an element of" or "is contained in". The phrase "where $\bar{Z}\,\varepsilon\left[Z_1, Z_1+\Delta Z\right]$" is equivalent to saying "where $Z_1 \leq \bar{Z} \leq Z_1 + \Delta Z$."

and using the midpoint rule, one obtains

$$\zeta_{x+\Delta x}^{t} = \zeta_{x}^{t} + \left.\frac{\partial \zeta}{\partial x}\right|_{x+\Delta x}^{t} \Delta x$$

or

$$\left.\frac{\partial \zeta}{\partial x}\right|_{x+\Delta x/2}^{t} = \frac{\zeta_{x+\Delta x}^{t} - \zeta_{x}^{t}}{\Delta x} \tag{3-29}$$

The value $(u\zeta)_{x+\Delta x/2}^{t}$ can be evaluated by simple averaging, as

$$(u\zeta)_{x+\Delta x/2}^{t} = \frac{1}{2}\left[(u\zeta)_{x}^{t} + (u\zeta)_{x+\Delta x}^{t}\right] \tag{3-30}$$

and similarly for $(u\zeta)_{x-\Delta x/2}^{t}$.

Using (3-29) and (3-30) in (3-28) gives

$$\left[\zeta_{x}^{t+\Delta t} - \zeta_{x}^{t}\right]\Delta x = -\left[\cancel{\frac{1}{2}(u\zeta)_{x}^{t}} + \frac{1}{2}(u\zeta)_{x+\Delta x}^{t} - \cancel{\frac{1}{2}(u\zeta)_{x}^{t}} - \frac{1}{2}(u\zeta)_{x-\Delta x}^{t}\right]\Delta t$$

$$+ \alpha\left[\frac{\zeta_{x+\Delta x}^{t} - \zeta_{x}^{t}}{\Delta x} - \frac{\zeta_{x}^{t} - \zeta_{x-\Delta x}^{t}}{\Delta x}\right]\Delta t$$

Dividing by $\Delta x \cdot \Delta t$ gives

$$\frac{\zeta_{x}^{t+\Delta t} - \zeta_{x}^{t}}{\Delta t} = -\frac{(u\zeta)_{x+\Delta x}^{t} - (u\zeta)_{x-\Delta x}^{t}}{2\Delta x} + \alpha\,\frac{\zeta_{x+\Delta x}^{t} + \zeta_{x-\Delta x}^{t} - 2\zeta_{x}^{t}}{\Delta x^{2}} \tag{3-31}$$

When converted to (i,n) index notation, this equation (3-31) is identical to equation (3-18), obtained by Taylor-series expansions. There is clearly a great deal of arbitrariness in deriving the equations in either method. If the time integration had been carried out from $(t-\Delta t)$ to $(t+\Delta t)$, instead of (t) to $(t+\Delta t)$, and the midpoint integration rule is used, equation (3-17) would have resulted. As we have already stated, this equation is unconditionally unstable.

The advantage of the integral method can be appreciated only after the conservation property is studied. The difference between the integral method and the Taylor-series method is most easily appreciated in non-rectangular coordinate systems.

III-A-2. Control Volume Approach

The control volume approach to deriving finite-difference equations is very similar to the integral approach, but is more physical in its basis. This approach highlights the "numerical simulation" process, best exemplified by the famous PIC (particle-in-cell) and FLIC (fluid-in-cell) codes developed at Los Alamos (Evans and Harlow, 1947; Gentry, Martin, and Daley, 1966). (See Section V-E-3.)

We picture a "control volume" in space about the location x as depicted in Figure 3-5. The node-point value of ζ will refer to the average over the control volume, CV. For a *specific* (i.e., volume averaged) quantity ζ, where ζ can now refer to any quantity, we write $\zeta = \Gamma/\text{volume}$.

For example, if ζ is the mass density ρ, then Γ_x is just the total mass in the control volume at x. If ζ is the vorticity, Γ is the "circulation" (see Lamb, 1945). We then write

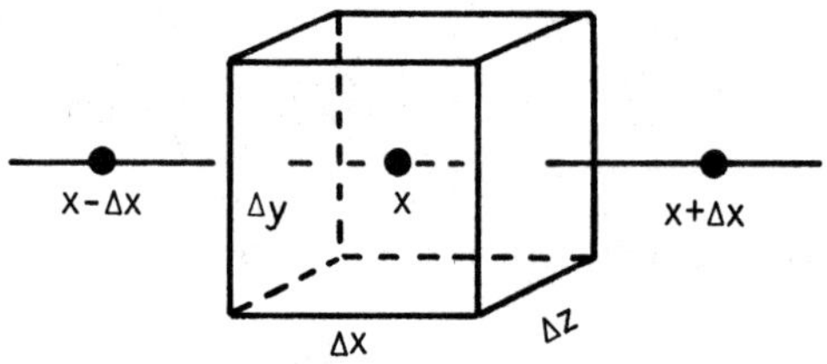

Figure 3-5. Control volume CV at point x.

the following verbal conservation law:

total increase of Γ in CV = net flux of Γ into CV by advection + net flux of Γ into CV by diffusion (3-32)

The total increase of $\Gamma = \zeta \times$(volume) in the CV over Δt is

$$\zeta\Big|_x^{t+\Delta t} \times (\Delta x \Delta y \Delta z) - \zeta\Big|_x^{t} \times (\Delta x \Delta y \Delta z)$$

The advective flux rate of Γ into CV from the left side is

$$(u\zeta)_{x-\Delta x/2} \times (\text{area}) = (u\zeta)_{x-\Delta x/2} \Delta y \Delta z$$

where u is possibly varying, and where the interface values at $(x-\Delta x/2)$, as yet to be determined, must represent some average over Δt. Due to this advective flux *rate* in, the total advective flux quantity of Γ into CV over Δt, through the interface at $(x-\Delta x/2)$, is

$$(u\zeta)_{x-\Delta x/2}\ \Delta y \Delta z \Delta t$$

Likewise, the total advective flux quantity of Γ out of CV at $(x+\Delta x/2)$ is

$$(u\zeta)_{x+\Delta x/2}\ \Delta y \Delta z \Delta t$$

and the net flux of Γ into CV is (total flux in) - (total flux out), or

$$\left[(u\zeta)_{x-\Delta x/2} - (u\zeta)_{x+\Delta x/2}\right] \Delta y \Delta z \Delta t$$

To calculate the flux into CV by diffusion, we need a diffusion-rate law. The simplest diffusion law (and the one compatible with the vorticity transport equation) is the linear law that says the diffusion flux rate of ζ, which we call q, is proportional to the gradient of ζ (i.e., Fick's law).

$$q = -\alpha \frac{\partial \zeta}{\partial x}$$

The minus sign indicates that a ζ which increases in x causes diffusion in the negative x-direction.

Then the diffusion flux rate into CV on the left is

$$q\big|_{x-\Delta x/2}\ \Delta y\,\Delta z = -\alpha\left.\frac{\partial\zeta}{\partial x}\right|_{x-\Delta x/2}\Delta y\,\Delta z$$

and the diffusion flux rate out of CV on the right is

$$q\big|_{x+\Delta x/2}\ \Delta y\Delta z = -\alpha\left.\frac{\partial\zeta}{\partial x}\right|_{x+\Delta x/2}\Delta y\Delta z$$

Again, the interface values at $(x\pm\Delta x/2)$ are some appropriate time averages, yet to be determined. The net diffusion flux quantity into CV over Δt is

$$+\ \alpha(\Delta y\Delta z\Delta t)\left[\left.\frac{\partial\zeta}{\partial x}\right|_{x+\Delta x/2} - \left.\frac{\partial\zeta}{\partial x}\right|_{x-\Delta x/2}\right]$$

Using these expressions, the verbal conservation law (3-32) for 1D advection-diffusion becomes

$$\zeta\big|_x^{t+\Delta t}\cdot(\Delta x\Delta y\Delta z) - \zeta\big|_x^{t}\cdot(\Delta x\Delta y\Delta z) = (u\zeta\big|_{x-\Delta x/2} - u\zeta\big|_{x+\Delta x/2})\Delta y\Delta z\Delta t$$

$$+\ \alpha\Delta y\Delta z\Delta t\left[\left.\frac{\partial\zeta}{\partial x}\right|_{x+\Delta x/2} - \left.\frac{\partial\zeta}{\partial x}\right|_{x-\Delta x/2}\right] \qquad (3\text{-}33)$$

Dividing by $\Delta x\Delta y\Delta z\Delta t$ gives

$$\frac{1}{\Delta t}\left[\zeta\big|_x^{t+\Delta t} - \zeta\big|_x^{t}\right] = \frac{1}{\Delta x}\left[u\zeta\big|_{x-\Delta x/2} - u\zeta\big|_{x+\Delta x/2}\right] + \frac{\alpha}{\Delta x}\left[\left.\frac{\partial\zeta}{\partial x}\right|_{x+\Delta x/2} - \left.\frac{\partial\zeta}{\partial x}\right|_{x-\Delta x/2}\right] \quad (3\text{-}34)$$

As in the integral approach, further latitude in deriving finite-difference expressions arises in the evaluation of the interface values. If we take each interface value to be the simple arithmetic mean between adjacent node values at time n, such as

$$(u\zeta)_{x\pm\Delta x/2} = \frac{1}{2}\left[(u\zeta)^n_{x\pm\Delta x} + (u\zeta)^n_x\right]$$

and

$$\left.\frac{\partial\zeta}{\partial x}\right|_{x\pm\Delta x/2} = \left.\frac{\delta\zeta}{\delta x}\right|^n_{x\pm\Delta x/2}$$

and, if we evaluate the gradients by centered differences, as

$$\left.\frac{\delta\zeta}{\delta x}\right|^n_{x+\Delta x/2} = \frac{\zeta\big|^n_{x+\Delta x} - \zeta\big|^n_x}{\Delta x}$$

then equation (3-34) becomes

$$\frac{1}{\Delta t}\left[\zeta\big|_x^{t+\Delta t} - \zeta\big|_x^{t}\right] = \frac{1}{\Delta x}\left\{\frac{1}{2}\left[(u\zeta)^t_x + (u\zeta)^t_{x-\Delta x}\right] - \frac{1}{2}\left[(u\zeta)^t_x + (u\zeta)^t_{x+\Delta x}\right]\right\}$$

$$+\ \frac{\alpha}{\Delta x}\left[\frac{1}{\Delta x}\left(\zeta\big|^t_{x+\Delta x} - \zeta\big|^t_x\right) - \frac{1}{\Delta x}\left(\zeta\big|^t_x - \zeta\big|^t_{x-\Delta x}\right)\right]$$

or

$$\frac{\zeta_x^{t+\Delta t} - \zeta_x^t}{\Delta t} = -\frac{(u\zeta)_{x+\Delta x}^t - (u\zeta)_{x-\Delta x}^t}{2\Delta x} + \alpha \frac{\zeta_{x+\Delta x}^t + \zeta_{x-\Delta x}^t - 2\zeta_x^t}{\Delta x^2} \tag{3-35}$$

When converted to (i,n) index notation, this equation is identical to equation (3-18), already obtained.

It is thus seen that all four approaches to deriving finite-difference analogs of partial differential equations -- Taylor-series expansions, polynomial curve fitting, the integral method, and the control-volume approach -- can lead to the same expressions. This is encouraging, and tends to build confidence in all the approaches. But there is a latitude present in all of them, so that the choice of an approach to deriving a finite-difference analog of a continuum equation does not uniquely determine the analog. In fact, there is a considerable variety of analogs used. Although most of them differ by what appears to the uninitiated to be minor points, they can differ greatly in behavior. On a personal basis, one of the fascinating aspects of computational fluid dynamics to the present author is the large number of plausible schemes which do *not* work, e.g., equation (3-17), already mentioned. This is true of both the "basic," or interior point differencing schemes, and of the methods of treating boundary points.

The control-volume approach has no unique claim to success, but it appears to have the best batting average. The advantage of the control-volume approach is that it is based on the macroscopic physical laws, rather than on the continuum mathematics. This is most clearly significant when we deal with rarefied gases or with inviscid continuum flows in which shock waves exist. In these cases, there are no continuous differential equations to which a Taylor-series expansion is everywhere applicable. Yet mass, for example, is still conserved, and the advection portion of equation (3-35) is still valid. And even when continuum equations are valid, the control-volume approach focuses attention on the actual satisfaction of the physical laws *macroscopically*, not merely in some academic limit as $\Delta x, \Delta t$ approach zero. This is the basic concept of the conservative property, which we now discuss.

III-A-3. The Conservative Property

A finite-difference method possesses the conservative property if it preserves certain integral conservation relations of the continuum equation.

Consider the vorticity transport equation (2-12), with $1/Re = \alpha$.

$$\frac{\partial \zeta}{\partial t} = -\nabla \cdot (\vec{V}\zeta) + \alpha \nabla^2 \zeta \tag{3-36}$$

We integrate this over some space region R, and obtain

$$\int_R \frac{\partial \zeta}{\partial t}\, dR = -\int_R \nabla \cdot (\vec{V}\zeta)\, dR + \int_R \alpha \nabla^2 \zeta\, dR \tag{3-37}$$

Because t is independent of the space variables, we have

$$\int_R \frac{\partial \zeta}{\partial t}\, dR = \frac{\partial}{\partial t} \int_R \zeta\, dR \tag{3-38}$$

Using the Gauss divergence theorem, we have

$$\int_R \nabla \cdot (\vec{V}\zeta)\, dR = \int_{\partial R} (\vec{V}\zeta) \cdot \vec{n}\, ds \tag{3-39}$$

where ∂R is the boundary of R, $\vec{n}$ is the unit normal surface vector (positive outward), and ds is the differential element of the boundary ∂R. Also, by the same theorem,

$$\int_R \alpha \nabla^2 \zeta \, dR = \alpha \int_{\partial R} (\nabla \zeta) \cdot \vec{n} \, ds \tag{3-40}$$

Then equation (3-37) becomes

$$\frac{\partial}{\partial t} \int_R \zeta \, dR = - \int_{\partial R} (\vec{V} \zeta) \cdot \vec{n} \, ds + \alpha \int_{\partial R} (\nabla \zeta) \cdot \vec{n} \, ds \tag{3-41}$$

Equation (3-41) states that the time rate of accumulation of ζ in R equals the net advective flux rate of ζ across ∂R into R, plus the net diffusion flux rate of ζ across ∂R into R.* The idea of the conservative property is to maintain this integral relation identically in the finite-difference system.

Consider for clarity the inviscid limit ($\alpha = 0$) in a one-dimensional model system. Then equation (3-36) becomes

$$\frac{\partial \zeta}{\partial t} = - \frac{\partial (u\zeta)}{\partial x} \tag{3-42}$$

(Alternately, the flux quantity ζ may be taken to represent mass density, making equation (3-42) the compressible continuity equation.) Using FTCS differencing, the finite-difference analog of equation (3-42) is

$$\frac{\zeta_i^{n+1} - \zeta_i^n}{\Delta t} = - \left(\frac{u_{i+1}\zeta_{i+1} - u_{i-1}\zeta_{i-1}}{2\Delta x} \right) \tag{3-43}$$

where we have dropped the superscript n for clarity. We now consider a one-dimensional region R running from $i = I_1$ to $i = I_2$, and we evaluate the summation

$$\frac{1}{\Delta t} \sum_{i=I_1}^{I_2} \zeta_i \Delta x$$

corresponding to the integration $\frac{\partial}{\partial t} \int_R \zeta \, dR$ in equation (3-41).

$$\frac{1}{\Delta t} \left[\sum_{i=I_1}^{I_2} \zeta_i^{n+1} \Delta x - \sum_{i=I_1}^{I_2} \zeta_i^n \Delta x \right] = \sum_{i=I_1}^{I_2} - \left(\frac{u_{i+1}\zeta_{i+1} - u_{i-1}\zeta_{i-1}}{2\Delta x} \right) \Delta x$$

$$= + \frac{1}{2} \sum_{i=I_1}^{I_2} \left[(u\zeta)_{i-1} - (u\zeta)_{i+1} \right] \tag{3-44a}$$

---

*We have derived (3-41) from (3-36) to show their relation, but (3-41) is actually more fundamental than (3-36). For example, with $\alpha = 0$ and ζ = mass density, these are continuity equations expressing conservation of mass. Then (3-41) remains valid, even though the derivatives in (3-36) may not exist at some points interior to R.

The summation is as follows:

$$\sum_{i=I_1}^{I_2} \left[(u\zeta)_{i-1} - (u\zeta)_{i+1} \right] \tag{3-44b}$$

$$\begin{array}{lr}
= + (u\zeta)_{I_1-1} - \cancel{(u\zeta)}_{I_1+1} & [i=I_1] \\
\quad + (u\zeta)_{I_1} - \cancel{(u\zeta)}_{I_1+2} & [i=I_1+1] \\
\quad\quad + \cancel{(u\zeta)}_{I_1+1} - \cancel{(u\zeta)}_{I_1+3} & [i=I_1+2] \\
\quad\quad\quad + \cancel{(u\zeta)}_{I_1+2} - \cancel{(u\zeta)}_{I_1+4} & [i=I_1+3] \\
\quad\quad\quad\quad + \cancel{(u\zeta)}_{I_1+3} - \cancel{(u\zeta)}_{I_1+5} & [i=I_1+4] \\
\quad\quad\quad\quad\quad + \ldots & \\
\cdots\cdots & \\
\quad \cdots\cdots & \\
+ \cancel{(u\zeta)}_{I_2-4} - \cancel{(u\zeta)}_{I_2-2} & [i=I_2-3] \\
\quad + \cancel{(u\zeta)}_{I_2-3} + \cancel{(u\zeta)}_{I_2-1} & [i=I_2-2] \\
\quad\quad + \cancel{(u\zeta)}_{I_2-2} - (u\zeta)_{I_2} & [i=I_2-1] \\
\quad\quad\quad + \cancel{(u\zeta)}_{I_2-1} - (u\zeta)_{I_2+1} & [i=I_2]
\end{array}$$

$$= (u\zeta)_{I_1-1} + (u\zeta)_{I_1} - (u\zeta)_{I_2} - (u\zeta)_{I_2+1}$$

Then equation (3-44a) becomes

$$\frac{1}{\Delta t} \left[\sum_{i=I_1}^{I_2} \zeta_i^{n+1} \Delta x - \sum_{i=I_1}^{I_2} \zeta_i^n \Delta x \right] = \frac{1}{2} \left[(u\zeta)_{I_1-1} + (u\zeta)_{I_1} \right] - \frac{1}{2} \left[(u\zeta)_{I_2} + (u\zeta)_{I_2+1} \right]$$

$$= (u\zeta)_{I_1-1/2} - (u\zeta)_{I_2+1/2} \tag{3-44c}$$

This equation states that the rate of accumulation of ζ_i in R is identically equal* to the flux of ζ into R across the boundary at $I_1-1/2$, analogous to the inviscid part of the continuum equation (3-41). Thus, the finite-difference analog has preserved the integral Gauss divergence property of the continuum equation, and is said to possess the conservative property.

---

*We are concerned here with algebraic equality, disregarding the machine roundoff errors.

The conservative property depends on the form of the continuum equation used, as well as the finite-difference scheme. For example, the non-conservative form of the 1D model equation (2-18), with $\alpha = 0$, is

$$\frac{\partial \zeta}{\partial t} = - u \frac{\partial \zeta}{\partial x} \tag{3-45}$$

Using the same differencing technique as in the previous example, i.e., forward time and centered space differences, we obtain

$$\frac{\zeta_i^{n+1} - \zeta_i^n}{\Delta t} = - u_i^n \left[\frac{\zeta_{i+1} - \zeta_{i-1}}{2\Delta x}\right] \tag{3-46}$$

The summation corresponding to equation (3-44a) is then

$$\frac{1}{\Delta t}\left[\sum_{i=I_1}^{I_2} \zeta_i^{n+1} \Delta x - \sum_{i=I_1}^{I_2} \zeta_i^n \Delta x\right] = \sum_{i=I_1}^{I_2} - u_i \left(\frac{\zeta_{i+1} - \zeta_{i-1}}{2\Delta x}\right) \Delta x$$

$$= \frac{1}{2} \sum_{i=I_1}^{I_2} u_i (\zeta_{i-1} - \zeta_{i+1}) \tag{3-47a}$$

Performing the summation,

$$\sum_{i=I_1}^{I_2} u_i (\zeta_{i-1} - \zeta_{i+1}) \tag{3-47b}$$

$$\begin{aligned}
&= u_{I_1} \zeta_{I_1-1} - u_{I_1} \zeta_{I_1+1} && [i=I_1] \\
&\quad + u_{I_1+1} \zeta_{I_1} - u_{I_1+1} \zeta_{I_1+2} && [i=I_1+1] \\
&\quad + u_{I_1+2} \zeta_{I_1+1} - u_{I_1+2} \zeta_{I_1+3} && [i=I_1+2] \\
&\quad + u_{I_1+3} \zeta_{I_1+2} - u_{I_1+3} \zeta_{I_1+4} && [i=I_1+3] \\
&\quad + \ldots
\end{aligned}$$

It is seen that, in this formulation, the pairs of terms corresponding to inter-cell fluxes do not cancel, e.g.,

$$u_{I_1+2} \zeta_{I_1+1} - u_{I_1} \zeta_{I_1+1} = \left(u_{I_1+2} - u_{I_1}\right) \zeta_{I_1+1} \neq 0 \tag{3-47c}$$

except in the particular case of u_i = constant, and that the finite-difference analog has failed to preserve the integral Gauss divergence property of the continuum equation. The meaning of calling equation (2-10) the "conservation" or "divergence" form of the equation is now clear.

The success of the first method in preserving the conservative property is pinpointed with reference to the control-volume approach to deriving the finite-difference expressions. Using the conservation form, the advective flux rate of ζ out of a control volume located at i across the interface at i+1/2, $1/2(u_i\zeta_i + u_{i+1}\zeta_{i+1})$, is identically equal to the advective flux rate into the (i+1) control volume across the same interface. This is not the case when the non-conservation form is used, as seen above.

Exercise: Demonstrate that the use of the centered difference expression (3-12) for $\partial^2\zeta/\partial x^2$ provides conservation of the diffusion flux terms, for $\alpha > 0$.

It is clear that, for $\alpha > 0$, the only way to assure conservation of the total flux in general (i.e., allowing for spatially varying u) is to conserve the *advective-flux* terms and the *diffusive-flux* terms independently; for higher dimensions, it is necessary to conserve in each *dimension* independently.

The significance of the conservative property is easily appreciated in terms of the compressible continuity equation. Consider the problem of computing natural convection in a completely closed container with impermeable walls. The calculation is begun with $\vec{V} = 0$ everywhere. Heat is applied to the bottom of the box, and natural convection ensues, perhaps developing to a final steady-state pattern. If some non-conservative method is used (see Problem 3-2), the total mass in the box will change as the calculation proceeds! If a conservative method is used, the mass will not change, except for machine roundoff errors. Some solace may be found in the fact that the conservation errors in the first case will vanish as $\Delta x \to 0$, etc., but this is little consolation in a practical computation with finite Δx.

We find this argument to be very compelling, and we do recommend the use of the conservation form. However, there exist pro and con arguments, and the numerical tests which have been reported in the literature do not present an entirely one-sided case. We now discuss the arguments and the test results.

Conservation does not *necessarily* imply accuracy, except in regard to this integral property. For example, unstable solutions of the conservative equations still maintain the conservative property. Moreover, a non-conservative method may be more accurate, in some sense, than a conservative one. For example, one could use one-dimensional curve-fitting routines to pass a high-order polynomial through mesh-point values, and perhaps evaluate space derivatives with lower order truncation error (see Thomas, 1954). But the resulting method may not be conservative and therefore not more accurate if one's criteria for accuracy include conservation.

Experience so far has indicated that conservative systems do *generally* give more accurate results. Cheng (1968) and Allen (1968) have shown that the conservation form gives substantially more accurate results for certain solutions of Burgers equation, (2-19) and (2-20). Cyrus and Fulton (1967) have shown that the conservation form ("half-station approximation") gives more accurate results than the non-conservation form ("whole-station approximation") for elliptic equations. Torrance et al.(1972) have shown that the use of conservation equations with only first-order accuracy is more accurate than the use of non-conservation equations with second-order accuracy, for the driven cavity problem. The advantage in shock wave calculations is well known (Gary 1964) and will be covered in Chapter V (but it is worth noting that Gary's calculation of a rarefaction wave is slightly more accurate in the non-conservation form). Also, the divergence form of the equations is more physically meaningful, and aids in the formation of boundary conditions for compressible flow.

Piacsek (1966) has shown how to obtain a conservative set of equations in axisymmetric coordinates. Roberts and Weiss (1966) discuss conservation of vector quantities. Lax (1954) was the first to use the conservation form of the compressible flow equations given by Courant and Friedrichs (1948) in a finite-difference calculation, and he elaborated on the concept.

Meteorologists have extended the conservation idea to moment quantities. Bryan (1963, 1966) has devised schemes which conserve not only vorticity but also kinetic energy. Arakawa's scheme (1966) [see also Lilly (1965) or Fromm (1967) and section III-A-2] conserves vorticity, vorticity squared, linear momentum, and kinetic energy. But the additional complexity is not *always* warranted or even beneficial. Bengtsson (1964) has shown that such forms give small improvements, insignificant with respect to actual data, and may actually give larger errors in wave speeds. But in the inviscid limit, conservation of kinetic energy avoids the "nonlinear" instability* discussed by Phillips (1959) and Sundqvist (1963). Bengtsson (1964) has devised a system that conserves the difference between the kinetic energy and the (meteorological) gross static stability, which is beneficial to his problem involving significant gravity-force gradients.

It usually costs little to use a differencing method which achieves conservation of the basic flux quantities such as vorticity, mass, momentum, or total energy. In the two-dimensional vorticity transport problem, the additional cost is in performing two more difference operations to obtain the additional velocity components from the stream-function solution, and two additional multiplications. For compressible flow, the cost is higher and, in certain cases, there may be reason not to use the conservation form (see Moretti's method, Chapter VI). There are many solutions obtained without the conservation form; for example, Allen and Southwell (1955), Hung and Macagno (1966), Mehta and Lavan (1968), and Pao and Daugherty (1969). Note that all method-of-characteristics solutions are non-conservative, and that finite-difference calculations of boundary-layer equations typically have not used the conservation form. In such cases, conservation checks can be performed as an indicator of truncation error (see Section III-D).

Finally, lest the use of the conservation forms becomes a religion, we note that the use of the non-conservation form for the variable-coefficient diffusion term $\frac{\partial}{\partial x}\left(\alpha \frac{\partial \zeta}{\partial x}\right)$ can produce more accurate results than the conservation form (see problems 3-3 and 3-4).

Exercise: Show that the first moment of the inviscid vorticity transport equation, obtained by multiplying equation (2-12) by ζ with $1/R_e = 0$, can be written in the conservation form

$$\frac{\partial E}{\partial t} = - \nabla \cdot (\vec{V} E)$$

where $E = \zeta^2$.

III-A-4. A Description of Instability

We consider the one-dimensional linear model equation for ζ to describe some of the phenomenology of numerical instability. In Figure 3-6, part (a) represents a steady-state solution, $\hat{\zeta}^n$, at time level n. Part (b) represents a perturbation ε on ζ^n. This disturbance may be thought to originate from machine roundoff error, or from transverse disturbances in an actual two-dimensional problem. Using forward time and centered space (FTCS) differences, we now trace the development of the disturbance. The linear model equation in conservation form is

$$\frac{\partial \zeta}{\partial t} = - \frac{\partial (u\zeta)}{\partial x} + \alpha \frac{\partial^2 \zeta}{\partial x^2}$$

and the FTCS difference equation is

$$\frac{\zeta_i^{n+1} - \zeta_i^n}{\Delta t} = - \frac{u\zeta_{i+1}^n - u\zeta_{i-1}^n}{2\Delta x} + \alpha \frac{\zeta_{i+1}^n + \zeta_{i-1}^n - 2\zeta_i^n}{\Delta x^2} \tag{3-48}$$

Separating ζ into its steady-state component $\hat{\zeta}$ and the perturbation ε, we have

$$\zeta_i^n = \hat{\zeta}_i^n + \varepsilon_i \tag{3-49}$$

*Actually due only to non-constant coefficients rather than true nonlinearity.

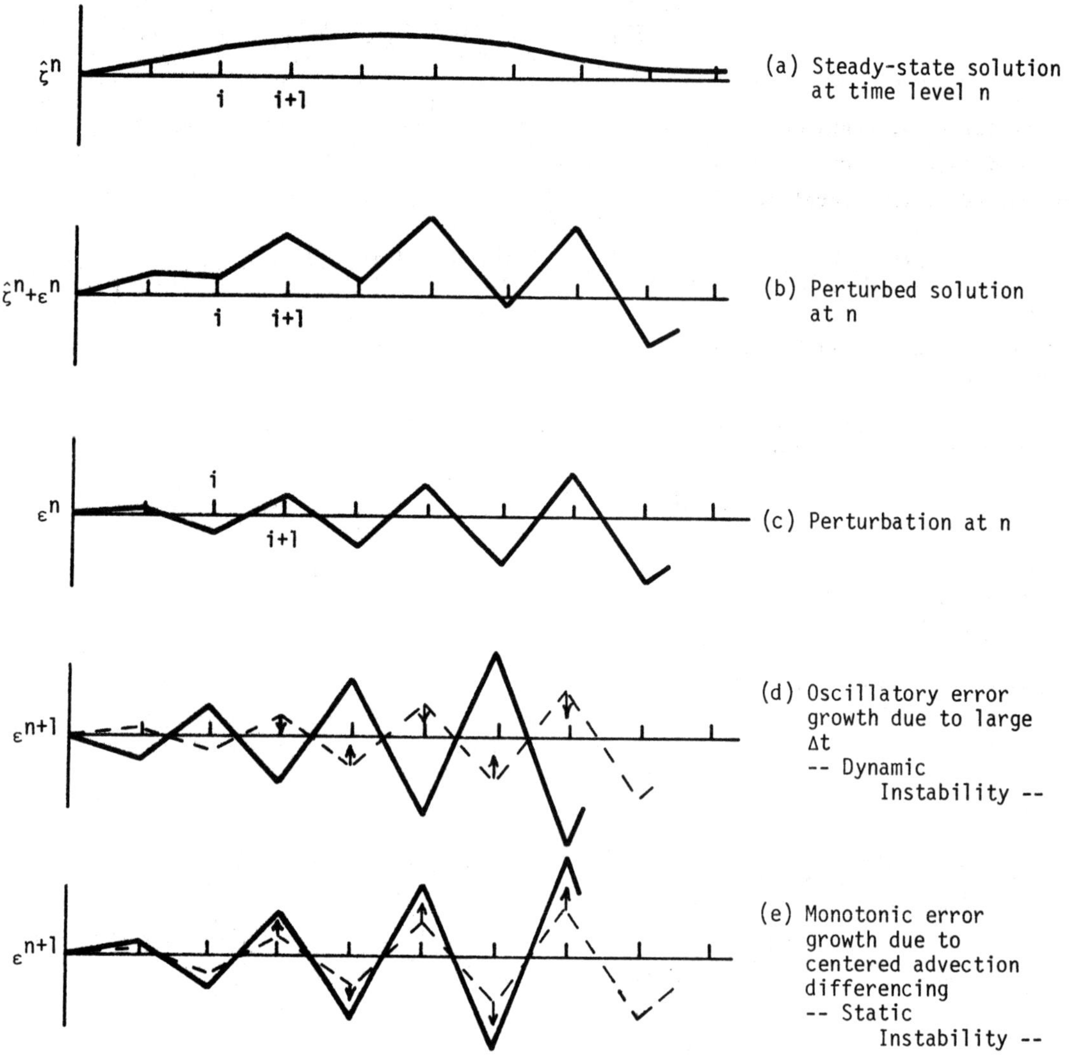

Figure 3-6. A description of error growth for FTCS differencing of the model advection-diffusion equation.

Then equation (3-48) becomes

$$\frac{\zeta_i^{n+1} - \zeta_i^n}{\Delta t} = - \frac{u\hat{\zeta}_{i+1}^n + u\hat{\zeta}_{i-1}^n}{2\Delta x} + \alpha \frac{\hat{\zeta}_{i+1}^n + \hat{\zeta}_{i-1}^n - 2\hat{\zeta}_i^n}{\Delta x^2}$$

$$- \frac{u\varepsilon_{i+1} - u\varepsilon_{i-1}}{2\Delta x} + \alpha \frac{\varepsilon_{i+1} + \varepsilon_{i-1} - 2\varepsilon_i}{\Delta x^2} \qquad (3\text{-}50)$$

The first two terms on the right of equation (3-50) equal the finite-difference version of $(\partial\hat{\zeta}/\partial t)_i^n$, which is zero by the assumption that a steady finite-difference solution existed at time level n. Then equation (3-50) reduces to

$$\zeta_i^{n+1} - \zeta_i^n \equiv \Delta\zeta_i = - \Delta t \frac{u\varepsilon_{i+1} - u\varepsilon_{i-1}}{2\Delta x} + \alpha\Delta t \frac{\varepsilon_{i+1} + \varepsilon_{i-1} - 2\varepsilon_i}{\Delta x^2} \qquad (3\text{-}51)$$

The first term on the right is the $\Delta\zeta$ due to the advection term;* the second is due to diffusion.

Consider now a problem with just the diffusion term alone, evaluated at point i in Part (c) of Figure 3-6. Since $\varepsilon_{i+1} > 0$, $\varepsilon_i < 0$ and $\varepsilon_{i-1} > 0$, we therefore have

$$\Delta\zeta_i = \alpha\Delta t \frac{\varepsilon_{i+1} + \varepsilon_{i-1} - 2\varepsilon_i}{\Delta x^2} > 0 \qquad (3\text{-}52)$$

That is, for all $\Delta t > 0$, the change $\Delta\zeta_i$ is > 0, thus tending to correct the negative perturbation ε_i.

Likewise, considering $\Delta\zeta$ at i+1, we have $\varepsilon_{i+2} < 0$, $\varepsilon_i < 0$, $\varepsilon_{i+1} > 0$; therefore,

$$\Delta\zeta_{i+1} = \alpha\Delta t \frac{\varepsilon_{i+2} + \varepsilon_i - 2\varepsilon_{i+1}}{\Delta x^2} < 0 \qquad (3\text{-}53)$$

and the positive perturbation ε_{i+1} is corrected by a negative $\Delta\zeta$ at i+1.

Note, however, that the change $\Delta\zeta_i = \zeta_i^{n+1} - \zeta_i^n$ (as well as $\Delta\zeta_{i+1}$, etc.) is proportional to Δt. If Δt is large enough, the correction will overshoot. For Δt too large, the magnitude of the new ζ_i^{n+1} will be greater than the original perturbation, as depicted in Part (d) of Figure 3-6.

$$|\zeta_i^{n+1}| > |\varepsilon_i| \qquad (3\text{-}54)$$

---

*In discussions of the stability problems of the nonlinear vorticity transport equations or the primitive equations, it is common practice to identify the advection term as the "nonlinear term." This phrase misses the point entirely. The most common stability problems arising from it do not result from nonlinearity, nor even from nonconstant coefficients. For example, all the examples here apply when ζ is taken to be the temperature in an incompressible problem, where u is regarded as fixed in time and may even be constant in space. The stability problems discussed here arise simply because it is a first-derivative term.

Likewise,

$$|\zeta_{i\pm1}^{n+1}| > |\varepsilon_{i\pm1}| \tag{3-55}$$

This type of oscillatory overshoot with increasing amplitude is a *dynamic* instability, which may be removed by lowering the time step below some "critical time step" Δt_{crit}.

Consider now a problem with just the advection term alone. We evaluate equation (3-51) at i, with $u > 0$. The form of the perturbation which we have assumed is oscillatory in i, with amplitude increasing with i. Therefore $u\varepsilon_{i+1} > 0$, $u\varepsilon_{i-1} > 0$, $u\varepsilon_{i+1} > u\varepsilon_{i-1}$, and

$$\Delta\zeta_i = -\Delta t \frac{u\varepsilon_{i+1} - u\varepsilon_{i-1}}{2\Delta x} < 0 \tag{3-56}$$

or $\Delta\zeta_i < 0$, due to advection, even though $\varepsilon_i < 0$. This means that the error grows monotonically, as shown in Part (e) of Figure 3-6. This error growth is a *static* instability which cannot be removed by lowering the time step Δt. It can be removed by using some other finite-difference method.

If the spatial direction of growth of ε with respect to u is changed from that shown in Figure 3-6, i.e., if either $u < 0$ or if the envelope of ε decreases in i, then the advection term becomes statically stable, but subject still to dynamic instability if Δt is too large. In any realistic problem, the initial errors are more or less randomly distributed, and we may be assured that, at some time and at some location, the distribution will resemble the disastrous one shown in Figure 3-6.

When both advection terms and diffusion terms are present, they interact. As we will see shortly, the result for this difference scheme is a Δt limitation due to the diffusion term, and another Δt limitation which depends on the relative importance of the statically unstable advection term compared with the statically stable diffusion term, i.e., on the Reynolds number. If we also require no overshoot at all (which is the behavior of the continuum equations) we also obtain a limitation on Reynolds number, independent of Δt. These points should become clear in the next section.

III-A-5. Stability Analyses

Now that we have presented a pedagogic description of stability, we will present three methods of stability analysis, followed by a discussion of their relations and relative merits. The methods will be illustrated by application to the forward time-centered space (FTCS) difference scheme applied to the linear model equation, as in equation (3-18).

III-A-5-a. Discrete Perturbation Stability Analysis

The method which we here call the discrete perturbation stability analysis is an extension of the method first used by Thom and Apelt (1961) and further developed by Thoman and Szewczyk (1966). This method closely follows our description of instability already presented. It is conceptually simple and direct, and lends insight both to the stability problem and to the transportive property, to be defined later. Briefly, a discrete perturbation in ζ is introduced into the equations at an arbitrary point, and its effect is followed; stability is indicated if the perturbation dies out.

For simplicity, we first consider only the diffusion term, and we assume that a steady-state solution $\zeta_i^n = 0$ for all i has been reached. We then introduce a disturbance ε in ζ_i^n, so the FTCS method (equation 3-18) gives

$$\frac{\zeta_i^{n+1} - (\zeta_i^n + \varepsilon)}{\Delta t} = \alpha\left[(\zeta_{i+1}^n + \zeta_{i-1}^n - 2(\zeta_i^n + \varepsilon)\right]/\Delta x^2 \tag{3-57}$$

or

$$\frac{\zeta_i^{n+1} - \varepsilon}{\Delta t} = -2\alpha\varepsilon/\Delta x^2 \tag{3-58}$$

$$\zeta_i^{n+1} = \varepsilon(1 - 2d) \tag{3-59}$$

where the diffusion number d is defined by

$$d = \frac{\alpha \Delta t}{\Delta x^2} \tag{3-60}$$

For stability, this disturbance must die out. For this first computational step, this requires

$$|\zeta_i^{n+1}/\varepsilon| \leq 1 \tag{3-61}$$

or

$$-1 \leq 1 - 2d \leq +1 \tag{3-62}$$

The right-hand inequality is the "static stability" requirement, and is automatically satisfied by the positivity of d for α, $\Delta t > 0$.* The left-hand inequality is the "dynamic stability" requirement, and is satisfied by $d \leq 1$. If, following Thoman and Szewczyk (1966), we also require that the numerics model the physics in that *no overshoot* is allowed, i.e.,

$$\zeta_i^{n+1}/\varepsilon \geq 0 \tag{3-63}$$

then we obtain the requirement

$$d \leq \frac{1}{2} \tag{3-64}$$

The inequality (3-63) is not a "stability requirement," however, in the sense of a disturbance decreasing in amplitude. But, interestingly enough, if we continue to later time planes, we find that equation (3-64) *is* required. We first calculate neighboring disturbances $\zeta_{i\pm1}^{n+1}$ from the FTCS method (equation (3-18) as

$$\zeta_{i\pm1}^{n+1} = d\varepsilon \tag{3-65}$$

The calculation at the next time plane then gives

$$\zeta_i^{n+2} = \zeta_i^{n+1} + d\left[\zeta_{i+1}^{n+1} + \zeta_{i-1}^{n+1} - 2\zeta_i^{n+1}\right] = \varepsilon(1 - 2d) + d\left[d\varepsilon + d\varepsilon - 2\varepsilon(1 - 2d)\right]$$

$$\zeta_i^{n+2} = \varepsilon(1 - 4d + 6d^2) \tag{3-66}$$

---

* Note that the diffusion equation cannot be solved backward in time ($\Delta t < 0$) nor with $\alpha < 0$, since these are mathematically and physically unstable. This point will be discussed later.

We again require

$$|\zeta^{n+2}/\varepsilon| \leq 1 \tag{3-67}$$

which gives

$$-1 \leq 1 - 4d + 6d^2 \leq 1 \tag{3-68}$$

The left inequality is always satisfied, while the right inequality requires $d \leq 2/3$.

Thus, consideration of just the first time level leads to the requirement $d \leq 1$, and the second time level to $d \leq 2/3$. Higher time levels may be traced out, with ever more restrictive conditions on d. Asymptotically, the initial single perturbation ε at i approaches the oscillatory distribution $\zeta_i = \pm\,\varepsilon'$, where ε' is some smaller perturbation amplitude, as shown in Figure 3-7.

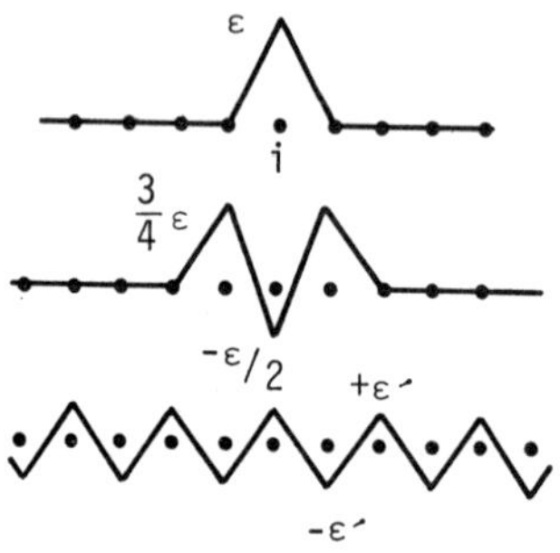

Figure 3-7. Asymptotic spreading of a single perturbation ε at i for the FTCS method applied to the diffusion equation.

Thus, it is seen that the most restrictive condition on d occurs with this type of distribution; if we start with this oscillatory distribution of ε' about $\zeta = 0$ and apply the FTCS method (3-18) we obtain

$$\zeta_i^{n+1} = \varepsilon' + d(-\varepsilon' - \varepsilon' - 2\varepsilon') = \varepsilon'(1 - 4d) \tag{3-69}$$

The stability requirement

$$|\zeta_i^{n+1}/\varepsilon'| \leq 1 \tag{3-70}$$

then gives

$$-1 \leq 1 - 4d \leq 1 \tag{3-71}$$

or

$$d \leq \frac{1}{2} \tag{3-72}$$

Succeeding time levels do not change equation (3-72). Thus, the long time stability requirement of equation (3-72) is equivalent to equation (3-64), the requirement for zero overshoot for an isolated disturbance.

From equation (3-60) we see that the requirement $d \leq 1/2$, for a fixed mesh space and fixed α, places a limitation on the time step of

$$\Delta t \leq \frac{1}{2}\frac{\Delta x^2}{\alpha} \tag{3-73}$$

Note the stringent requirement implied by equation (3-73) in terms of calculation time for a diffusion equation. Suppose we make a calculation with some mesh size Δx_1 up to some dimensionless time $T = N_1 \Delta t_1$, where $\Delta t_1 = \frac{1}{2}\Delta x_1^2/\alpha$, the maximum possible time step. If we wish to repeat the calculation with a halved mesh size $\Delta x_2 = \Delta x_1/2$ (in order to check for truncation-error convergence, for example) then $\Delta t_2 = \frac{1}{2}\ \Delta x_2^2/\alpha = \frac{1}{4}\Delta t_1$. Thus to reach the same dimensionless time T, 4 times as many time levels are required, i.e., $T = N_2 \Delta t_2$ and $N_2 = 4N_1$. Furthermore, each time-level calculation takes twice as long, since $\Delta x_2 = \Delta x_1/2$ means there are twice as many node points in the field. Thus, for the one-dimensional case, *halving* the mesh size increases the computer time *8-fold*! In a two-dimensional problem,* halving Δx and Δy increases the number of node points fourfold, thus increasing the computer time by a factor of 16. In a three-dimensional diffusion problem, halving all three mesh sizes increases the computer time by a factor of 32. Generally, reducing the mesh size from Δx_1 to Δx_2 in a D-space-dimensional diffusion problem, using the FTCS explicit method, increases the computer time by the factor $(\Delta x_1/\Delta x_2)^{(2+D)}$ It is clear that methods which can overcome the stability requirement (3-73) are highly desirable.

It is not essential in the above development to assume a steady state. The entire unperturbed unsteady equation can be subtracted from the perturbed equation, giving the equation for just the error growth,

$$\varepsilon_i^{n+1} = d\left(\varepsilon_{i-1}^n + \varepsilon_{i-1}^n - 2\varepsilon_i^n\right) + \varepsilon_i^n \tag{3-74}$$

with stability requirements that $|\varepsilon_i^{n+1}/\varepsilon_i^n| \leq 1$, etc. The results are identical.

We now consider the advection and diffusion terms in equation (3-18). Without loss of generality, we assume $u > 0$. [If $u < 0$, the roles of the indexes (i+1) and (i-1) are reversed.] We again apply the FTCS method at i with a perturbation ε to ζ_i^{n+1} giving

$$\frac{\zeta_i^{n+1} - (\zeta_i^n + \varepsilon)}{\Delta t} = -\frac{u\zeta_{i+1}^n - u\zeta_{i-1}^n}{2\Delta x} + \alpha\left[\frac{\zeta_{i+1}^n + \zeta_{i-1}^n - 2(\zeta_i^n + \varepsilon)}{\Delta x^2}\right] \tag{3-75}$$

This step gives no more additional information than the previously analyzed diffusion-only equation, since the advection terms at (i) do not affect the perturbation at (i). If we also apply FTCS at (i+1), we obtain

$$\zeta_{i+1}^{n+1} = -\frac{\Delta t}{2\Delta x}\left[u\zeta_{i+2}^n - u(\zeta_i^n + \varepsilon)\right] + \frac{\alpha\Delta t}{\Delta x^2}\left[\zeta_{i+2}^n + (\zeta_i^n + \varepsilon) - 2\zeta_{i+1}^n\right] + \zeta_{i+1}^n \tag{3-76}$$

---

*In two- and three-dimensional problems, the time-step restrictions are $\Delta t \leq \frac{1}{2}\frac{A}{\alpha}$, where $A = 1/(\Delta x^{-2} + \Delta y^{-2})$ or $A = 1/(\Delta x^{-2} + \Delta y^{-2} + \Delta z^{-2})$.

or

$$\zeta_{i+1}^{n+1} = \frac{c}{2}\varepsilon + d\varepsilon \tag{3-77}$$

where $c = u\Delta t/\Delta x$ is called the "Courant number",* and $d = \alpha\Delta t/\Delta x^2$, as before.

For stability, we again require

$$|\zeta_{i+1}^{n+1}/\varepsilon| \leq 1 \tag{3-78}$$

or

$$-1 \leq \frac{c}{2} + d \leq 1 \tag{3-79}$$

The left inequality is automatically satisfied by $u > 0$. The right-hand inequality, the static stability requirement, gives another necessary condition for stability.

$$u\Delta t/\Delta x + 2\alpha\Delta t/\Delta x^2 \leq 2$$

or

$$\Delta t \leq \frac{2}{(2\alpha/\Delta x^2) + \frac{u}{\Delta x}} \tag{3-80}$$

Proceeding similarly at (i-1), we obtain

$$\zeta_{i-1}^{n+1} = -\frac{c}{2}\varepsilon + d\varepsilon \tag{3-81}$$

and the stability requirement $|\zeta_{i-1}^{n+1}/\varepsilon| \leq 1$ gives

$$-1 \leq -\frac{c}{2} + d \leq 1 \tag{3-82}$$

Considering first the right-hand inequality of (3-82) (static instability) gives

$$\Delta t \left[\frac{2\alpha}{\Delta x^2} - \frac{u}{\Delta x}\right] \leq 2 \tag{3-83}$$

If the bracketed term is < 0, this inequality is satisfied for all $\Delta t > 0$. If the bracketed term is > 0, we obtain

$$\Delta t \leq \frac{2}{2\alpha/\Delta x^2 - u/\Delta x} \tag{3-84}$$

Since the denominator is greater than zero, equation (3-84) is less restrictive than (3-80) and is therefore superseded by it.

---

*After Richard Courant (1888-1972), the mathematician whose inspiring work in the analysis of numerical methods and nonlinear partial differential equations laid much of the groundwork for modern computational fluid dynamics.

Considering next the left-hand inequality of (3-82) (dynamic instability) gives

$$-2 \leq \Delta t\left[2\alpha/\Delta x^2 - u/\Delta x\right] \tag{3-85}$$

If the bracketed term is > 0, the inequality is satisfied by all $\Delta t > 0$. If it is < 0, we obtain

$$\Delta t \leq \frac{2}{u/\Delta x - 2\alpha/\Delta x^2} \tag{3-86}$$

where the denominator is > 0. This (3-86) is also less restrictive than (3-80), and is therefore superseded by it.

So from this discrete perturbation stability analysis, the two necessary conditions, equations (3-73) and (3-80), for stability of the advection-diffusion equation arise. If the analysis were extended to higher time levels, other more restrictive conditions might arise, but the method becomes very clumsy. In fact, the von Neumann analysis does indicate another condition (3-112), as we shall see.

Also, if we follow Thoman and Szewczyk (1966) in additionally requiring zero overshoot, at (i-1) we require

$$\zeta_{i-1}^{n+1}/\varepsilon > 0 \tag{3-87}$$

and equation (3-81) then gives

$$-\frac{c}{2} + d \geq 0 \tag{3-88}$$

or

$$-u\Delta t/\Delta x + 2\alpha\Delta t/\Delta x^2 \geq 0 \tag{3-89}$$

or

$$u\Delta x/\alpha \leq 2 \tag{3-90}$$

For $\alpha = 1/R_e$ in the vorticity transport equation, the term $u\Delta x/\alpha$ is a *cell Reynolds number,* R_c. That is, R_c is the Reynolds number based on local velocity and a characteristic length equal to the cell mesh size Δx. The zero-overshoot criterion then requires

$$R_c \leq 2 \tag{3-91}$$

independent of Δt.* If this zero-overshoot requirement (3-90) is combined with the diffusion restriction (3-73), the restrictions that the "Courant number" $c \equiv u\Delta t/\Delta x \leq 1$ and that $\Delta t \leq 2\alpha/u^2$ result. As we shall see, these are the correct restrictions.

---

*This R_c limitation should not be confused with the R_c stability limitation given by Thom and Apelt (1961), page 136. In that case, iteration of the steady equations was performed with no "underrelaxation." This is analogous to running the unsteady equations with a fixed Δt (see Section III-A-2), and their R_c limitation really corresponds to a Δt limitation.

Exercise: Apply the discrete perturbation stability analysis to the following "upwind differencing form" (see Sections III-A-7, 8, 9) of the inviscid model equation,

$$\frac{\zeta_i^{n+1} - \zeta_i^n}{\Delta t} = -\frac{u\zeta_i^n - u\zeta_{i-1}^n}{\Delta x}, \quad u \geq 0 \tag{3-92}$$

Show that the stability requirement is that the Courant number $c = u\Delta t/\Delta x < 1$, and that this also meets the zero-overshoot criterion.

III-A-5-b. von Neumann Stability Analysis

The most commonly used method of stability analysis was originated about 1944 by J. von Neumann at Los Alamos. The method was circulated privately among the relatively small group of interested workers at that time (Eddy, 1949). Short descriptions of the method were first published by Crank and Nicolson (1947), and later by Charney, Fjortoft, and von Neumann (1950). The earliest complete treatment was given by O'Brien, Hyman, and Kaplan (1950). In this method, a finite Fourier series expansion of the solution to a model equation is made, and the decay or amplification of each mode is considered separately to determine stability or instability, as we now demonstrate.

Consider first the linear model equation with diffusion only, again using the FTCS method (3-18).

$$\frac{\zeta_i^{n+1} - \zeta_i^n}{\Delta t} = \alpha\left(\frac{\zeta_{i+1}^n + \zeta_{i-1}^n - 2\zeta_i^n}{\Delta x^2}\right)$$

or

$$\zeta_i^{n+1} = \zeta_i^n + d\left(\zeta_{i+1}^n + \zeta_{i-1}^n - 2\zeta_i^n\right) \tag{3-93}$$

wherein $d = \alpha\Delta t/\Delta x^2$. Each Fourier component of the solution is written

$$\zeta_i^n = V^n e^{Ik_x(i\Delta x)} \tag{3-94}$$

where V^n is the amplitude function at time-level n of the particular component whose wave number is k_x (wavelength $\Lambda = 2\pi/k_x$) and $I = \sqrt{-1}$. The spatial domain is considered infinite in extent.* Define the phase angle $\theta = k_x\Delta x$, giving

---

*There are several confusing aspects of this approach which are best ignored on first reading, but which come to the fore when other literature is consulted.

Equation (3-94) is actually a solution of the FDE (3-93), provided that zero boundary conditions are used. The general solution is (3-94), with V^n replaced by $A\xi^n$, where n is algebraically interpreted as an exponent. This FDE solution may be exploited to demonstrate very nicely some convergence properties of (3-93); see Richtmyer and Morton (1967), p. 9ff. However, equation (3-94) is not a solution of the combined advection-diffusion equation, and we wish shortly to treat the combined equation while avoiding questions of boundary influence. This is readily accomplished (after an additional approximation) by analyzing stability for an infinite region, following von Neumann.

For the infinite spatial domain considered here, k_x takes on all integer values $k_x = 1, 2, 3 \ldots$. If it is desired to analyze boundary effects, then max k_x is finite, depending on max $i = I$. The literature is tiresomely confused by the use of two different "normalizing" systems for length. In one system, max $k_x = (I - 1)$ is taken as $\pi/\Delta x$. This requires that $X (\equiv$ max x of the mesh) be given the value $X = \pi$. With phase angle θ defined by $\theta = k_x\Delta x$ and $k_x \varepsilon [1, I-1]$, this gives min $\theta = \Delta x = \pi(I - 1)$ and max $\theta = (I - 1)\Delta x = \pi$. In the second system, the max x of the mesh is given the more natural normalized value $X = 1$. Then $\Delta x = 1/(I - 1)$, and the phase angle θ is now defined by $\theta = \pi k_x\Delta x$. This again properly gives min $\theta = \pi\Delta x = \pi/(I - 1)$ and max $\theta = \pi(I - 1)\Delta x = \pi$. See?

$$\zeta_i^n = V^n e^{Ii\theta} \tag{3-95}$$

Similarly

$$\zeta_{i\pm1}^{n+1} = V^{n+1} e^{I(i\pm1)} \tag{3-96}$$

Substituting these into (3-93) gives

$$V^{n+1} e^{Ii\theta} = V^n e^{Ii\theta} + d\left[V^n e^{I(i+1)\theta} + V^n e^{I(i-1)\theta} - 2V^n e^{Ii\theta}\right] \tag{3-97}$$

Canceling the common term $e^{Ii\theta}$ gives

$$V^{n+1} = V^n \left[1+d(e^{I\theta} + e^{-I\theta} -2)\right] \tag{3-98}$$

We use the identity

$$e^{I\theta} + e^{-I\theta} = 2 \cos \theta \tag{3-99}$$

and define the *amplification factor* G from the equation

$$V^{n+1} = GV^n \tag{3-100}$$

From (3-98) we evaluate G as

$$G = 1 - 2d(1 - \cos \theta) \tag{3-101}$$

Note that $G = G(\theta)$, i.e., the amplitude factor varies for each Fourier component, in this case.

Equation (3-100) shows clearly that, if solutions are to remain bounded, we must have

$$|G| \leq 1 \tag{3-102}$$

for all θ. This is the stability criterion for the diffusion equation (3-93).

From equations (3-101) and (3-102), we have

$$-1 \leq 1 - 2d(1 - \cos \theta) \leq 1 \tag{3-103}$$

which must be satisfied for all possible θ, that is, all possible Fourier modes. The right-hand inequality is satisfied for all θ. The left-hand inequality is critical at the maximum value of $(1 - \cos \theta) = 2$, which then gives the stability requirement on d of $d \leq 1/2$, or

$$\Delta t \leq \frac{1}{2} \frac{\Delta x^2}{\alpha} \tag{3-104}$$

This is identical to equation (3-73), found by the discrete perturbation analysis.

We now consider the FTCS method for the entire advection-diffusion model equation (3-18).

$$\zeta_i^{n+1} = \zeta_i^n + \left[-\frac{c}{2}\left(\zeta_{i+1}^n - \zeta_{i-1}^n\right) + d\left(\zeta_{i+1}^n + \zeta_{i-1}^n - 2\zeta_i^n\right)\right] \qquad (3\text{-}105)$$

Substituting (3-95) and (3-96) and canceling the term $e^{Ii\theta}$ again gives (3-101), but with the amplification factor G given by

$$G = 1 - \frac{c}{2}\left(e^{I\theta} - e^{-I\theta}\right) + d\left(e^{I\theta} + e^{-I\theta} - 2\right) \qquad (3\text{-}106)$$

Using the identities (3-99) and

$$e^{I\theta} - e^{-I\theta} = 2I \sin\theta \qquad (3\text{-}107)$$

gives

$$G = 1 - 2d(1 - \cos\theta) - Ic \sin\theta \qquad (3\text{-}108)$$

As contrasted with the previous case, we see that the advection-diffusion equation gives rise to a *complex* amplification factor (3-108). This complex G reduces to the real G given in (3-100) as $c \to 0$, i.e., as the advection-diffusion equation reduces to just the diffusion equation.

The stability requirement is that

$$|G| \leq 1 \qquad (3\text{-}109)$$

where $|G|$ is now the modulus of complex G, which is represented in Figure 3-8. Equation (3-108) may be written as

$$G = (1 - 2d) + (2d)\cos\theta - I(c)\sin\theta \qquad (3\text{-}110)$$

which can be recognized as the equation of an ellipse centered on (1-2d) on the real axis, with axis half-lengths of c and 2d. Stability ($|G| \leq 1$) is indicated when the ellipse lies entirely within the unit circle.

Clearly, it is necessary for stability that $c \leq 1$ and $d \leq 1/2$. The most general restriction is found by solving for the modulus of G (with the complex conjugate denoted by $\bar{G}$) as

$$|G|^2 = G\bar{G} = \left[1 + 2d(\cos\theta - 1)\right]^2 + c^2(1 - \cos^2\theta) \qquad (3\text{-}111)$$

Using elementary methods to solve for the maximum $|G|^2$ over $\cos\theta$, we find* that, for a cell Reynolds number

$$R_c = u\Delta x/\alpha \leq 2 \qquad (3\text{-}112)$$

*The methods are elementary, but the calculations are difficult. The results for the two-dimensional case were presented by Fromm (1964).

no maximum occurs over the interior range of $-1 < \cos\theta < +1$. The maximum occurs at $\cos\theta = -1$, and simply gives the diffusion restriction $d \le 1/2$. For $R_c > 2$, a maximum over $-1 < \cos\theta < +1$ does occur, and $|G|^2_{max} > 1$ always. Therefore, equation (3-112) is necessary for stability.

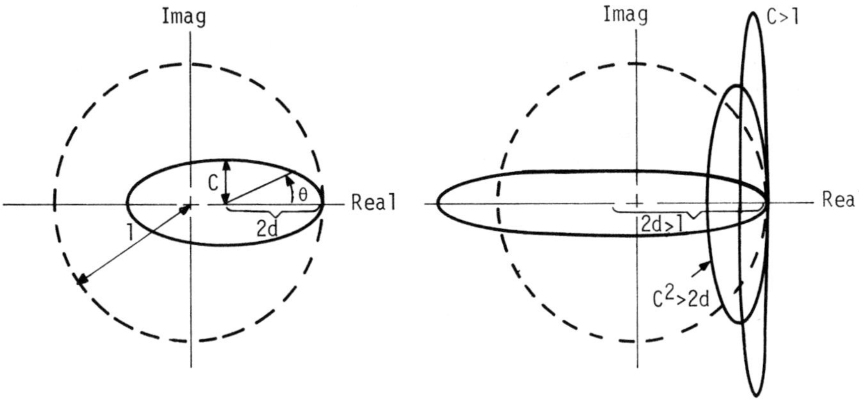

Figure 3-8. Polar diagram of the amplification factor G from equation (3-96). The ellipse with $c < 1$, $d < 1/2$, and $c^2 < 2d$ falls inside the unit circle, representing stability.

If (3-112) is combined with the diffusion restriction $d \le 1/2$, the result is $c^2 \le 2d$ or*

$$\Delta t \le 2\alpha/u^2 \tag{3-113}$$

It is seen from equation (3-113) or the more restrictive (3-112) that FTCS of the inviscid equation ($\alpha = 0$) is unstable for *all* Δt.

The two conditions $d \le 1/2$ and $R_c \le 2$ (which together include $c \le 1$ and $c^2 \le 2d$) are necessary and sufficient to guarantee stability, in the case of the linear, constant u equation in an infinite domain. The effects of spatially varying u cannot be ascertained by this method.

In the case of more general finite-difference methods involving three or more time levels, the correspondent of (3-101) is a *matrix* equation. The stability condition then requires that $|\lambda_m| \le 1$, where λ_m are all the eigenvalues** of the matrix G. This condition is equivalent to equation (3-102) when G is simply a number. An example of this case will be presented in Section III-A-6.

In addition to stability, the von Neumann analysis also provides information on dispersion errors, which will be covered in Section III-A-14.

Exercise: Repeat the previous exercise of determining the stability conditions for the upwind differencing method, this time using the von Neumann analysis.

---

*This restriction, which does not imply the $R_c \le 2$ limitation, can also be obtained by requiring the curvature $|d^2y/dx^2|$ along G at (1,0) to be greater than the curvature of the unit circle (W. D. Sundberg, private communication.)

**The problem of determining the eigenvalues is non-trivial for complex methods and for systems of equations (which arise in compressible flow). The eigenvalue problem may itself be attacked numerically. References on eigenvalue problems include Walden (1967) and Westlake (1968).

III-A-5-c. Hirt's Stability Analysis

A third method of stability analysis was given by Hirt (1968). In this interesting method, the terms of the finite-difference equations are expanded in a Taylor series in order to develop a continuum partial differential equation. Stability is then determined from the known stability properties of the continuum equations.* (This same turn-around, of studying the finite-difference equations by generating continuum equations, was used by Cyrus and Fulton (1967) to study the accuracy, rather than stability, of finite-difference methods for elliptic equations.)

We again consider the FTCS method (3-18) for the advection-diffusion model equation, assuming constant u.

$$\frac{\zeta_i^{n+1} - \zeta_i^n}{\Delta t} = -u\left(\frac{\zeta_{i+1}^n - \zeta_{i-1}^n}{2\Delta x}\right) + \alpha\left(\frac{\zeta_{i+1}^n + \zeta_{i-1}^n - 2\zeta_i^n}{\Delta x^2}\right) \qquad (3\text{-}114)$$

We now evaluate each term of equation (3-114) by expanding in a Taylor series about (x,t), that is, about the value of ζ_i^n.

$$\zeta_i^{n+1} = \zeta_i^n + \Delta t \left.\frac{\partial\zeta}{\partial t}\right|_i^n + \frac{1}{2}\Delta t^2 \left.\frac{\partial^2\zeta}{\partial t^2}\right|_i^n + O(\Delta t^3) \qquad (3\text{-}115)$$

$$\zeta_{i\pm1}^n = \zeta_i^n \pm \Delta x \left.\frac{\partial\zeta}{\partial x}\right|_i^n + \frac{1}{2}\Delta x^2 \left.\frac{\partial^2\zeta}{\partial x^2}\right|_i^n \pm O(\Delta x^3) \qquad (3\text{-}116)$$

Substituting these into (3-114) and canceling gives

$$\frac{1}{\Delta t}\left[\Delta t \left.\frac{\partial\zeta}{\partial t}\right|_i^n + \frac{1}{2}\Delta t^2 \left.\frac{\partial^2\zeta}{\partial t^2}\right|_i^n + O(\Delta t^3)\right] = -\frac{u}{2\Delta x}\left[2\Delta x \left.\frac{\partial\zeta}{\partial x}\right|_i^n + O(\Delta x^3)\right] + \frac{\alpha}{\Delta x^2}\left[\Delta x^2 \left.\frac{\partial^2\zeta}{\partial x^2}\right|_i^n + O(\Delta x^4)\right] \qquad (3\text{-}117)$$

or, dropping the indexes i and n,

$$\frac{\partial\zeta}{\partial t} + \left(\frac{\Delta t}{2}\right)\frac{\partial^2\zeta}{\partial t^2} = -u\frac{\partial\zeta}{\partial x} + \alpha\frac{\partial^2\zeta}{\partial x^2} + O(\Delta t^2, \Delta x^2) \qquad (3\text{-}118)$$

For Δt, $\Delta x \rightarrow 0$, this equation reverts to the original PDE (3-99). But for $\Delta t > 0$, we drop HOT and rearrange equation (3-118) as

$$\left(\frac{\Delta t}{2\alpha}\right)\frac{\partial^2\zeta}{\partial t^2} - \frac{\partial^2\zeta}{\partial x^2} + \frac{1}{\alpha}\frac{\partial\zeta}{\partial t} + \frac{u}{\alpha}\frac{\partial\zeta}{\partial x} = 0 \qquad (3\text{-}119)$$

This form, obtained by retaining all the first-order terms in the Taylor series expansion, displays a *hyperbolic* character (see, for example, Weinberger, 1965). Such equations exhibit a domain of influence of an arbitrary point (x,t) demarcated by the characteristic lines of slope $\pm\sqrt{\Delta t/2\alpha}$ through (x,t), as indicated in Figure 3-9a. It is only within this domain that a perturbation at (x,t) is felt. The region on the (x,t) plane outside of the characteristics is sometimes called the zone of silence.

---

*It is fortunate that this process generates continuum equations with known stability properties. Otherwise, one might be tempted to determine the stability of the generated continuum equations by a numerical method, necessitating a stability analysis of *this* numerical method, and so on in a dizzying spiral.

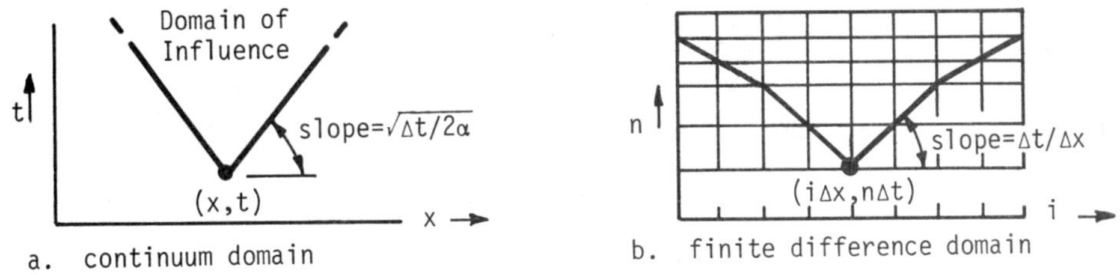

Figure 3-9. Domain of influence of the point (x,t) for the hyperbolic equation (3-118).

The finite-difference equation (3-114) also exhibits a domain of influence. Every new calculated point ζ_i^{n+1} depends on previous values at neighboring points, $\zeta_{i\pm1}^n$. Equivalently, each value ζ_i^n propagates its effect to neighboring points at the next time level, $\zeta_{i\pm1}^{n+1}$. These in turn propagate to $\zeta_{i\pm2}^{n+2}$, etc. Thus, the domain of influence of the discretized equation (3-118) is demarcated by the finite-difference "characteristic" lines of slope $\Delta t/\Delta x$, as shown in Figure 3-9b. The Courant condition (Courant, Friedrichs and Lewy, 1928) for stability of finite-difference analogs of such hyperbolic equations states that the finite-difference domain of influence at least include the continuum domain of influence.* With reference to Figure 3-9, this requires $\Delta t/\Delta x \leq \sqrt{\Delta t/2\alpha}$, or

$$\Delta t \leq \frac{1}{2}\frac{\Delta x^2}{\alpha} \tag{3-120}$$

This is the same diffusion limit on Δt as obtained previously by both the discrete perturbation stability analysis and the von Neumann analysis.

To determine another necessary stability condition, we evaluate the term $\partial^2\zeta/\partial t^2$ in equation (3-119) by differentiating the original equation,** recalling that u = constant has been assumed.

$$\frac{\partial\zeta}{\partial t} = -u\frac{\partial\zeta}{\partial x} + \alpha\frac{\partial^2\zeta}{\partial x^2} \tag{3-121}$$

$$\frac{\partial^2\zeta}{\partial t^2} = -u\frac{\partial^2\zeta}{\partial t\partial x} + \alpha\frac{\partial^3\zeta}{\partial t\partial x^2} \tag{3-122}$$

Interchanging the order of chain differentiation and substituting for $\frac{\partial\zeta}{\partial t}$ from the original PDE gives

$$\frac{\partial^2\zeta}{\partial t^2} = -u\frac{\partial}{\partial x}\left(-u\frac{\partial\zeta}{\partial x} + \alpha\frac{\partial^2\zeta}{\partial x^2}\right) + \alpha\frac{\partial^2}{\partial x^2}\left(-u\frac{\partial\zeta}{\partial x} + \alpha\frac{\partial^2\zeta}{\partial x^2}\right) \tag{3-123}$$

---

*The physical interpretation for the acoustic wave equation is particularly appealing. The Courant condition simply states that a sound wave cannot travel more than one cell length in one time increment.

**Alternately, the next result can be obtained by expanding the terms in equation (3-114) in double Taylor series centered about $\zeta_i^{n+1/2}$.

$$\frac{\partial^2 \zeta}{\partial t^2} = u^2 \frac{\partial^2 \zeta}{\partial x^2} - 2u\alpha \frac{\partial^3 \zeta}{\partial x^3} + \alpha^2 \frac{\partial^4 \zeta}{\partial x^4} \tag{3-124}$$

Substituting this into (3-119) and rearranging gives

$$\frac{\partial \zeta}{\partial t} = -u \frac{\partial \zeta}{\partial x} + \left(\alpha - \frac{u^2 \Delta t}{2}\right) \frac{\partial^2 \zeta}{\partial x^2} + u\alpha\Delta t \frac{\partial^3 \zeta}{\partial x^2} - \frac{\alpha^2 \Delta t}{2} \frac{\partial^4 \zeta}{\partial x^4} \tag{3-125}$$

Following Hirt (1968), we obtain a useful approximation by dropping the higher order derivatives in equation (3-125), retaining the lowest-ordered even and odd derivative terms in each independent variable x and t. This may be rationalized in two ways. First, the higher order derivatives are generally smaller. Second, we know *a posteriori* that the stability condition resulting from this analysis will dominate the diffusion time-step limitation only for slightly viscous flow, i.e., for $\alpha << u$, making the coefficients of the higher derivatives small in equation (3-125). The result of this approximation is

$$\frac{\partial \zeta}{\partial t} = -u \frac{\partial \zeta}{\partial x} + \alpha_e \frac{\partial^2 \zeta}{\partial x^2} \tag{3-126}$$

where

$$\alpha_e = \alpha - u^2 \Delta t/2 \tag{3-127}$$

Because the form of (3-126) is equivalent to the original model equation, we refer to α_e as an "effective viscosity coefficient."

It is the mathematical (and physical) nature of a diffusion coefficient to smear out (diffuse) a disturbance in ζ, tending toward a uniform ζ distribution. A negative diffusion coefficient is a physical impossibility. It would tend to concentrate any slight perturbation from a uniform distribution, thus constituting a monotonic instability.* Stability thus requires $\alpha_e \geq 0$, or

$$\Delta t \leq 2\alpha/u^2 \tag{3-128}$$

This is the same as equation (3-113) obtained by the von Neumann analysis. Together with equation (3-120), it includes the condition of Courant number $c = u\Delta t/\Delta x \leq 1$. This analysis does not uncover the cell Reynolds-number limitation (3-112), and is therefore seen to provide necessary but not sufficient conditions for stability of the model advection-diffusion equation.

Exercise: Repeat the previous two exercises of determining the stability conditions for the upwind differencing method, this time using Hirt's stability analysis.

III-A-5-d. Summary and Evaluation of Stability Criteria

We have presented examples of three different methods of stability analysis: the discrete perturbation stability analysis, the von Neumann stability analysis, and Hirt's stability analysis. In Hirt's analysis, we also use the criterion of Courant, Friedrichs, and Lewy (1928) for hyperbolic systems. There are at least three other methods which are more or less popular, and still others that are less popular. The boundedness of the solution can be tested directly by the Friedrich's criterion on the positivity of coefficients

---

*Applied to the heat-conduction equation, the requirement $\alpha_e \geq 0$ may be interpreted as a statement that the finite-difference equation must not violate the second law of thermodynamics.

(Richtmyer and Morton, 1967, p. 12; Hahn, 1958) and by the "energy" methods* of Keller and of Lax (see Richtmyer and Morton, 1967, p. 13, et seq.). These methods appear to be applicable in practice only to the simplest difference schemes for the differential equations. Like these two, the method of Eddy (1949) also treats the amplification properties of the finite-difference equations directly, rather than the discrete Fourier components. It appears to give the same results as the Neumann method for the simple methods tested in the original paper. It is more difficult to apply and has not been used in the open literature.

These three methods all depend on boundedness of the solution as a criterion for stability, as does the von Neumann analysis,** which requires the amplification factor $|G| \leq 1$. (But the von Neumann criterion may be modified to $|G| \leq 1 + O(\Delta t)$ to allow for cases wherein the true solutions of the continuum equations are unbounded.) These various criteria are not identical. Hahn (1958) has shown that the von Neumann, the Friedrichs, and the Courant-Friedrichs-Lewy criteria are equivalent only for the simplest finite-difference schemes where the linear equations have constant coefficients. For variable coefficients, the von Neumann condition is only necessary, and Friedrich's condition is only sufficient. For finite-difference analogs of the wave equation which use values at i farther than $(i\pm 1)$, the Courant-Friedrichs-Lewy condition is generally no longer sufficient. Mitchell (1969) says that the von Neumann condition is sufficient as well as necessary for (constant coefficient) two-level schemes with only one dependent variable and any number of independent variables; otherwise, it is *necessary* only. Further, these criteria are clearly not equivalent to those used in the discrete perturbation analysis or in Hirt's analysis, and we really should not expect the same answers in all cases.

For other insights into the definitions and criteria of stability, the following references are recommended. Hildebrand (1968, p. 205) discusses stepwise stability (concerned with time behavior as $t \to \infty$ with fixed Δt, as in the three methods given here***) versus pointwise stability (concerned with the behavior of the finite-difference equations as the mesh size approaches zero). Gustafsson (1969) defines the "A-stable" criterion. Rogers (1967) treats the stability of difference operators in the same way as transition functions in the Lyapunov theory familiar to control systems analysts. Kusic and Lavi (1968) and Lavi (1969) discuss stability of various methods, and present a non-iterative method of evaluating stability during the course of the calculation. Other works important to the mathematical foundations of stability theory are the papers of Kreiss (1964, 1968) and Osher (1969B) and the books by Forsythe and Wasow (1960) and by Keller (1968). Karplus (1958) presented an approach to FDE stability based on electric circuit theory; although sometimes successful, there appears to be an inherent ambiguity in the method which limits its utility. Another concept of numerical instability is instability in the sense of Hadamard (see, for example, Weinberger, 1965), i.e., sensitivity to initial conditions in an initial-value problem. A method for experimentally testing for this type of instability was given by Miller (1967). Cheng (1970) uses a stability analysis similar to Hirt's, but applies it only in the limit of Δx, $\Delta t \to 0$ in some specified relationship.

Thus, it is seen that stability of even linear systems is not defined with universal applicability. Frankel (1956) avoided trying to give a precise definition of stability. Richtmyer (1963) points out that the concept of stability depends on the choice of a norm in the function space of the dependent variable, and that the use of Fourier analysis as in the von Neumann method implies an L_2 or root-mean-square norm, which is somewhat arbitrary. The reader is also directed to the discussion on alternate definitions of stability in Richtmyer and Morton (1967), p. 95.

---

*Not necessarily related to physical energy, but generally the square of the dependent variable.

**In the paper most often credited with open literature presentation of the von Neumann method (O'Brien, Hyman, and Kaplan, 1950) stability was actually defined in terms of growth or decay of machine roundoff errors. This is conceptually different from boundedness, but the distinction is of no practical importance since the von Neumann analysis is the same. But boundedness is to be preferred as a criterion, since it emphasizes that stability still must be considered even in the limit of algebraically perfect computations, for general initial conditions (Lax and Richtmyer, 1956).

***The stability properties are similar to those obtained with $\Delta t \to 0$ with fixed time interval (Lax and Richtmyer, 1956).

The concept of stability is intimately related to the concepts of consistency and convergence.* A finite-difference analog of the differential equation is consistent if the finite-difference *equation* (FDE) approaches the partial differential *equation* (PDE) as $\Delta x, \Delta t \to 0$. Although this might seem to be automatically satisfied by the Taylor-series method of developing the FDE, it is not; other restrictions on the relative rates of convergence of Δx and Δt may be required (see Section III-A-7). The FDE is convergent if the *solution* of the FDE approaches the *solution* of the PDE as Δx, $\Delta t \to 0$. Two obvious necessary requirements for convergence are that the FDE be stable, in some sense, and consistent.

For linear systems such as our model equation with constant coefficients, the equivalence theorem of Lax (Lax and Richtmyer, 1956) establishes the equivalence of stability and convergence, provided that the following conditions apply: the initial-value problem must be well posed, in the sense of Hadamard (Weinberger, 1965), so that the PDE solution depends continuously on the initial data; the FDE must be consistent with the PDE; stability must be defined in the L_2 norm (as in the von Neumann analysis). Under these conditions, the von Neumann necessary condition for stability also becomes sufficient. F. John (1952) has shown that a mildly strengthened form of the von Neumann condition is sufficient for stability of linear parabolic equations, even with variable coefficients. Lax (see Lax and Richtmyer, 1956) obtained a similar result for variable-coefficient hyperbolic equations.

This equivalence theorem of Lax is certainly important, but unfortunately its significance tends to be over-emphasized. In particular, some authors have based arguments for the convergence of nonlinear FDE's on the Lax equivalence theorem for linear systems, apparently out of desperation. While we grant the utility of studying linear systems as guidelines to the nonlinear systems, it is obvious that Lax's equivalence theorem is not simply applicable to nonlinear equations. The single fact of possible non-uniqueness of solutions of the nonlinear equations, discussed in Chapter I, ought to dissuade such misuse. And even for linear systems, it is not correct to apply the Lax theorem when stability is not defined in the L_2 norm.

Even more fundamentally, a precise stability criterion is really not required mathematically. For the study of nonlinear equations, Hicks (1969) suggests skipping over the problems of stability criteria and going directly to the heart of the matter which is, after all, convergence (Lax and Richtmyer, 1956). Fundamentally, we want our FDE solution to approach the PDE solution, and stability definitions are of secondary interest. In this light, Lax's equivalence theorem may be regarded as a direct investigation of convergence, with stability so defined that the definitions are equivalent.

None of these criteria or analyses are really adequate for practical computations. In actual fluid dynamics problems, the stability restrictions are applied locally. The mesh is scanned for the most restrictive values of the stability limitations, and the resulting minimum Δt max is used throughout the mesh. The typical and prudent practice is to use some fraction, usually 0.8 or 0.9, of this analytically-indicated maximum time step. In early time calculations when transients are large, smaller fractions may be needed (e.g., Torrance, 1968).

The shortcomings of this approach are clear. Several authors (Phillips, 1959; Richtmyer, 1963; Hirt, 1968; and Gourlay and Morris, 1968A) have reported instabilities due to nonlinearity, or at least to non-constant coefficients. Others (Lilly, 1965) have reported the phenomena of time splitting of solutions (see Section III-A-6) which, though not an instability in the sense of producing unbounded solutions, is an instability in a practical sense of preventing iterative convergence. It is of fundamental import to realize that it may be impossible to distinguish between what we might call a "true" instability and just a very poor rate of convergence.

In fact, preoccupation with tidy definitions of consistency, convergence, and stability as Δx, $\Delta t \to 0$ is sometimes effete, since computations are not run under these conditions. (Occasionally, practical guidelines do result, as in the consistency requirement for the Dufort-Frankel method; see Section III-A-7.)

---

*The "convergence" discussed here is truncation-error convergence. For a discussion of the problems of iteration convergence, see Section III-D.

The final evaluation of the three methods presented in the previous sections is as follows. The commonly used von Neumann method is generally the easiest, the most straightforward, and the most dependable. An important feature is its straightforward formal extension to multidimensional problems (see next section). However, for more complicated FDE, the algebra of solving for $|G| \leq 1$ (or the eigenvalue inequality, Section III-A-6) may become intractable.* Also, the smallest disturbance accounted for is a periodic disturbance with $\lambda = 2\Delta x$; point disturbances are not resolved. It can be extended to predict the stability effects of certain boundary conditions (Campbell and Keast, 1968; P. J. Taylor, 1968).

The discrete perturbation stability analysis is much less dependable. The application of this method is a hit-or-miss affair, compared with the systematic and formalized von Neumann analysis. It did produce the same answers as the von Neumann analysis for the upwind differencing method (see last three exercises). The additional requirement of zero-overshoot, though not obviously a stability requirement in the sense of requiring a bounded solution, also produced results identical to those of the von Neumann analysis for the FTCS method, with considerably less work. But it is not at all clear that this criterion will provide the correct restrictions for more complicated schemes, so its general applicability is questionable at this time. However, it has provided insight into the stability problems associated with boundary and mesh problems, where the von Neumann analysis did not.

Hirt's method is not as formal as von Neumann's method, and the implications of some of the assumptions are not quite clear at this time. Although it has been generally successful in predicting stability limitations of simple FDE (in some cases, with less algebraic work than the von Neumann analysis), it did not disclose the cell Reynolds-number limitation of the FTCS method. It would appear that, like the discrete perturbation analysis, it as yet cannot be confidently applied to more complex methods. But it has been carefully extended to provide insight into some stability problems associated with nonlinear and nonconstant coefficients (Hirt, 1968) which cannot be accomplished readily with the von Neumann analysis.

In summary, all three methods provide information of value. The von Neumann analysis appears to be the most dependable, but none of the methods is entirely adequate. If we are concerned with actually obtaining computational solutions, and not merely with the analysis of numerical methods for its own sake, we must face up to the need for numerical experimentation, with any or preferably all of these stability analyses providing nothing more than clues to practical stability.

However, the utility of these analyses does not end with the prediction of stability conditions. The discrete perturbation analysis has the virtue of focusing attention on the discrete, individual calculations which are actually performed, rather than on a related abstraction. This provides insight in formulating boundary condition treatments, and in defining the transportive property (see Section III-A-9). The von Neumann analysis provides information not only on damping (i.e., stability) but also on the phase relationships of the FDE and the resulting dispersion errors (see Section III-A-13). The approach used in Hirt's analysis also sheds light on dispersion errors, and on the finite-difference behavior referred to as an "artificial viscosity" effect. So all three analyses have their uses, and will be applied in succeeding sections of this book.

III-A-5-e. The von Neumann Analysis in Multidimensional Problems

The extension to multi-dimensional problems of the discrete perturbation stability analysis (Thoman and Szewczyk, 1966) and Hirt's analysis (Hirt, 1968) is possible. We present here the more straightforward extension of the von Neumann analysis for the example method. Using FTCS differencing on the linearized, constant-coefficient, two-dimensional vorticity transport equation (2-12) with $\alpha = 1/R_e$ gives

---

*Numerical searches for the stability boundaries of the amplification matrix may then be required

$$\frac{\zeta_{i,j}^{n+1} - \zeta_{i,j}^{n}}{\Delta t} = -u\,\frac{\zeta_{i+1,j}^{n} - \zeta_{i-1,j}^{n}}{2\Delta x} - v\,\frac{\zeta_{i,j+1}^{n} - \zeta_{i,j-1}^{n}}{2\Delta y}$$

$$+ \alpha\left[\frac{\zeta_{i+1,j}^{n} + \zeta_{i-1,j}^{n} - 2\zeta_{ij}}{\Delta x^2} + \frac{\zeta_{i,j+1}^{n} + \zeta_{i,j-1}^{n} - 2\zeta_{i,j}^{n}}{\Delta y^2}\right] \quad (3\text{-}129)$$

Each Fourier component of the solution is now written

$$\zeta_{ij}^{n} = V^n e^{I(k_x i\Delta x + k_y j\Delta y)} \quad (3\text{-}130)$$

where V^n is again the amplitude function at time level n of the particular component whose wave numbers in the x and y directions are k_x and k_y (wavelengths $\Lambda_x = 2\pi/k_x$, $\Lambda_y = 2\pi/k_y$), and $I = \sqrt{-1}$. Defining the x and y phase angles as $\theta_x = k_x\Delta x$ and $\theta_y = k_y\Delta y$, equation (3-130) becomes

$$\zeta_{ij}^{n} = V^n e^{I(i\theta_x + j\theta_y)} \quad (3\text{-}131)$$

Similarly,

$$\zeta_{i+1,j+1}^{n+1} = V^{n+1} e^{I\left[(i+1)\theta_x + (j+1)\theta_y\right]} \quad (3\text{-}132)$$

and so forth. The dimensional counterparts of the Courant number c are defined as $c_x = u\Delta t/\Delta x$ and $c_y = v\Delta t/\Delta y$, and the counterparts of d are $d_x = \alpha\Delta t/\Delta x^2$ and $d_y = \alpha\Delta t/\Delta y^2$. Substituting into equation (3-130), canceling the common term $e^{I(i\theta_x + j\theta_y)}$, and using exponential-trigonometric identities again gives $V^{n+1} = GV^n$, with

$$G = 1 - 2(d_x + d_y) + 2d_x \cos\theta_x + 2d_y \cos\theta_y - I(c_x \sin\theta_x + c_y \sin\theta_y) \quad (3\text{-}133)$$

The obvious necessary conditions for $|G| \leq 1$ are

$$d_x + d_y \leq \frac{1}{2} \quad (3\text{-}134)$$

and

$$c_x + c_y \leq 1 \quad (3\text{-}135)$$

For the special case of $d_x = d_y = d$, equation (3-134) gives

$$d \leq \frac{1}{4} \quad (3\text{-}136)$$

which is twice as restrictive as the one-dimensional diffusion limitation. For the special case of $c_x = c_y = c$, equation (3-135) gives

$$c \leq \frac{1}{2} \quad (3\text{-}137)$$

which is again twice as restrictive as the corresponding one-dimensional necessary condition. Fromm (1964) shows that, for the special case of $\Delta x = \Delta y = \Delta$ and $\theta_x = \theta_y$, the cell Reynolds number limitation on $\bar{R}_c = (|u| + |v|)\Delta/\alpha$ is

$$\bar{R}_c \leq 4 \tag{3-138}$$

which is less restrictive than the one-dimensional case.

Exercise: Apply the von Neumann analysis to the two-dimensional upwind differencing method for the inviscid transport equation,

$$\frac{\zeta_{ij}^{n+1} - \zeta_{ij}^{n}}{\Delta t} = -u\,\frac{\zeta_{ij}^{n} - \zeta_{i-1,j}^{n}}{\Delta x} - v\,\frac{\zeta_{ij}^{n} - \zeta_{i,j-1}^{n}}{\Delta y} \quad \text{where } u > 0 \;,\; v > 0 \tag{3-139}$$

Show that the stability restriction is

$$c_x + c_y \leq 1 \tag{3-140}$$

Exercise: Apply the von Neumann analysis to the FTCS version of the 3D diffusion equation,

$$\frac{\partial \zeta}{\partial t} = \alpha\left(\frac{\partial^2 \zeta}{\partial x^2} + \frac{\partial^2 \zeta}{\partial y^2} + \frac{\partial^2 \zeta}{\partial z^2}\right) \tag{3-141}$$

and show that

$$dx + dy + dz \leq \frac{1}{2} \tag{3-142}$$

is necessary and sufficient for stability. For the special case of $dx = dy = dz = d$, equation (3-142) gives

$$d \leq \frac{1}{6} \tag{3-143}$$

which is three times as restrictive as the one-dimensional case.

III-A-6. One Step Explicit Methods: The Midpoint Leapfrog Method

The crude FTCS method we have been considering for the linear model equation, using forward time differences and centered space differences, is a one-step, explicit, two-time-level method. It is "one-step" because only one calculation step is required to advance to a new time level. It is "explicit" because all the values on the right-hand side of (3-44c) needed to calculate the advance time-level values of ζ^{n+1} are known; that is, values of $\zeta_{i\pm1}^{n+1}$ do not appear on the right-hand side of the equation.* It is "two-time-level" because only two time levels are involved in the calculation; the new values at (n+1) are calculated solely in terms of old values at (n).**

In equation (3-17) we presented a method using centered space and centered time derivatives, which we said was completely unstable for all $\alpha > 0$ and $\Delta t > 0$. But when applied

---

*Methods wherein ζ_{i+1}^{n+1} and ζ_{i-1}^{n+1} appear on the right-hand side also are called implicit and generally require matrix inversion to calculate a new time level.

***Some authors have used the term "one step" for what we here refer to as "one level" methods.

to the advection terms alone, (i.e., $\alpha = 0$) this method, called the "midpoint rule" (Lilly, 1965), or the "midpoint leapfrog" method, or more frequently just "leapfrog" method (Courant, Friedrichs, and Lewy, 1928) has some very desirable stability properties.

$$\frac{\partial \zeta}{\partial t} = -\frac{\partial (u\zeta)}{\partial x} \qquad (3\text{-}144)$$

$$\frac{\zeta_i^{n+1} - \zeta_i^{n-1}}{2\Delta t} = -\frac{u\zeta_{i+1}^{n} - u\zeta_{i-1}^{n}}{2\Delta x} \qquad (3\text{-}145)$$

This method is second-order accurate in space and time, is one-step, explicit, and three-time-level. That is, values at (n) and at (n-1) are needed to calculate new values at (n+1). Note that the new ζ at an *even* time step is calculated as ζ at the previous *even* time step plus a change, skipping over the previous *odd* time step; hence, the term "leap-frog."

The von Neumann stability analysis for this and other multi-level methods proceeds as follows. Using the same definitions and assumptions as in the previous examples, equation (3-145) becomes

$$\zeta_i^{n+1} = \zeta_i^{n-1} - c\left(\zeta_{i+1}^{n} - \zeta_{i-1}^{n}\right) \qquad (3\text{-}146)$$

where c is the Courant number, $u\Delta t/\Delta x$. Substituting the Fourier components into (3-146), the amplitude relation is obtained as

$$V^{n+1} = aV^{n} + V^{n-1} \qquad (3\text{-}147)$$

where

$$a = -2Ic \sin \theta \qquad (3\text{-}148)$$

To obtain the needed matrix equation, we supplement equation (3-147) with the identity

$$V^{n} = (1)V^{n} + 0(V^{n-1}) \qquad (3\text{-}149)$$

Combined with equation (3-147), this gives

$$\begin{bmatrix} V^{n+1} \\ V^{n} \end{bmatrix} = G \begin{bmatrix} V^{n} \\ V^{n-1} \end{bmatrix} \qquad (3\text{-}150)$$

where the amplification factor G is now the *matrix*

$$G = \begin{bmatrix} a & 1 \\ 1 & 0 \end{bmatrix} \qquad (3\text{-}151)$$

For the previously studied one-level method, G was simply a number, and the stability criterion was $|G| \leq 1$. For the present case where G is a matrix, the stability criterion (following von Neumann) becomes

$$|\lambda| \leq 1 \qquad (3\text{-}152)$$

where λ represents all possible *eigenvalues* of G.* An eigenvalue λ of a matrix is defined as the number which, when subtracted from all the diagonal elements, zeros the determinant of the matrix. The defining equation for λ of G is then

$$\begin{vmatrix} (a-\lambda) & 1 \\ 1 & (0-\lambda) \end{vmatrix} = 0 \tag{3-153}$$

[When G is simply a number, as in previous examples, it is viewed as a degenerate, one-element matrix. The eigenvalue definition then becomes $G - \lambda_{-} = 0$, or $\lambda = G$, so equation (3-152) reduces to our previous criterion (3-102) of $|G| \leq 1$.] Expanding the determinant and solving the resultant quadratic equation for λ gives the two solutions

$$\lambda\pm = \frac{1}{2}\left[a \pm \sqrt{a^2 + 4}\right] \tag{3-154}$$

With $a = -2Ic \sin\theta$ and $a^2 = -4c^2 \sin^2\theta$, this becomes

$$\lambda\pm = -Ic \sin\theta \pm \sqrt{1 - c^2 \sin^2\theta} \tag{3-155}$$

For the cases when $c^2 \sin^2\theta > 1$, the square-root term is imaginary, and we get

$$\lambda\pm = I\left[-c \sin\theta \pm \sqrt{c^2 \sin^2\theta - 1}\right] \tag{3-156}$$

The modulus $|\lambda| > 1$, indicating instability. For the case when $c^2 \sin^2\theta \leq 1$, which generally requires

$$c \leq 1 \tag{3-157}$$

we solve for the modulus as

$$|\lambda\pm|^2 = c^2 \sin^2\theta + (1 - c^2 \sin^2\theta) \tag{3-158}$$

$$|\lambda\pm| = 1 \text{ for } c \leq 1 \tag{3-159}$$

This obviously satisfies the stability requirement (3-152) marginally. The same result accrues in two dimensions, where $c_x + c_y \leq 1$ is required for stability.

It might be thought that this marginal satisfaction of the stability requirement $|\lambda\pm| \leq 1$ is marginally acceptable, but actually it is highly desirable, since the continuum equation itself has no damping. The inviscid advection equation (3-144) for constant u expresses the fact that an arbitrary initial distribution function $\zeta(x,0)$ is simply translated at the advection speed u; for any elapsed time τ, the solution is

$$\zeta(x,t + \tau) = \zeta(x - u\tau, t) \tag{3-160}$$

*A frequently used statement, of some semantic convenience, is that the "spectral radius" of G must be less than or equal to one; that is, $\rho(G) \leq 1$, where $\rho(G) = \max|\lambda_p|$ and where λ_p are the p eigenvalues of G. The spectral radius $\rho(G)$ is obviously the radius of a circle in the complex plane with origin at (0,0) inside of which all eigenvalues lie.

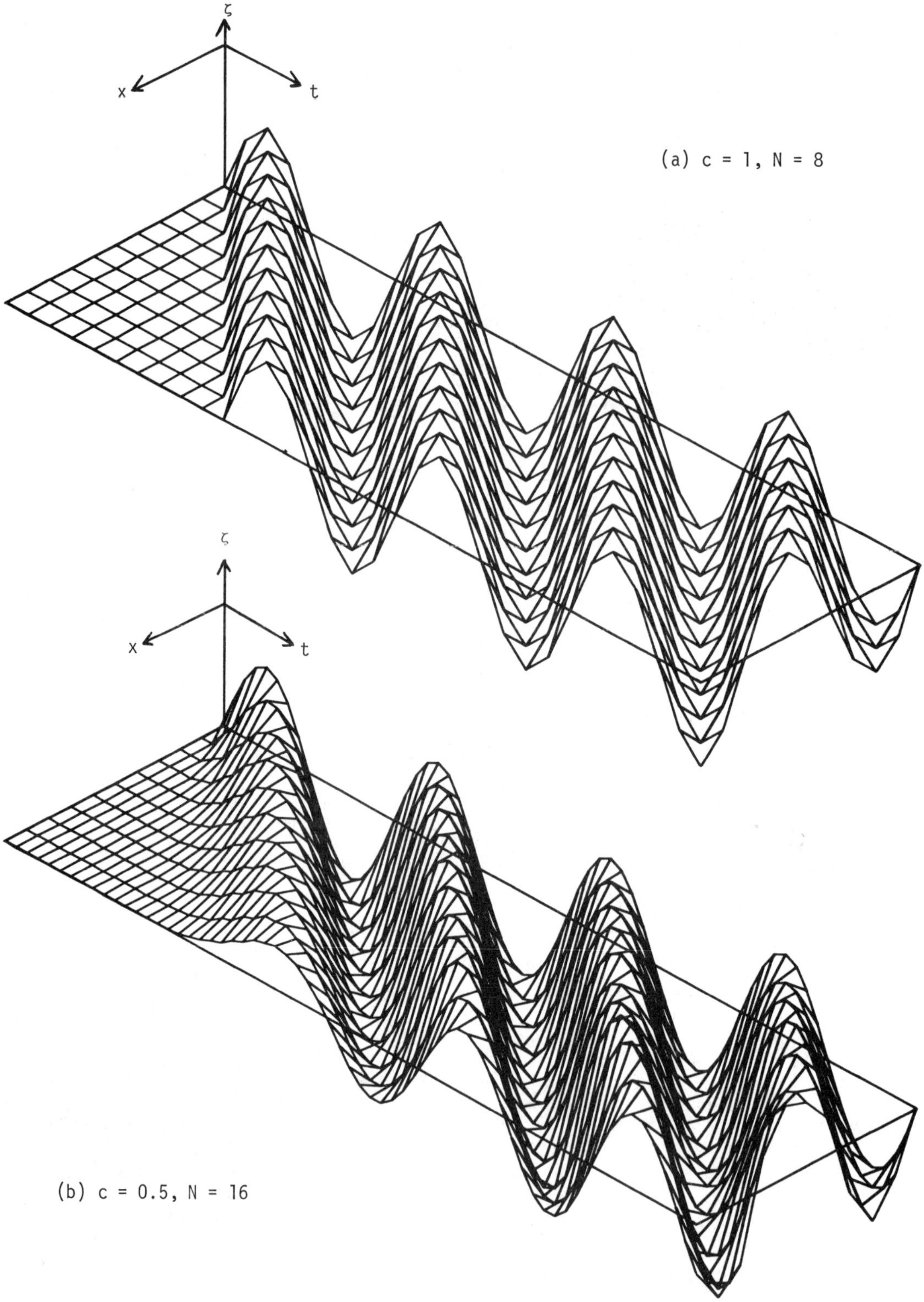

Figure 3-10. Solutions to $\partial\zeta/\partial t = -u\partial\zeta/\partial x$ obtained with the midpoint leapfrog method. c = Courant number and N = sampling frequency. (Plots courtesy W. Sundberg, Sandia Laboratories.)

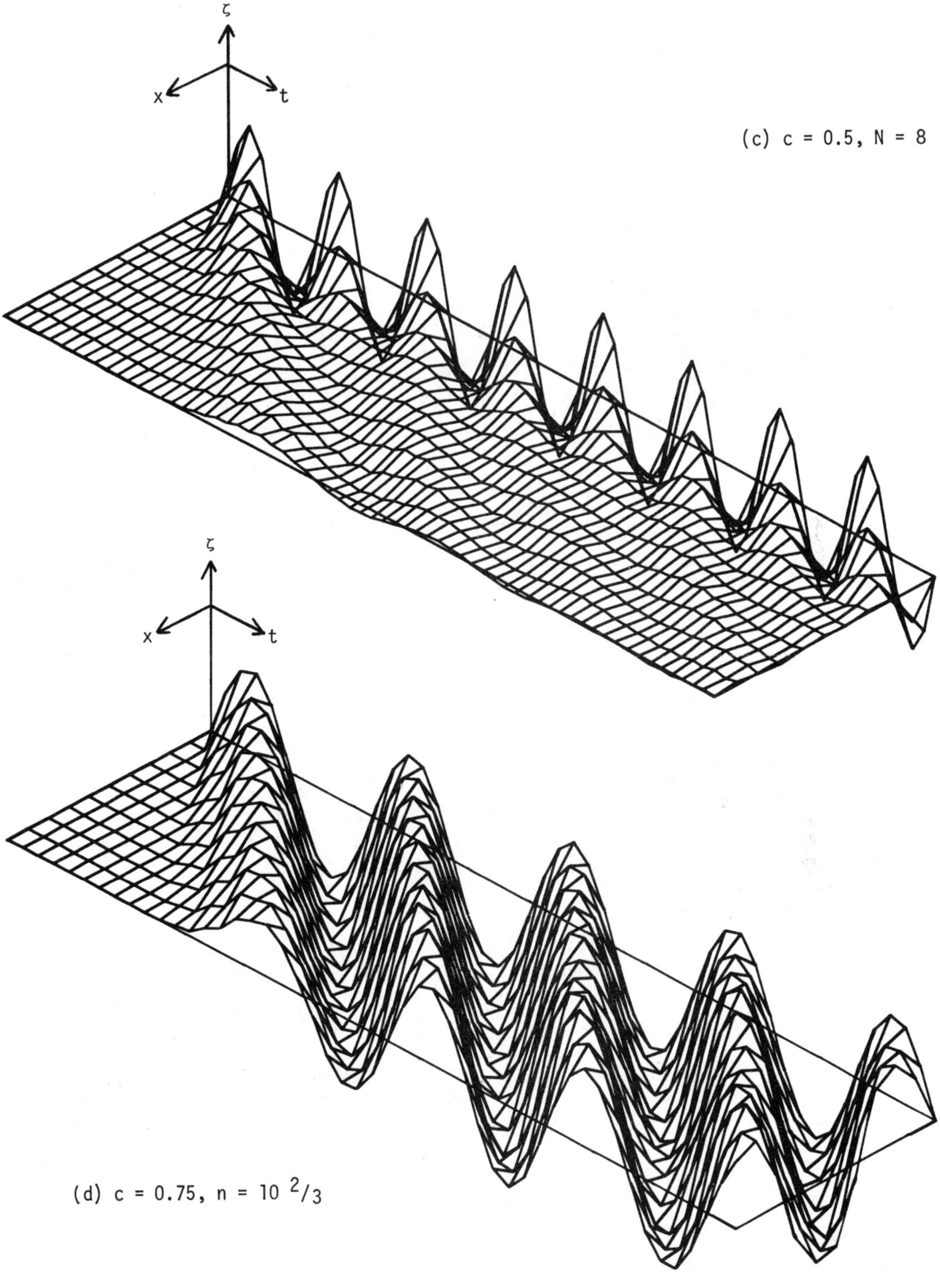

Figure 3-10. (continued)

Thus, for u = constant, the von Neumann analysis indicates that the leapfrog method identically models one feature of the continuum equation, in that no damping occurs. Any numerical method for the inviscid equation which has $|G| < 1$ exhibits an artificial or numerical damping error. For any convergent method, the numerical damping error must, of course, vanish as Δx, $\Delta t \to 0$, but the von Neumann analysis indicates that the leapfrog method has zero damping error even for finite Δx and Δt, for the case u = constant and $c \leq 1$.

In fact, for the special case c = 1, the method gives exactly the correct results. With $\tau = \Delta t$ in (3-160) we have $(x - u\tau) = (x - c\Delta x)$ so for c = 1, (3-160) can be written

$$\zeta_i^{n+1} = \zeta_{i-1}^n \tag{3-161}$$

That is, the exact continuum solution evaluated at the location of the finite-difference mesh is just a point-to-point transfer of ζ values, from (i-1) at (n) to (i) at (n+1). The advection velocity u carries ζ the distance $u\Delta t$ in one time step, and the distance $u\Delta t$ is equal to Δx for c = 1. Over two time steps, the exact solution is

$$\zeta_i^{n+1} = \zeta_{i-2}^{n-1} \tag{3-162}$$

Application of the leapfrog method, (3-146), with c = 1, gives

$$\zeta_i^{n+1} = \zeta_i^{n-1} - \zeta_{i+1}^n + \zeta_{i-1}^n \tag{3-163}$$

Given the correct starting values according to equation (3-161), $\zeta_{i+1}^n = \zeta_i^{n-1}$ and $\zeta_{i-1}^n = \zeta_{i-2}^{n-1}$, then (3-163) becomes identical to the exact solution (3-162). Thus, the leapfrog method produces the exact solution for constant u and c = 1.

However, in a practical fluid dynamics calculation, the problem of spatially varying u (ignoring the additional complications of nonlinearity) means that Δt will be limited by the largest u in the mesh. Thus, we cannot generally have c = 1 everywhere. For c < 1, errors occur.

First of all, despite the results of the von Neumann analysis, there can be a kind of numerical "damping" in the method, although this point has not been generally admitted. Figure 3-10 presents three-dimensional plots $\zeta(x,t)$ computed by the leapfrog method for the case of sinusoidally varying input $\zeta(o,t) = \sin(t)$. For c = 1, the exact solution is obtained, with the sinusoidal input being advected without damping, as shown in part a of Figure 3-10. Part b was produced at half the time step, giving c = 1/2; the maximum amplitude of the first pulse is clearly seen to decrease as it travels downstream. Part c is also produced at c = 1/2, but with the advection speed and time step adjusted so as to give the same Δx as part a; the "damping" is very strong in this case.

A problem of semantics exists. The von Neumann analysis has become so well-known that "damping" is commonly thought of in terms of the Fourier components, with $|\lambda| < 1$. Clearly, if "damping" is taken to mean $|\lambda| < 1$ by definition, then the midpoint leapfrog method has no damping, by definition. The decrease in extreme amplitudes shown in Figure 3-10b and c is correctly attributed to dispersion error,* which will be discussed shortly, and which does manifest itself in the von Neumann analysis. This terminology is precise and is even recommended, as long as we keep in mind that the method can indeed produce a decrease in extreme amplitudes of waves. In ordinary language, this would certainly be described as a "damping"; we would say that the wave in Figure 3-10c "damps out" as it is advected downstream.

Another very interesting point is to be made from Figure 3-10. It is not usually recognized that there are *two* characteristic parameters for the FDE solution to problems like that of Figure 3-10. The first parameter is the Courant number, which is the only

---

*See pages 80 and 81 for a discussion of dispersion error. Briefly, the error in wave speed is different for each Fourier component, so each has a different advection speed. Thus the Fourier components of the original distribution tend to spread out or *disperse* as the solution proceeds. The error is usually larger for the shortest wavelength components, and of course depends on the Courant number, with no error occuring at c = 1.

parameter which appears in the interior-point FDE. The second parameter is the "sampling rate" $N = 2\pi/\Delta t$, i.e., the number of time levels per cycle at which inflow boundary values are defined.

By comparing parts b and c of Figure 3-10, we see that, for a fixed $c < 1$, the "damping" (decrease in extreme amplitudes) diminishes with increased sampling rate, when the sampling rate N is integer. But when N is not integer, as is the case in part d, "undamping" occurs; due to phase errors, amplitudes as much as 15% greater than the peak inflow amplitude are evidenced in Figure 3-10d. Nor is this effect entirely attributable to the outflow boundary condition; before any outflow effect is felt, an 8% "undamping" occurs.

There are other errors and anomalies in the leapfrog method. The continuum equation (3-144) is first order in space and time; the solution for all $x > 0$, $t > 0$ is completely specified by an initial condition function, $\zeta(x,0)$, and a boundary-condition function, $\zeta(0,t)$. But the discrete analog (3-145) requires two sets of initial data to start the calculations, since values at (n-1) and (n) are needed to calculate the (n+1) level. Thus, the FDE is really a second-order equation in time, requiring initial conditions of ζ_i^1 and ζ_i^2 ; this is analogous to specifying $\zeta(x,0)$ and $\frac{\partial\zeta}{\partial t}(x,0)$ in the continuum equations, which would *over*-specify the continuum problem. Another starting FDE must be used to generate $\zeta_i^{n=2}$ after which the leapfrog method may be used. If the starting method gives exact results, as assumed after equation (3-163), the subsequent leapfrog solution with $c = 1$ is correct. If the starting method produces an error in $\zeta^{n=2}$, this error persists in the subsequent leapfrog calculations. It is thus more precise to say that the leapfrog method with $c = 1$ *perpetuates*, rather than produces, the exact solution for all time, given an exact first time-level solution.

Another type of error in the leapfrog method (and all other methods) for $c < 1$ is the *phase error*. In the continuum solution, the entire initial distribution $\zeta(x,0)$ is propagated at the advection speed u. In the finite-difference calculations, different Fourier components have different advection speeds, with the speed of the longest wavelength Λ approaching the correct value u, and shorter wavelengths being propagated at slower speeds. A more complete analysis of phase error will follow in Section III-A-13, but the phenomenon is easily demonstrated by considering the smallest possible wavelength $\Lambda = 2\Delta x$, in a spatial mesh of infinite extent, as shown in Figure 3-11. For generality, the disturbances are shown with different amplitudes at (n-1) and (n), a situation representative of using a starting method which does not have $G = 1$.

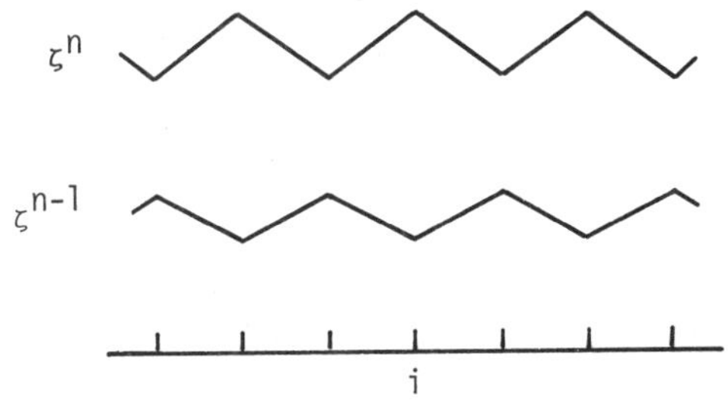

Figure 3-11. Fourier component of wavelength $\Lambda = 2\Delta x$ in a mesh of infinite spatial extent.

Considering Figure 3-11, it is clear that

$$\left.\frac{\delta\zeta}{\delta t}\right|_i^n = \frac{\zeta_{i+1}^n - \zeta_{i-1}^n}{2\Delta x} = 0 \qquad (3\text{-}164)$$

for all i. Thus, $\zeta_i^{n+1} = \zeta_i^{n-1}$, $\zeta_i^{n+2} = \zeta_i^n$, etc. The calculations alternate between two false and arbitrary-amplitude initial distributions. The Fourier component with $\Lambda = 2\Delta x$ is *completely stationary,* with a complete phase speed error.

This weird behavior is resolvable with the seemingly contradictory fact that the leapfrog method with c = 1 perpetuates the exact solution. If started with the exact solution $\zeta_i^n = \zeta_{i-1}^n$, the next correct solution is indeed $\zeta_i^{n+1} = \zeta_{i-2}^{n-1} = \zeta_i^{n-1}$ for the $\Lambda = 2\Delta x$ component. It is clear that, although the method is formally of second-order accuracy, *in fact the accuracy is determined by the accuracy of the starting method.**

Computational experience shows that the phenomena indicated by this model problem with constant advection velocity does occur in nonlinear problems also. Thus, there is always the possibility in a practical computation of *time-step splitting* (Lilly, 1965) wherein two unrelated, disjoint solutions develop which alternate at each time step. Notice that, since $\delta\zeta/\delta t = 0$, changing the time step does not change the two disjoint solutions! Lilly (1965) has pointed out that this time-splitting "instability" is likely to develop as a steady-state solution is approached. The present author has also encountered the phenomenon in two-dimensional flows, even with viscous terms present. In a calculation of backstep flow, the viscous terms (which may not be treated by leapfrog differencing; see Section III-A-7) did tend to bring the two disjoint solutions together, but even at Re as low as 100, the split solutions developed as a steady state was approached.**

The leapfrog method for the advection terms is still useful in low Re flows (Hung and Macagno, 1966), and in inviscid flows provided that an accurate starting solution is separately computed and that a steady state is not approached.

An additional error occurs in the leapfrog method, and indeed in all methods which use second-order-accurate centered-space derivatives. Consider a finite computational mesh in which the largest value of i is IL. Application of leapfrog differencing (3-145) to calculate ζ_{IL}^{n+1} would require knowledge of ζ_{IL+1}^{n}, which is outside the computational mesh. Thus, the computation cannot be performed at IL, and the evaluation of ζ_{IL}^{n+1} must be accomplished by a computational boundary condition at IL. These conditions are the subject of section III-C. We note here that this requirement is again analogous, as in the need for two sets of initial conditions, to overspecifying the problem in the continuum case. We also note here that the commonly used zero-gradient method for specifying the outflow boundary condition (to be covered in section III-C-7) by setting $\zeta_{IL}^{n+1} = \zeta_{IL-1}^{n+1}$ does have the effect of moving the otherwise stationary Fourier component with $\Lambda = 2\Delta x$, but this motion has nothing to do with advection. As time progresses, the $\Lambda = 2\Delta x$ component decays from the right (outflow) to left in the mesh, instead of the correct advection proceeding from left to right.

This false requirement for an additional outflow boundary condition is the result of yet another type of error, called a transportive error, which will be discussed in later sections.

Exercise: By hand computation and sketching, verify that the midpoint leapfrog method does produce the correct behavior at an inflow boundary. Given an initial condition comprised of only the $\Lambda = 2\Delta x$ component and the inflow boundary condition fixed at its initial value for all time, start the leapfrog method with the correct solution at the second time step, for c = 1. Show that the initial profile correctly advects out of the mesh, for c = 1.

---

*To obtain a good starting solution, one may resort to smaller Δt for 10 or 20 steps, using some two-level method. But this question of a starting solution is not trivial. For example, Polger (1971) has shown the important effects of the starting solution on long-term nonlinear instability of inviscid equations using leapfrog time differencing.

**Fromm (1967) said that Mintz (1965) overcame this disjoint solution (which he terms a "phase instability," page 97) by "periodically setting the solution at time (n-1) equal to that at time (n)." Just how often to do this, when to start, and what effects will appear in the transient and in the steady-state solution are not clear. The property of disjoint solutions remains a problem.

It also should be noted that the term "leapfrog" has been applied to many methods which differ in the evaluation of spatial derivatives, but which all use the three-level centered time-differencing of the presently described method.

III-A-7. The DuFort-Frankel Leapfrog Method

As we have seen, the midpoint leapfrog method for the inviscid model equation, using time- and space-centered derivatives, has some desirable features, including second-order accuracy in space and time, and an amplification factor $|G| = 1$. Unfortunately, when this differencing method is applied to the diffusion equation, it gives Richardson's method, which is unconditionally unstable. Since $|G| = 1$ for the advection term alone and the diffusion equation gives $|G| > 1$, it is not surprising that the advection-diffusion equation (3-17) also is unconditionally unstable.

Several workers have combined leapfrog advection differences, giving errors of $O(\Delta t^2, \Delta x^2)$, with forward-time, centered-space differencing (over $2\Delta t$) of the diffusion terms, giving errors of $O(\Delta t, \Delta x^2)$.

$$\frac{\zeta_i^{n+1} - \zeta_i^{n-1}}{2\Delta t} = - \frac{u\zeta_{i+1}^n - u\zeta_{i-1}^n}{2\Delta x} + \alpha \frac{\zeta_{i+1}^{n-1} + \zeta_{i-1}^{n-1} - 2\zeta_i^{n-1}}{\Delta x^2} \tag{3-165}$$

Formally, the "order" of the truncation error refers to the error magnitude as $\Delta x, \Delta t \to 0$, so the combined error of equation (3-165) is $O(\Delta t, \Delta x^2)$. Practically speaking, the "size" of the error can be less. For a small but non-zero Δt, we may have $\alpha = 1/Re = O(\Delta t)$. In such a case, the first error term in the Taylor series expansion for the diffusion term will be $O\,[\alpha(\Delta t, \Delta x^2)]$, and the "size" of the total error of equation (3-165) will be $(\Delta t^2, \Delta x^2)$.

The stability of equation (3-165) will be limited by the smallest of the advection limit, $c \equiv u\Delta t/\Delta x \leq 1$, and the diffusion limit, $d \equiv \alpha\Delta t/\Delta x^2 \leq 1/2$. (The stability of the advection and diffusion terms may be analyzed separately in many cases, but not always. See Kasahara (1965), for example.) A well-known explicit method for removing the diffusion restriction is the method of DuFort and Frankel (1953). This method has been used successfully in a great number of fluid-dynamics problems: for example, Payne (1958), Fromm and Harlow (1963), Fromm (1963, 1965, 1967), Amsden and Harlow (1964), Hung and Macagno (1966), and Torrance (1968).

The center node value ζ_i^n in the diffusion term of equation (3-17) is replaced by its average at times (n-1) and (n+1), giving

$$\frac{\zeta_i^{n+1} - \zeta_i^{n-1}}{2\Delta t} = - \frac{u\zeta_{i+1}^n - u\zeta_{i-1}^n}{2\Delta x} + \alpha \frac{\zeta_{i+1}^{n} + \zeta_{i-1}^{n} - \zeta_i^{n+1} - \zeta_i^{n-1}}{\Delta x^2} \tag{3-166}$$

Although (n+1) values appear on the right-hand side, they appear only at the space position (i). Hence, equation (3-166) may be solved explicitly as

$$\zeta_i^{n+1} = \left[\zeta_i^{n-1} - \frac{\Delta t}{\Delta x}\left(u\zeta_{i+1}^n - u\zeta_{i-1}^n\right) + \frac{2\alpha\Delta t}{\Delta x^2}\left(\zeta_{i+1}^n + \zeta_{i-1}^n - \zeta_i^{n-1}\right)\right] \Big/ \left(1 + \frac{2\alpha\Delta t}{\Delta x^2}\right) \tag{3-167}$$

It may be verified by the von Neumann stability analysis that the only stability restriction on equation (3-167) is the inviscid requirement of $c \leq 1$. But in multidimensions, large Re may reduce this criterion by more than 50%. (See Schumann, J.Comp.Phys.,Vol.18,No.4,1975,pg 465.)

The cost of this device is peculiar. If we expand equation (3-167) in a Taylor series expansion (as in Section III-A-5-c), we obtain a *hyperbolic* equation

$$\alpha \left(\frac{\Delta t}{\Delta x}\right)^2 \frac{\partial^2 \zeta}{\partial t^2} + \frac{\partial \zeta}{\partial t} = - u \frac{\partial \zeta}{\partial x} + \alpha \frac{\partial^2 \zeta}{\partial x^2} \tag{3-168}$$

The stability of the DuFort-Frankel method "may be regarded as arising from the introduction of this hyperbolic term in the approximating differential equation" (DuFort and Frankel, 1953). Thus, the FDE (3-167) is a consistent analog (that is, equation (3-168) approaches the model advection-diffusion (2-18) as Δx, $\Delta t \to 0$) *only* if the limits Δx, $\Delta t \to 0$ are taken such that $\Delta t/\Delta x \to 0$. If we instead take Δt, $\Delta x \to 0$ with $\Delta t/\Delta x \neq 0$, we find that we have a consistent analog for the hyperbolic equation (3-168).

In practice, the limits Δx, $\Delta t \to 0$ are not taken, and the idea of consistency becomes vague. But equation (3-168) does give a practical guide to computation. Richtmyer (1957) suggests that if we are numerically checking convergence of an unsteady solution by the usual practice (from ordinary differential equations) of recomputing with halved Δx, then we should reduce Δt by more than 1/4. Also, if we bring the second time derivative of equation (3-168) to the right-hand side, we have

$$\frac{\partial \zeta}{\partial t} = -u\frac{\partial \zeta}{\partial x} + \alpha \frac{\partial^2 \zeta}{\partial x^2} + 0\left[\Delta t^2, \Delta x^2, \alpha(\Delta t/\Delta x)^2\right] \tag{3-169}$$

This shows that the method is second-order accurate only if $\alpha(\Delta t/\Delta x)^2$ is small. In fact, solving for the condition of only first-order accuracy in time gives $0[\alpha(\Delta t/\Delta x)^2] = 0(\Delta t)$, or $(\alpha\Delta t/\Delta x^2) = 0(1)$. This is the same ordering required by the stability limitation of (3-73), that $d \leq 1/2$, for the simple first-order accurate FTCS method!

We have noted that finite-difference analogs may duplicate certain properties of the continuum equations without taking the limits Δx, $\Delta t \to 0$. These include the order-of-differentiation property $\delta^2 f/\delta x \delta y = \delta^2 f/\delta y \delta x$, the conservative property, and the unity amplification factor of the midpoint leapfrog method; we will shortly discuss the transportive property. The diffusion equation above possesses a boundedness property, in that $\zeta(x,t)$ will never exceed the maximum of the initial and boundary conditions* applied to $\partial \zeta/\partial t = \alpha \partial^2 \zeta/\partial x^2$. This is also true for the FTCS analog, provided the calculation is stable. (This property may be inferred from the zero-overshoot behavior, as discussed in Section III-A-5-2.) Gordon (1968) has shown that the DuFort-Frankel method misses this behavior by terms of $0(\Delta t, \Delta x)$, which is an additional shortcoming of the method.

Taylor (1970) has shown that the Neumann boundary conditions (specified values of ζ gradients) may destabilize DuFort-Frankel calculations of the diffusion equation, unless the near-boundary differencing is carefully handled in a manner consistent with the interior point method. We would expect that this behavior would not be important in high Re flows, although it would in low Re flows and in pure diffusion problems. Allen (1968) did experience some boundary difficulty in attempting to apply the method to the compressible flow equations.

Although the DuFort-Frankel method is thus seen to have several shortcomings, it is still advantageous to have an explicit unconditionally stable method. And in a practical computation with fixed Δx and small α, the equation may be "second-order" accurate, as argued above, in the sense of small "size" of the errors rather than true "order." Also, Pearson (1965A,B) has demonstrated that the method actually is more accurate than the FTCS method in a particular practical computation; see also Fromm (1964).

DuFort-Frankel differencing of diffusion terms may also be used in conjunction with other three-level methods for the advection terms, but stability must be checked each time for the total equation. The only other one-step, explicit, unconditionally stable method for the diffusion equation is one of the Saul'yev methods (Saul'yev, 1964; Richtmyer and Morton, 1967; Carnahan et al., 1969; see also Section III-A-17). In unpublished analyses by the present author, this approach has not proved adaptable to the entire advection-diffusion method. Using any of several treatments of the advection terms, the result gives back the simple FTCS diffusion time limit $d < 1/2$ whenever the Courant number $c > 0$. Also, the Saul'yev method is actually implicit in its boundary conditions, which requires special treatment in a fluid dynamics calculation.

---

*Applied to the heat-conduction equation, this property is associated with the second law of thermodynamics. The temperature in a block of material controlled by simple diffusion can never exceed its highest initial value and its highest boundary value.

It is instructive to consider a two-level method similar to the DuFort-Frankel method. Allen (1968) has pointed out the subtle pitfall of the approach. Considering FTCS differencing of the diffusion equation,

$$\frac{\zeta_i^{n+1} - \zeta_i^n}{\Delta t} = + \alpha\left(\frac{\zeta_{i+1}^n + \zeta_{i-1}^n - 2\zeta_i^n}{\Delta x^2}\right) \tag{3-170}$$

we proceed in a manner analogous to the DuFort-Frankel method, and replace the center value ζ_i^n of the diffusion term by ζ_i^{n+1}, giving

$$\frac{\zeta_i^{n+1} - \zeta_i^n}{\Delta t} = \alpha\left(\frac{\zeta_{i+1}^n + \zeta_{i-1}^n - 2\zeta_i^{n+1}}{\Delta x^2}\right) \tag{3-171}$$

Then ζ_i^{n+1} can be solved explicitly. A von Neumann stability analysis then shows, as hoped, that the method is unconditionally stable. But the method can be rewritten, with a little algebra, as

$$\frac{\zeta_i^{n+1} - \zeta_i^n}{\Delta t'} = \alpha\left(\frac{\zeta_{i+1}^n + \zeta_{i-1}^n - 2\zeta_i^n}{\Delta x^2}\right) \tag{3-172}$$

wherein

$$\Delta t' = \Delta t \Big/ \left(1 + \frac{2\alpha\Delta t}{\Delta x^2}\right) \tag{3-173}$$

Now equation (3-172) is limited by $\Delta t' \leq 1/2\ \Delta x^2/\alpha$. But from equation (3-173), as $\Delta t \to \infty$, we have $\Delta t' \to 1/2\ \Delta x^2/\alpha$. In other words, the method of equation (3-171) is nothing more than an obscure way of using the simple FTCS method, with an erroneously small time step (erroneous because the user thinks the result of n calculations applies to the time $n\Delta t$, when it actually should apply to the time $n\Delta t'$).*

It is instructive to consider the DuFort-Frankel method applied to a stationary (steady-state) flow. In that case,

$$\zeta_i^n = \frac{1}{2}\left(\zeta_i^{n+1} + \zeta_i^{n-1}\right) \tag{3-174}$$

and it may easily be seen that the DuFort-Frankel method (3-166) is algebraically equivalent to the Richardson method (midpoint leapfrog on both advection and diffusion terms). Yet the analyses indicate that the DuFort-Frankel method is stable, while the Richardson method is unconditionally unstable. The resolution of this paradox is connected with the idea of a "well-posed" problem, or the sensitivity of the solution to initial data. Certain special cases are not significant to a practical computation, and we do not define stability in terms of these special cases. That is, if

$$\zeta_i^n = \frac{1}{2}\left(\zeta_i^{n+1} + \zeta_i^{n-1}\right) + \varepsilon_i \tag{3-175}$$

where the ε_i are arbitrarily small but not all are identically zero, then the FDE calculation with Richardson's method will diverge, but the DuFort-Frankel calculation will be stable. In any true fluid dynamics problem, ε_i will not be identically zero.

---

*Although this method is in error, the error is a truncation error. Strictly speaking, the method could be mathematically consistent if, for example, the limits Δx, $\Delta t \to 0$ were taken such that $\Delta t/\Delta x^3$ was a constant, giving $t' \to t$ in the limit.

Also note that, for $d \equiv \alpha\Delta t/\Delta x^2 = 1/2$, the DuFort-Frankel method applied to the diffusion equation with no advection is algebraically equivalent to the FTCS method. But, for $d \neq 1/2$, the method, the answers, and the stability behavior are different from the FTCS method.

III-A-8. The First Upwind Differencing Method; Artificial Viscosity Errors

A one-step, explicit, two-time-level method which achieves static stability of the advection terms involves the use of one-sided, rather than space-centered, differencing. Backward differences are used when the velocities are positive, and vice-versa.* Thus, the one-sided difference is always on the "upwind" or "upstream" side of the point at which $\delta\zeta/\delta t$ is evaluated. The method has a truncation error $E = O(\Delta t, \Delta x)$.

$$\frac{\zeta_i^{n+1} - \zeta_i^n}{\Delta t} = -\frac{u\zeta_i^n - u\zeta_{i-1}^n}{\Delta x} \quad \text{for } u > 0$$

$$= -\frac{u\zeta_{i+1}^n - u\zeta_i^n}{\Delta x} \quad \text{for } u < 0 \qquad (3\text{-}176)$$

This method has often been used in the literatures under various names and with different rationales. Meteorologists have long known of the stabilizing effect of "upwind" (Lilly, 1965; Forsythe and Wasow, 1960) or "weather" (Frankel, 1956) differencing,** and have applied it in incompressible and Boussinesq problems. Mathematicians just refer to the difference equations "with positive coefficients" (Forsythe and Wasow, 1960; Motzkin and Wasow, 1953). The term "skew differencing" has also been used (Lomas, et al., 1970). Richtmyer (1957) first credited Lelevier for applying it to inviscid compressible flow with slab symmetry. Richtmyer (1963) later credited Courant, Isaacson, and Rees (1952) and called the method "upstream differencing." Roberts and Weiss (1966), Kurzrock (1966), and Crocco (1965) also credit Lelevier, apparently following Richtmyer (1957). But Stone and Brian (1963) and others (Richtmyer, 1963) credit Courant, Isaacson, and Rees (1952) for their paper on compressible flow in which they first demonstrate the essential link of this scheme to the theory of characteristics, and also apply it to inviscid compressible two-dimensional flow. Longley's (1960) "Type II" scheme and Filler and Ludloff's (1961) first method were applications of this method to one-dimensional compressible flow, including viscosity effects. In the FLIC code (Fluid-In-Cell) of Gentry, Martin, and Daly (1966), the method*** is called "donor cell differencing." Kurzrock (1966) applied it to two-dimensional viscous compressible flow and derived the critical time step for this case. Many other applications exist in the literature.

Rewriting equation (3-176) in terms of the Courant number $c = u\Delta t/\Delta x$ gives, for u = positive constant,

$$\zeta_i^{n+1} = \zeta_i^n - c\left(\zeta_i^n - \zeta_{i-1}^n\right) \qquad (3\text{-}177)$$

For c = 1, the method gives $\zeta_i^{n+1} = \zeta_{i-1}^n$, which is the exact solution (as discussed in Section III-A-6). The condition c = 1 is also the stability limit (see previous exercises). For c < 1, the method introduces an artificial damping, in that the von Neumann stability analysis shows that the amplification matrix has eigenvalues $\lambda < 1$. Any method for the inviscid advection equation which has $\lambda < 1$ introduces such an artificial "damping," but a Taylor series expansion, as in the application of Hirt's stability analysis, shows that equation (3-176) is equivalent to

---

*The term "backward" differencing is sometimes used to describe the method, but it is clear that the only meaningful reference for "backward" or "forward" is the velocity.

**The opposite of "weather differencing" is "lee differencing" (Frankel, 1956) and is unconditionally unstable.

***Their donor cell differencing is what we later will refer to as the second upwind differencing method.

$$\frac{\partial \zeta}{\partial t} = -\frac{\partial (u\zeta)}{\partial x} + \alpha_e \frac{\partial^2 \zeta}{\partial x^2} + \text{HOT} + \text{HOD} \tag{3-178}$$

where

$$\alpha_e = 1/2\ u\Delta x(1 - c) \tag{3-179}$$

Since the method has introduced a non-physical coefficient α_e of $\partial^2\zeta/\partial x^2$, we are justified in referring not only to the artificial damping but, more specifically, to artificial or numerical *diffusion* (Noh and Protter, 1963) or to numerical or *artificial viscosity* of the upwind differencing method.*

The interpretation of α_e in multidimensional and viscous problems is not as straightforward as it might appear. For example, consider a situation where a steady state has developed. Then the left-hand side of equation (3-176) is zero, and we can reduce Δt without changing the FDE solution. Yet equation (3-179) would indicate that a reduction in Δt increases α_e (through c). If the concept of artificial diffusion α_e means anything, it would appear that the FDE solution should depend on α_e. If, instead of analyzing the transient equation, we instead set $\partial\zeta/\partial t = 0$ in equation (3-176) and expand in a Taylor series, we obtain

$$\alpha_e = 1/2\ u\Delta x \tag{3-180}$$

In this formulation, $\alpha_e \neq \text{fcn}(\Delta t)$, and the steady-state independence of Δt is not suspect.

The resolution of the discrepancies between the two different expressions [equations (3-179) and (3-180)] for α_e is accomplished by recognizing that, for the inviscid model equation, the only possible steady-state solution with u = constant is the trivial solution $\zeta_i^n = \zeta_1^n$ = constant. In this case, $\delta^2\zeta/\delta x^2 = \partial^2\zeta/\partial x^2 = 0$, permitting an arbitrary form for α_e.

Now, consider the application of upwind differencing with (physical) diffusion terms in two-dimensional flow. For constant $u_i, v_i, > 0$, this gives

$$\frac{\zeta_{ij}^{n+1} - \zeta_{ij}^n}{\Delta t} = -\frac{u\zeta_{ij}^n - u\zeta_{i-1,j}^n}{\Delta x} - \frac{v\zeta_{ij}^n - v\zeta_{i,j-1}^n}{\Delta y} + \alpha\left(\frac{\zeta_{i+1,j}^n + \zeta_{i-1,j}^n - 2\zeta_{ij}^n}{\Delta x^2} + \frac{\zeta_{i,j+1}^n + \zeta_{i,j-1}^n - 2\zeta_{ij}^n}{\Delta y^2}\right) \tag{3-181}$$

for which stability requires

$$\Delta t \leq \frac{1}{2\alpha\left(\frac{1}{\Delta x^2} + \frac{1}{\Delta y^2}\right) + \frac{|u|}{\Delta x} + \frac{|v|}{\Delta y}} \tag{3-182}$$

---

*The numerical viscosity α_e of the equations is not unique to upwind differencing, as we will see later.

The Taylor series expansion now gives

$$\frac{\partial \zeta}{\partial t} = -\frac{\partial (u\zeta)}{\partial x} - \frac{\partial (v\zeta)}{\partial y} + (\alpha + \alpha_{ex}) \frac{\partial^2 \zeta}{\partial x^2} + (\alpha + \alpha_{ey}) \frac{\partial^2 \zeta}{\partial y^2} + \text{HOT} + \text{HOD} \tag{3-183}$$

where, in the transient analysis,

$$\alpha_{ex} = 1/2\, u\Delta x(1 - c_x) \;,\; \alpha_{ey} = 1/2\, v\Delta y(1 - c_y) \tag{3-184}$$

but in the steady-state analysis,

$$\alpha_{ex} = 1/2\, u\Delta x \;,\; \alpha_{ey} = 1/2\, v\Delta y \tag{3-185}$$

The present author (see Appendix B) has shown that the steady-state form (3-185) is indeed appropriate for steady-state solutions. For transient solutions, we note from (3-184) that the artificial viscosity effect will be minimized for c_x and c_y as close to 1 as possible. But in a practical problem, it will be impossible to simultaneously have both near 1 in all parts of the flow field, so some artificial viscosity will necessarily be present.

Note also that the artificial viscosity coefficients depend on the velocity components u and v, which are measured with respect to the fixed Eulerian reference frame. This results in the equations failing to be "Galilean invariant", i.e., the "wind-tunnel transformation" of the continuum equations does not apply to these finite difference equations, except as Δx and $\Delta y \to 0$.

The terms "artificial viscosity" and "first-order method" have often been used as though synonymous, but in fact they are not. For example, one could simply add to a second-order method an additional, artificial viscosity term $\alpha_x\, \partial^2\zeta/\partial x^2$, with $\alpha_x \propto \Delta x^2$. This permits a second-order method with artificial viscosity. This method is, in fact, the basis of the von Neumann-Richtmyer shock method (see Section V-D-1).

Arguments have been made to the effect that accurate solutions are not possible unless $\alpha_e \ll \alpha$. From the steady-state analysis, equation (3-185) then indicates that $u\Delta x/\alpha$ and $v\Delta y/\alpha \ll 2$, or that the directional cell Reynolds numbers must be much less than 2. This is the condition for the formal accuracy requirement, but the practical situation is not quite this bad. Consider any region where the boundary-layer approximations apply (see Schlichting, 1968). Then $\partial^2\zeta/\partial x^2$ will be small and the effect of $(\alpha + \alpha_{ex})$ will be small in equation (3-183). Also, v will be small, so α_{ey} in (3-184) & (3-185) may be less than α (see Problem 3-9).

The analysis of Runchal and Wolfshtein (1969) (see also Wolfshtein, 1968), indicates that, for two-dimensional flow, $\alpha_e \simeq 1/3\, u\Delta x \sin(2\theta)$, where θ is the angle that the streamlines make with the coordinate system. They use the second upwind differencing method* (to be covered shortly) to compute driven cavity flows. At Re = 100, their nonuniform 13 x 13 mesh gives a maximum-cell Reynolds number of about 20. Yet their results compare very favorably with a second-order solution in a 51 x 51 mesh. Similarly, the calculations by Torrance (1968) of natural convection, also using the second upwind differencing method, at a high Grashof number (an equivalent Re $\simeq$ 300) differed by < 5% from a second-order solution. Campbell and Mueller (1968) and Mueller and O'Leary (1970) demonstrate good comparisons

*Their analysis of artificial viscosity also applies to the first upwind differencing method of this section.

with physical experiments in several separated flows at high Reynolds numbers. In calculations of driven cavity flow, Torrance, et al. (1972) have shown that the second upwind differencing method applied to the conservative equations is considerably more accurate than second-order differencing applied to the non-conservative equations.

It is also important to recognize that viscosity affects the flow-field not only by way of the diffusion terms in the vorticity transport equation, but also through the enforcement of no-slip conditions at the wall. The latter may provide the more important distinction between viscid and inviscid flows. Thus, Kentzner (1970A) finds that numerical Reynolds numbers (i.e., based on α_e) as low as 300 can very well approximate the inviscid solutions ($\alpha = 0$) provided that slip boundary conditions are used. The actual cut-off numerical Re will of course be problem-dependent. (It is obvious that for Re-independent problems such as Poiseuille and Couette flows, the artificial viscosity will have no effect.)

It appears then, that usable solutions are obtainable using upwind differencing methods, although the artificial viscosity effect must be considered when assessing the accuracy of the results. The upwind differencing methods also have several advantages. Compared with the forward-time, centered-space method, upwind differencing is not stability limited by a cell Re. Compared with the midpoint leapfrog method (previous section) and several others, upwind differencing does not lead to the time-step splitting behavior and requires one less array of ζ values. Compared with all methods which use second-order-accurate centered-space differences, upwind differencing does not leave the $\Lambda = 2\Delta x$ Fourier component stationary, and requires no more boundary and initial conditions than the original PDE.

Related to these last two points, it possesses a desirable property of some importance. Three sets of authors have given some physical significance to the success of one-sided differencing. Gentry, Martin, and Daley (1966) used "donor-cell mass differencing" in their FLIC (fluid-in-cell) technique to avoid "empty cells" and for its good stability properties. Thoman and Szewczyk (1966), for their incompressible problem, "define cell boundary vorticity values according to the velocity component directions at these boundaries," and thus "employ the average vorticity of the cells from which the vorticity is transported." Frankel (1956) speaks of the "unidirectional flow of information." All these descriptions relate closely to the concept of the "transportive property," which we now define.

III-A-9. The Transportive Property

We will say that a finite-difference formulation of a flow equation possesses the *transportive property* (Roache and Mueller, 1970) if the effect of a perturbation in a transport property is advected only in the direction of the velocity.

Innocuous and obvious as this definition may read, the fact is that the most frequently used methods do not possess this property. All methods which use centered-space derivatives for the advection terms do not possess this property.

The emphasis is on the word "advected." A physical perturbation in vorticity will spread in all directions due to diffusion. But it should be *carried* along only in the direction of the velocity. Consider the inviscid model equation,

$$\frac{\partial \zeta}{\partial t} = -\frac{\partial u\zeta}{\partial x} \tag{3-186}$$

The one-dimensional finite-difference form of equation (3-186) using FTCS differences is

$$\frac{\zeta_i^{n+1} - \zeta_i^n}{\Delta t} = -\frac{u\zeta_{i+1}^n - u\zeta_{i-1}^n}{2\Delta x} \tag{3-187}$$

Consider a perturbation $\varepsilon_m = \delta$, all other $\varepsilon = 0$, and $u > 0$. Then at (m+1), downstream of the perturbation,

$$\frac{\zeta_{m+1}^{n+1} - \zeta_{m+1}^n}{\Delta t} = -\frac{(0 - u\delta)}{2\Delta x} = +\frac{u\delta}{2\Delta x} \tag{3-188}$$

which is acceptable. But, at the point where the perturbation occurs,

$$\frac{\zeta_m^{n+1} - \zeta_m^n}{\Delta t} = - \frac{(0 - 0)}{2\Delta x} = 0 \tag{3-189}$$

which is not reasonable. Most significantly, at i = (m-1) *upstream* of the disturbance,

$$\frac{\zeta_{m-1}^{n+1} - \zeta_{m-1}^n}{\Delta t} = - \frac{(u\delta - 0)}{2\Delta x} = - \frac{u\delta}{2\Delta x} \tag{3-190}$$

Thus, a perturbation effect appears *upstream* of the perturbation and the transportive property is violated. At the next time step, a positive disturbance will appear at ζ_{m-2}, and so forth.

Compare this to the upwind difference for u > 0.

$$\frac{\zeta_i^{n+1} - \zeta_i^n}{\Delta t} = - \frac{u\zeta_i^n - u\zeta_{i-1}^n}{\Delta x} \tag{3-191}$$

Then for $\varepsilon_m = \delta$ as before, at (m+1), downstream of the disturbance,

$$\frac{\zeta_{m+1}^{n+1} - \zeta_{m+1}^n}{\Delta t} = - \frac{(0 - u\delta)}{\Delta x} = + \frac{u\delta}{\Delta x} \tag{3-192}$$

which is reasonable. At the point m of the disturbance,

$$\frac{\zeta_m^{n+1} - \zeta_m^n}{\Delta t} = - \frac{(u\delta - 0)}{\Delta x} = - \frac{u\delta}{\Delta x} \tag{3-193}$$

which means that the perturbation is being transported out of the affected region, as it should. (The connection between this observation and the previously mentioned property of upwind-difference schemes, that the Fourier mode with $\Lambda = 2\Delta x$ is not stationary, is clear.) Finally, at (m-1) upstream of the disturbance point,

$$\frac{\zeta_{m-1}^{n+1} - \zeta_{m-1}^n}{\Delta t} = \frac{(0 - 0)}{\Delta x} = 0 \tag{3-194}$$

which indicates that no perturbation effect is carried upstream. Thus, this method possesses the transportive property. It maintains the "unidirectional flow of information" (Frankel, 1956).

"Upwind differencing" is not quite synonymous with the transportive property. Consider the following two-dimensional problem, using a contrived upwind differencing approach in which the fluxes are defined by space averages in both directions. For clarity, we assume that the velocity components are constant.

$$\frac{\partial\zeta}{\partial t} = - \nabla\cdot(\vec{V}\zeta) \tag{3-195}$$

$$\frac{\zeta_{i,j}^{n+1} - \zeta_{i,j}^n}{\Delta t} = - u \frac{\hat{\zeta}_R - \hat{\zeta}_L}{\Delta x} - v \frac{\hat{\zeta}_T - \hat{\zeta}_B}{\Delta y} \tag{3-196}$$

The upwind differencing has been written for positive velocity components; the average vorticities are then defined as follows:

$$\hat{\zeta}_R = \frac{\zeta_{i,j+1} + 2\zeta_{i,j} + \zeta_{i,j-1}}{4}$$

$$\hat{\zeta}_L = \frac{\zeta_{i-1,j+1} + 2\zeta_{i-1,j} + \zeta_{i-1,j-1}}{4}$$

$$\hat{\zeta}_T = \frac{\zeta_{i-1,j} + 2\zeta_{i,j} + \zeta_{i+1,j}}{4} \qquad (3\text{-}197)$$

$$\hat{\zeta}_B = \frac{\zeta_{i-1,j-1} + 2\zeta_{i,j-1} + \zeta_{i+1,j-1}}{4}$$

In this method, the upwind cell property $\hat{\zeta}_R$ is the parabolic weighted average over the column from which ζ is transported. We now consider a steady-state solution with all v = 0, and introduce a vorticity perturbation $\varepsilon_{a,b} = \delta$, all other $\varepsilon = 0$. Then, at (a,b-1),

$$\frac{\zeta^{n+1}_{a,b-1} - \zeta^{n}_{a,b-1}}{\Delta t} = -u\frac{\left(\frac{\delta}{4}\right) - 0}{\Delta x} - 0 = -u\delta/4 \qquad (3\text{-}198)$$

Thus, the perturbation effect at (a,b) is transported in the v direction to (a,b-1), even though the v-velocity is identically zero. The method violates the transportive property, although it is based on a kind of upwind differencing scheme.

Space-centered differences are more accurate than the upwind one-sided differences insofar as the formal Taylor series expansion indicates. As discussed in Section III-A-3 on the conservative property, it is also possible to more accurately represent a derivative by using a non-conservative method, but the whole system is not more accurate if one's criteria for accuracy include the conservative property. The transportive property appears to be as physically significant as the conservative property. Upwind differencing schemes which possess the transportive property are more accurate in this sense, although not in the sense of order of the truncation error, than schemes with space-centered first derivatives.

To emphasize the significance of the transportive property, consider the opposite of upwind differencing -- downwind, or lee differencing (Frankel, 1965). This method is unstable, but we hypothesize that it may be made stable by some time differencing. From the aspect of accuracy of the differentials only, this method and that of upwind differencing are equally acceptable. Yet in downwind differencing, a disturbance will be advected only upstream, and not at all in the direction of the velocity! This is a physical absurdity,* and serves to point out again what we said concerning the conservative property; accuracy in the finite-difference representation of the differentials is not equivalent to accuracy in representation of the differential equation.

Roberts and Weiss (1966) term the Lelevier (upwind) method "inadequate," yet admit in a footnote, "Nevertheless, the one-sided scheme preserves the sign of positive definite quantities, as do Lagrangian methods also; this property is not shared by space-centered Eulerian schemes."

It is of interest to note that, in this finite-difference scheme, each node point in the mesh is analogous to a finite stage in the finite-stage reactor-model computations sometimes used by chemical engineers (Crider and Foss, 1966). Its description as a "simulation" method is clearly justified.

---

*D. B. Spalding is purported to have referred to upwind differencing as the "pigpen method," the idea being that if ζ is taken as a concentration fraction, then we should smell the pigpen when we're on the downwind side, and not on the upwind side (except for diffusion effects).

III-A-10. Transportive and Conservative Differencing.

The first upwind differencing method (3-176) is conservative, as well as transportive, as long as the velocity components do not reverse. For one-dimensional flow with all $u_i > 0$, we demonstrate the conservative property. Proceeding as in Section III-A-3, we have

$$\frac{\Delta\zeta_i}{\Delta t} = -\frac{u_i\zeta_i - u_{i-1}\zeta_{i-1}}{\Delta x} \qquad (3\text{-}199)$$

$$\sum_{i=I_1}^{I_2} \frac{\Delta\zeta_i}{\Delta t} = \sum_{i=I_1}^{I_2} \frac{u\zeta|_{i-1} - u\zeta|_i}{\Delta x}\,\Delta x \qquad (3\text{-}200)$$

$$\begin{aligned}
&= u\zeta|_{I_1-1} - u\zeta|_{I_1} && i=I_1\\
&\quad + u\zeta|_{I_1} - u\zeta|_{I_1+1} && i=I_1+1\\
&\quad + u\zeta|_{I_1+1} - u\zeta|_{I_1+2} && i=I_1+2\\
&\quad + \ldots\\
&\quad \ldots\\
&\quad + u\zeta|_{I_2-3} - u\zeta|_{I_2-2} && i=I_2-2\\
&\quad + u\zeta|_{I_2-2} - u\zeta|_{I_2-1} && i=I_2-1\\
&\quad + u\zeta|_{I_2-1} - u\zeta|_{I_2} && i=I_2
\end{aligned}$$

or

$$\sum_{i=I_1}^{I_2} \frac{\Delta\zeta_i}{\Delta t} = u\zeta|_{I_1-1} - u\zeta|_{I_2} \qquad (3\text{-}201)$$

These are the consistent upwind values of flux rate into region R (from I_1 to I_2) over I_1, and out of R over I_2. The conservative property is maintained.

However, in a region where the velocity reverses, the direction of upwind one-sided differencing also reverses and the conservative property is violated. We demonstrate with a one-dimensional case, where $u_i > 0$ for $i \leq \ell$, but $u_i < 0$ for $i > \ell$.

Upwind differencing then gives

$$\frac{\Delta\zeta}{\Delta t} = -\frac{u\zeta|_i - u\zeta|_{i-1}}{\Delta x} \quad \text{for } i \leq \ell \quad \text{where } u_i > 0 \qquad (3\text{-}202)$$

and

$$\frac{\Delta\zeta}{\Delta t} = -\frac{u\zeta|_{i+1} - u\zeta|_i}{\Delta x} \quad \text{for } i > \ell \quad \text{where } u_i < 0 \qquad (3\text{-}203)$$

Then

$$\sum_{i=I_1}^{I_2} \frac{\Delta\zeta_i}{\Delta t} = \sum_{i=I_1}^{\ell} \frac{u\zeta|_{i-1} - u\zeta|_i}{\Delta x} \Delta x + \sum_{i=\ell+1}^{I_2} \frac{u\zeta|_i - u\zeta|_{i+1}}{\Delta x} \Delta x \tag{3-204}$$

$$= u\zeta|_{I_1-1} - u\zeta|_{I_1}$$

$$+ u\zeta|_{I_1} - u\zeta|_{I_1+1}$$

$$+ u\zeta|_{I_1+1} - u\zeta|_{I_1+2}$$

$$\cdots$$

$$+ u\zeta|_{\ell-2} - u\zeta|_{\ell-1}$$

$$+ u\zeta|_{\ell-1} - u\zeta|_{\ell}$$

$$+ u\zeta|_{\ell+1} - u\zeta|_{\ell+2}$$

$$+ u\zeta|_{\ell+2} - u\zeta|_{\ell+3}$$

$$+ \cdots$$

$$\cdots$$

$$+ u\zeta|_{I_2-2} - u\zeta|_{I_2-1}$$

$$+ u\zeta|_{I_2-1} - u\zeta|_{I_2}$$

$$+ u\zeta|_{I_2} - u\zeta|_{I_2+1}$$

or

$$\sum_{i=I_1}^{I_2} \frac{\Delta\zeta_i}{\Delta t} = u\zeta|_{I_1-1} - u\zeta|_{I_2+1} - u\zeta|_{\ell} + u\zeta|_{\ell+1} \tag{3-205}$$

The first two terms are the consistent upwind values of flux rate into R; the third and fourth terms are non-conservative error terms. Since $u_\ell > 0$ and $u_{\ell+1} < 0$, they may be written as

$$- u\zeta|_{\ell} + u\zeta|_{\ell+1} = - \left[|u_\ell|\zeta_\ell + |u_{\ell+1}|\zeta_{\ell+1} \right] \tag{3-206}$$

and

$$\sum_{i=I_1}^{I_2} \frac{\Delta\zeta_i}{\Delta t} = u\zeta\Big|_{I_1-1} - u\zeta\Big|_{I_2+1} - \Big[|u_\ell|\zeta_\ell + |u_{\ell+1}|\zeta_{\ell+1}\Big] \tag{3-207}$$

Consider the case where ζ does not change sign* from ℓ to $\ell+1$. Then the error term may be regarded as an *artificial sink* of ζ. Or, if $u_\ell < 0$ and $u_{\ell+1} > 0$, then the error appears as an *artificial source* of ζ.**

Since $u = 0$ somewhere between ℓ and $\ell+1$, then u_ℓ and $u_{\ell+1}$ are generally small and the error approaches to zero as the mesh decreases.

The artificial source can be eliminated in two modifications of upwind differencing. The first modification uses the first upwind difference method (3-176) when no velocity reversals occur between (i-1) and (i), nor between (i) and (i+1). When a reversal does occur, the finite-difference method is formulated from a control volume approach about point (i).***

$$\frac{\Delta\zeta_i}{\Delta t} = \left(f_{i-1} + f_i + f_{i+1}\right)/\Delta x \tag{3-208}$$

where

$$f_i = -|u_i|\zeta_i \tag{3-209}$$

$$\begin{aligned} f_{i-1} &= +|u_{i-1}|\zeta_{i-1} && \text{for } u_{i-1} > 0 \\ &= 0 && \text{for } u_{i-1} \leq 0 \end{aligned} \tag{3-210}$$

and

$$\begin{aligned} f_{i+1} &= 0 && \text{for } u_{i+1} \geq 0 \\ &= +|u_{i+1}|\zeta_{i+1} && \text{for } u_{i+1} < 0 \end{aligned} \tag{3-211}$$

That this modification makes the first upwind differencing method both conservative and transportive may be easily verified. This modification requires fewer arithmetic operations but is not as accurate as the second upwind differencing method, which we now present.

---

*This is always the case when ζ is density, a positive definite quantity, in the compressible continuity equation.

**In a two-dimensional problem, this error occurs only if a velocity component changes sign as one moves in the direction of the component. That is, if $v_{i,j} > 0$ and $v_{i,j+1} < 0$, the non-conservative error appears. But if $v_{i,j} > 0$ and $v_{i+1,j} < 0$, there is no error. The first case occurs in separated flow problems. The second case occurs as the v-component of velocity over a cylinder changes sign from the forward to the aft region, with no separation. Thus, the solutions of Scala and Gordon (1967) are conservative.

***This special case is actually equivalent to equation (3-176) when no velocity reversals occur. It is not used at all points to replace equation (3-176) only because the calculations and FORTRAN "IF" statements required make it slightly more time consuming than (3-176).

III-A-11. The Second Upwind Differencing Method

In the second upwind differencing method, or "donor cell" method (Gentry, Martin, and Daly, 1966) some sort of average interface velocities on each side of the mesh point are defined; the sign of these velocities determines, by upwind differencing, which cell values of ζ to use. In one-dimensional notation,

$$\frac{\Delta\zeta_i}{\Delta t} = -\frac{u_R\zeta_R - u_L\zeta_L}{\Delta x} \tag{3-212}$$

where

$$u_R = 1/2(u_{i+1} + u_i)$$
$$u_L = 1/2(u_i + u_{i-1}) \tag{3-213}$$

or perhaps some other averaging scheme, and

$$\zeta_R = \zeta_i \text{ for } u_R > 0 \quad , \quad \zeta_R = \zeta_{i+1} \text{ for } u_R < 0$$
$$\zeta_L = \zeta_{i-1} \text{ for } u_L > 0 \quad , \quad \zeta_L = \zeta_i \text{ for } u_L < 0 \tag{3-214}$$

That this method is both conservative and transportive is easily verified. It is easily interpreted from the control-volume point of view, with interface velocites determined by averaging and interface ζ values determined by the flow direction. (Note: If interface ζ's also are defined by averaging, we obtain the FTCS method, which is not transportive.)

Compared with the first upwind differencing method, this one involves additional computations for velocities: an additional numerical differentiation of Ψ to obtain u and the additional averaging calculation of equation (3-213). However, it is more accurate than the first method, since it maintains something of the second-order accuracy of $\partial(u\zeta)/\partial x$ possessed by the centered difference schemes. Consider ζ constant with $\zeta_{i-1} = \zeta_i = \zeta_{i+1} = \hat{\zeta}$, but spatially varying u. Then (3-212) becomes

$$\frac{\Delta\zeta_i}{\Delta t} = -\frac{u_R\zeta_R - u_L\zeta_L}{\Delta x} = -\frac{\hat{\zeta}\frac{1}{2}(u_{i+1} + u_i) - \hat{\zeta}\frac{1}{2}(u_i + u_{i-1})}{\Delta x} \tag{3-215A}$$

or

$$\frac{\Delta\zeta_i}{\Delta t} = -\hat{\zeta}\,\frac{u_{i+1} - u_{i-1}}{2\Delta x} \tag{3-215B}$$

which is second-order accurate for the advection field. Thus, the second upwind differencing method possesses both the conservative and transportive properties, and retains something of the second-order accuracy of centered-space derivatives. That it is indeed superior to the first upwind difference method was demonstrated in an actual two-dimensional computation by Torrance (1968). The surprising agreement of this first-order method with Torrance's second-order calculations of the driven cavity problem (less than 5% deviation at Re = 100) can now be explained. Along the mid-section of the walls, the boundary-layer approximations make the effect of the numerical viscosity term α_e small(see discussion in Section III-A-8). Near the corners, the velocities are small, so the physical $\alpha >> \alpha_e$. And in the rotating core region, the spatial variation of ζ is small, so the method is nearly second-order accurate, in accordance with equation (3-216).

The method of Runchal et al. (1969) is algebraically equivalent to this second upwind differencing method. At the cost of some added complexity and arithmetic operations, it avoids the necessity of FORTRAN "IF" statements to test for the signs of u and v.

A non-conservative second-*order* upwind method was used in 1D by Price et al.(J. Mathematics and Physics, Vol. 45, 1966, pp.301-311.) The advection derivative for $u > 0$ is evaluated by

$$\delta\zeta_i/\delta x = (3\zeta_i - 4\zeta_{i-1} + \zeta_{i-2})/2\Delta x \tag{3-216}$$

III-A-12. Adams-Bashforth and Crocco Methods

The Adams-Bashforth differencing scheme as applied by Lilly (1965) to the advection equation only is an explicit, one-step, 3-time-level, forward-time scheme with error of $O(\Delta t^2, \Delta x^2)$. It can be interpreted as a finite-difference approximation to the *second time derivative*.

We write a Taylor series expansion for ζ_i^{n+1} in time.

$$\zeta_i^{n+1} = \zeta_i^n + \left.\frac{\partial\zeta}{\partial t}\right|_i^n \Delta t + 1/2 \left.\frac{\partial^2\zeta}{\partial t^2}\right|_i^n \Delta t^2 + O(\Delta t^3) \tag{3-217}$$

We now evaluate the second derivative by one-sided finite differences in time.

$$\left.\frac{\partial^2\zeta}{\partial t^2}\right|_i^n = \frac{\partial}{\partial t}\left[\frac{\partial\zeta}{\partial t}\right]_i^n = \frac{\left.\frac{\partial\zeta}{\partial t}\right|_i^n - \left.\frac{\partial\zeta}{\partial t}\right|_i^{n-1}}{\Delta t} + O(\Delta t) \tag{3-218}$$

Combining with the previous equation gives

$$\zeta_i^{n+1} = \zeta_i^n + \left.\frac{\partial\zeta}{\partial t}\right|_i^n \Delta t + \frac{1}{2}\left[\frac{\left.\frac{\partial\zeta}{\partial t}\right|_i^n - \left.\frac{\partial\zeta}{\partial t}\right|_i^{n-1}}{\Delta t} + O(\Delta t)\right]\Delta t^2 + O(\Delta t^3)$$

or

$$\zeta_i^{n+1} = \zeta_i^n + \left[\frac{3}{2}\left.\frac{\partial\zeta}{\partial t}\right|_i^n - \frac{1}{2}\left.\frac{\partial\zeta}{\partial t}\right|_i^{n-1}\right]\Delta t + O(\Delta t^3) \tag{3-219}$$

This is the basic form of the Adams-Bashforth differencing scheme for the advection terms. It can be combined with evaluation of the diffusion terms by centered-space derivatives at time n, giving, for a two-dimensional problem,

$$\zeta_i^{n+1} = \zeta_i^n + \Delta t\left[-\frac{3}{2}\left.\frac{\delta u\zeta}{\delta x}\right|_i^n + \frac{1}{2}\left.\frac{\delta u\zeta}{\delta x}\right|_i^{n-1} - \frac{3}{2}\left.\frac{\delta v\zeta}{\delta x}\right|_i^n + \frac{1}{2}\left.\frac{\delta v\zeta}{\delta x}\right|_i^{n-1} + \alpha\left(\delta^2\zeta^n/\delta x^2 + \delta^2\zeta^n/\delta y^2\right)\right] \tag{3-220}$$

However, this method with the viscous terms is only first-order accurate.

The basic form of the difference expansion of equation (3-219) was given in 1833 by Bashforth and Adams. Thomas (1954) used higher order versions of it in one-sided spatial expansions approaching a shock front, and referred to it as the Adams method. Lilly (1965) applied the time expansion as in equation (3-219) to the inviscid vorticity transport equation, and called it the "Adams-Bashforth method," without historical reference.

As it stands, equation (3-219) is unconditionally unstable, with a weak divergence caused by an amplification factor $G = 1 + O(\Delta t^2)$; see Lilly, 1965. Because the instability is weak, the method can be used for unsteady calculations of inviscid flows, provided the total time is not large. Lilly (1965) found it more accurate than the Lax-Wendroff (1964) method (see Section V-E-5). Also, the presence of the viscous terms in equation (3-220) stabilize the equation, giving a usable $\Delta t = \text{fcn}(Re)$. (See problem 3-11.)

Crocco (1965) has presented a quasi-one-dimensional compressible flow method using the following method, which is a consistent method *only* in the steady-state limit.

$$\zeta_i^{n+1} = \zeta_i^n - u\Delta\zeta_i + \alpha\Delta^2\zeta_i \tag{3-221a}$$

$$\Delta\zeta_i = (1 + \Gamma)\frac{\zeta_{i+1}^n - \zeta_{i-1}^n}{2\Delta x} - \Gamma\frac{\zeta_{i+1}^{n-1} - \zeta_{i-1}^{n-1}}{2\Delta x} \tag{3-221b}$$

$$\Delta^2\zeta_i = \frac{\zeta_{i+1}^n - 2\zeta_i^{n+1} + \zeta_{i-1}^n}{\Delta x^2} \tag{3-221c}$$

He investigated the weighting factor* Γ and, for inviscid flow ($\alpha = 0$), found the lower limit on Γ for stability to be $\Gamma = 1/2$, which is precisely the first-order Adams-Bashforth method, equation (3-219). The algebra of the von Neumann stability analysis is clumsy. Crocco presents the numerical results of the stability analysis graphically, showing which combinations of Γ, Rc, c, and Δt can give a stable, long-term calculation. The actual flow calculations were done with $\Gamma = 1$. Application of Hirt's stability analysis (see Problem 3-12) gives a transient $\alpha_e = u^2\Delta t(\Gamma - 1/2)$, which also shows that $\Gamma \geq 1/2$ is required for stability.

The Adams-Bashforth and Crocco methods have second-order-accurate advection space derivatives, as do the midpoint leapfrog and the DuFort-Frankel leapfrog methods. Like the leapfrog methods, they use data at two time levels to predict the third, but they are not "leapfrog" methods since the value ζ_i^{n+1} is calculated as the old value ζ_i^n at the immediately previous time step, plus a change. Consequently, they do not display the disjoint solution behavior of leapfrog methods. They neither are transportive, nor give exact solutions for $c = 1$.

The method of Miyakoda (1962) (see also Lilly, 1965) is somewhat similar to the Adams-Bashforth method. It is a four-level method, using data at (n-2), (n-1), and (n) to predict values at (n+1). It is also second-order accurate and does not develop disjoint solutions. It is also weakly unstable, and it appears to offer no advantages over the simpler Adams-Bashforth method.

III-A-13. Leith's Method; Phase Errors, Aliasing Errors and Time-Splitting.

Leith's (1965) method is an exceedingly interesting one-step, two-level, second-order-accurate method. The 1D version was presented earlier by Noh and Protter (1963). We first construct the 1D method by reverting to a Lagrangian point of view, following the particle motion.

For a one-dimensional problem with constant u, the trajectories (time-space curves) of the fluid particles are represented by heavy arrows in Figure 3-12a. In traveling from (t) to $(t + \Delta t)$, the particles move an x-distance of $u\Delta t$. Now we "mark" each particle with a value of ζ. That is, ζ can be any intrinsic property associated with a particular fluid particle. In the absence of diffusion, each fluid particle will retain its value of ζ. Thus, the trajectories of Figure 3-12a are lines of constant ζ.

The inviscid advection equation $\partial\zeta/\partial t = -u\partial\zeta/\partial x$ just says that ζ is some property of the fluid particle, which does not change as the flow proceeds. This is the meaning of the "substantive derivative" or Lagrangian notation $D\zeta/Dt = \partial\zeta/\partial t + u\partial\zeta/\partial x$. The $D\zeta/Dt$ is written following the fluid particle and the inviscid advection equation just says

*This is a well-known technique; see Richtmyer (1957).

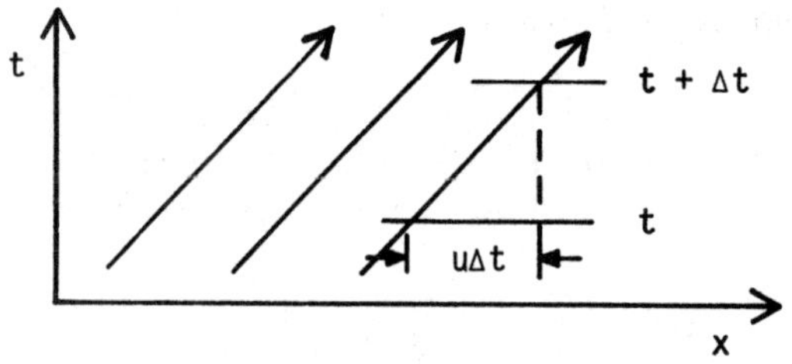

(a) time-space particle trajectories in a constant u - field

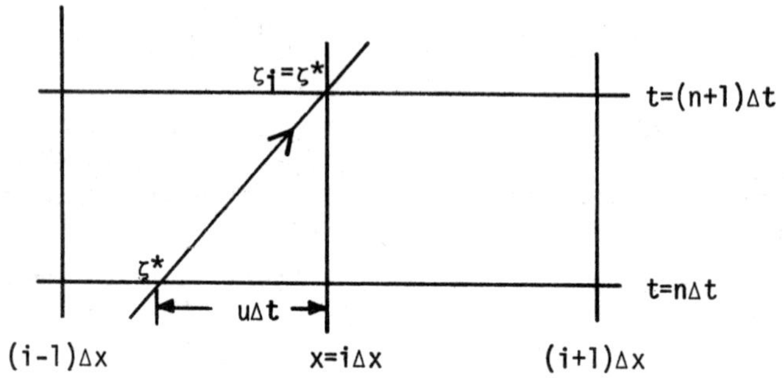

(b) location of ζ^*

Figure 3-12. Construction of Leith's method in one dimension.

$D\zeta/Dt = 0$, i.e., the ζ-value of any particle does not change.†

Referring to Figure 3-12b, the inviscid advection equation just says that $\zeta_i^{n+1} = \zeta^*$. The construction of FDE's then reduces to the problem of approximately determining ζ^* by some kind of interpolation between known values of ζ^n.

We first note that, for a particular combination of parameters, the trajectory passes through the lattice point at (i-1) and (n). In this case, we have $\zeta^* = \zeta_{i-1}^n$ exactly, with no interpolation error. The necessary condition is seen from the geometry of Figure 3-12b to be $u\Delta t = \Delta x$, or $c = 1$; for a unity Courant number, we have $\zeta_i^{n+1} = \zeta_{i-1}^n$, an exact solution (as in Section III-A-6).

We now consider a more general condition of $c \neq 1$. If we have $u\Delta t < \Delta x$, ζ^* is located between (i) and (i+1), as in the figure. If we use linear interpolation to estimate ζ^*,

---

†For students new to this interpretation, it is sometimes helpful to think of placing a drop of ink on the surface of smoothly flowing water, and regarding ζ as a color concentration, which is clearly attached to a (finite) fluid "particle." Neglecting diffusion, the Eulerian description of color at a point is given by $\partial\zeta/\partial t = -u(\partial\zeta/\partial x)$, and the Lagrangian description is given by $D\zeta/Dt = 0$. The explanation given by Bird, Stewart, and Lightfoot (1960) is also recommended.

we obtain the first-order-accurate estimate of

$$\zeta^* = \zeta_i^n - \left(\zeta_i^n - \zeta_{i-1}^n\right) \frac{u\Delta t}{\Delta x} \qquad (3\text{-}222)$$

Setting $\zeta_i^{n+1} = \zeta^*$ then gives

$$\frac{\zeta_i^{n+1} - \zeta_i^n}{\Delta t} = - u \left(\frac{\zeta_i^n - \zeta_{i-1}^n}{\Delta x}\right) \qquad (3\text{-}223)$$

which is just the upwind differencing method of Section III-A-8. If instead we use linear interpolation over (i+1) and (i-1), we obtain the FTCS method. If we use a quadratic polynomical fit (see Section III-A-1) to interpolate over (i-1), (i), and (i+1), we obtain Leith's method

$$\zeta_i^{n+1} = \zeta_i^n - \frac{1}{2}\left(\frac{u\Delta t}{\Delta x}\right)\left(\zeta_{i+1}^n - \zeta_{i-1}^n\right) + \frac{1}{2}\left(\frac{u\Delta t}{\Delta x}\right)^2\left(\zeta_{i+1}^n - 2\zeta_i^n + \zeta_{i-1}^n\right) \qquad (3\text{-}224)$$

The Courant number $c = u\Delta t/\Delta x$ can now be regarded as an *interpolation parameter*. The stability restriction $c \leq 1$, which can be shown to apply to equation (3-224) as well as (3-223) can now be interpreted as a statement that ζ^* must be determined by *interpolation* rather than *extrapolation*.

An alternate derivation of Leith's method is obtained by considering a forward-time Taylor series expansion to $O(\Delta t^3)$, as in the Adams-Bashforth method, but now evaluating the second time derivative from the original advection equation, as

$$\frac{\partial \zeta}{\partial t} = - u \frac{\partial \zeta}{\partial x} \qquad (3\text{-}225)$$

$$\frac{\partial^2 \zeta}{\partial t^2} = - u \frac{\partial^2 \zeta}{\partial x \partial t} = - u \frac{\partial}{\partial x}\left(- u \frac{\partial \zeta}{\partial x}\right) = u^2 \frac{\partial^2 \zeta}{\partial x^2} \qquad (3\text{-}226)$$

Then the Taylor series gives

$$\zeta_i^{n+1} = \zeta_i^n + \Delta t \frac{\partial \zeta}{\partial t} + \frac{1}{2}\Delta t^2 \frac{\partial^2 \zeta}{\partial t^2} + O(\Delta t^3) \qquad (3\text{-}227)$$

Using the space-centered second-order forms, $\delta\zeta/\delta x$ and $\delta^2\zeta/\delta x^2$, for the derivatives in equations (3-225) and (3-226), and substituting in (3-227), gives

$$\zeta_i^{n+1} = \zeta_i^n - u\Delta t \frac{\delta \zeta^n}{\delta x} + \frac{1}{2} u^2 \Delta t^2 \frac{\delta^2 \zeta^n}{\delta x^2} + O(\Delta t^3, \Delta x^2) \qquad (3\text{-}228)$$

which is identical to (3-224), and shows the second-order accuracy of the method.

The two derivations, one based on quadratic space interpolation and one based on second-derivative time expanisons give identical results because of equation (3-226), relating $\partial^2\zeta/\partial t^2$ to $\partial^2\zeta/\partial x^2$. This relation holds *only* for the inviscid equation with constant u. In this case, Leith's method is the same as the Lax-Wendroff method and the various two-step Lax-Wendroff methods which are based on the time expansion (see Chapter V).

There is a most interesting point, important to the interpretation of other methods, to be made about the artificial damping of Leith's method. When we evaluate the various terms in equation (3-224) from a Taylor series expansion about (i) and (n), as in Hirt's stability analysis, we find

$$\frac{\partial \zeta}{\partial t}\,\Delta t + \frac{1}{2}\frac{\partial^2 \zeta}{\partial t^2}\,\Delta t^2 + O(\Delta t^3) = -\frac{u\Delta t}{\Delta x}\frac{\partial \zeta}{\partial x}\,\Delta x - \left(\frac{u\Delta t}{\Delta x}\right)\frac{1}{6}\frac{\partial^3 \zeta}{\partial x^3}\,\Delta x^3$$

$$+ \left(\frac{u^2\Delta t^2}{\Delta x^2}\right)\frac{1}{2}\left[\frac{\partial^2 \zeta}{\partial x^2}\,\Delta x^2 + O(\Delta x^4)\right] \tag{3-229}$$

or

$$\frac{\partial \zeta}{\partial t} = -u\frac{\partial \zeta}{\partial x} + \frac{1}{2}\Delta t\left[-\frac{\partial^2 \zeta}{\partial t^2} + u^2\frac{\partial^2 \zeta}{\partial x^2}\right] + O(\Delta t^2, \Delta x^2) \tag{3-230}$$

The bracketed term in equation (3-230) is identically zero, for a constant u-field, by virtue of (3-226). The coefficient of $\partial^2\zeta/\partial x^2$ is zero, and Leith's method has no "artificial viscosity." Thus, a method which has the same form (3-224) as the $O(\Delta t, \Delta x^2)$ FTCS method for the entire viscous equation with $\alpha = 1/2\, u^2\Delta t$ (compare with 3-18) is actually an analog of the inviscid equation, with error of $O(\Delta t^2, \Delta x^2)$.

However, the analysis again becomes confused if we examine only the steady-state equations, in which case we find $\alpha_e = 1/2\, u^2\Delta t$. This indicates that a steady-state solution will depend on Δt, and will be only first-order accurate. This anomaly arises because of the need for an extraneous outflow boundary condition, necessitated by the use of centered space derivatives. Practically speaking, high accuracy is maintained if gradient outflow boundary conditions (section III-C-7) are used. This confusing subject can only be discussed along with boundary conditions; for details, see Roache (1971c). We do note an important point, often overlooked. When viscous terms are included, with Leith's method used for the advection term, the artificial viscosity is unequivocally given by $\alpha_e = 1/2\, u^2\Delta t$ except for the singular case of $c = 1$. (See Appendix B.)

The von Neumann stability analysis can be taken directly from the example of Section III-A-5-b, worked out for the FTCS method, by noting that d in equation (3-105) is now replaced by $c^2/2$. Then equation (3-108) gives the amplification factor for Leith's method as

$$G = 1 - c^2(1 - \cos\theta) - Ic\sin\theta \tag{3-231}$$

Using the trigonometric identities, this can be rearranged as

$$|G|^2 = 1 + c^2(c^2 - 1)(1 - \cos\theta)^2 \tag{3-232}$$

or

$$-1 \le 1 + c^2(c^2 - 1)(1 - \cos\theta)^2 \le +1 \tag{3-233}$$

It is easily verified that the left-hand inequality is always satisfied, and that the right-hand inequality requires only $c \le 1$, as usual.

Equation (3-232) also gives the *damping error* and the *phase error* of the method. Leith (1965) has analyzed these for the conditions of interest to him in meteorological calculations, $c \ll 1$ and θ small. The first condition arises because the meteorological calculations will be carried out at small Δt, and the second condition indicates that the long wavelength disturbances (compared with Δx) are of primary interest, with $\theta = k_x \Delta x = 2\pi\Delta x/\Lambda << 1$. Then, with $c^4 << c^2$ and $1 - \cos\theta \simeq \theta^2/2$, equation (3-232) becomes

$$|G|^2 \simeq 1 - c^2\theta^4/4 \tag{3-234}$$

Using the usual approximation $(1 - \varepsilon)^{1/2} \simeq 1 - 1/2\varepsilon$ for $\varepsilon << 1$ gives

$$|G| \simeq 1 - c^2\theta^4/8 \tag{3-235}$$

Since $|G| = 1$ in the continuum equation, the term $c^2\theta^4/8$ is an approximation to the truncation damping error of the method. Note that the damping error is fourth order in Δx.

The phase error is similarly approximated. As discussed previously in Section III-A-6, the correct phase speed of the continuum equation is u, for all θ. The continuum solution (neglecting boundary influence) can be written as in equation (3-160), or equivalently as

$$\zeta(x,t) = \zeta(x - u\tau,\ t - \tau) \tag{3-236}$$

for any time shift τ. In terms of the Fourier components with wave number k_x and corresponding amplitude function V, the continuum solution (3-236) is written as*

$$\zeta_{k_x} = V\, e^{Ik_x(x-ut)} \tag{3-237}$$

or

$$\zeta_{k_x}(x,t) = V\, e^{I(\theta - k_x ut)} \tag{3-238}$$

The correct continuum phase shift over τ is then $\Delta\theta = -k_x u\tau$. We wish now to compare this to the numerical phase shift. We evaluate the continuum (PDE) phase shift over $\tau = \Delta t$, and write

$$[\Delta\theta]_{PDE} = -k_x u\Delta t = -k_x c\Delta x \tag{3-239}$$

or

$$[\Delta\theta]_{PDE} = -c\theta \tag{3-240}$$

The actual phase shift of the finite difference equation (FDE) is found from the geometry of the polar diagram (see Figure 3-8, for example) which gives

$$\sin[(\Delta\theta)_{FDE}] = \frac{I_{mag}(G)}{|G|} \tag{3-241}$$

From equation (3-231), we have $I_{mag}(G) = -c \sin\theta$. Taking $\sin(\Delta\theta) \simeq \Delta\theta$ and $|G|$ from

---

*Differentiating (3-237) gives $\partial\zeta/\partial t = -Iuk_x\zeta$ and $\partial\zeta/\partial x = +Ik_x\zeta$, and thus $\partial\zeta/\partial t = -u\partial\zeta/\partial x$, which is correct.

(3-235) gives

$$[\Delta\theta]_{FDE} \simeq - c \sin\theta \left(\frac{1}{1 - c^2\theta^4/8}\right) \tag{3-242}$$

or, using the approximation $(1-\varepsilon)^{-1} \simeq 1 + \varepsilon$ for $\varepsilon << 1$,

$$[\Delta\theta]_{FDE} \simeq - c \sin\theta \; (1 + c^2\theta^4/8) \tag{3-243}$$

To compare directly to (3-240), we write (3-243) as

$$[\Delta\theta]_{FDE} = - c\theta r \tag{3-244}$$

where

$$r = \frac{\sin\theta}{\theta}\left(1 + \frac{c^2\theta^4}{8}\right) \simeq \frac{\sin\theta}{\theta} \tag{3-245}$$

Expanding $\sin\theta$ for small θ gives

$$r = \frac{\theta - \theta^3/3! + 0(\theta^5)}{\theta} \tag{3-246}$$

$$r \simeq 1 - \theta^2/6 < 1 \tag{3-247}$$

Comparing equations (3-244) and (3-240), we see that each Fourier component is carried along too slowly, by the factor $r(\theta) < 1$. In the exact continuum PDE solution, all components are advected at the speed u; in the FDE solution, all Fourier components lag the true solution, and the FDE advection speed, ur, differs for each component. The error is larger for the larger θ, i.e., shorter wavelength Λ. Thus, different Fourier components will spread apart, or disperse, as the numerical solution proceeds, and the phenomenon is frequently referred to as *dispersion* error. (For an early study of dispersion error, see Stone and Brian, 1963.)

The phase error E_θ per time step is $(\Delta\theta)_{FDE} - (\Delta\theta)_{PDE}$. Using equations (3-240) and (3-243), $E_\theta \simeq -c \sin\theta - (-c\theta) = -c[\theta - \theta^3/3! + \ldots - \theta]$, or

$$E_\theta \simeq c\theta^3/6 \tag{3-248}$$

Thus the dispersion error is third order in θ, and therefore third order in Δx.

The short wavelength limit can be examined without approximations. For $c < 1$, equation (3-232) shows that $|G|^2 < 1$ except for $\theta = 0$. From (3-231), $\mathrm{Imag}(G) = -c \sin\theta$. In the limit of $\theta \to 0$, $\mathrm{Imag}(G) \to 0$ and from (3-241), $\sin(\Delta\theta) \to 0$, or $(\Delta\theta)_{FDE} \to 0$. Thus, the phase error is complete, with the *smallest wavelength being completely stationary*. This also occurred for the midpoint leapfrog method of Section III-A-6 and is typical of methods which use space-centered $\delta\zeta/\delta x$ (Fromm, 1968).

For $c = 1$, equation (3-232) shows $|G|^2 = 1$, which is correct. The phase error also disappears, since (3-241) becomes

$$[\sin(\Delta\theta)]_{FDE} = \frac{-\sin\theta}{(1)} \qquad (3\text{-}249)$$

or

$$[\Delta\theta]_{FDE} = \theta \qquad (3\text{-}250)$$

which agrees with the exact relation (3-240) for c = 1. This behavior is verified by setting c = 1 in the FDE (3-224), giving

$$\zeta_i^{n+1} = \zeta_i^n - \frac{1}{2}\left(\zeta_{i+1}^n - \zeta_{i-1}^n\right) + \frac{1}{2}\left(\zeta_{i+1}^n + \zeta_{i-1}^n - 2\zeta_i^n\right) \qquad (3\text{-}251)$$

or $\zeta_i^{n+1} = \zeta_{i-1}^n$, the exact solution.

Contour plots of the phase and amplitude errors of Leith's method, with c and θ as parameters, are given by Fromm (1968). The limits as $c \to 1$ and/or $\theta \to 0,\ 2\pi$ are shown to be well behaved. Another fundamental reference or phase errors is Crowley (1968A).

The phase errors are obviously of no concern if only a steady-state solution is desired. But for the transient solution, their importance may dominate. Leith (1965) solves equation (3-248) for the number of calculations N to give a phase error of 1 radian, and shows that $N = 6/c\theta^3$. Leith says that at this point in the calculations the phase error is relatively more important than the amplitude error.

Exercise: Evaluate the phase error and dispersion error of the upwind differencing method.

Another type of error which affects the accuracy of transient and steady-state calculations, using Leith's or any other method, is called *aliasing error*. Aliasing error was first delineated and analyzed by Phillips (1959). Other important papers on this subject are Grammeltvedt (1969) and Robert et al., (1970). Here, we merely describe the phenomenon. Aliasing is associated with energy exchange between Fourier components. (We use "energy" here in the general sense of the first moment of the transport property, ζ^2.) As we have already noted, the shortest wavelength component which can be discriminated by the computational mesh is $\Lambda = 2\Delta x$. We are generally more interested in accuracy of the long wavelengths, so this lack of short-wavelength discrimination might seem to be of little consequence. Indeed, if we assume that the advection field u is everywhere constant, no aliasing occurs. But, in the nonlinear problem, it is well known that the components interact in such a way that energy cascades down from the long wavelengths to the short. (Physically, we know that the energy of turbulence generally moves from large eddies to small; beyond small eddies, it is dissipated, or degraded into internal energy, via friction.) For this interaction of modes to take place, true nonlinearity is not necessary; spatially varying u suffices. The question is then, what happens when there is no dissipative mechanism to remove energy from the short wavelengths, and when the computation cannot resolve $\Lambda < 2\Delta x$? The surprising answer is that the energy flips over and reappears in the long wavelengths, distorting those very components of greatest interest, and even resulting in a kind of computational instability (Phillips, 1959). The presence of any damping, physical or numerical, in the computation will alleviate this problem by providing a dissipative mechanism. Thus, the computation of low Re flows and the computation of inviscid flows using numerical methods which introduce artificial damping of short-wavelength modes will have less trouble from aliasing errors. The word *aliasing* is also used to describe the fact that a finite grid simply cannot distinguish between certain frequencies. (This is why the energy can flip over into the long wavelengths.) That is, two distributions like $\zeta_i(\pm) = \cos\{\pi(m \pm n)i\}$ have exactly the same values at the node points i, a result of simple trigonometric identities. The same discretization or "sampling" phenomenon in *time* causes stroboscopic effects in motion pictures, e.g. wagon wheels which appear to slow down and reverse direction. This confusion of frequencies is an *inevitable consequence of discretization*. See Hamming (1962),pp.276, 303.

A final interesting point about Leith's method is its adaptation to two dimensions. The obvious extension of equation (3-224) would be

$$\zeta_{ij}^{n+1} = \zeta_{ij}^{n} - \frac{1}{2} c_x \left(\zeta_{i+1,j}^{n} - \zeta_{i-1,j}^{n}\right) + \frac{1}{2} c_x^2 \left(\zeta_{i+1,j}^{n} - 2\zeta_{ij}^{n} + \zeta_{i-1,j}^{n}\right)$$

$$- \frac{1}{2} c_y \left(\zeta_{i,j+1}^{n} - \zeta_{i,j-1}^{n}\right) + \frac{1}{2} c_y^2 \left(\zeta_{i,j+1}^{n} - 2\zeta_{ij}^{n} + \zeta_{i,j-1}^{n}\right) \tag{3-252}$$

or, in an obvious shorthand notation,

$$\zeta_{ij}^{n+1} = \zeta_{ij}^{n} - \frac{1}{2} c_x \delta_x(\zeta^n) + \frac{1}{2} c_x^2 \delta_x^2(\zeta^n) - \frac{1}{2} c_y \delta_y(\zeta^n) + \frac{1}{2} c_y^2 \delta_y^2(\zeta^n) \tag{3-253}$$

where $c_x = u\Delta t/\Delta x$ and $c_y = v\Delta t/\Delta y$. This method is *not* second-order accurate. A second-order scheme would have to include cross-derivative terms $\partial^2\zeta/\partial x\partial y$ in the Taylor series expansion. More importantly, the method is *unstable* (Leith, 1965). Thus (3-252) is an example of a two-dimensionally unstable method, arising from the interaction of two one-dimensional methods, each of which is stable.

Leith (1965) gives a stable two-dimensional method based on the concept of "fractional time steps" of Marchuk (1965). The method is now also commonly called "time splitting". It consists of applying each one-dimensional method separately and in succession, with no significance attached to the result of just the first step.*

$$\zeta_{ij}^{n+1/2} = \zeta_{ij}^{n} - \frac{1}{2} c_x \delta_x(\zeta^n) + \frac{1}{2} c_x^2 \delta_x^2(\zeta^n) \tag{3-254A}$$

$$\zeta_{ij}^{n+1} = \zeta_{ij}^{n+1/2} - \frac{1}{2} c_y \delta_y(\zeta^{n+1/2}) + \frac{1}{2} c_y^2 \delta_y^2(\zeta^{n+1/2}) \tag{3-254B}$$

This method is consistent, and does provide the full second-order accuracy missing from equation (3-253), and is stable provided that $c_x, c_y < 1$.

It is apparently true that any stable 1D method may be applied in 2D by using time-splitting with no change in the 1D stability condition. For problems of flow about bodies, such time-split methods introduce a difficulty in the calculation of boundary values for the intermediate step. As we shall see, the calculation of surface boundary values of ζ involves the stream function ψ at internal points. It seems to be meaningless to calculate a provisional $\psi^{n+1/2}$. For small-enough time steps, the original value of ζ^n at walls could probably be used in the second step of equation (3-254). The effects on stability and accuracy of this boundary treatment have not been ascertained at this time for any of the time-split methods. (The difficulty does not exist in Leith's meteorological problem.) Fromm (1971) achieved stability, in a method which was similarly stable in 1D and unstable in 2D, by explicitly including the cross-derivative terms in a one-step method.

In addition to improving stability, time-splitting may also improve accuracy, as shown in the following exercise.

Exercise: Apply upwind differencing in 2D to the inviscid, constant-coefficient transport equation.

$$\zeta_t = -u\zeta_x - v\zeta_y$$

Demonstrate that the time-split method produces the exact answer when the directional Courant numbers $u\Delta t/\Delta x = 1$ and $v\Delta t/\Delta y = 1$, whereas the non-time-split method does not.

---

*This concept is similar to, but not equivalent to, the splitting used earlier by Peaceman and Rachford (1955) in their ADI methods (see Section III-A-16). A recent book by Yanenko (1971) covers other methods of fractional time steps.

Other methods similar to Leith's are discussed by Kasahara (1965) and Fischer (1965A). These and Leith's are all similar to the Lax-Wendroff method and two-step versions thereof, to be treated in Chapter 5. Although the Lax-Wendroff methods were developed for compressible flow, they are of interest in incompressible flows as well (Lilly, 1965), although by design they do strongly damp out the short-wavelength disturbances.

Other methods which involve the fractional-time-step concept and have improved dispersion properties will be discussed in later sections. We continue now with a discussion of some simpler methods.

III-A-14. Implicit Methods

The methods previously described are explicit, in that only known values at time levels (n), (n-1), ..., are needed to advance the calculation to time level (n+1). We now discuss implicit methods, which use advance values in the spatial derivatives, thereby requiring the simultaneous solution of equations at (n+1) in order to advance the calculation.

We write a general equation for the inviscid model equation as

$$\zeta^{n+1} = \zeta^{n} - (u\Delta t)\delta\zeta/\delta x \qquad (3\text{-}255)$$

We will consider here only centered-space evaluations, so that

$$\frac{\delta\zeta_i}{\delta x} = \frac{\zeta_{i+1} - \zeta_{i-1}}{2\Delta x} \qquad (3\text{-}256)$$

where the time level n is as yet unspecified. Likewise, for the diffusion equation

$$\zeta^{n+1} = \zeta^{n} + (\alpha\Delta t)\delta^2\zeta/\delta x^2 \qquad (3\text{-}257)$$

If we evaluated the space derivatives in equation (3-255) or (3-257) at time level (n), we would have the previously discussed FTCS explicit method, with error of $O(\Delta t, \Delta x^2)$. If we instead evaluate the term $\delta\zeta/\delta x$ in the advection equation (3-255) entirely at the new time level (n+1), we have the so-called *fully* implicit method.

$$\zeta^{n+1} = \zeta^{n} - (u\Delta t)\ \delta\zeta^{n+1}/\delta x \qquad (3\text{-}258)$$

The error is still of order $(\Delta t, \Delta x^2)$, but this method has a great stability advantage. Using the von Neumann stability analysis, we have, with $c = u\Delta t/\Delta x$,

$$V^{n+1} = V^{n} - cV^{n+1}\left(e^{I\theta} - e^{-I\theta}\right) \qquad (3\text{-}259)$$

$$V^{n+1}(1 + cI\sin\theta) = V^{n} \qquad (3\text{-}260)$$

$$G = \frac{1}{1 + cI\sin\theta} = \frac{1 - cI\sin\theta}{1 + c^2\sin^2\theta} \qquad (3\text{-}261)$$

$$|G|^2 = \frac{1 + c^2\sin^2\theta}{\left[1 + c^2\sin^2\theta\right]^2} = \frac{1}{1 + c^2\sin^2\theta} \qquad (3\text{-}262)$$

Thus for the fully implicit method, we have $|G| \leq 1$ regardless of c. The method is unconditionally stable, allowing an arbitrarily large time step, which is of great advantage.

The first order accurate, fully implicit method is also unconditionally stable for the diffusion equation* (Laasonen, 1949; Richtmyer and Morton, 1967). With $d = \alpha\Delta t/\Delta x^2$, we have

$$\zeta^{n+1} = \zeta^n + (\alpha\Delta t)\delta^2\zeta^{n+1}/\delta x^2 \tag{3-263}$$

$$V^{n+1} = V^n + V^{n+1}\, d(e^{I\theta} + e^{-I\theta} - 2) \tag{3-264}$$

$$V^{n+1}\,[1 - 2d(\cos\theta - 1)] = V^n \tag{3-265}$$

$$G = \frac{1}{1 + 2d(1 - \cos\theta)} \tag{3-266}$$

Since $(1 - \cos\theta) \geq 0$ for all θ, we have $|G| \leq 1$ for all d or all Δt. Note also that $G > 0$ for all θ; this models the zero-overshoot behavior of the continuum equation, as previously discussed.

Exercise: Verify by a von Neumann stability analysis that the fully implicit method applied to the combined advection-diffusion equation is unconditionally stable.

Exercise: Using the Taylor series expansions as in Hirt's stability analysis, show that the artificial viscosity of the fully implicit method in the transient case gives $\alpha_e = u^2\Delta t/2$. Hint: to simplify the algebra, expand about (i) and (n+1).

If, in the advection equation (3-255), we evaluate $\delta\zeta/\delta x$ as the average of the values at (n) and (n+1), we have

$$\zeta^{n+1} = \zeta^n - (u\Delta t)\,\frac{1}{2}\,[\delta\zeta^n/\delta x + \delta\zeta^{n+1}/\delta x] \tag{3-267}$$

Lilly (1965) calls this "Euler's modified method" (the forward-time centered-space, FTCS, method being the "Euler" method).** The method is still implicit, but because the averaging effectively centers the spatial derivative at (n+1/2), similar to the midpoint leapfrog method of Section III-A-6, the error is now of $O(\Delta t^2, \Delta x^2)$. Also, it gives an amplification factor $|G| = 1$ identically, as we now demonstrate.

$$V^{n+1} = V^n - \frac{c}{4}\,(e^{I\theta} - e^{-I\theta})(V^{n+1} + V^n) \tag{3-268}$$

$$V^{n+1} = V^n - Ia(V^{n+1} + V^n) \tag{3-269}$$

where

$$a = \frac{c}{2}\sin\theta \tag{3-270}$$

Then

$$G = \frac{1 - aI}{1 + aI}\left(\frac{1 - aI}{1 - aI}\right) = \frac{1 - 2aI - a^2}{1 - a^2} \tag{3-271}$$

---

*It is generally true, as may be observed in previous examples, that for explicit methods, a time differencing scheme which is successful for the advection-only equation will not be successful for the diffusion equation. But this rule of thumb does not apply to implicit methods.

**The method is also referred to as the Crank-Nicolson method. As we will note presently, this name is more precisely applied to the method for the diffusion equation, as in (3-273).

$$|G|^2 = G\tilde{G} = \frac{(1 - a^2)^2 + (2a)^2}{(1 - a^2)^2} = 1 \tag{3-272}$$

Other advantages of the Euler modified method are that, unlike the midpoint leapfrog method, it does not require two sets of initial data, and it does not develop a "time splitting" instability.

This same type of averaging applied to the diffusion equation (3-257) gives the well-known method of Crank and Nicolson (1947), also with error of $O(\Delta t^2, \Delta x^2)$.

$$\zeta^{n+1} = \zeta^n + (\alpha\Delta t)\frac{1}{2}\left[\delta^2\zeta^n/\delta x^2 + \delta^2\zeta^{n+1}/\delta x^2\right] \tag{3-273}$$

The von Neumann analysis gives

$$V^{n+1} = V^n + \frac{d}{2}(e^{+I\theta} + e^{-I\theta} - 2)(V^n + V^{n+1}) \tag{3-274}$$

$$V^{n+1} = V^n + d(\cos\theta - 1)(V^n + V^{n+1}) \tag{3-275}$$

$$G = \frac{1 - d(1 - \cos\theta)}{1 + d(1 - \cos\theta)} \tag{3-276}$$

For $(1 - \cos\theta) = 0$, we have $G = 1$. Now we consider $(1 - \cos\theta) > 0$. For $d \to 0$, $G \to 1$ (as it must for $\Delta t \to 0$). For $d \to \infty$, $G \to -1$. Thus the Crank-Nicolson method is unconditionally stable in that $|G| < 1$ for large Δt, but large Δt will cause some Fourier modes to overshoot. This suggests that the Crank-Nicolson second-order method is less accurate than the explicit FTCS method for large enough Δt. Indeed, this is the simple-minded and correct interpretation of the ordering of error terms: for small enough Δt, an $O(\Delta t^2)$ term will be smaller than an $O(\Delta t)$ term; for large enough Δt, the advantage is reversed.

Exercise: From equation (3-276), show that the $\Lambda = 2\Delta x$ component will overshoot when the of the Crank-Nicolson method exceeds the critical Δt of the explicit FTCS method.

The "Euler modified" treatment of the advection equation (3-267) and the Crank-Nicolson treatment of the diffusion equation (3-273) can be combined for the entire advection-diffusion equation, and the method is still unconditionally stable.

In these "partially implicit" methods, the present level $\delta\zeta^n/\delta x$ and the future level $\delta\zeta^{n+1}/\delta x$ are equally weighted in the averaging to give errors of $O(\Delta t^2, \Delta x^2)$. The most refined version of this approach has been given by Stone and Brian (1963), based partly on the earlier work of Brian (1961). They generalized the separate weighting factors and attempted to optimize them. Their best results were obtained by using combinations of different weighting factors at different time steps.

Another implicit method was considered by Ivanov, et al., (1970).

One disadvantage of implicit methods applied to the inviscid advection equation is that they result in an infinite signal propagation speed. For the continuum inviscid model equation, the effect of a disturbance in ζ is propagated (carried) a distance $\ell = u\Delta t$ over a time Δt. For a simple explicit finite-difference method, the disturbance is always propagated to a neighboring node point, a distance $\ell = \Delta x$, over any Δt. But for an implicit method, since all ζ^{n+1} are solved simultaneously, the disturbance is propagated a distance

$\ell = \infty$ (or to the boundaries of the computational mesh).* Note, however, that this is a desirable property for the diffusion equation, which has such an infinite signal propagation speed in the continuum. This property is not shared by the simple explicit finite-difference analogs for the diffusion equation. In order to model the correct qualitative behavior of the combined advection-diffusion equation in regard to signal speeds, one could consider using an explicit method for the advection term and an implicit method for the diffusion term (see Pracht, 1971A). Of course, this behavior is not significant if only steady-state solutions are sought.

The other disadvantage of these implicit methods is the obvious one of requiring the simultaneous solution of the N algebraic equations (where N = number of i locations not specified by known boundary values) at the new time step. If the partially implicit methods are to be really of $O(\Delta t^2,\Delta x^2)$ for the nonlinear problem, the velocity field u^{n+1} must also be solved implicitly. The simultaneous nonlinear equations are presently intractable and, in practice, the advection field solution is not treated implicitly. The N linear equations are certainly not as easy to solve as the simple explicit methods, but neither are they extremely difficult or time consuming (for the one-dimensional case), as we now demonstrate.

The equation at each node point i involves unknowns only at the neighboring points ($i\pm1$). For illustrative purposes, we indicate how the solution of such a system might proceed. (In fact, the actual method as outlined is not often used.) Say we know the (n+1) values of ζ_1 and ζ_I. The node point equation is of the form

$$\zeta_{i-1} + a\zeta_i + b\zeta_{i+1} = c' \tag{3-277}$$

where we have omitted the (n+1) superscript for clarity. With the known value of ζ_1, we can solve equation (3-277) at i = 2, giving

$$\zeta_3 = \text{fcn}(\zeta_2,\zeta_1) \tag{3-278}$$

where this notation indicates that ζ_3 is a function of the values ζ_2 and ζ_1. Proceeding to i = 3,

$$\zeta_4 = \text{fcn}(\zeta_3,\zeta_2) \tag{3-279A}$$

Using (3-278) to relate ζ_3 to ζ_2 and ζ_1, we get

$$\zeta_4 = \text{fcn}(\zeta_2,\zeta_1) \tag{3-279B}$$

Continuing, we will have at i = I - 2,

$$\zeta_{I-1} = \text{fcn}(\zeta_{I-2},\zeta_{I-3}) = \text{fcn}(\zeta_2,\zeta_1) \tag{3-280}$$

and finally at i = I - 1,

$$\zeta_I = \text{fcn}(\zeta_2,\zeta_1) \tag{3-281}$$

---

*This infinite signal propagation speed also means that at least some Fourier components have a leading phase error, contrasted to the lagging phase error of simple explicit methods.

With the known boundary values of ζ_1 and ζ_I, we can solve for ζ_2. Then, in a second "sweep" of equation (3-277), we can solve for the final values. Different types of boundary conditions can easily be accommodated.

This method as outlined is not often used because "double-sweep" methods of this type sometimes suffer from the accumulation of roundoff errors on a computer. This is overcome in several "tridiagonal algorithms", one of which is given in Appendix A.

The system is called tridiagonal because, in the matrix equation of the system,

$$[A][\zeta] = [B] \qquad (3\text{-}282)$$

the matrix [A] which is to be "inverted" is of tridiagonal form, i.e., all elements more than one position off the diagonal are zero. The tridiagonal form is schematically illustrated as

$$[A] = \begin{bmatrix} \diagdown & & 0 \\ & \diagdown & \\ 0 & & \diagdown \end{bmatrix} \qquad (3\text{-}283)$$

This tridiagonal system is fairly easy to solve. However, the application of implicit methods to the two-dimensional problem results in a *block*-tridiagonal form,

$$[A] = \begin{bmatrix} d & t & 0 \\ t & \diagdown & \\ 0 & & \diagdown \end{bmatrix} \qquad (3\text{-}284)$$

with the elements d and t themselves being tridiagonal matrices (see Mitchell, 1969, pg 120). This matrix is not so easily solved, and the most popular methods of attack are iterative.

So, although the use of implicit methods allows a large time step, it generally takes many iterations to solve that step. There is then no gain over just using the explicit method many times. Consequently, these implicit methods do not find many direct applications in multi-dimensional fluid-dynamics problems.* The one exception is in the boundary-layer equations (Section VI-D); diffusion is neglected in the streamwise coordinate, and the Crank-Nicolson method has been used for the diffusion in the other coordinate, since only a tridiagonal system is involved.**

The implicit formulation does find frequent use in the Alternating Direction-Implicit or ADI methods, which make use of the fractional time-step concept (Section III-A-13) to generate tridiagonal matrices even for the multi-dimensional problems. These ADI methods will be covered shortly, but first we will consider some multi-step explicit methods.

III-A-15. *Multi-Step Explicit Methods*

The methods previously described for the one-dimensional linear model equation are "one-step" methods, in that only one computational step is required to advance to a new time level.*** We now consider a few multi-step methods.

_____

*For the simple time-dependent diffusion equation in 2D or 3D, $\zeta_t = \alpha \nabla^2 \zeta$, the implicit Crank-Nicolson equation is a Poisson equation with the non-homogeneous or "source" term being just $(2\zeta_t/\alpha - \nabla^2 \zeta^n)$. In this case, non-iterative Poisson solvers (Section III-B-1,8,9) can be used to solve the implicit diffusion equation, and the method is both accurate and efficient.

**An instability can result with the Crank-Nicolson procedure when a gradient boundary condition is involved (F. G. Blottner, private communication).

***Leith's method (Section III-A-13) requires two steps only when extended to two dimensions.

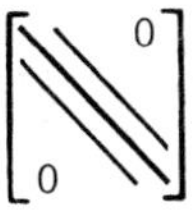

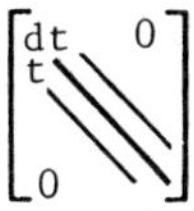

Heun's method is an explicit two-step method for the inviscid advection equation. It was used by Lorenz (1963) and was considered by Lilly (1965). It is a first iterative approximation to "Euler's modified method", equation (3-267). Instead of the term $\delta\zeta^{n+1}/\delta x$ being evaluated implicitly, it is approximated in a preliminary calculation of the term $\overline{\zeta^{n+1}}$ using forward-time, centered-space differences, as follows.

$$\overline{\zeta^{n+1}} = \zeta^n - (u\Delta t)\delta\zeta^n/\delta x \qquad (3\text{-}285A)$$

$$\zeta^{n+1} = \zeta^n - (u\Delta t)\frac{1}{2}\left[\delta\zeta^n/\delta x + \delta\overline{\zeta^{n+1}}/\delta x\right] \qquad (3\text{-}285B)$$

This method retains the second-order accuracy of equation (3-267), but is weakly unstable (Lilly, 1965), with $G = 1 + 0(\Delta t^2)$. It can be used for some early transient calculations, but not for long-term or steady calculations.

A first iterative approximation to the fully implicit treatment of the inviscid advection equation was given by Matsuno (see Lilly, 1965). Brailovskaya (1965) used the same approach for the compressible flow equations including viscosity.

$$\overline{\zeta^{n+1}} = \zeta^n - (u\Delta t)\delta\zeta^n/\delta x \qquad (3\text{-}286A)$$

$$\zeta^{n+1} = \zeta^n - (u\Delta t)\,\delta\overline{\zeta^{n+1}}/\delta x \qquad (3\text{-}286B)$$

The truncation error is $0(\Delta t,\Delta x^2)$.

The von Neumann stability analysis of this two-step method proceeds by first re-writing it as a one-step method. Applying equation (3-286A) at $(i\pm1)$ gives

$$\overline{\zeta^{n+1}_{i+1}} = \zeta^n_{i+1} - \frac{c}{2}\left(\zeta^n_{i+2} - \zeta^n_i\right) \qquad (3\text{-}287)$$

$$\overline{\zeta^{n+1}_{i-1}} = \zeta^n_{i-1} - \frac{c}{2}\left(\zeta^n_i - \zeta^n_{i-2}\right) \qquad (3\text{-}288)$$

Substituting these into equation (3-286B) gives

$$\zeta^{n+1}_i = \zeta^n_i - \frac{c}{2}\left[\overline{\zeta^{n+1}_{i+1}} - \overline{\zeta^{n+1}_{i-1}}\right]$$

$$= \zeta^n_i - \frac{c}{2}\left[\zeta^n_{i+1} - \frac{c}{2}\left(\zeta^n_{i+2} - \zeta^n_i\right) - \zeta_{i-1} + \frac{c}{2}\left(\zeta^n_i - \zeta^n_{i-2}\right)\right] \qquad (3\text{-}289)$$

or

$$\zeta^{n+1}_i = \zeta^n_i - \frac{c}{2}\left(\zeta^n_{i+1} - \zeta^n_{i-1}\right) + \left(\frac{c}{2}\right)^2\left(\zeta^n_{i+2} + \zeta^n_{i-2} - 2\zeta^n_i\right) \qquad (3\text{-}290)$$

The von Neumann analysis then proceeds as

$$V^{n+1} = V^n\left[1 - \frac{c}{2}\left(e^{I\theta} - e^{-I\theta}\right) + \left(\frac{c}{2}\right)^2\left(e^{I2\theta} + e^{-I2\theta} - 2\right)\right] \qquad (3\text{-}291)$$

$$G = 1 - \frac{c}{2}\,(I2\sin\theta) + \left(\frac{c}{2}\right)^2(2\cos 2\theta - 2) \qquad (3\text{-}292)$$

With $(1 - \cos 2\theta) = 2 \sin^2 \theta$, this gives

$$G = 1 - Ic \sin\theta - c^2 \sin^2\theta \tag{3-293}$$

With $a = c \sin \theta$, this gives

$$|G|^2 = (1 - a^2)^2 + a^2 \tag{3-294}$$

$$|G|^2 = 1 - a^2 + a^4 \tag{3-295}$$

The stability requirement of $|G|^2 \leq 1$ will be satisfied for $|a| \leq 1$, which gives $c \leq 1$. Thus, this first iterative approximation to the fully implicit method gives the usual stability condition of explicit methods, rather than the unconditional stability of the implicit method.

The two-step equations (3-286) have the same stability analysis as the one-step equation (3-290), but they are not computationally equivalent. The two-step equations can be applied at node points adjacent to a boundary, but the one-step equation would require non-physical node values inside the boundary. Also, in the compressible flow applications of Brailovskaya (1965), nonlinearity makes a difference.

We determine the artificial viscosity of the method for the transient solution by expanding the one-step equation (3-240) in a Taylor series, as in Hirt's stability analysis. Dropping subscript i and superscript n,

$$\begin{aligned}
\zeta + \frac{\partial\zeta}{\partial t}\Delta t + \frac{1}{2}\frac{\partial^2\zeta}{\partial t^2}\Delta t^2 + O(\Delta t^3) = \zeta - \frac{u\Delta t}{2\Delta x}\Bigg\{&\left[\zeta + \frac{\partial\zeta}{\partial x}\Delta x + \frac{1}{2}\frac{\partial^2\zeta}{\partial x^2}\Delta x^2 + O(\Delta x^3)\right] \\
&- \left[\zeta - \frac{\partial\zeta}{\partial x}\Delta x + \frac{1}{2}\frac{\partial^2\zeta}{\partial x^2}\Delta x^2 + O(\Delta x^3)\right]\Bigg\} \\
&+ \frac{u^2\Delta t^2}{4\Delta x^2}\Bigg\{\left[\zeta + \frac{\partial\zeta}{\partial x}(2\Delta x) + \frac{1}{2}\frac{\partial^2\zeta}{\partial x^2}(2\Delta x)^2 + \frac{1}{6}\frac{\partial^3\zeta}{\partial x^3}(2\Delta x)^3\right. \\
&\left. + O(2\Delta x)^4\right] + \left[\zeta - \frac{\partial\zeta}{\partial x}(2\Delta x) + \frac{1}{2}\frac{\partial^2\zeta}{\partial x^2}(2\Delta x)^2\right. \\
&\left. - \frac{1}{6}\frac{\partial^3\zeta}{\partial x^3}(2\Delta x)^3 + O(2\Delta x)^4\right] - 2\zeta\Bigg\}
\end{aligned} \tag{3-296}$$

Canceling and dividing by Δt gives

$$\frac{\partial\zeta}{\partial t} + \frac{\Delta t}{2}\frac{\partial^2\zeta}{\partial t^2} + O(\Delta t^2) = -u\frac{\partial\zeta}{\partial x} + O(\Delta x^2) + u^2\Delta t\frac{\partial^2\zeta}{\partial x^2} \tag{3-297}$$

Evaluating $\partial^2\zeta/\partial t^2$ from the inviscid advection equation, as in equation (3-226), we have

$$\frac{\partial^2\zeta}{\partial t^2} = u^2\frac{\partial^2\zeta}{\partial x^2} \tag{3-298}$$

So equation (3-297) becomes

$$\frac{\partial\zeta}{\partial t} = -u\frac{\partial\zeta}{\partial x} + \frac{u^2\Delta t}{2}\frac{\partial^2\zeta}{\partial x^2} + O(\Delta t^2,\Delta x^2) \qquad (3\text{-}299)$$

The artificial viscosity of the method for the transient solution then gives an effective diffusion coefficient of

$$\alpha_e = \frac{1}{2}u^2\Delta t = \frac{1}{2}c(u\Delta x) \qquad (3\text{-}300)$$

If we now apply the diffusion time-step limitation of the explicit method to equation (3-300), requiring

$$\Delta t \le \frac{1}{2}\frac{\Delta x^2}{\alpha_e} \qquad (3\text{-}301)$$

we obtain

$$\Delta t \le \frac{1}{2}\frac{\Delta x^2}{\frac{1}{2}u^2\Delta t} \qquad (3\text{-}302A)$$

$$u^2\Delta t^2 \le \Delta x^2 \qquad (3\text{-}302B)$$

or $c \le 1$, as in the von Neumann analysis.

Contrasted to the artificial viscosity of upwind differencing (3-179), the viscosity of the Matsuno method for the transient solution decreases with Δt. For the steady-state solution, the artifical diffusion coefficient is zero. This is obtained by noting that when a steady-state solution is obtained, the results of each of the two steps are identical (Allen, 1968; Allen and Cheng, 1970), since the two formulas are identical. (This is a desirable feature, not shared by all other two-step methods.) In that case, the use of centered-space derivatives for $\delta\zeta/\delta x$ gives $\alpha_e = 0$.

This method is also applicable to the complete advection-diffusion equation, either by using provisional values only for the advection term but keeping old values for the diffusion term, or by using provisional values for both. In either case, the von Neumann analysis then shows (Brailovskaya, 1965; Allen, 1968) that sufficient conditions for stability are $c \le 1$ and $d \le 1/4$. The second condition is a factor of 2 *more* restrictive than the usual diffusion limitation of the FTCS method. Allen and Cheng (1970)(see also Allen, 1968) have removed the diffusion limitation by the following method, similar in spirit and simplicity to the DuFort-Frankel adaptation of the leapfrog method given in Section III-A-7.

$$\overline{\zeta_i^{n+1}} = \zeta_i^n - \frac{c}{2}\left(\zeta_{i+1}^n - \zeta_{i-1}^n\right) + d\left(\zeta_{i+1}^n + \zeta_{i-1}^n - 2\overline{\zeta_i^{n+1}}\right) \qquad (3\text{-}303A)$$

$$\zeta^{n+1} = \zeta_i^n - \frac{c}{2}\left(\overline{\zeta_{i+1}^{n+1}} - \overline{\zeta_{i-1}^{n+1}}\right) + d\left(\overline{\zeta_{i+1}^{n+1}} + \overline{\zeta_{i-1}^{n+1}} - 2\zeta_i^{n+1}\right) \qquad (3\text{-}303B)$$

They have successfully applied the method to Burger's equation (2-20) and to the compressible flow equations in two dimensions (Allen, 1968; Allen and Cheng, 1970). The method is still of $O(\Delta t,\Delta x^2)$.

If the iteration of the Matsuno-Brailovskaya method (3-286) for the inviscid equation were continued, we would have an approximation to the fully implicit method. Using outer superscript k to indicate the iteration level, we would have

$$\zeta^{(n+1)^1} = \zeta^n - (u\Delta t)\frac{\delta}{\delta x}(\zeta^n)$$

$$\zeta^{(n+1)^2} = \zeta^n - (u\Delta t)\frac{\delta}{\delta x}\left(\zeta^{(n+1)^1}\right)$$

$$\vdots$$

$$\zeta^{(n+1)^{k+1}} = \zeta^n - (u\Delta t)\frac{\delta}{\delta x}\left(\zeta^{(n+1)^k}\right) \tag{3-304}$$

This approach could also be extended to Heun's method (3-285), giving a stable method with an error of $0(\Delta t^2,\Delta x^2)$, as in

$$\zeta^{(n+1)^0} = \zeta^n - (u\Delta t)\ \delta\zeta^n/\delta x$$

$$\zeta^{(n+1)^1} = \zeta^n - (u\Delta t)\ \frac{1}{2}\left[\delta\zeta^n/\delta x + \frac{\delta}{\delta x}\left(\zeta^{(n+1)^1}\right)\right]$$

$$\zeta^{(n+1)^2} = \zeta^n - (u\Delta t)\ \frac{1}{2}\left[\delta\zeta^n/\delta x + \frac{\delta}{\delta x}\left(\zeta^{(n+1)^2}\right)\right]$$

$$\zeta^{(n+1)^{k+1}} = \zeta^n - (u\Delta t)\ \frac{1}{2}\left[\delta\zeta^n/\delta x + \frac{\delta}{\delta x}\left(\zeta^{(n+1)^k}\right)\right] \tag{3-305}$$

Neither of these methods has been applied to the vorticity transport equation, but Veronis (1968) has applied equation (3-305) to a set of ordinary differential equations. To achieve convergence, he found k = 3 was sufficient when Δt was chosen "optimally." (However, only for large k would the gain in stability be significant.)

A three-level, two-step method proposed by Kurihara (1965)(see also Polger, 1971) has some interesting properties. For the inviscid equation, it may be written as

$$\overline{\zeta^{n+1}} = \zeta^{n-1} - (2u\Delta t)\frac{\delta\zeta^n}{\delta x} \tag{3-306A}$$

$$\zeta^{n+1} = \zeta^n - (u\Delta t)\ \frac{1}{2}\left[\frac{\delta\zeta^n}{\delta x} + \frac{\delta\overline{\zeta^{n+1}}}{\delta x}\right] \tag{3-306B}$$

The first step is a leapfrog predictor, and the second step is that of Heun's method (3-285). The method has **several** interesting characteristics (see Problems 3-16,a through g). Like the leapfrog method, it has the advantages of second order accuracy $E = 0(\Delta x^2,\Delta t^2)$, $|\lambda| = 1$ for $c \leq 1$ in the von Neumann stability analysis, and has zero artificial viscosity in both the transient and steady-state analyses. It also shares the leapfrog method disadvantages, in that extra outflow conditions and an extra line of initial conditions are required, and the $\Lambda = 2\Delta x$ component is stationary. Unlike the leapfrog method, it has the disadvantage of not giving the exact answer to the model equation for c = 1, but has the considerable advantage of not being subject to a time-splitting instability.

III-A-16. ADI Methods

The alternating direction implicit methods, or ADI methods, were introduced in companion papers by Peacemen and Rachford (1955) and Douglas (1955). Also known as the method of variable direction (Kuskova, 1968), this method makes use of a splitting of the time step to obtain a multi-dimensional implicit method which requires only the inversion of a tridiagonal matrix. Early applications to fluid dynamics problems were given by Wilkes and Churchill (1966), Samuels and Churchill (1967), Pearson* (1964, 1965A, 1965B), and Aziz and Hellums (1967). ADI methods are currently the most popular approach to viscous problems.

For the linearized problem, the Peaceman-Rachford ADI method is as follows. We use the notation $\delta\zeta/\delta x$ and $\delta^2\zeta/\delta x^2$ to indicate the space-centered approximations to $\delta\zeta/\delta x$ and $\delta^2\zeta/\delta x^2$ at i. The advancement of the advection-diffusion equation

$$\frac{\partial\zeta}{\partial t} = -u\frac{\partial\zeta}{\partial x} - v\frac{\partial\zeta}{\partial y} + \alpha\frac{\partial^2\zeta}{\partial x^2} + \alpha\frac{\partial^2\zeta}{\partial y^2} \tag{3-307}$$

over Δt is accomplished in two steps, as

$$\frac{\zeta^{n+1/2} - \zeta^n}{\Delta t/2} = -u\frac{\delta\zeta^{n+1/2}}{\delta x} - v\frac{\delta\zeta^n}{\delta y} + \alpha\frac{\delta^2\zeta^{n+1/2}}{\delta x^2} + \alpha\frac{\delta^2\zeta^n}{\delta y^2} \tag{3-308A}$$

$$\frac{\zeta^{n+1} - \zeta^{n+1/2}}{\Delta t/2} = -u\frac{\delta\zeta^{n+1/2}}{\delta x} - v\frac{\delta\zeta^{n+1}}{\delta y} + \alpha\frac{\delta^2\zeta^{n+1/2}}{\delta x^2} + \alpha\frac{\delta^2\zeta^{n+1}}{\delta y^2} \tag{3-308B}$$

or the x-y permutation thereof.

The advantage of this approach over the fully implicit methods is that each equation, although implicit, is only tridiagonal. Equation (3-308A) contains implicit unknowns $\zeta_{ij}^{n+1/2}$, $\zeta_{i\pm1,j}^{n+1/2}$. Equation (3-308B) contains implicit unknowns $\zeta_{i,j}^{n+1}$, $\zeta_{i,j\pm1}^{n+1}$. The stability of this two-dimensional method is unconditional, as in the fully implicit method [equations (3-258) and (3-263)]. But the method requires only the solution of a tridiagonal system (see Appendix A) which occurs only for usual implicit methods in one dimension. (The other disadvantage of implicit methods, that of infinite signal propagation speed for the advection term, is retained in ADI methods.)

Furthermore, the method as applied to the linear equation has a formal error of $O(\Delta t^2, \Delta x^2, \Delta y^2)$. The plausibility of the second-order time accuracy, which may not be obvious at first glance, can be recognized by writing the contributions of the x and y distributions separately. Ignoring the y-dependence, equation (3-308) can be written as

$$\frac{\zeta^{n+1} - \zeta^n}{\Delta t} = -u\frac{\delta\zeta^{n+1/2}}{\Delta x} + \alpha\frac{\delta^2\zeta^{n+1/2}}{\delta x^2} \tag{3-309}$$

which is clearly $O(\Delta t^2)$. Similarly, ignoring the x-dependence in equation (3-308) gives

$$\frac{\zeta^{n+1} - \zeta^n}{\Delta t} = -v\,\frac{1}{2}\left[\frac{\delta\zeta^n}{\delta y} + \frac{\delta\zeta^{n+1}}{\delta y}\right] + \alpha\,\frac{1}{2}\left[\frac{\delta^2\zeta^n}{\delta_y{}^2} + \frac{\delta^2\zeta^{n+1}}{\delta_y{}^2}\right] \tag{3-310}$$

which is also obviously $O(\Delta t^2)$. The second-order accuracy, and the consistency and convergence of the method applied to the diffusion equation, were shown formally by Douglas (1955, 1957) for rectangular and non-rectangular regions.

---

*Pearson (1964), pg. A31 et seq., showed that advection terms do not alter the unconditional stability, as for the fully implicit method. However, Houston and De Bremaecker (J.Comp. Phys., 1974,Vol.16,pg.230) state that the error eigenfunctions used by Pearson are real only for small advection terms, and that the proof therefore lacks generality. It may readily be verified that the required smallness is measured, as might be guessed, by cell $R_c \leq 2$.

The full second-order accuracy of the method can be deteriorated by the nonlinear terms, which should properly be evaluated as $u^{n+1/2}$, v^n in equation (3-308A) and $u^{n+1/2}$, v^{n+1} in equation (3-308B). Since u and v are determined from ψ, which is determined from the elliptic equation $\nabla^2\psi = \zeta$, this procedure would require the implicit coupled solution of ζ and ψ at both (n+1/2) and (n+1), which is out of the question. If just the old values of u^n and v^n are used throughout as in the work of Son and Hanratty (1969), the formal accuracy is now $O(\Delta t, \Delta x^2)$, but something of the second-order accuracy of the linearized system is retained if the velocity field is slowly varying.* Briley (1970) calculated u^n and v^n and, from the previously calculated u^{n-1} and v^{n-1}, he linearly extrapolated forward to $u^{n+1/2}$ and $v^{n+1/2}$. The procedure was stable and would appear to be second-order accurate. It requires the additional storage of ψ^{n-1}. It is also possible (Pearson, 1965; Aziz and Hellums, 1967) to iterate the entire procedure, using one application of equation (3-308) with u^n and v^n to obtain a first estimate ζ^{n+1} and, from a $\Delta^2\psi = \zeta^{n+1}$ solution, to obtain a first estimate u^{n+1} and v^{n+1}. In the second iteration, the advection terms in equation (3-308) can be taken as the averages $u = 1/2\ (u^n + u^{n+1})$, etc. The iteration can be stopped here or continued to (k+1) iterations, until $(u^{n+1})^{k+1} \simeq (u^{n+1})^k$; in either case, the error is $O(\Delta t^2)$, as in the Heun method (3-285) of Section III-A-15. The work per step is obviously doubled for a single iteration, and an additional storage array is required for ψ^{n+1}. Another possibility would be to calculate a $\psi^{n+1/2}$ after equation (3-308A) and then use $u^{n+1/2}$ and $v^{n+1/2}$ in (3-308B). This moves the linearization up $1/2\ \Delta t$. It may be more accurate than linearizing at u^n and v^n, and does not require the additional storage of the true second-order method as ψ^n and $\psi^{n+1/2}$ need not be distinguished in storage. (That is, ψ^n and $\psi^{n+1/2}$ can have the same FORTRAN name.) Aziz and Hellums (1967) examined these alternatives. As expected, the true second-order method (ψ^n and ψ^{n+1}) was more accurate, but they used the last method (ψ^n and $\psi^{n+1/2}$) in their three-dimensional problem wherein storage was critical.

The iteration of these schemes, required to attain second-order accuracy of the nonlinear terms, may not represent additional work at all, since some iteration is desirable for the boundary values. ADI methods, and all other implicit methods, suffer from the requirements for boundary values of ζ^{n+1}. Along some boundaries we can specify conditions on ζ^{n+1} which allow the implicit solution. But along no-slip wall boundaries, the wall values of ζ_w depend on the internal point values of ψ, in one of several possible ways to be discussed in Section III-C-2. So the implicit solution of $\nabla^2\psi^{n+1} = \zeta^{n+1}$ is required to determine ζ_w^{n+1} on the wall. The complete implicit problem is thus not practically computable for problems with no-slip walls, even if the velocities are linearized at u^n and v^n values.

The alternative treatments for boundary values ζ_w^{n+1} are identical to the u and v alternatives. We can evaluate the wall values as $\zeta_w^{n+1} = \zeta_w^n$. In this case, the wall values of ζ lag the internal values by Δt. This method was used by Wilkes and Churchill (1966). For small Δt, the approximation is accurate, but large Δt is of course a major motivation for using the ADI methods. For large Δt, the method can be not only inaccurate, but destabilizing. Iteration, as described for the ψ^{n+1} solution, is obviously preferable.

---

*This is similar to the formal $O(\Delta x)$ accuracy of the second upwind differencing method, which retains something of the second-order accuracy of the advection field if ζ is slowly varying in space (see Section III-A-11).

Many workers found computational instability (or perhaps just very slow iteration convergence) using ADI methods for high Re flows (or high Grashof number free-convection flows). They either did not calculate high Re flows (e.g., Paris and Whitaker, 1965; Torrance, 1968) or they "weighted" the results of the old and new time steps (Pearson, 1965A) which is equivalent to reducing Δt, or they reverted to upwind differencing for the advection terms (Pao and Daugherty, 1969) which is something like reducing the Re (see Section III-A-8). It was not clear whether the lack of convergence was due to poor convergence of the linearized problem, to nonlinear instability of the interior point equations, to the lagging of ζ_w^n by Δt in the one-step procedure or insufficient convergence of ζ_w^{n+1} in the iterative procedure, or to the equation used to evaluate ζ_w from internal point values of ψ. The last two have emerged as the culprits and, with the illuminating recent work of Briley (1970), the difficulty can be considered to be resolved.

The difficulty of Wilkes and Churchill (1966) was due both to the time lag of ζ_w and to the particular second-order equation used for ζ_w. Briley's results for the ζ_w equation will be covered in Section III-C-2. Samuels and Churchill (1967) reverted to a first-order equation for ζ_w, which is not destabilizing, and were thereby able to extend the computations of Wilkes and Churchill to higher Grashof numbers before the Δt lag of ζ_w caused instability.

The degree of convergence required for ζ_w to obtain stability will be problem dependent. For large Δt, convergence may be prohibited entirely by nonlinear effects. It is clear that, for a fixed number of iterations, a smaller Δt will be required to attain the same iteration convergence, i.e., the same ε in a convergence criterion line $(\zeta_n^{n+1})^{k+1} = (\zeta_w^{n+1})^k + \varepsilon$. Torrance (1968) (see also Briley, 1970 and Briley and Walls, 1971) found that the convergence of wall values of vorticity actually imposed a time-step restriction of the form $\Delta t \leq a/\Delta x^2$, where a is some number dependent on the problem and the convergence requirements. Thus, although the von Neumann analysis indicates unconditional stability, we see that the problem of the implicit wall-boundary values of vorticity actually gives a time-step restriction similar to that of the simplest explicit FTCS method. This behavior is not limited to ADI methods, but occurs with all implicit methods.

Although the practical advantage of ADI methods over explicit methods is not anything like that indicated by the von Neumann analysis, the experience of many researchers indicates that ADI methods do allow larger time steps and faster overall computation, by a factor of two or more, and furthermore allow second-order accuracy in time. It seems assured that they will continue to be used extensively for simple rectangular regions. For irregularly shaped regions, the programming can become complicated, and explicit methods may be more appealing.

The successful extension of the basic ADI method (3-308) to three dimensions is a little subtle. In the most obvious method, three calculations would be performed, with two intermediate calculations at $(t + \Delta t/3)$ and $(t + 2\Delta t/3)$. In this method, second-order accuracy in time and unconditional stability are both lost (Richtmyer and Morton, 1967), and the method goes unstable for $d > 3/2$ (Carnahan et al., 1969). We illustrate a successful extension for the three-dimensional diffusion equation

$$\frac{\partial \zeta}{\partial t} = \alpha\left(\frac{\partial^2 \zeta}{\partial x^2} + \frac{\partial^2 \zeta}{\partial y^2} + \frac{\partial^2 \zeta}{\partial z^2}\right) \tag{3-311}$$

Douglas (1962) gives the following three-step method, where the superscripts * and ** refer to intermediate values calculated via the tridiagonal algorithm.

$$\frac{\zeta^* - \zeta^n}{\Delta t} = \alpha \frac{\delta^2}{\delta x^2} \frac{1}{2}\left[(\zeta^* + \zeta^n)\right] + \alpha \frac{\delta^2 \zeta^n}{\delta y^2} + \alpha \frac{\delta^2 \zeta^n}{\delta z^2} \tag{3-312A}$$

$$\frac{\zeta^{**} - \zeta^{n}}{\Delta t} = \alpha \frac{\delta^2}{\delta x^2}\left[\frac{1}{2}(\zeta^{*} + \zeta^{n})\right] + \alpha \frac{\delta^2}{\delta y^2}\left[\frac{1}{2}(\zeta^{**} + \zeta^{n})\right] + \alpha \frac{\delta^2 \zeta^n}{\delta z^2} \qquad (3\text{-}312B)$$

$$\frac{\zeta^{n+1} - \zeta^{n}}{\Delta t} = \alpha \frac{\delta^2}{\delta x^2}\left[\frac{1}{2}(\zeta^{*} + \zeta^{n})\right] + \alpha \frac{\delta^2}{\delta y^2}\left[\frac{1}{2}(\zeta^{**} + \zeta^{n})\right] + \alpha \frac{\delta^2}{\delta z^2}\left[\frac{1}{2}(\zeta^{n+1} + \zeta^{n})\right] \qquad (3\text{-}312C)$$

The method is of accuracy $0(\Delta t^2, \Delta x^2)$ and is unconditionally stable. This procedure may be generalized to higher dimensions. Douglas and Rachford (1956), Douglas and Gunn (1964), and Brian (1961) give other ADI methods in three dimensions (see also Carnahan et al., 1969). We again note that plausible extensions frequently fail. As Richtmyer and Morton (1967) point out, if the most recent approximation ζ^{**} is used in the $\delta^2/\delta x^2$ term of equation (3-312C) instead of ζ^*, the unconditional stability is lost.

Aziz and Hellums (1967) have successfully used three-dimensional ADE methods on the entire advection-diffusion equations. McKee and Mitchell (1970) have considered ADI methods for problems with mixed spatial derivatives $\partial^2\zeta/\partial x \partial y$. Kellog (1969) has considered an ADI method for a nonlinear diffusion equation with a nonlinear boundary condition. A discussion of ADI methods applied to the diffusion equation with variable mesh spacing and general boundary conditions was given by Spanier (1967). A general discussion was given by Widlund (1967). Gustafsson (1971) developed an ADI method for the shallow-water equations. Gourlay and Mitchell (1969A) establish the equivalence of certain ADI and "locally one-dimensional methods" (see also Mitchell, 1969). Richards (1970) used ADI on the vorticity transport equations in cylindrical coordinates. See also Piacsek (1968, 1969B).

III-A-17. ADE Methods

Mainly for pedagogic reasons, we consider now the Alternating Direction Explicit or ADE methods. This is the class of methods first considered by Saul'yev (1957). [See also the texts of Saul'yev (1964), Richtmyer and Morton (1967), and Carnahan et al., (1969).] Applied to the one-dimensional diffusion equation, the simplest Saul'yev method is the following one-step version. It is more easily understood if we write the diffusion equation as

$$\frac{\partial \zeta}{\partial t} = \alpha \frac{\partial^2 \zeta}{\partial x^2} = \alpha \frac{\partial}{\partial x}\left(\frac{\partial \zeta}{\partial x}\right) \qquad (3\text{-}313)$$

and the Saul'yev finite-difference form as

$$\frac{\zeta_i^{n+1} - \zeta_i^{n}}{\Delta t} = \alpha \frac{\delta}{\delta x}\left(\frac{\delta \zeta}{\delta x}\right) = \alpha \frac{\left.\frac{\delta\zeta}{\delta x}\right|_{i+1/2}^{n} - \left.\frac{\delta\zeta}{\delta x}\right|_{i-1/2}^{n+1}}{\Delta x} \qquad (3\text{-}314)$$

or with $d = \alpha\Delta t/\Delta x^2$,

$$\zeta_i^{n+1} = \zeta_i^{n} + d\left(\zeta_{i+1}^{n} - \zeta_i^{n} - \zeta_i^{n+1} + \zeta_{i-1}^{n+1}\right) \quad \text{with } i\uparrow \qquad (3\text{-}315)$$

Note that the advance time value ζ_{i-1}^{n+1} is already known, provided that the calculation sweeps in the direction of increasing i, indicated by i↑ in equation (3-315). Although it is a two-level method, it requires only one storage array, since the same FORTRAN name can be used for ζ^{n+1} and ζ^{n} in the replacement statement (3-315). Since (3-315) can be solved for ζ^{n+1}, the method is *explicit*.

If the direction of the calculation sweep is alternated from increasing i to decreasing i at each successive time step, the resulting alternating-direction explicit (ADE) method is more nearly symmetric.

$$\zeta_i^{n+1} = \zeta_i^n + d\left(\zeta_{i+1}^n - \zeta_i^n - \zeta_i^{n+1} + \zeta_{i-1}^{n+1}\right) \text{ with } i\uparrow \tag{3-316A}$$

$$\zeta_i^{n+2} = \zeta_i^n + d\left(\zeta_{i+1}^{n+2} - \zeta_i^{n+2} - \zeta_i^{n+1} + \zeta_{i-1}^{n+1}\right) \text{ with } i\downarrow \tag{3-316B}$$

The von Neumann stability analysis of equation (3-316) proceeds in two steps. For the first step, equation (3-316A) gives

$$V^{n+1} = V^n + d\left[V^n(e^{+I\theta} - 1) + V^{n+1}(e^{-I\theta} - 1)\right] \tag{3-317}$$

$$G_A = \frac{1 - d + d\,e^{+I\theta}}{1 + d - d\,e^{-I\theta}} \tag{3-318}$$

For the second step, equation (3-316B) gives

$$V^{n+2} = V^{n+1} + d\left[V^{n+2}(e^{+I\theta} - 1) + V^{n+1}(e^{-I\theta} - 1)\right] \tag{3-319}$$

$$G_B = \frac{1 - d + d\,e^{-I\theta}}{1 + d - d\,e^{+I\theta}} \tag{3-320}$$

Now for the total two-step process we have

$$V^{n+2} = G_B V^{n+1} = G_B(G_A V^n) \tag{3-321}$$

or

$$V^{n+2} = GV^n \tag{3-322}$$

where

$$G = G_A G_B \tag{3-323}$$

Writing $e^{\pm I\theta} = \cos\theta \pm I \sin\theta$, these equations give

$$G = \left(\frac{1 - d + d\cos\theta - Id\sin\theta}{1 + d - d\cos\theta - Id\sin\theta}\right)\left(\frac{1 - d + d\cos\theta + Id\sin\theta}{1 + d - d\cos\theta + dI\sin\theta}\right) \tag{3-324}$$

Using the identity $(a + Ib)(a - Ib) = a^2 + b^2$ to simplify numerator and denominator gives

$$G = \frac{[1 - d(1 - \cos\theta)]^2 + d^2\sin^2\theta}{[1 + d(1 - \cos\theta)]^2 + d^2\sin^2\theta} \le 1 \tag{3-325}$$

which indicates unconditional stability. See Saul'yev (1964) for the proof that the one-sweep method (3-315) is also unconditionally stable.

The ADE method applied to the diffusion equation is unconditionally stable, as are ADI methods, has a formal truncation error $E = O(\Delta t^2, \Delta x^2)$ (see Saul'yev, 1964), and is likewise adaptable to higher dimensions, as in

$$\frac{\zeta_{ij}^{n+1} - \zeta_{ij}^{n}}{\Delta t} = \alpha\left(\frac{\zeta_{i+1,j}^{n} - \zeta_{ij}^{n} - \zeta_{ij}^{n+1} + \zeta_{i-1,j}^{n+1}}{\Delta x^2}\right)$$

$$+ \alpha\left(\frac{\zeta_{i,j+1}^{n} - \zeta_{ij}^{n} - \zeta_{ij}^{n+1} + \zeta_{i,j-1}^{n+1}}{\Delta y^2}\right) \quad \text{with } i\uparrow,\ j\uparrow \qquad (3\text{-}326)$$

(See Larkin, 1964, for various cyclic sweep permutations.) It has the further advantage over ADI of not requiring the implicit tridiagonal solution. Other versions of ADE methods are given by Saul'yev (1964), Larkin (1964), and Barakat and Clark (1966); see also the text of Carnahan, et al., (1969). The ADE adaptation to the nonlinear diffusion equation

$$\frac{\partial \zeta}{\partial t} = \frac{\partial}{\partial x}\left(\alpha \frac{\partial \zeta}{\partial x}\right) \quad , \quad \alpha = \alpha(x,\zeta) \qquad (3\text{-}327)$$

is particularly simple (see Quon et al., 1966).

However, the application of ADE methods to fluid dynamics is limited for two reasons. First, although the interior point equation (3-326) is explicit, the entire method is actually implicit in the boundary conditions. In the first sweep (3-326A), the advance value ζ_1^{n+1} must be known; in the second sweep, ζ_I^{n+1} is needed, where I = max i in the field. This is not troublesome in heat conduction problems, where temperatures or temperature gradients are typically known on boundaries for all time. But wall vorticity values are not known and cause difficulty, as already described for ADI methods. Second, and more fundamental, when the method is combined with any of several treatments of the advection terms, including variants of upwind differencing, FTCS, midpoint leapfrog, and ADE differencing, the result is either unconditional instability or a return to the simple explicit limits of $c \leq 1$ and $d \leq 1/2$.* The only combination that has been moderately successful is a two-sweep upwind differencing combined with an averaged ADE of the diffusion terms following Larkin (1964).

$$\zeta_i^{*} = \zeta_i^{n} - c\left(\zeta_i^{n} - \zeta_{i-1}^{n}\right) + d\left(\zeta_{i+1}^{n} - \zeta_i^{n} - \zeta_i^{*} + \zeta_{i-1}^{*}\right) \qquad (3\text{-}328A)$$

$$\zeta_i^{**} = \zeta_i^{n} - c\left(\zeta_i^{n} - \zeta_{i-1}^{n}\right) + d\left(\zeta_{i+1}^{**} - \zeta_i^{**} - \zeta_i^{n} + \zeta_{i-1}^{n}\right) \qquad (3\text{-}328B)$$

$$\zeta_i^{n+1} - \frac{1}{2}\left(\zeta_i^{*} + \zeta_i^{**}\right) \qquad (3\text{-}328C)$$

The von Neumann stability analysis gives

$$G^{*} = \frac{1 - c - d + de^{+I\theta} + ce^{-I\theta}}{1 + d - de^{-I\theta}} \qquad (3\text{-}329A)$$

$$G^{**} = \frac{1 - c + ce^{-I\theta} - d + de^{-I\theta}}{1 + d - de^{+I\theta}} \qquad (3\text{-}329B)$$

---

*Unpublished work by the present author.

$$G = \frac{1}{2}\left(G^{*} + G^{**}\right) \tag{3-329C}$$

Equations (3-329) are algebraically difficult. It was determined by numerical solution of equation (3-329) for a full range of c, d, and θ, that stability requires $c \leq 1$, but no restriction on d. Neither (3-328) nor any other ADE method has been used in an actual fluid dynamics calculation with non-zero advection terms, with one exception: the 1D shock structure calculated by Sakurai and Iwasaki (1970), which has no boundary problems.

Roberts and Weiss (1966) gave a successful ADE treatment of the inviscid transport equation, which they call their "angled derivative method." This method uses time and space differences centered at half-steps.

$$\frac{\zeta_i^{n+1} = \zeta_i^{n}}{\Delta t} = -u\,\frac{\zeta_{i+1/2}^{n+1/2} - \zeta_{i-1/2}^{n+1/2}}{\Delta x} \tag{3-330}$$

The right-hand terms are evaluated by diagonal averaging in the space-time mesh. Referring to Figure 3-13, we have for the sweep proceeding in the increasing i direction,

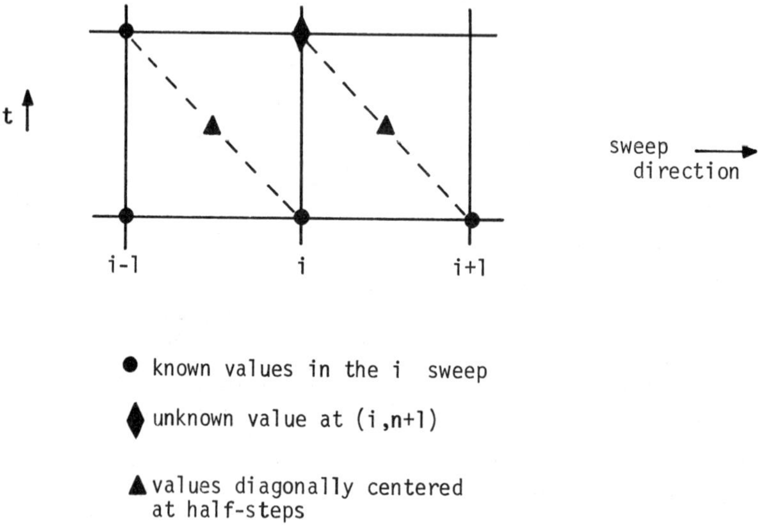

Figure 3-13. Angled derivative method.

$$\zeta_{i+1/2}^{n+1/2} = \frac{1}{2}\left(\zeta_i^{n+1} + \zeta_{i+1}^{n}\right) \quad \text{with } i\uparrow \tag{3-331A}$$

$$\zeta_{i-1/2}^{n+1/2} = \frac{1}{2}\left(\zeta_{i-1}^{n+1} + \zeta_{i}^{n}\right) \quad \text{with } i\uparrow \tag{3-331B}$$

Substituting these into equation (3-330) gives

$$\zeta_i^{n+1} = \zeta_i^{n} - \xi\left(\zeta_{i+1}^{n} - \zeta_{i-1}^{n+1}\right) \tag{3-332A}$$

where

$$\xi = \frac{c/2}{1 + c/2} \tag{3-332B}$$

Exercise: For a sweep proceeding in the decreasing i direction, show that the Roberts and Weiss angled derivative method gives

$$\zeta_i^{n+1} = \zeta_i^n - \xi'\left(\zeta_{i+1}^{n+1} - \zeta_{i-1}^n\right) \tag{3-333A}$$

where

$$\xi' = \frac{c/2}{1 - c/2} \tag{3-333B}$$

Because of the space-time centering, the method appears to be second-order accurate. However, this is difficult to show formally because of the presence of mixed space-time derivatives in the Taylor series expansion. Piacsek and Williams (1970) found the method to be accurate in an actual calculation.

Like the Saul'yev method, this method is explicit at interior points but implicit in boundary conditions. To approach symmetry, the sweep direction may be alternated from i↑ to i↓ . The von Neumann stability analysis of equation (3-332) gives

$$V^{n+1} = V^n - \xi\left[V^n e^{+I\theta} - V^{n+1} e^{-I\theta}\right] \tag{3-334}$$

$$G = \frac{1 - \xi e^{+I\theta}}{1 - \xi e^{-I\theta}} \tag{3-335}$$

which gives $|G|^2 = 1$ identically. This ADE method is unconditionally stable for the interior point equations according to the von Neumann analysis and gives $|G| = 1$, as did the midpoint leapfrog of Section III-A-6; but it does not exhibit the time-splitting instability of the leapfrog method. It is a two-level method, but requires only one FORTRAN storage array for ζ^n and ζ^{n+1}.

However, any error in a boundary calculation at the beginning of the sweep may be magnified, not as the time proceeds forward, but as the *spatial* sweep proceeds forward. Consider all $\zeta^n = 0$, and introduce an error ε in the boundary value ζ_i^{n+1}. Then equation (3-332) gives $\zeta_{i+1}^{n+1} = \xi^i \varepsilon$. In order to avoid *spatial* amplification of (say) a machine round-off error, it is necessary that $|\xi| \le 1$ for i↑ and $|\xi'| \le 1$ for i↓. For the alternating-direction method, both requirements dictate $|c| \le 1$ (see Problem 3-19). This is one instance where the usual von Neumann stability analysis fails to give any clue about a true computational instability.

Like the implicit methods, the ADE methods for the inviscid advection equation give an infinite signal propagation speed, which is not a property of the continuum equation.

This ADE method has not been combined with an ADE method for the diffusion equation to give an unconditionally stable explicit method for the entire advection-diffusion equation, nor has it been used in an actual fluid-dynamics calculation.

Exercise: Demonstrate that the Saul'yev ADE method (3-315) for the diffusion equation does not cause spatial amplification of errors.

III-A-18. The Hopscotch Method.

The hopscotch method was first used for the iterative solution of the Poisson equation by Sheldon (1962)(see Section III-B-7) and for the compressible flow equations by Scala and Gordon (1966, 1967). Gourlay (1970A) generalized the approach, proved stability for quite general equations in multi-dimensions, applied it to the transient heat conduction and the elliptic equations, and gave it the accurately picturesque name of "hopscotch."

This name comes about from the pattern of the sweeps, shown in Figure 3-14. Each time-step n is calculated in two sweeps of the mesh. In the first and subsequent odd-numbered

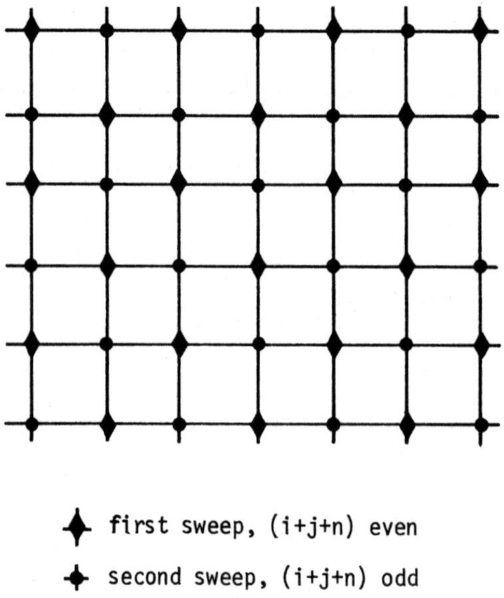

Figure 3-14. Hopscotch method.

time steps n, the nodes marked with an "◆" in Figure 3-14, defined by (i+j) odd, are calculated based on current values of neighboring points. For the following simple diffusion equation

$$\frac{\partial \zeta}{\partial t} = \frac{\partial^2 \zeta}{\partial x^2} + \frac{\partial^2 \zeta}{\partial y^2} \tag{3-336}$$

this gives

$$\zeta_{ij}^{n+1} = \zeta_{ij}^{n} + \Delta t \left[\frac{\zeta_{i+1,j}^{n} - 2\zeta_{i,j}^{n} + \zeta_{i-1,j}^{n}}{\Delta x^2} + \frac{\zeta_{i,j+1}^{n} - 2\zeta_{ij}^{n} + \zeta_{i,j-1}^{n}}{\Delta y^2} \right], (i+j) \text{ odd} \tag{3-337}$$

This first sweep consists of FTCS differencing for (i+j) odd. For the second sweep at the same time level, the same calculation is used at nodes marked with "•" defined by (i+j) even, but now using the known advance values of neighboring points calculated in the first sweep:

$$\zeta_{ij}^{n+1} = \zeta_{ij}^{n} + \Delta t \left[\frac{\zeta_{i+1,j}^{n+1} - 2\zeta_{ij}^{n+1} + \zeta_{i-1,j}^{n+1}}{\Delta x^2} + \frac{\zeta_{i,j+1}^{n+1} - 2\zeta_{ij}^{n+1} + \zeta_{i,j-1}^{n+1}}{\Delta y^2} \right], (i+j) \text{ even} \tag{3-338}$$

This second sweep is fully implicit in the sense that (n+1) values are required at (i,j), (i±1,j) and i,j±1), but this "implicitness" involves no simultaneous algebraic solutions, since the neighboring advance values are already known from the first sweep and the ζ_{ij}^{n+1} value can be eliminated by algebra as

$$\zeta_{ij}^{n+1} = \left[\zeta_{ij}^{n} + \frac{\zeta_{i+1,j}^{n+1} + \zeta_{i-1,j}^{n+1}}{\Delta x^2} + \frac{\zeta_{i,j+1}^{n+1} - \zeta_{i,j-1}^{n+1}}{\Delta y^2} \right] \Bigg/ \left[1 + \frac{2}{\Delta x^2} + \frac{2}{\Delta y^2} \right] \tag{3-339}$$

In the second and subsequent even-numbered time steps, the role of the ♦ and • nodes are interchanged. This is succintly expressed by noting that equation (3-337) is applied at (i+j+n) even, and (3-338) at (i+j+n) odd.

The speed of the computation is considerably increased by the use of the following relation, which follows from the algebra of the method (Gourlay, 1970A).

Exercise: Show that the hopscotch equations (3-337) and (3-338) give

$$\zeta_{ij}^{n+2} = 2\zeta_{ij}^{n+1} - \zeta_{ij}^{n} \text{ for } (i+j+n) \text{ even} \tag{3-340}$$

The original equations are used at steps where print-out is required.

The method has a formal truncation error $E = O(\Delta t, \Delta x^2)$. The method is also applicable to the advection-diffusion equation, and equation (3-340) still holds. Gourlay (1970A) has shown unconditional stability for the diffusion equation, and for the advection-diffusion equation when upwind differencing is used for the advection terms. Scala and Gordon (1966, 1967) used upwind differencing in hopscotch for their compressible viscous flow calculations. However, it is possible that hopscotch methods could be used with centered-difference advection terms, as ADI methods have been. Gourlay (1970A) reports that hopscotch is 3 to 4 times as fast, per computational step, as the Peaceman-Rachford ADI method, due to the absence of tridiagonal inversions and to the efficiency of equation (3-340). It is also considerably easier to program than ADI, especially for complex geometries, and is readily extendable to three dimensions. However, for calculations of the vorticity transport equations, it shares with ADI and ADE methods the difficulty of implicit boundary values. The close relationship between the hopscotch method and both the ADI and the Dufort-Frankel methods is explored by Gourlay (1970A,B).

The "explicitness" of equations (3-338) depends on the fact that the usual five-point analog for the Laplacian operator involves values only at (ij), (i±1,j), and i,j±1). Gourlay (1970A) calls such finite-difference expressions "E-operators." The centered difference form for the advection terms is also an E-operator, but the forms for cross-derivative terms and the nine-point forms for the Laplacian (see Section III-B-10) are not. These forms require implicit evaluation of the equations corresponding to equation (3-338). The application of a hopscotch method to cross derivatives has been considered by Gourlay and McKee (1971) and to hyperbolic systems with shocks by Gourlay and Morris (1971). The latter paper actually included 1D shock calculations. Morris (1971) has applied the method to the parabolic heat-conduction (diffusion) equation in 3D, with the mixed (Robbin's) boundary conditions which arise from the use of Newtonian convective-heating equations. The method has not yet been applied to multi-dimensional incompressible flow equations with some gradient boundary conditions, but extensive further use is anticipated.

III-A-19. THE FOURTH-ORDER METHODS OF ROBERTS AND WEISS AND OF CROWLEY

In the next three sections, we return to the consideration of methods which, although explicit, are more complicated than implicit methods. The complexity arises because the motivation for their development was the reduction of the phase error, discussed in Section III-A-13 in connection with Leith's method.

The paper of Roberts and Weiss (1966) appeared at about the same time as Leith's (1965) paper. Both served to illuminate the importance of phase error. Roberts and Weiss put forth four different classes of explicit methods and examined others. Their work, and that of Fromm (1968) and Grammeltvedt (1969), should be read in the original by anyone seriously concerned with phase errors. We have already presented the Roberts and Weiss second-order accurate "angled derivative" method in the previous section. We present here their most interesting and, from the viewpoint of reduced phase error, most successful methods which are accurate to $O(\Delta t^2, \Delta x^4)$. The methods were derived via the integral approach (Section III-A-1-c), assuring strict conservation. Following Roberts and Weiss (1966), we define two interlaced or staggered mesh systems,* as shown in Figure 3-15. The symbols

*Their method can also be described in terms of a single mesh in two dimensions. Their second-order method on a staggered mesh is actually equivalent to the midpoint leapfrog method, Section III-A-6.

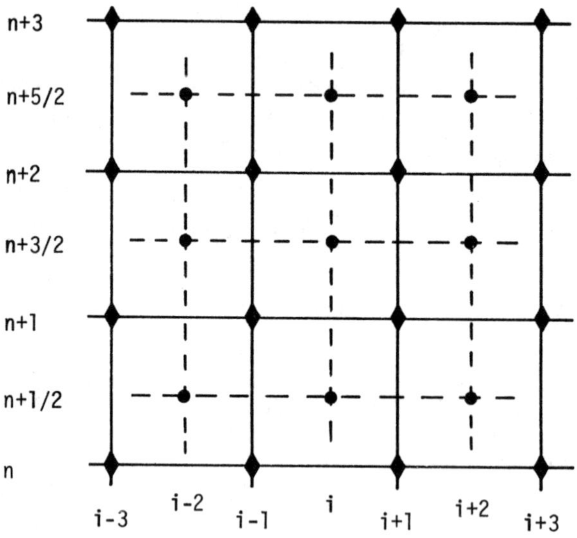

Figure 3-15. Staggered mesh system for the 4th-order method of Roberts and Weiss in one space dimension.

◆ and ● represent nodes in the two meshes. Using four values at (n+1/2) to evaluate the spatial derivative for the linearized equation, the first fourth-order method of Roberts and Weiss is obtained as

$$\zeta_i^{n+1} = \zeta_i^n - \frac{c}{2}\left(\zeta_{i+1}^{n+1/2} - \zeta_{i-1}^{n+1/2}\right)$$

$$+ \frac{1}{24}\left(1 - \frac{1}{4}c^2\right)\left(\zeta_{i-3}^{n+1/2} - 3\zeta_{i-1}^{n+1/2} + 3\zeta_{i+1}^{n+1/2} - \zeta_{i+3}^{n+1/2}\right) \tag{3-341}$$

The first bracketed term in equation (3-341) is just the midpoint leapfrog method; the second is a higher order correction. The method gives $|\lambda| = 1$ for $c \leq 2$. (Note that $c = u\Delta t/\Delta x$, where $\Delta t = t^{n+1} - t^n$. But, in one computational step, we progress only $\Delta t/2$, from a known solution at $t^{n+1/2}$ to t^{n+1}. So the effective Courant number should really be $c' = (u\Delta t/2)/\Delta x$, with $c' \leq 1$ required for stability.) The phase shift $\Delta\theta/2$ per time step, using the notation of Section III-A-13, is given by Roberts and Weiss (1966) as

$$\sin\left(\frac{\Delta\theta}{2}\right)_{FDE} = \frac{c}{2}\sin\theta\left[6 + \left(1 - \frac{c^2}{4}\right)\sin^2\theta\right] \tag{3-342}$$

The second fourth-order method of Roberts and Weiss is obtained by a combination of their two second-order methods, the "angled derivative" ADE method (equations 3-332 and 3-333) of Section III-A-17, and the midpoint leapfrog, written in the staggered mesh.

$$\zeta_i^{n+1} = \zeta_i^n - \frac{c/2}{1 - M\,c/2}\left[\left(1 + 2M\right)\left(\zeta_{i+1}^{n+1/2} - \zeta_{i-1}^{n+1/2}\right) - M\left(\zeta_{i+2}^n - \zeta_{i-2}^{n+1}\right)\right] \text{ with } i\uparrow \tag{3-343A}$$

wherein

$$M = \frac{1 - \frac{c}{2}}{6\left(1 + \frac{c}{2}\right)} \qquad (3\text{-}343B)$$

Since the right side of equation (3-343A) involves ζ_{i-2}^{n+1}, this second method is an ADE method written for i↑. Like the other ADE methods of Section III-A-17, it is implicit in the boundary condition, i.e., ζ_1^{n+1} must be known in order to start the i↑ sweep.

For $c \to 0$, we have $M \to 1/6$ and $\frac{c/2}{1 - M\,c/2} \to c/2$. Then equation (3-343A) may be approximated by

$$\zeta_i^{n+1} = \zeta_i^n - \frac{c}{2}\left[\frac{4}{3}\left(\zeta_{i+1}^{n+1/2} - \zeta_{i-1}^{n+1/2}\right) - \frac{1}{6}\left(\zeta_{i+2}^{n} - \zeta_{i-2}^{n+1}\right)\right] \qquad (3\text{-}344)$$

The phase shift is given by Roberts and Weiss (1966). For $c \to 0$, the phase error may be found from

$$\frac{(\Delta\theta)_{FDE}}{(\Delta\theta)_{PDE}} \simeq \frac{c \sin\theta\,(4 - \cos\theta)}{3\theta} \qquad (3\text{-}345)$$

This second method gives slightly less phase error than the first, and both are significantly better than second-order methods. All fail to move the $\Lambda = 2\Delta x$ component, i.e., the phase error is complete, as is true of all methods except the upwind differencing methods.

As expected, the method is not without drawbacks. Where applied to two space dimensions, four staggered meshes are required: two interlaced in space at each of the integer and half-integer time levels (see Roberts and Weiss, 1966). The computational work is severe. Molenkamp (1968) says the staggered mesh method requires 45 times as much computation time, as well as 4 times the core storage, as the upwind differencing method. The formal truncation error $E = (\Delta t^2, \Delta x^4)$ will not hold globally unless the boundary evaluation is also $O(\Delta x^4)$, which is generally not true (see Section III-C-2). The time-splitting instability of the midpoint leapfrog method (Section III-A-6) allows the appearance of two disjoint solutions; the present method allows *four* disjoint solutions to develop. This obviously requires diffusion terms to couple the solutions and, if experience with the midpoint leapfrog is any guide, a low Re is required if there is a likelihood of a steady state being approached. The staggered mesh system also causes difficulty in boundary conditions; the treatment suggested by Roberts and Weiss (1966) causes the control volume interpretation of node values at boundaries to be inconsistent with interior node interpretations, diminishing accuracy near boundaries.

Another fourth-order method with improved phase error is given by Crowley (1967). The first step predicts provisional values $\zeta^{n+1/2}$ for i, (i±1) using Leith's method (3-224) with $\Delta t/2$. The second step is a leapfrog calculation

$$\zeta_i^{n+1} = \zeta_i^n - \frac{c}{2}\left(\zeta_{i+1}^{n+1/2} - \zeta_{i-1}^{n+1/2}\right) \qquad (3\text{-}346)$$

The phase-error properties may be found in Crowley (1967) and Fromm (1968).

III-A-20. Fromm's Method of Zero Average Phase Error.

Most explicit methods exhibit only a lagging phase error, i.e., the computational advection speed of the θ-component is $u \cdot r(\theta)$, where $r(\theta) \leq 1$. Fromm (1968) has designed a method based on a combination of leading and lagging phase-error methods, the design objectives being to produce a composite method which is (1) conditionally stable, and (2) has a zero phase error, on an average.

The idea of the method is to project the exact solution (for the linearized constant u equation) at an advance time with c = 1. The desired solution with c < 1 is then obtained using Leith's method with *backward* differencing in time from the c = 1 projected solution. For clarity, we first illustrate the idea using FTCS differencing rather than Leith's method.

With a unity Courant number for the projected solution, we have a $\Delta t' = (\Delta x/u)$. The point-to-point transfer of grid values for the exact solution then gives

$$\zeta_i^{t+\Delta t'} = \zeta_{i-1}^{t} \equiv \zeta_{i-1}^{n} \qquad (3\text{-}347A)$$

$$\zeta_{i+1}^{t+\Delta t'} = \zeta_{i}^{t} \equiv \zeta_{i}^{n} \qquad (3\text{-}347B)$$

$$\zeta_{i-1}^{t+\Delta t'} = \zeta_{i-2}^{t} \equiv \zeta_{i-2}^{n} \qquad (3\text{-}347C)$$

The backward solution in time, from $(t+\Delta t')$ to $(t+\Delta t)$, where $\Delta t < \Delta t'$, then proceeds as

$$\frac{\zeta_i^{t+\Delta t'} - \zeta_i^{t+\Delta t}}{(\Delta t' - \Delta t)} = -u\,\frac{\delta\zeta^{t+\Delta t'}}{\delta x} = \frac{-u}{\Delta x}\left(\zeta_{i+1}^{t+\Delta t'} - \zeta_{i-1}^{t+\Delta t'}\right) \qquad (3\text{-}348)$$

Solving for $\zeta_i^{t+\Delta t}$ and noting that

$$\frac{u(\Delta t' - \Delta t)}{\Delta x} = 1 - c \qquad (3\text{-}349)$$

gives

$$\zeta_i^{t+\Delta t} = \zeta_i^{t+\Delta t'} + (1 - c)\left(\zeta_{i+1}^{t+\Delta t'} - \zeta_{i-1}^{t+\Delta t'}\right) \qquad (3\text{-}350)$$

The right-hand members are now evaluated using the exact solution (3-347) and the notational identity $\zeta_i^{t+\Delta t} \equiv \zeta_i^{n+1}$

$$\zeta_i^{n+1} = \zeta_{i-1}^{n} + (c - 1)\left(\zeta_i^{n} - \zeta_{i-2}\right) \qquad (3\text{-}351)$$

Equation (3-351) illustrates the rationale of obtaining a one-dimensional method with a leading phase error. Instead of the second-order centered-space expression for $\delta\zeta^{t+\Delta t'}/\delta x$ as in equation (3-348), Fromm uses Leith's method, applied in two space dimensions in the fractional time-step form (3-254). This method, with a leading phase error, is then averaged with Leith's method (3-254), which has a lagging phase error. The result is Fromm's "method of zero average phase error." We use a tilde to represent the fractional time-step calculation at (n+1/2) to which no physical significance is attached. Note $c_x = u\Delta t/\Delta x$, and $c_y = v\Delta t/\Delta y$. Fromm's method of zero average phase error is then

$$\tilde{\zeta}_{ij} = \zeta_{ij}^{n} + \frac{c_x}{4}\left(\zeta_{i-1,j}^{n} - \zeta_{i+1,j}^{n} + \zeta_{i-2,j}^{n} - \zeta_{ij}^{n}\right) + \frac{c_x^2}{4}\left(\zeta_{i-1,j}^{n} - 2\zeta_{ij}^{n} + \zeta_{i+1,j}^{n}\right)$$
$$+ \left(\frac{c_x^2 - 2c_x}{4}\right)\left(\zeta_{i-2,j}^{n} - 2\zeta_{i-1,j}^{n} + \zeta_{ij}^{n}\right) \qquad (3\text{-}352A)$$

$$\zeta_{ij}^{n+1} = \tilde{\zeta}_{ij} + \frac{c_y}{4}\left(\tilde{\zeta}_{i,j-1} - \tilde{\zeta}_{i,j+1} + \tilde{\zeta}_{i,j-2} - \tilde{\zeta}_{ij}\right) + \frac{c_y^2}{4}\left(\tilde{\zeta}_{i,j-1} - 2\tilde{\zeta}_{ij} + \tilde{\zeta}_{i,j+1}\right)$$

$$+ \left(\frac{c_y^2 - 2c_y}{4}\right)\left(\tilde{\zeta}_{i,j-2} - 2\tilde{\zeta}_{i,j-1} + \tilde{\zeta}_{ij}\right) \qquad (3\text{-}352B)$$

The method is stable for $c_x + c_y \leq 1$. Fromm (1968) presents contour plots of the modulus $|G|$ and the phase error, with c_x, c_y and θ as parameters. Although the method is formally second-order accurate, the phase properties are considerably improved over the fourth-order method of Roberts and Weiss (1966) and Crowley (1967) presented in the previous section. Like these, the computational time is considerably increased over simpler methods. As in Leith's method and all fractional time-step methods, there is some problem with boundary values for the provisional step (3-252B). These can possibly be handled by lagging wall values of ζ by $\Delta t/2$ or by iterating (see Section III-A-16). Fromm* recommends switching to a simpler centered-space or upwind-difference form adjacent to the boundary. Methods of treating the diffusion terms in conjunction with equation (3-352) have not yet appeared in the open literature.

Exercise: Design a method of zero-average phase error based on upwind differencing.

In a later work, Fromm (1971) used the upwind and centered difference forms at alternate time steps, rather than as a two-step method, thus reducing computation time. But upwind differencing was used for directional Courant numbers < 1/2. Special considerations are also required near boundaries; see Fromm (1971).

III-A-21. Arakawa's Method

A method frequently used in meteorological calculations for the inviscid vorticity transport equation is the method of Arakawa (1966). This is an essentially two-dimensional method, using a 9-point formulation involving nodes like (i+1,j-1) etc. It intrinsically involves the stream function relations $u = \partial\psi/\partial y$, $v = -\partial\psi/\partial x$. We cannot present a derivation of the method; we merely convert it from the original operator notation to our index notation.

$$\zeta_{ij}^{n+1} = \zeta_{ij}^{n-1} + \frac{\Delta t}{6\Delta x\Delta y}\Big\{\left(\psi_{i+1,j} - \psi_{i-1,j}\right)\left(\zeta_{i,j+1} - \zeta_{i,j-1}\right)$$

$$- \left(\psi_{i,j+1} - \psi_{i,j-1}\right)\left(\zeta_{i+1,j} - \zeta_{i-1,j}\right) + \psi_{i+1,j}\left(\zeta_{i+1,j+1} - \zeta_{i+1,j-1}\right)$$

$$- \psi_{i-1,j}\left(\zeta_{i-1,j+1} - \zeta_{i-1,j-1}\right) - \psi_{i,j+1}\left(\zeta_{i+1,j+1} - \zeta_{i-1,j+1}\right)$$

$$+ \psi_{i,j-1}\left(\zeta_{i+1,j-1} - \zeta_{i-1,j-1}\right) + \zeta_{i,j+1}\left(\psi_{i+1,j+1} - \psi_{i-1,j+1}\right)$$

$$- \zeta_{i,j-1}\left(\psi_{i+1,j-1} - \psi_{i-1,j-1}\right) - \zeta_{i+1,j}\left(\psi_{i+1,j+1} - \psi_{i+1,j-1}\right)$$

$$+ \zeta_{i-1,j}\left(\psi_{i-1,j+1} - \psi_{i-1,j-1}\right)\Big\}^n \qquad (3\text{-}353)$$

Although obviously complex, the method has advantages. The formal truncation error E is $O(\Delta t^2, \Delta x^4, \Delta y^4)$. It is single-step and has no boundary condition problems. It gives

*Personal communication.

$|G| = 1$ identically, and identically conserves ζ, ζ^2 and kinetic energy $u^2 + v^2$; these properties make it especially well suited to hydrodynamic stability problems. Since it conserves ζ^2, it is not subject to the nonlinear instability of Phillips (1959) which arises from aliasing errors. (Aliasing errors are present but remain bounded, since ζ^2 remains bounded.) Its good phase-error properties and a generalization to the meteorological "β-plane" equations are given by Grammeltvedt (1969). Festa (1970) included viscous diffusion terms using the Dufort-Frankel approach (Section III-A-7).

Unfortunately, this method does share the shortcoming of other methods which use leap-frog time differencing (see Section III-A-6), in that it is susceptible to the time-splitting instability (see Williams, 1969). Also, Festa (1970) used occasional averaging of time levels to achieve steady-state stability.

III-A-22. Remarks on Steady Flow Methods.

Many features of the methods for the vorticity transport equation described above are applicable only to the transient solution (e.g., phase error). Even though the philosophy of the time-dependent approach is attractive (Harlow and Fromm, 1965; Macagno, 1965), it is natural to ask why should one bother with the transient solution in those cases where the only interest is in the eventual steady-state solution? Why not set $\partial\zeta/\partial t = 0$ and just work with the steady flow equations?

This approach has, in fact, been used successfully by many workers. Yet we generally recommend the transient approach. As the first crucial step in the argument, we illustrate how the simplest steady-flow method is equivalent to an unsteady flow method.

The steady-flow, one-dimensional model transport equation is obtained from the parabolic model transport equation (2-18) as

$$u \frac{\partial \zeta}{\partial x} = \alpha \frac{\partial^2 \zeta}{\partial x^2} \tag{3-354}$$

which can be discretized, using centered-space differencing, to

$$u_i \left(\frac{\zeta_{i+1} - \zeta_{i-1}}{2\Delta x} \right) = \alpha \left(\frac{\zeta_{i+1} - 2\zeta_i + \zeta_{i-1}}{\Delta x^2} \right) \tag{3-355}$$

This elliptic equation may be solved by some iteration procedure, such as those which will be described in Section III-B. The simplest method (Richardson or Jacobi iteration) involves solving equation (3-355) for ζ_i. From some initial guess ζ_i^1 for all i, new (k+1) iterative values of ζ_i^{k+1} are determined by using old (k) values on the right-hand side.

$$\zeta_i^{k+1} = - \left(\frac{u\Delta x}{4\alpha} \right) \left(\zeta_{i+1}^k - \zeta_{i-1}^k \right) + \frac{1}{2} \left(\zeta_{i+1}^k + \zeta_{i-1}^k \right) \tag{3-356}$$

Iteration is continued until some convergence criterion (Section III-D) is reached. If we subtract ζ_i^k from both sides of equation (3-356), the iteration computational step is not changed.

$$\zeta_i^{k+1} - \zeta_i^k = - \left(\frac{u\Delta x}{4\alpha} \right) \left(\zeta_{i+1}^k - \zeta_{i-1}^k \right) + \frac{1}{2} \left(\zeta_{i+1}^k - 2\zeta_i^k + \zeta_{i-1}^k \right) \tag{3-357}$$

or

$$\zeta_i^{k+1} - \zeta_i^k = - \left(\frac{u\Delta x^2}{2\alpha} \right) \left(\frac{\zeta_{i+1}^k - \zeta_{i-1}^k}{2\Delta x} \right) + \frac{\Delta x^2}{2} \left(\frac{\zeta_{i+1}^k - 2\zeta_i^k + \zeta_{i-1}^k}{\Delta x^2} \right) \tag{3-358}$$

We now note that (3-358) is equivalent to (3-18), the *unsteady* forward-time, centered-space FDE, if we define an iterative time step in equation (3-18) of $\Delta t^I = 1$, an iterative advection speed $u^I = u\Delta x^2/2\alpha$, and an iterative diffusion coefficient $\alpha^I = \Delta x^2/2$. From our analysis of (3-18), we know that convergence requires an iterative Courant number $C^I \leq 1$, or

$$C^I = \frac{u^I \Delta t^I}{\Delta x} = \frac{u\Delta x^2(1)}{2\alpha\Delta x} = \frac{u\Delta x}{2\alpha} \leq 1 \qquad (3\text{-}359A)$$

or

$$Rc \leq 2 \qquad (3\text{-}359B)$$

For a cell Reynolds number $u\Delta x/\alpha > 2$, the method will diverge.

This example illustrates how the Richardson iteration method is *strictly equivalent* to an unsteady method, and a poor one at that. Other iterative methods for the elliptic equation are equivalent or at least analogous to unsteady methods for the parabolic equation. The analogy was first perceptively discussed by Frankel (1950).

More recently, Hodgkins (1966) established the correspondence between the Chebyshev semianalytic method and a time-dependent hyperbolic equation. The relation between the solution of the steady flow equations and the limit of the unsteady solution was explored by Heywood (1970).

Under- and over-relaxation methods essentially involve multiplying the right-hand side of equation (3-357)by a factor r, where $r < 1$ implies under-relaxation and $r > 1$ implies over-relaxation. For the example method (3-358), reducing r is clearly equivalent to reducing the time step. As we can anticipate from our analysis of the unsteady equations, under-relaxation will be required as Re is increased. Textor (1968) and Tejeira (1966) both found experimentally that $r \propto 1/Re$ was required, which is in accordance with our Courant-number restriction. But above some Rc, convergence will not be achieved even for r arbitrarily small.* This has been the common experience of researchers who have (successfully) used this type of method, e.g., Thom and Orr (1931), Thom (1933), Thom and Apelt (1961), Kawaguti (1953, 1961, 1965), Burggraf (1966), Michael (1966), Hamielec, et al., (1967A, 1967B), Friedman, et al., (1968), Dennis, et al., (1968), Friedman (1970), and Lee and Fung (1970). The Rc limitation persists, even when some (not all) of the boundary-layer approximations are used (Plotkin, 1968). The Rc limitation can be removed in an iterative steady-state approach by reverting to an upwind-difference formulation for the advection terms, as done by Runchal and Wolfshtein (1969), Greenspan (1969A, 1969B), and Gosman and Spalding (1971). Other iterative schemes are possible, and can be quite successful; in addition to the above, see Allen and Southwell (1955), who used Southwell's residual relaxation method (Section III-B-3) and attained a cylinder flow solution at Re = 1000; Griffiths, et al., (1969), who used line SOR (Section III-B-4) in cylindrical coordinates; and Katsanis (1967) and Brady (1967), who iterated potential flow steady-state equations.

It is an interesting historical fact that most workers who have used the steady-flow iteration approach have not analyzed their methods for stability and for convergence rates, but have determined these empirically. Yet the stability considerations for the unsteady flow equations have been well known since the early work of von Neumann. A possible explanation is that the steady-flow approach evolved from that branch of numerical analysis concerned with the solution of the Poisson equation, for which the simplest iterative methods have no stability restrictions.

The treatment of the nonlinear advection terms can alter the strict equivalence of Richardson iteration to the time-dependent approach. In the time-dependent approach, the elliptic Poisson equation, $\nabla^2\psi = \zeta$, is usually iterated to convergence at each time step of the vorticity transport equation. In the steady-flow approach, each equation may (though not necessarily) be iterated upon in succession. In this "combined iteration" approach,

---

*While it is true that this Rc limitation may be linked to a linear instability in the time-like iteration, it may also be linked to the boundary phenomenon which will be described in Section III-C-8.

iteration convergence, if it occurs, can thereby be accomplished in fewer steps. A. D. Gosman (private communication; see Gosman and Spalding, 1971) reports computer time savings of 40%. However, because $\nabla^2\psi = \zeta$ is not converged at each vorticity iteration, a poor choice of the initial conditions may lead to nonlinear instability in the steady method (Tejeira, 1966; Leal, 1969; Richards, 1970) through the advection terms and/or the boundary conditions. Experience shows that this difficulty does not occur with the use of explicit, time-dependent methods, nor with implicit methods if the boundary values are properly iterated. Further, a FORTRAN computer program may be written for the unsteady equations and then be readily adapted to the steady approach. The "combined iteration" of the Poisson equation and the vorticity transport equation may be achieved by using an easy convergence criterion for the Poisson iteration. (This is, in fact, the recommended procedure; see Section III-D.) And the advantage of Liebman (Gauss-Seidel) or SOR iteration methods (to be covered in Section III-B), which are analogous to the time-dependent ADE methods (Section III-A-17), may then be achieved in a FORTRAN program by the simple addition of an EQUIVALENCE statement between the array names for ζ^{n+1} and ζ^n. Even the practice of using a smaller under-relaxation factor near boundaries [Friedman (1970) used r boundary ≈ 1/3r interior] can be achieved with a spatially-varying* Δt.

An early work in which both the steady-state and the time-dependent approaches were used was that of Hung and Macagno (1966). They found that the time-dependent equations were easier to handle and more stable, a conclusion rediscovered by many workers since that time. This conclusion obviously depends on the simplicity of the time-dependent method used. It is not recommended that an elaborate method such as Fromm's (Section III-A-19) be used when only steady-state solutions are desired. For reasonably simple time-dependent methods, the flexibility of being able to achieve the transient solution, if desired, is also attractive. More important, the time-dependent approach does not presume the existence of a steady-state solution, which indeed may not exist.

Here, a word of caution is needed. If the time-dependent FDE's do converge to a stable steady state, we cannot casually assume that the PDE's have a stable steady state. As we have seen, discretization sometimes introduces an "artificial viscosity" effect. This and other truncation errors can cause the FDE to be more stable than the PDE; the distinction of hydrodynamic stability from numerical over-stability is a difficult problem (see Section VI-E for references).

Another approach to the incompressible problem which deserves mention is the single-variable, fourth-order equation approach. Simply substituting the Poisson equation (2-13) and the velocity relations (2-7) into the vorticity transport equation (2-12) gives

$$\frac{\partial(\nabla^2\psi)}{\partial t} = -\frac{\partial}{\partial x}\left(\frac{\partial\psi}{\partial y}\nabla^2\psi\right) + \frac{\partial}{\partial y}\left(\frac{\partial\psi}{\partial x}\nabla^2\psi\right) + \frac{1}{Re}\nabla^4\psi \tag{3-360A}$$

Though applicable in principal to the unsteady-flow equation, this approach is usually considered for the left-hand side of (3-360A) set = 0. Genuinely successful applications were achieved by Bourcier and Francois (1969), who used ADI methods (see Section III-B-7), and Powe, et al., (1971), using Liebman iteration. Most workers (Thom and Apelt, 1961; Paris and Whitaker, 1965; Pearson, 1964, 1965A) found difficulties with boundary conditions and slow convergence rates. Pearson (1964, 1965A) performed calculations for the limit of zero advection terms, in which case equation (3-360B) reduces, in steady flow, to

$$\nabla^4\psi = 0 \tag{3-360B}$$

and for simple Dirichlet boundary conditions. Even in this easy case, Pearson found convergence rates reduced by an order of magnitude compared with the two-equation approach. In fact, it is known that the most efficient iterative method of solving the biharmonic equation (3-360B), common in elasticity problems, is to decompose it into two Poisson equations whenever

---

*That is, a local $\Delta t(x,y)$ can be calculated as a function of the local critical Δt, and then be applied to advance ζ through ζ_t locally. The time-dependent solution will obviously not be meaningful, but the iterative convergence to a steady-state might be accelerated.

the boundary specifications permit. See, e.g., Thom and Apelt (1961) and Pal'tsev (1970). Dias (1970) used fourth- and fifth-order polynomial fits to evaluate $\nabla^2\psi$, but encountered problems at boundaries.

A really conclusive advantage of the steady-flow method occurs when the steady equations display a parabolic behavior in space, i.e., a "marching" solution in a spatial coordinate is possible or required. These cases include the boundary layer equations, flows with finite chemical reaction rates, or thixotropic effects (Section VI-D) which depend on the Lagrangian particle history, and the inverse approach to the detached shock problem (Section V-A-1). Kyriss (1970) attempted to attain the best features of each approach in a compressible flow detached-shock problem, using time-dependent equations for the continuity and momentum equations and steady-state equations for the temperature and chemical composition.

Davis (1971) has developed a method for the Navier-Stokes equations which is time-like but which has several steady-state features. The calculation is split into two directions; in one direction, the calculation is based on the unsteady boundary layer equations with second-order curvature corrections and, in the other direction, the equations are linear. Because of the transformations used, initial conditions of the *steady-state* boundary layer solution are obtained in a natural way. The initial solution also fixes the boundary conditions at infinity at their steady-state values. (The transformations also remove the leading-edge singularity for the limiting case of the flat plate.) For high-Re problems in which the boundary layer equations are accurate, convergence is attained in one iteration. For lower Re flows without separation, convergence is still obtained in comparatively few iterations. The method is expected to be used extensively in future work.

The present author (Roache, 1972) has also formulated several methods which are not time-like in their iterations. These methods, the Numerical Oseen (NOS) and Laplace Driver (LAD) methods, make use of recent advances in solving *linear* second-order FDE's (see Sections III-B-1, 8, 9). The evaluation of these methods is far from complete, but at least for some flows they are clearly superior to time-dependent methods, even for separated flows.

Non-iterative methods are not applicable for the steady-state flow equations because of the nonlinearities. But methods such as those of Sections III-B-1, 8, 9 may be the most efficient for the linearized equations (e.g., the temperature equation, Section III-F).

We summarize the comparison between the iterative steady-state and time-dependent methods as follows. (1) Some steady-state methods are strictly equivalent to time-dependent methods, with under- and over-relaxation adjustments being equivalent to Δt changes. Most steady-state iterative methods are at least analogous to time-dependent methods. In any case, the analogy illustrates that (2) steady-state iterative methods cannot be presumed to be stable, and should be analyzed for stability via the von Neumann analysis. (3) A FORTRAN computer program written for the time-dependent approach can (a) be used to achieve the "combined iteration" approach of the steady-state method by merely adjusting the convergence criterion of the Poisson equation, and (b) be converted quickly to a Liebman or SOR-type iteration through the use of an EQUIVALENCE statement. (4) The explicit time-dependent methods, in which the Poisson equation is converged after each vorticity transport iteration, are less susceptible to nonlinear instabilities and are thereby less sensitive to initial conditions. (5) The time-dependent formulation offers the flexibility of obtaining the transient solution if it should be desired; more important, it does not presume the existence of a steady-state solution, which indeed may not exist. (6) There is a philosophical and even aesthetic attraction, nebulous though widespread, in modeling the actual physical process which is, after all, fundamentally time-dependent.

III-A-23. Remarks on Evaluating Methods; Behavioral Errors

In the preceding sections, we have covered a representative sampling of computational methods for treating the vorticity transport equation. The potential user is now faced with the task of evaluating these methods for applicability to his fluid dynamics problems, and of devising new methods if he finds these lacking.

Consideration of the error of a computational analog is obviously a major factor in evaluation and design. In the usual textbook classification, FDE errors are either *round-off* errors or *truncation* errors. Round-off errors are due to the finite floating-point word length of electronic computers. This word length, or the number of significant figures carried, is integer only in the arithmetic base of the computer (usually 2, 8, 10, or 16). For current U. S. computers, the equivalent decimal word length ranges from less than 7 to more than 14 significant decimal figures, for single-precision arithmetic. Round-off error is difficult to analyze because it introduces qualitative aberrant behavior; e.g., floating-point addition and multiplication are cummutative, but not associative or distributive. [See Forsythe (1970) for an excellent and readable presentation of round-off error effects.] Round-off error can be a major concern in highly refined solutions of ordinary differential equations (ODE) because the Δx can become very small and because of the use of high-order methods which are sensitive to round-off error. For PDE's in one space dimension, the Δx and Δt can also be small enough that round-off error would be important.* Also, round-off error will be important in some matrix inversion problems (see Section III-B-8), will affect the choice of convergence criteria (see Section III-D), and will obviously limit the smallest Δt for which a computation is meaningful. Although awareness of the round-off problem is important, it is generally true that, for PDE solutions of multi-dimensional problems, the mesh spacings $\Delta\vec{x} = (\Delta x, \Delta y, \Delta z)$ and Δt are necessarily coarse enough for truncation error to dominate round-off error.

"Truncation error" refers to the error incurred by not retaining all terms in the infinite Taylor series expansions or, equivalently, by using finite $\Delta\vec{x}$, Δt. Neglecting round-off error, it is true that all the rest of the error is truncation error. Although accurate, this classification is not adequate. In the preceding sections, we have discussed another classification of errors, which may be described as a *behavioral* classification.

These behavioral errors include conservative errors (or lack of the conservative property), transportive errors, numerical damping errors and numerical viscosity errors, lack of Galilean invariance ("wind tunnel transformation"), boundedness (or overshoot) errors, phase errors, and aliasing errors. All of these are *truncation errors* in the sense that they vanish as $\Delta\vec{x}$, $\Delta t \to 0$, but this is a coarse description indeed. For example, conservative errors may be eliminated regardless of truncation error (although some contribution of round-off error persists). Likewise, some methods possess the transportive property, others have no damping error for the constant-coefficient equation, etc. It seems that this behavioral classification is more meaningful than the simple "round-off error" or "truncation error" classification.

In selecting a method, the user must weigh the importance of these errors to his problem. For example, the conservative error of a non-conservative method may be used as a convergence indicator; phase errors are of no consequence to a steady-state solution; Buneman (1967) considers the time-reversibility of time-symmetric methods to be a desirable behavior; etc. It is worthwhile to evaluate a method with an eye to this behavioral classification of errors, rather than to focus exclusively on the "order" of the truncation error, i.e. $E = O(\Delta t^2, \Delta x^2)$, etc., although the order is also important.

Persons familiar only with numerical methods for ODE's are invariably surprised at the low-order approximations which generally have been used in the past for PDE's. The reason is simply that it is difficult to achieve uniformly high-order *performance* in non-trivial fluid dynamics problems. In the total problem, the accuracy of the vorticity transport equation will be restricted by the Poisson solution (see Section III-B) and by the *boundary formulation* (see Section III-C-1). The latter especially makes it difficult to achieve a uniformly high-order formulation of the entire problem using conventional multi-point high-order equations such as those in Section III-B-10. (Near a straight boundary conveniently aligned with a coordinate direction, an $O(\Delta x^4)$ method will involve the boundary value and the next five interior points, for example. See Southwell, 1946.) Stability analyses become very difficult, although the concept of *time-splitting* (Section III-A-13) can help. Another reason for the often disappointing performance of higher-order methods in PDE's is that the "order" of the method is significant only as $\Delta\vec{x}, \Delta t \to 0$. Thus "order" is more significant to ODE's, where the less stringent storage and time limitations allow for much smaller $\Delta\vec{x}$. In many test calculations, lower order methods have produced as accurate or more accurate calculations for coarse Δx and Δt; see Cyrus and Fulton (1967), Cheng (1968), page 210 of Hamming (1962), and the examples cited in Section III-A-8. (Cheng, 1970A feels that second-order methods are, roughly speaking, optimal.) In fact, for the frequently desireable large Δt, we may expect a *reversal* of advantage in methods which are higher-order in time. See the examples of Section III-A-14, in which the Crank-Nicolson method with $E = O(\Delta t^2, \Delta x^2)$ for the diffusion equation gives an erroneous overshoot $(G < 0)$ for large Δt, whereas the fully implicit method with only $E = (\Delta t, \Delta x^2)$ gives the qualitatively correct behavior. Even among the successful applications, there is a wide range of accuracy of second-order methods, and an even wider range for fourth-order methods. The differences in accuracy of competing fourth-order methods is often due to *conservation* forms (e.g., see Williamson, 1969) and to their focus on the *phase error* problem (see Sections III-A-18,19,20). Also, the *aliasing error* of some $O(\Delta x^4)$ methods is worse than the $O(\Delta x^2)$ methods, especially for the shorter wavelength components (Grammeltvedt, 1969).

Nevertheless, current research is being pursued on higher-order methods, and we now feel that they will become more commonly used in computational fluid dynamics. In the first printing of this book (copyright 1972) we repeated the "conventional wisdom" about higher-order methods, that one should expect only moderate accuracy for PDE's as compared to ODE's because of the information lost in the discretization of the boundary data. We now seriously question the validity of this argument. It is true that the practical difficulties which affect the accuracy of second-order methods - difficulties like boundary conditions, complicated boundary shapes, consistent solutions of the Poisson equation, and the cell Reynolds number problem - all seem to become more delicate for higher-order solutions, and it is true that the convergence is not always uniformly rapid over a 2D mesh. But higher-order performance on multidimensional problems can and has been obtained.

We have presented some higher-order methods in Sections III-A-18,19,20. Additional references using "conventional" higher-order methods follow. Fairweather (1969) used an ADI method for the diffusion equation which is $O(\Delta t^2, \Delta x^2)$. (We note that some of the methods tabulated by Richtmyer and Morton, 1967, for the diffusion equaiton achieve a higher "order" of accuracy for certain parameter combinations, but these conditions are not generally significant in fluid dynamics problems.) Gunaratnam and Perkins (1970) derived high-order schemes by the method of weighted residues. Dawson and Marcus (1970) have used a modified Runge-Kutta-Gill scheme for time integration only. Lomax et al. (1970) used fourth-order Runge-Kutta time integration for the 1D model inviscid equation. Friedman (1970) used fourth-order forms for second derivatives normal to the wall (the dominant direction for diffusion) but second-order forms parallel to the wall. A similar advantage may be gained by employing a rectangular mesh with $\Delta x \neq \Delta y$; that the advantage can be real was shown by Hung and Macagno (1966). Rybicki and Hopper(1970) treated the 2D diffusion equation using fully implicit $O(\Delta t)$ differencing in time and 36 degree-of-freedom finite elements in space.

There have been other applications of conventional higher-order finite differences since 1972, but a more promising development is the use of "compact differences". Orszag and Israeli[1] and Hirsh[2] have credited this scheme to Kreiss[3] and refer to it as compact differencing. But according to Rubin and Khosla[4] the following methods are equivalent, i.e. any one can be derived from any other: Kreiss compact, Hermitian, Pade´ approximation, Mehrstelleng, and Rubin and Khosla's own 4th-order spline-on-spline. The compact scheme can be described as follows, using the notation of Hirsh[2].

Consider a discrete function ζ_i for which we want to evaluate an approximation F_i to the first partial derivative , and an approximation S_i to the second partial derivative. To evaluate F_i, we first evaluate the usual second-order centered difference approximation for the first derivative and store it in an array called f_i. That is,

$$f_i = \frac{\zeta_{i+1} - \zeta_{i-1}}{2\Delta} \tag{3-361A}$$

where Δ is the mesh spacing in either x or y. Then, a fourth-order approximation F_i is obtained by solving

$$\frac{1}{6}(F_{i+1} + 4F_i + F_{i-1}) = f_i \tag{3-361B}$$

Conventional fourth-order approximations are explicit formulae involving 5 local points, i and its neighbors i±1 and i±2. The compact scheme involves only 3 points, i and i±1, in the formula, but the formula is implicit, i.e. non-local. The F_i are solved by a tridiagonal solution (see Appendix A) of equation (3-361B), so that values of F at all i are dependent on one another, and thus on f_i and ζ_i, globally rather than locally. (In this global dependence, the compact scheme is like spectral and pseudo-spectral methods[1].) Also, the compact scheme has a lower coefficient of the truncation error term of $O(\Delta^4)$ than does the conventional fourth-order method. In a similar way, the second-order approximation to the second derivative is calculated explicitly and stored in s_i. That is,

$$s_i = \frac{\zeta_{i+1} - 2\zeta_i + \zeta_{i-1}}{\Delta^2} \qquad (3\text{-}361C)$$

Then a fourth-order approximation S_i is obtained from the tridiagonal solution of

$$\frac{1}{12}(S_{i+1} + 10S_i + S_{i-1}) = s_i \qquad (3\text{-}361D)$$

By itself, the equation for the F's would require a boundary condition on the first derivative, and similarly for the S's. This is a common difficulty with higher-order methods . However, Hirsh[2] has shown that the boundary values for the coupled system can be obtained to fourth-order accuracy by using the second diagonal Padé approximant,

$$\zeta_i - \zeta_{i+1} + \frac{\Delta}{2}(F_i + F_{i+1}) + \frac{\Delta^2}{12}(S_i - S_{i+1}) = 0 \qquad (3\text{-}362)$$

This equation allows the coupled system for the F's and S's to retain both its fourth-order accuracy and its tridiagonal form right up to the boundary. This form is thus simpler than conventional 5-point expressions, as well as more accurate.

Hirsh[2] has solved 2D low Reynolds number viscous steady flows using this compact scheme applied in an ADI manner (see Section III-A-16). For roughly the same accuracy, the compact fourth-order scheme showed a savings over a second-order scheme of a factor of 20 in computer time and a factor of 3 in storage. The boundary conditions on vorticity were lagged, as is commonly done when only the steady-state solution is of interest, so that temporal accuracy was lost. Three-point compact differences can also be developed for $O(\Delta^6)$ and higher methods (Hirsh, private communication). The spline-on-spline formulation of Rubin and Khosla[4] includes *variable mesh spacing*, in which case the accuracy of F remains $O(\Delta^4)$ but the accuracy of S deteriorates to $O(\Delta^3)$.

The monograph by Kreiss and Oliger[3] and the review by Orszag and Israeli[1] are recommended for introductions to higher-order finite difference methods, and to spectral and pseudo-spectral methods.

It seems important here to emphasize that higher-order methods, even when correctly and uniformly applied, do not solve the cell Re problem described in Section III-C-8. In fact, the "wiggles" that arise when $Rc > 2$ are often aggravated by higher-order differencing. Apparently, the unfavorable evaluation of many early studies on high-order methods can be traced to the lack of awareness of the role of the Rc limitation, which represents possibly the most difficult problem in computational fluid dynamics[5]. Finally, we note that accuracy is not the only consideration in choosing a method. The total cost, both in computer time and *people time*, is often the dominant consideration. The total cost must include both production-run costs and development costs, and the latter is strongly dependent on the simplicity of the method. Complex methods are more difficult to program, to test, and especially to debug than simple methods, which becomes more important for complicated boundaries and additional physical laws, e.g. chemical reactions. Clearly, there is no clear choice for a "best method".

Additional references for higher-order methods.

1. Orszag, S.A. and Israeli, M. (1974), "Numerical Solution of Viscous Incompressible Flows", in Annual Review of Fluid Mechanics, Vol. 6, Annual Reviews, Inc., Palo Alto.
2. Hirsh, R.S. (1975), "Higher Order Accurate Difference Solutions of Fluid Mechanics Problems by a Compact Differencing Scheme", J. of Computational Physics, Vol.19,pp.90-109.
3. Kreiss, H. and Oliger, J. (1973), Methods for the Approximate Solution of Time Dependent Problems, GARP Publication Series No. 10, World Meteorological Organization, Feb. 1973. Available from UNIPUB, Inc., P.O. Box 433, New York, NY 10016.
4. Rubin, S.G. and Khosla, P.K. (1975), "Higher Order Numerical Solutions Using Cubic Splines", pp. 55-66 of Proc. Second AIAA Computational Fluid Dynamics Conference, Hartford, Conn., June 19-20, 1975.
5. Roache, P.J. (1975), "A Review of Numerical Techniques", in Proc. First International Conference on Numerical Ship Hydrodynamics", to be published by U.S. Government Printing Office.

III-B. METHODS FOR SOLVING THE STREAM FUNCTION EQUATIONS

In the previous sections, we dealt only with 1/3 of the incompressible fluid dynamics problem, that of solving the parabolic vorticity transport equation. This is an initial value of "marching" problem in time. We now consider the second 1/3 of the total problem, methods for solving the elliptic Poisson equation (2-13) for stream function ψ.

$$\nabla^2\psi \equiv \frac{\partial^2\psi}{\partial x^2} + \frac{\partial^2\psi}{\partial y^2} = \zeta \tag{3-363}$$

This represents a boundary-value or "jury" problem, and requires different methods. For the present, we will consider solving the Poisson equation with two types of boundary conditions along parts of the boundary: either Dirichlet conditions, in which values of ψ are known, or the Neumann conditions, in which values of the normal gradient $\partial\psi/\partial n$ are known. Just *when* these conditions are appropriate is the final 1/3 of the problem, and will be covered in Section III-C.

The discretized form of the Poisson equation, using second-order differences, is the "5-point formula" (Thom and Apelt, 1961).

$$\frac{\delta^2\psi}{\delta x^2} + \frac{\delta^2\psi}{\delta y^2} = \zeta \tag{3-364}$$

$$\frac{\psi_{i+1,j} - 2\psi_{ij} + \psi_{i-1,j}}{\Delta x^2} + \frac{\psi_{i,j+1} - 2\psi_{ij} + \psi_{i,j-1}}{\Delta y^2} = \zeta_{ij} \tag{3-365}$$

where ζ_{ij} is known.

III-B-1. Direct Methods

In a rectangular domain where max i = I and max j = J, equation (3-365) and the boundary equations give a system of $N = (I-2) \times (J-2)$ simultaneous linear algebraic equations. Except for the "source" or non-homogeneous term, ζ_{ij}, it is the same block-tridiagonal system which results from a fully implicit formulation of the 2D diffusion equation (see Section III-A-14) so the tridiagonal algorithm is not applicable. The most elementary methods of solving such a system are Cramer's rule and the various forms of Gaussian elimination (see Crandall, 1956). For the problems of interest, N is a large number, and these methods are inadequate. Cramer's rule involves quite an unbelievable number of operations - approximately $(N+1)!$ multiplications. Even if time were available, the accuracy would be destroyed by round-off error.* The multiplications in Gaussian methods vary as N^3, and the accuracy may be expected to deteriorate for N much greater than 50 (Hamming, 1962) depending on the details of the method and the computer word length. These and older methods are evaluated in the book by Westlake (1968).

In recent years, highly efficient direct methods have been developed. A recent paper by Dorr (1970) reviews the "block" methods, cyclic reduction methods, tensor product methods, the Fourier series methods, and a few others. Lancaster (1970) also presents a review of direct methods. Other recently published direct methods are the discrete invariant imbedding method of Angel (1968A), his dynamic programming approach (Angel, 1968B), the method of summary representation of Polozhii (1965) and Didenko and Liashenko (1964) (see also Chalenko, 1970), the method of Yee (1969), the method of odd/even reduction by Buzbee, et al., (1969, 1970B), and the EVP method of Roache (1971A, B). Hayes (1970) used Green's integral to compute the Laplace solution ($\zeta = 0$) directly. Angel and Kababa (1970) have established the formalism of one-sweep method based on invariant embedding. Other direct methods are treated by Swift (1971) and Tsao (1970).

---

*The time is not available. Using the figures cited by Forsythe (1970), the multiplications for the solution of 26 equations using Cramer's rule on a CDC 6600 would require 10^{16} years, or about 10^6 times the current estimates of the age of the universe.

A particularly attractive method based on "odd-even reduction" is that of Buneman (1969). The computer program is fairly short and self-contained, being independent of fast Fourier-transform routines, and it gives answers to essentially machine accuracy. But it is restricted in its direct application to a rectangular region with Dirichlet boundary conditions everywhere, and (I-1) and (J-1) must each be powers of 2. (However, see the procedures for extending such methods in Section III-B-9). The method was used by Fromm (1971) in a large problem, 128 × 128. The program is excellent, but the (M,N) indexing system used therein is not compatible with the (I,J) system used in this book, i.e., (M,N) cannot be rotated into (I,J).

All these methods have one or more of the following disadvantages: limited* to rectangular domains like L- or T-shapes; limited* to $\psi = 0$ boundary conditions; large storage requirements; not adaptable to non-Cartesion coordinates; field size limited (restricting the magnitude of I or J) due to accumulation of round-off error; field specification limited (e.g., I and J limited to the form of 2^k, k integer); elaborate preliminary calculations specific to the mesh required; complex to program; difficult to understand. However, the direct methods, particularly the Fourier-series methods, are increasing in use and will become more commonly used for large problems. The EVP method will be considered in Section III-B-8 because of its flexibility and simplicity compared with other direct methods, and the Fourier-series methods will be described in Section III-B-9 because of their growing importance.

By comparison, the various iterative methods are very easy to understand and to program, and are quite flexible. They are much faster than the older direct methods (Westlake, 1968) because they make use of the "sparseness" of the matrix, i.e., all terms more than two positions off the diagonal are zero. The iterative methods have historically been used most often in computational fluid dynamics and will undoubtedly continue to be important. The historical development of iterative methods will be followed in this presentation.

III-B-2. Richardson's Method and Liebman's Method

As discussed in Section III-A-22, the solution of (steady) elliptic equations by iteration is analogous to solving a time-dependent problem to an asymptotic steady state (Frankel, 1950). Suppose we consider a time-dependent diffusion equation for ψ with a source term, ζ, and with a diffusion coefficient of unity.

$$\frac{\partial\psi}{\partial t} = \nabla^2\psi - \zeta \qquad (3\text{-}366)$$

We are not here interested in the physical significance of the transient solution, but as a steady state solution of this diffusion equation is approached, the solution approaches the desired solution for the Poisson equation, (3-363).

In some cases, the analogy is exact. To illustrate the equivalence, we shall proceed to derive Richardson's iterative method for the elliptic Poisson equation from the FTCS time-dependent method for the parabolic diffusion equation.

Applying FTCS differencing to (3-366) we obtain

$$\frac{\psi_{ij}^{k+1} - \psi_{ij}^{k}}{\Delta t} = \frac{\delta^2\psi^k}{\Delta x^2} + \frac{\delta^2\psi^k}{\Delta y^2} - \zeta_{ij} \qquad (3\text{-}367)$$

To avoid confusion, we momentarily restrict ourselves to $\Delta x = \Delta y = \Delta$. Then equation (3-367) gives

$$\psi_{ij}^{k+1} = \psi_{ij}^{k} + \frac{\Delta t}{\Delta^2}\left[\psi_{i+1,j}^{k} + \psi_{i-1,j}^{k} + \psi_{i,j+1}^{k} + \psi_{i,j-1}^{k} - 4\psi_{ij}^{k} - \Delta^2\zeta_{ij}\right] \qquad (3\text{-}368)$$

---

*In direct application; see Section III-B-9.

We first demonstrate that ζ_{ij} does not affect the stability properties of equation (3-368). Following Shortley and Weller (1938), we consider Dirichlet boundary conditions and denote by ψ^{∞} the exact FDE solution* of the FDE Poisson equation (3-364). The errors e^k_{ij} in the iterative values ψ^k_{ij} are then

$$e^k_{ij} = \psi^{\infty}_{ij} - \psi^k_{ij} \tag{3-369}$$

Substituting this into (3-364) gives

$$\frac{\delta^2\psi^{\infty}}{\delta x^2} - \frac{\delta^2 e^k}{\delta x^2} + \frac{\delta^2\psi^{\infty}}{\delta y^2} - \frac{\delta^2 e^k}{\delta y^2} = \zeta \tag{3-370}$$

Since ψ^{∞} exactly satisfies (3-364), then equation (3-370) reduces to the Laplace equation,

$$\frac{\delta^2 e^k}{\delta x^2} + \frac{\delta^2 e^k}{\delta y^2} = 0 \tag{3-371}$$

Since the boundary values are known, $\psi^k = \psi^{\infty}$ or $e = 0$ on all boundaries. Then the FDE iteration (3-368) can be written for the error e as

$$e^{k+1}_{ij} = \frac{\Delta t}{\Delta^2}\left[e^k_{i+1,j} + e^k_{i-1,j} + e^k_{i,j+1} + e^k_{i,j-1}\right] \tag{3-372}$$

So the iteration (3-368) for ψ is equivalent to the iteration (3-372) for e, which clearly does not depend on ζ.

The stability restriction on equation (3-368) or (3-372)(see Section III-A-5-e) with $\alpha=1$ is just $d = \alpha\Delta t/\Delta^2 \leq 1/4$ or $\Delta t \leq \Delta^2/4$. Since we wish to approach the asymptotic condition as rapidly as possible, we consider taking the largest possible $\Delta t = \Delta^2/4$. Substituting into (3-368) then gives

$$\psi^{k+1}_{ij} = \psi^k_{ij} + \frac{1}{4}\left[\psi^k_{i+1,j} + \psi^k_{i-1,j} + \psi^k_{i,j+1} + \psi^k_{i,j-1} - 4\psi^k_{ij} - \Delta^2\zeta_{ij}\right] \tag{3-373}$$

The ψ^k_{ij} terms inside and outside the brackets cancel, giving

$$\psi^{k+1}_{ij} = \frac{1}{4}\left[\psi^k_{i+1,j} + \psi^k_{i-1,j} + \psi^k_{i,j+1} + \psi^k_{i,j-1} - \Delta^2\zeta_{ij}\right] \tag{3-374}$$

This is *Richardson's method*** for $\Delta x = \Delta y$. For the special case of Laplace's equation ($\zeta_{ij} = 0$). the method simply involves setting the new iterative value equal to the arithmetic average of its four neighbors.

---

*The notational implication that the exact solution accrues as $k \to \infty$ is more suggestive than precise. The $\lim_{k \to \infty} \psi^k$ will differ from the exact solution of equation (3-365) because of round-off error.

**Also known as the Jacobi method (Salvadori and Baron, 1961) as "iteration by total steps" (Crandall, 1956), and as the method of "simultaneous displacements" (Young, 1954), since each ψ^{k+1} is calculated independent of the sequence in (i,j) and therefore, in a sense, simultaneously. This distinction is important to the consideration of techniques for parallel-processing computers.

Equation (3-374) is the same result that is obtained from simply solving the (steady) elliptic equation (3-365) for ψ_{ij} and evaluating this term on the left side of the equation at (k+1) and all terms on the right-hand side at k. Defining the mesh aspect ratio $\beta = \Delta x/\Delta y$, this gives

$$\psi_{ij}^{k+1} = \frac{1}{2(1+\beta^2)}\left[\psi_{i+1,j}^{k} + \psi_{i-1,j}^{k} + \beta^2\psi_{i,j+1}^{k} + \beta^2\psi_{i,j-1}^{k} - \Delta x^2\zeta_{ij}\right] \tag{3-375}$$

This is *Richardson's method* for $\Delta x \neq \Delta y$.

The analysis of the convergence rate can proceed as in the analysis of stability for the vorticity equation, now writing

$$V^{k+1} = GV^{k} \tag{3-376}$$

for the error equation (3-372) or alternately by expanding the discretized Laplacian operator in its eigenfunctions; identical results are obtained provided boundary conditions are correctly accounted.* We present here the results of Frankel (1950) obtained in this manner, with a different notation.

The highest and lowest wavelength error components damp most slowly (i.e., have largest $|G(\theta)|$. Thus, regardless of the initial error distribution in θ, these components will dominate for k asymptotically large. Both have the same $|G(\theta)|$, the short wavelength ($\Lambda = 2\Delta x$) oscillating in sign with $G < 0$, which is, in a sense, optimal (Frankel, 1950).

Equation (3-375) is a two-level equation, requiring storage of ψ^{k+1} and ψ^{k}. If we sweep in i↑, j↑ and use new values wherever available in equation (3-375), we obtain

$$\psi_{ij}^{k+1} = \frac{1}{2(1+\beta^2)}\left[\psi_{i+1,j}^{k} + \psi_{i-1,j}^{k+1} + \beta^2\psi_{i,j+1}^{k} + \beta^2\psi_{i,j-1}^{k} - \Delta x^2\zeta_{ij}\right] \tag{3-377}$$

which is *Liebman's method*.** This method may be programmed with only one storage level, interpreting (3-377) as a FORTRAN replacement statement. Further, Frankel (1950) showed that, for the most resistant high and low wavenumber error components,

$$G(\text{Liebman}) = [G(\text{Richardson})]^2 \tag{3-378}$$

Asymptotically, k Liebman iterations are worth 2k Richardson iterations, and only require half the core storage.***

The most sophisticated version of what was still essentially Liebman's method was given by Shortley and Weller (1938).

---

*The more elegant matrix methods are described in Forsythe and Wasow (1960) and Ames (1969).

**Also known as the Gauss-Seidel method (Salvadori and Baron, 1961), as "iteration by single steps" (Crandall, 1956), and as the method of "successive replacements" (Young, 1954). In the time-step analogy, this method is similar but not identical to the Saul'yev methods (see Section III-A-17).

***Actually, the core storage requirement of methods like Richardson's can be reduced with a little programming effort and at the cost of a few more replacement statements.

III-B-3. Southwell's Residual Relaxation Method.

Southwell's residual relaxation method (Southwell, 1946) was used for many years to obtain numerical solutions by hand calculation of important engineering and scientific problems, including one of the earliest fluid dynamics solutions at a high Re (Allen and Southwell, 1955). The original name for the method was just the "relaxation method," but here we use the term "residual relaxation" to distinguish it from Liebmann's method and other iterative procedures, which today are sometimes called relaxation methods.

The simplest form of Southwell's residual relaxation method uses the same equation as Richardson's method (3-375) to evaluate new ψ_{ij}^{k+1}. The difference is that equation (3-375) is not applied indiscriminately to all mesh points in a sweep of the matrix. Rather, the residual $r_{i,j}$ is defined by

$$r_{ij} = \frac{\psi_{i+1,j} + \psi_{i-1,j} - 2\psi_{ij}}{\Delta x^2} + \frac{\psi_{i,j+1} + \psi_{i,j-1} - 2\psi_{ij}}{\Delta y^2} - \zeta_{ij} \qquad (3\text{-}379)$$

When $r_{ij} = 0$, the Poisson equation (3-365) is satisfied, but only at the point (i,j). Thus, $|r_{ij}|$ is indicative of how much the present estimate for all ψ_{ij} is in error at (i,j). One then scans the field for the largest $|r_{ij}|$ and sets this $r_{ij} = 0$ by calculating a new ψ_{ij} from (3-375). This in turn changes r at all neighboring points, and the scan is repeated.

This method is not used in today's electronic computers because the time required to scan for max $|r_{ij}|$ and recalculate neighboring r's is not sufficiently shorter than the time needed to apply equation (3-375). Thus, on an electronic computer, it is more efficient to eliminate each residual in turn and use new values whenever available, giving Liebmann's method. Southwell's method was historically important because a more refined version which evolved suggested the "extrapolated Liebmann method", more commonly known as the successive over-relaxation method.

III-B-4. Successive Over-Relaxation (SOR) Method

Continued work (Fox, 1948) with Southwell's method showed that, for fastest convergence, not the *largest* residual $|r_{ij}|$ should be eliminated, but rather that r_{ij} which required the largest "displacement" $|\psi_{ij}^{k+1} - \psi_{ij}^{k}|$ to eliminate it. This technique is clearly applicable only to a skilled human computer who can quickly make an approximate calculation of maximum displacement in a visual scan of the residuals. A second technique then evolved. It was found that optimum ultimate convergence was obtained, not by identically setting the residuals to zero, but by "over-relaxing" or "under-relaxing," depending on whether neighboring residuals were of the same or of opposite sign (Fox, 1948). (The general concept of over-relaxation had been suggested by Richardson in 1910.) The application of this concept in Southwell's method was successful, but the method now required even more the skill and intuition of a human computer. (This requirement was actually advantageous (Fox, 1948) in human computations, since the computer was less likely to get bored!)

Frankel (1950) and independently Young (1954) developed a method of applying over-relaxation to Liebmann's method in a methodical manner suited to electronic computers. Frankel called it the "extrapolated Liebmann method" (see Problem 3-20), and Young called it "successive over-relaxation."

Adding $0 = \psi_{ij}^{k} - \psi_{ij}^{k}$ to equation (3-377) and re-grouping gives

$$\psi_{ij}^{k+1} = \psi_{ij}^{k} + \frac{1}{2(1+\beta^2)}\Big[\psi_{i+1,j}^{k} + \psi_{i-1,j}^{k+1} + \beta^2\psi_{i,j+1}^{k} + \beta^2\psi_{i,j-1}^{k+1}$$

$$- \Delta x^2\zeta_{ij} - 2(1+\beta^2)\,\psi_{ij}^{k}\Big] \qquad (3\text{-}380)$$

Now as a solution is approached, $\psi^{k+1} \to \psi^{k}$ for all (i,j); the bracketed term becomes zero identically by equation (3-365), and (3-380) becomes a statement of convergence, $\psi_{ij}^{k+1} = \psi_{ij}^{k}$. If Liebmann's method is used, the bracketed term is set identically to zero and at the point (i,j), $\psi_{ij}^{k+1} = \psi_{ij}^{k}$. That is, the residual $r_{ij} = 0$. In the SOR method, the bracketed term in (3-380) is multiplied by a relaxation factor ω where $\omega \neq 1$; thus, $r_{ij} \neq 0$, but $r_{ij} \to 0$ as $\psi^{n+1} \to \psi^{n}$, as before.

$$\psi_{ij}^{k+1} = \psi_{ij}^{k} + \frac{\omega}{2(1+\beta^2)}\Big[\psi_{i+1,j}^{k} + \psi_{i-1,j}^{k+1} + \beta^2\psi_{i,j+1}^{k} + \beta^2\psi_{i,j-1}^{k+1} - \Delta x^2\zeta_{ij} - 2(1+\beta^2)\psi_{ij}^{k}\Big] \qquad (3\text{-}381)$$

For convergence, it is required that $1 \leq \omega < 2$. Frankel and Young both determined an "optimum" value, ω_o, basing their optimality criterion on the asymptotic reduction of the most resistant error. The optimum value ω_o depends on the mesh, the shape of the domain, and the type of boundary conditions. Using the approach of Frankel (1950) for the Dirichlet problem in a rectangular domain of size $(I-1)\Delta x$ by $(J-1)\Delta y$ with constant Δx and Δy, it may be shown that

$$\omega_o = 2\left(\frac{1 - \sqrt{1-\xi}}{\xi}\right) \qquad (3\text{-}382A)$$

where

$$\xi = \left[\frac{\cos\left(\frac{\pi}{I-1}\right) + \beta^2 \cos\left(\frac{\pi}{J-1}\right)}{1+\beta^2}\right]^2 \qquad (3\text{-}382B)$$

With $\omega = \omega_o$, the k required to reduce ξ to some specified level varies directly with the total number of equations $N = (I-2) \times (J-2)$; for the Liebmann method, $k \propto N^2$. So the SOR method with the optimum ω, sometimes referred to as the optimum over-relaxation method, is much better for large problems.

Analytic evaluation of ω_o exists for only slightly more general problems (Young, 1954; Mitchell, 1969; and Warlick and Young, 1970). Miyakoda (1962) has shown that ω_o increases for the case of the Neumann conditions at all boundaries. For Dirichlet conditions along some boundaries and the Neumann conditions along others, for varying Δx or Δy, and for L-shaped and most other non-rectangular regions, no analytic evaluation exists. In such cases, ω_o may be determined experimentally, by solving successions of Laplace equations with zero boundary conditions using different values of ω over $1 < \omega < 2$, and monitoring convergence toward $\psi_{ij}^{k} = 0$ for large k. (The value of ω_o does not change with the source term, ζ.) It is important to assure that all error components are present in the initial condition. This is readily met by choosing $\psi_{ij}^{o} = 1$ at all interior points. The process of experimentally finding ω_o is tedious because the convergence rate is usually very sensitive to ω near ω_o. An example is given in Figure 3-16. The curvature near ω_o shows that it is usually best to slightly overestimate ω_o than to underestimate it. The use of some guessed value, say $\omega = 1.1$, is seen to have very *little* good effect. The experimental determination of an approximation to ω_o is almost always worthwhile, since the Poisson equation must be solved at every iteration of the vorticity transport equation. Jain (1967) gives another method for estimating ω_o from numerical computations which may be useful in some cases. Carré (1961), Strawbridge and Hooper (1968), and Hoeper and Wenstrup (1970) also estimate ω_o from the early behavior of the iteration in their axially symmetric problems, following Carré (1961). Other papers on estimating ω_o have been presented by Young and Eidson (1970) and Zitko (1970).

The SOR method described here is the original "point" SOR of Frankel and Young. It uses advance (k+1) values at 2 neighboring points, (i-1,j) and (i,j-1). It is possible to slightly improve the convergence rate further by "line" SOR, which uses advance (k+1) values at 3 neighboring points. The sweep proceeds (say) in the j↑ direction. When the sweep is at row j1, the preceding row at (j1-1) has been solved already for (k+1) values. The row j1 is then solved using these (k+1) values along (j1-1), and implicitly solving along i in the j1 row using the tridiagonal algorithm (see Appendix A). Ames (1969, page 147) states that line SOR will converge in fewer iterations than point SOR by the factor $1/\sqrt{2}$ for $\Delta x = \Delta y \to 0$. However, each iteration takes longer, because of the implicit tridiagonal solution itself. According to Pao and Daugherty (1969), their numerical experiments showed a small gain in computing speed and was not worth the added complexity of the tridiagonal solution. Dorr (1969) gave an estimate for the ω_o using line SOR for the Poisson equation with the Neumann boundary conditions.

Because of the simplicity and effectiveness, the point SOR method has been the most popular of the iterative methods for solving the Poisson equation in computational fluid dynamics problems. The slightly more complicated ADI methods, which we will present shortly, are becoming more popular in recent years.

A simple modification of SOR that is very easy to program is to use the Liebmann method for only the first iteration by setting $\omega = 1$; thereafter, $\omega = \omega_o$ is used (Sheldon, 1959; Carré, 1961; Young and Kincaid, 1969). Chu (1970) considers alternating the sweep direction, which had beneficial effects in a more general problem than the simple Poisson equation.

III-B-5. Tactics and Strategy

Forsythe (1956) pointed out early that the ω_o given by equations (3-382) is optimum in the sense of a "strategic" or long-range maneuver. That is, the total error $\sum_{i,j} e_{ij}^k \to 0$ asymptotically (as $k \to \infty$) the fastest for $\omega = \omega_o$. But for a finite k, the "optimum" may be something less than ω_o, depending on the initial error distribution. In fact, if we limit k = 1, then $\omega = 1$ (Liebmann's method) gives the greatest reduction in total error in a single sweep; thus, Liebmann's method is the optimum "tactical" or short-range maneuver (Forsythe, 1956). (Another comparison of Liebmann and SOR was given by Keller, 1958.)

As always, the choice of an optimality criterion is seen to affect the determination of an "optimum" parameter value, and the precise mathematics obscures the arbitrarily defined standards.* Young and Kincaid (1969) point out that the relative comparisons of different iterative methods strongly depend on different optimality criteria, even for $k \to \infty$. Numerical experiments by the present author in a 21 x 21 grid indicated that the advantage of $\omega = 1$ over $\omega = \omega_o$, based on total error, could persist to k = 6 or 8. Further, $\omega = \omega_o$ gave errors highly skewed along the diagonal, compared with the results of $\omega = 1$; if the direction were changed at each iteration (say, from i↑, j↑, to i↓, j↓, or others), convergence in the average is not affected, but the skewness is decreased. (The programming and computing time penalties of this device are machine-dependent.) The more efficient iterative methods do not produce a truly symmetric calculation for symmetric boundary conditions, a situation which may be viewed as another "behavioral error." Note that the crude Richardson method has the advantage of giving truly symmetric calculations. This symmetry might be desirable for hyperstable computations of a hydrodynamically unstable flow, for example. Also, we note that, in the SOR method, the dynamic overshoot of $\omega = \omega_o$ is greater than that of $\omega = 1$ or of the Richardson method.

---

*And this arbitrariness exists, even for the relatively well-defined problem of convergence for the discretized Poisson equation. How suspect are "optimal" solutions of defense planners, economists, sociologists, urban planners, etc.?

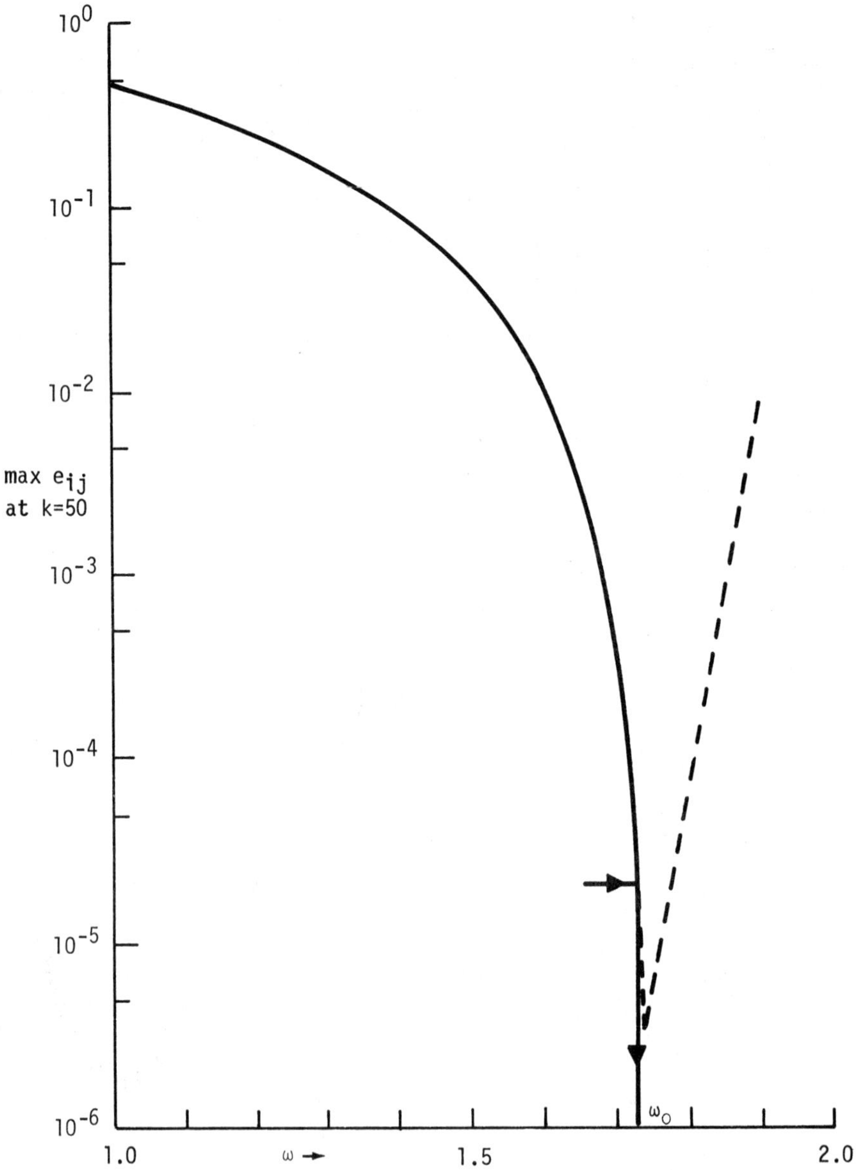

a. Values of max e_{ij} at the 50th iteration

Figure 3-16. Behavior of SOR iteration for values of the relaxation parameter ω. $I = J = 21$, $\Delta x = \Delta y$, which gives $\omega_0 = 1.7295$.

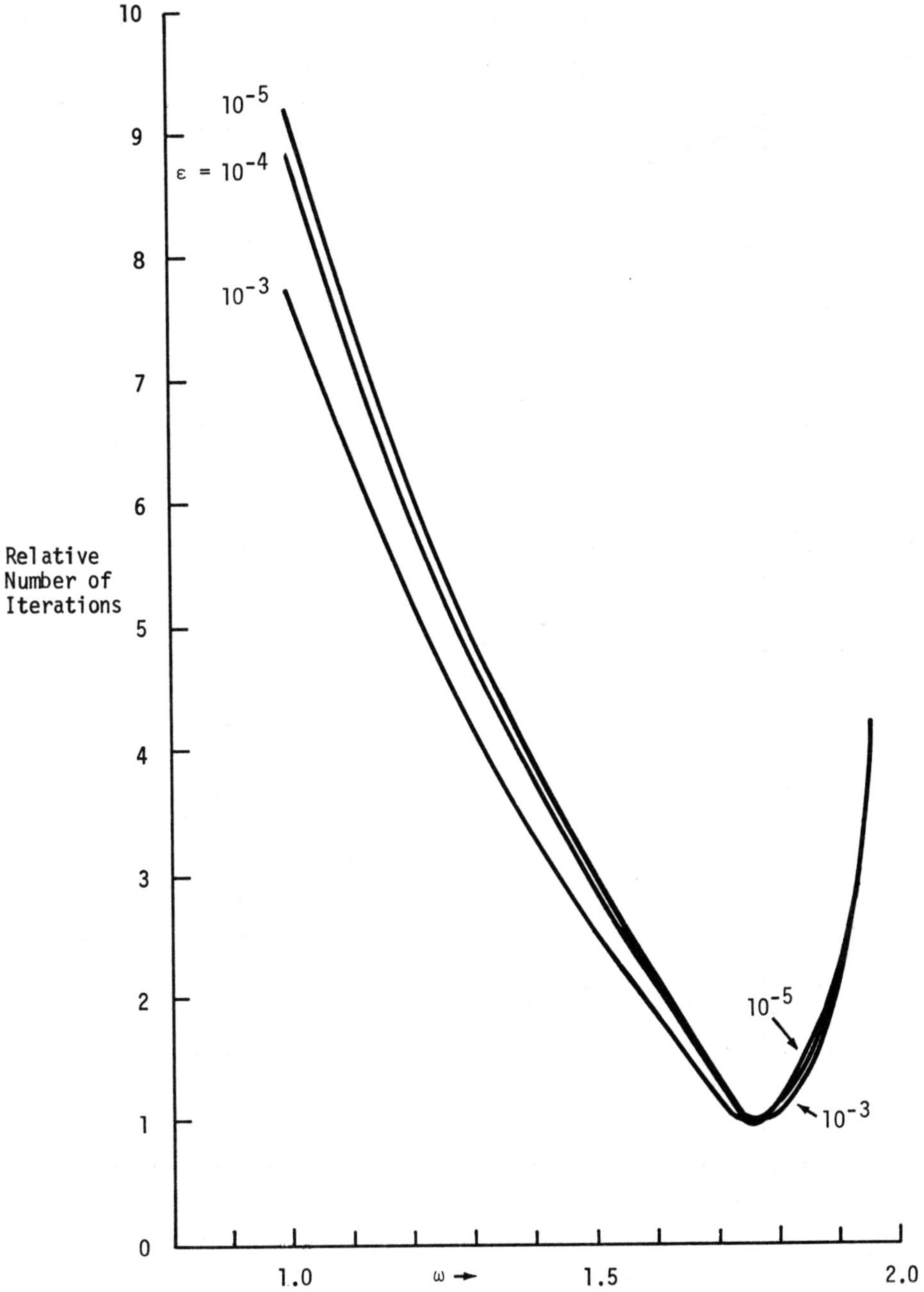

b. Relative number of iterations, based on $\psi_{max}^{k+1} - \psi_{max}^{k} = \varepsilon$.

Figure 3-16. (continued)

Finally, we note that the use of the ψ solution in fluid dynamics is in determining u and v velocity components by numerical differentiation. Thus, the errors in $\delta\psi/\delta x$ and $\delta\psi/\delta y$ ought to be the proper indexes of convergence but, to this author's knowledge, have not been so used in any study.

III-B-6. ADI Methods

Continuing the analogy between the iterative solution of the Poisson equation and the asymptotic time solution of the time-dependent diffusion equation, it is natural to consider the ADI methods of Section III-A-16. Indeed, Peaceman and Rachford (1955) considered both problems in their original paper. Analogizing to equation (3-308), with $\beta = \Delta x/\Delta y$, we have

$$\psi^{k+1/2} = \psi^k + \frac{\alpha\Delta t}{2\Delta x^2}\left[\delta_x^2\,\psi^{k+1/2} + \beta^2\delta_y^2\psi^k + \Delta x^2\zeta\right] \qquad (3\text{-}383\text{A})$$

$$\psi^{k+1} = \psi^{k+1/2} + \frac{\alpha\Delta t}{2\Delta x^2}\left[\delta_x^2\psi^{k+1/2} + \beta^2\delta_y^2\psi^{k+1} + \Delta x^2\zeta\right] \qquad (3\text{-}383\text{B})$$

wherein $\zeta_x^2\psi = \psi_{i+1,j} - 2\psi_{ij} + \psi_{i-1,j}$, and similarly for δ_y^2. The first equation is implicit in x with a tridiagonal form (see Appendix A). The second is implicit in y. Convergence is assured, provided the same Δt is used in both half-steps (see references in Section III-A-16).

Equations (3-383) may be written as follows, after multiplying by $\rho = 2\Delta x^2/\alpha\Delta t$.

$$\psi_{i+1,j}^{k+1/2} - (2+\rho)\psi_{ij}^{k+1/2} + \psi_{i-1,j}^{k+1/2} = -\beta^2\left[\psi_{i,j+1}^k - (2-\rho)\psi_{ij}^k + \psi_{i,j-1}^k\right] - \Delta x^2\zeta_{ij} \qquad (3\text{-}384\text{A})$$

$$\psi_{i,j+1}^{k+1} - (2+\rho)\psi_{ij}^{k+1} + \psi_{i,j-1}^{k+1} = -\frac{1}{\beta^2}\left[\psi_{i+1,j}^{k+1/2} - (2-\rho)\psi_{ij}^{k+1/2} + \psi_{i,j+1}^{k+1/2}\right] - \Delta y^2\zeta_{ij} \qquad (3\text{-}384\text{B})$$

It might be thought that very large Δt (small ρ) would hasten the asymptotic "time" convergence but actually an optimum Δt, or ρ, exists. The optimum ρ gives convergence in slightly fewer iterations than the optimim SOR. This faster convergence may appear plausible because the implicitness makes the effect of the elliptic boundary conditions felt along a whole line at a time. However, since each iteration of ADI takes longer, the optimum SOR method actually takes less computer time than this "single parameter" ADI method (Birkoff et al., 1962; Westlake, 1968).

The real strength of the ADI method comes in choosing a *sequence* of iteration parameters, ρ_k, which replaces ρ in equation (3-384). Obviously, the determination of an optimum sequence is more difficult than the determination of a single optimum value. Additional restraints are required, such as the desired fractional reduction of initial errors or the desired number of iterations. The determination of sequences ρ_k has been and continues to be an applied mathematics research subject in its own right. As an example, we give here the sequence used in the original work of Peaceman and Rachford (1955). The problem was a quarter-plane symmetric heat conduction problem on a square domain with $\Delta x = \Delta y$; in equation (3-384), $\zeta = 0$ and $\beta = 1$. The boundary conditions were

$$\frac{\partial\psi}{\partial x}(0,y) = 0 \quad , \quad \frac{\partial\psi}{\partial y}(x,0) = 0 \qquad (3\text{-}385\text{A})$$

$$\psi(1,y) = 0 \quad , \quad \psi(x,1) = 1 \qquad (3\text{-}385\text{B})$$

For fourteen mesh intervals along each edge ($\Delta x = \Delta y = 1/14$), the problem has 14 x 14 unknowns, for $i,j = 1,2 \ldots 14 = I-1 = J-1$. The ρ sequence used is

$$\rho_k = 4 \sin^2\left[\frac{(2k+1)\pi}{4(I-1)}\right] \quad (3\text{-}386)$$

This gives $\rho_o = 0.012576$, $\rho_{13} = 3.9874$. The corresponding Δt starts off large and drops off: $\Delta t_o = 0.40571$, $\Delta t_1 = 0.04546$, $\Delta t_{13} = 0.00128$.

The "optimum" sequence is generally not attainable. Methods of obtaining "good" sequences are given by Douglas (1962) and Wachpress (1966). These methods are described in the texts of Ames (1969) and Mitchell (1969); both of these latter authors say that the Wachpress parameters are preferable, but the evidence is not altogether convincing. Both methods require the estimation of maximum and minimum matrix eigenvalues, which is itself generally a non-trivial problem. The theories apply only for Dirichlet boundary conditions on a square region with $\Delta x = \Delta y$, but Briley (1970) has successfully used the Wachpress parameters in a cycle of 4 in a non-square mesh. Hadjidimos (1969) discusses the selection of the ρ_k sequence and points out some misinterpretations of theory made in the past. Westlake (1968) notes that it is generally better to over-estimate this cycle number than to under estimate it. No guidelines are available for non-rectangular geometries, although the method is known to be convergent for all ρ. (Note that the same ρ_k must be applied to both half-steps.) Mouradoglou (1968) and Crowder and Dalton (1971B) have done some preliminary work on ADI convergence properties in a non-rectangular mesh.

In SOR methods, the number of iterations required for convergence increases with N. For ADI methods applied to square regions, k_{max} is almost independent of N, so that for large enough N, ADI methods are preferable. In the numerical experiments of Birkoff, et al. (1962), ADI methods with Wachpress parameters were nearly four times faster than the optimum SOR method in a 40 x 40 grid. But for non-rectangular regions the ADI methods are not certainly known to be faster, and the SOR methods are certainly easier to program for complicated regions.

Lynch and Rice (1968) considered the effect of smoothness of initial errors on the optimum sequence of ρ_k. They also considered the effect of errors in estimating the matrix eigenvalues, recommended the use of the Wachpress parameters in practice, and pointed out that successive cycles of ρ_k are less effective than the first cycle. Other papers on the ρ_k sequence were given by Gaier and Todd (1967) and Wachpress (1968). ADI and similar methods were also considered by Young (1954), Samarski (1962) and by Fairweather (1969). ADI methods were applied to the biharmonic forms by Fairweather, et al., (1967), Hadjidimas (1969) and Bourcier and Francois (1969). Caspar (1968) considered ADI methods for mildly nonlinear elliptic equations. Widlund (1967) reported achieving some success for certain problems by generalizing the time-dependent asymptotic solution to the equation

$$c(x,y)\frac{\partial\psi}{\partial t} = \nabla^2\psi - \zeta \quad (3\text{-}387)$$

where $c(x,y) > 0$ is chosen to speed convergence.

The understanding of current research in ADI methods requires a familiarity with the notation and semantics of linear algebra. The précis given in Chapter 1 of Mitchell (1969) is recommended.

III-B-7. Other Iterative Methods

There are many variations of iterative methods for the Poisson equation. An early presentation of classes of iterative methods was given by Geiringer (1949). Although many of these methods have some kind of advantage for some problems or restrictions, the published comparisons must be viewed with some suspicion. As pointed out in Section III-B-5 above, and by Young and Kincaid (1969), a comparison of methods is often dependent on norm of the error, and on the geometry and boundary conditions of the problem. In evaluating a new method, one must weigh complexity, flexibility, proven adaptability (e.g., has the method ever been used with the Neumann boundary conditions? in a non-square mesh?) and the expected gain in the entire fluid dynamics problem for the effort put forth. In the end, personal familiarity with a method and the clarity of an author's presentation will weigh heavily in one's selection of a method.

In addition to the methods presented above, Westlake (1968) evaluated the conjugate gradient methods (see also Simeonov, 1967) and gradient methods which are faster than the Liebmann method but require excessive storage; the Newton-Raphson procedure, which requires too many iterations and too much storage; stationary linear iteration; and Monte-Carlo methods. The Monte Carlo methods are known to be effective for solving ψ at just one or a few points in the mesh, but this is of no value in a fluid dynamics problem.* Westlake also tested a two-line block SOR with cyclic Chebyshev acceleration. This method is superior to ADI for a square mesh spacing larger than some indeterminate, problem-dependent value, but ADI is better for fine-mesh problems. Martin and Tee (1961) conducted a comparison of iterative methods, including gradient methods. Pearson and Kaplan (1970) tested different scanning patterns for SOR. They found that convergence may be obtained in fewer iterations, but that the computer time may increase because of added program complexity.

Young and Kincaid (1969) compared the methods presented above with several other methods. These included an SOR method with varying ω, as in the Douglas-Rachford ADI method (see also McDowell, 1967); another modified SOR method; Sheldon's over-relaxation factors and a modification thereof; the cyclic Chebyshev semi-iterative method and a variation thereof (see also Hodgkins, 1966, and Rigler, 1969); and Sheldon's semi-iterative method and a variation. It is here that Young and Kincaid (1969) recommended a first SOR sweep with $\omega = 1$, and also discussed in some detail how the norm of the error affects comparisons.

Apelt (1969) and Son and Hanratty (1969) used an SOR variation devised by Russell (1962). One of the easiest SOR improvements to program is due to Sheldon (1962); see also Jenssen and Straede (1968). Sheldon splits the SOR sweep into two parts over staggered mesh points, the first part for (i+j) odd, the second for (i+j) even. For the first part, no new information is used (for the usual 5-point analog of the Laplacian considered here). For the second part, all four neighboring points are at new values (see Section III-A-18 on the Hopscotch method).

Stone (1968), Weinstein, et al., (1969), and DuPont, et al., (1968) considered methods for the diffusion equation (applicable here by the time-iteration analogy) which are more strongly implicit than ADI, but are still not "fully" implicit. These methods involve preliminary matrix factorization (as do many direct methods) and the solution of a sparse-matrix subproblem by direct Gaussian elimination.

Rushton and Laing (1968) used the "dynamic relaxation" method to solve Laplace's equation in 3D (see also Wood, 1971).

Winograd (1969) has considered a "chaotic relaxation" class of methods which, like the Richardson method, are suitable for programming on a parallel processor machine. That is, new computations are made "simultaneously" at many node points. He obtained convergence results assuming no pattern at all in either the (i,j) location processed or in the number of iterations of that equation. This intriguing result of Winograd also suggests the study of a Monte Carlo selection of ADI relaxation parameters, which would be at least of theoretical interest. Lick and Tunstall (1968) ahd Ahamed (1970) considered iterative methods for the equation $\nabla^2\psi = f(\psi)$. Other references for the Poisson solution are Aleksidze and Pertaia (1969) and Vonka (1970).

Because of its simplicity and acceptable speed, the basic SOR method (with $\omega = 1$ for the first iteration) will probably continue to be the most popular iterative method for non-rectangular regions, and the Douglas-Rachford ADI method (and perhaps SOR) will probably be the most popular for rectangular regions.

III-B-8. EVP Method

A relatively simple and flexible *direct* method for solving the Poisson equation was given by the present author (Roache, 1971A,B). The method is one of a class of "influence

*Gopalsamy and Aggarwala (1970) considered Monte Carlo methods for the biharmonic equation.

coefficient" or "double sweep" methods (e.g., Ishizaki, 1957; Lucey and Hansen, 1964; Hirota, et al., 1970). It makes use of only the most elementary principles of linear algebra, but it has the disadvantage of being limited by the field size of the problem due to round-off error.

We introduce the method via the one-dimensional problem (which will also have application to a problem of a boundary condition at outflow, considered in Section III-C-7). We first consider the one-dimensional problem in y, with Dirichlet boundary conditions, using the usual second-order form for $\delta^2\psi/\delta y^2$.

$$\frac{\psi_{j+1} - 2\psi_j + \psi_{j-1}}{\Delta y^2} = \zeta_j \tag{3-388}$$

$$\psi_1 = a \quad , \quad \psi_J = b \tag{3-389}$$

We pick an arbitrary value of ψ_2', where the prime denotes a provisional value, say $\psi_2' = \psi_1 = a$. This ψ_2' is in error from the true value, ψ_2, by the unit error, e. That is,

$$\psi_2 = \psi_2' + e \tag{3-390}$$

The remaining provisional values of ψ' up through J are now marched out in one sweep, starting at j = 3, by rearrangement of equation (3-388).

$$\psi_{j+1}' = \Delta y^2\zeta_j + 2\psi_j - \psi_{j-1}' \tag{3-391}$$

These provisional values, ψ_j', are in error by e_j. That is,

$$\psi_j = \psi_j' + e_j \tag{3-392}$$

Substituting equation (3-392) into (3-388) and using (3-391), we obtain the recursion relation for the error propagation as

$$e_{j+1} = 2e_j - e_{j-1} \tag{3-393}$$

which is seen to be independent of the non-homogeneous term ζ_j. For the boundary conditions (3-389), we have $e_1 = 0$ and $e_2 = e$; it may be verified by induction that equation (3-393) then gives

$$e_j = (j - 1)\cdot e \tag{3-394}$$

At the end of the first sweep, the unit error e is calculated from the known boundary value, $\psi_J = b$, as

$$e = \frac{b - \psi_J'}{J - 1} \tag{3-395}$$

With e so determined, the provisional values are now corrected to the final values in a second sweep, using

$$\psi_j = \psi_j' + (j - 1)\cdot e \tag{3-396}$$

Alternately, the final values may be marched out by using (3-391) and the *correct* values of

$$\psi_1 = a \quad , \quad \psi_2 = \psi_2' + e \tag{3-397}$$

This latter procedure will be used in the two-dimensional problem.

If a Neumann condition is used at the second boundary, where j = J, with either

$$\frac{\psi_J - \psi_{J-1}}{\Delta y} = U \quad \text{or} \quad \frac{\psi_{J+1} - \psi_{J-1}}{2\Delta y} = U \tag{3-398}$$

then equation (3-395) for determining e is replaced by

$$e = U\cdot\Delta y - (\psi_J' - \psi_{J-1}') \quad \text{or} \quad e = U\Delta y - \frac{1}{2}(\psi_{J+1}' - \psi_{J-1}') \tag{3-399}$$

which is easily verified. If a Neumann condition is used at j = 1, with

$$\frac{\psi_2 - \psi_1}{\Delta y} = C \tag{3-400}$$

then, instead of $e_1 = 0$ and $e_2 = e$, we have

$$e_1 = e_2 = e \tag{3-401}$$

and the error propagation equation (3-394) is replaced by

$$e_j = e + (j - 2)\cdot C\cdot\Delta y \tag{3-402}$$

Since the error propagation equation (3-394) or (3-402) is linear in j, there is no practical danger of generating excessively large ψ_j' and thereby destroying accuracy due to machine round-off error. The 2D problem is not free from this shortcoming.

The geometry of the reference two-dimensional problem is shown in Figure 3-17. We first consider Dirichlet conditions at all boundaries. Analogous to the one-dimensional case in which we picked a single provisional value, ψ_2', we now pick a provisional *vector* of values $\psi_{i,2}'$, where i runs from i = 2 to i = I - 1. This $\psi_{i,2}'$ vector, lying just inside boundary B1 in Figure 3-17, is in error by the unit error vector, $e_{i,2}$. That is

$$\psi_{i,2} = \psi_{i,2}' + e_{i,2} \tag{3-403}$$

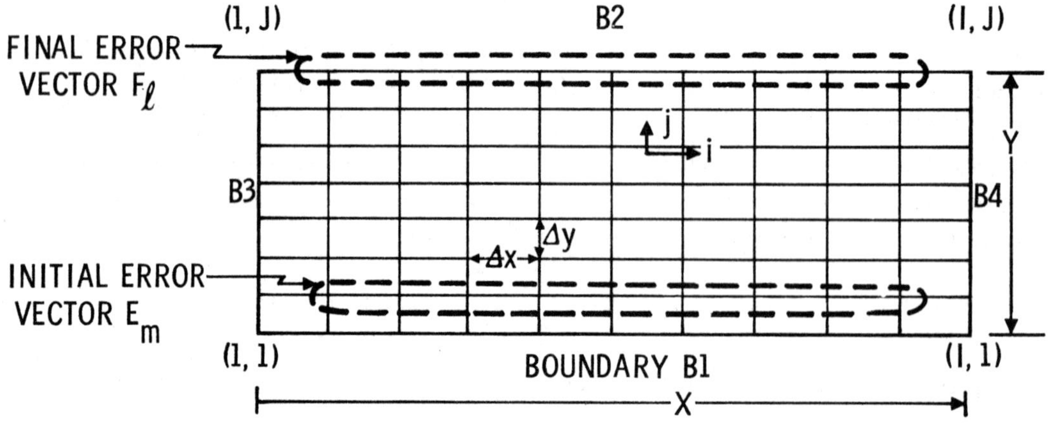

Figure 3-17. Geometry of the reference two-dimensional problem for the EVP method.

With $\psi'_{i,2}$ so chosen, the remaining provisional values for internal points up to and including $j = J$ are calculated in one sweep by rearranging equation (3-365). With $\beta = \Delta x/\Delta y$, the result is

$$\psi'_{i,j+1} = \Delta y^2 \zeta_{ij} + 2(1 + \beta^2)\psi'_{ij} - \beta^2 \left[\psi'_{i+1,j} + \psi'_{i-1,j}\right] - \psi'_{i,j-1} \tag{3-404}$$

The correct boundary values $\psi_{1,j}$ at B3 and $\psi_{I,j}$ at B4 are used in equation (3-404) when needed. The error propagation equation, corresponding to equation (3-393), is

$$e_{i,j+1} = 2(1 + \beta^2)\, e_{ij} - \beta^2 e_{i+1,j} - \beta^2 e_{i-1,j} - e_{i,j-1} \tag{3-405}$$

which has boundary values along B1, B3, and B4 of

$$e_{i,1} = e_{1,j} = e_{I,j} = 0 \tag{3-406}$$

After the first sweep, the final error vectors at B2 are calculated from the correct boundary values, $\psi_{i,J}$, as

$$e_{i,J} = \psi_{i,J} - \psi'_{i,J} \tag{3-407}$$

Then the initial error vector $e_{i,2}$ is solved from $e_{i,J}$, using a pre-established linear relation (below). The correct values of $\psi_{i,2}$ are then solved from equation (3-402); then a second sweep, using the march equation (3-404) with ψ replacing ψ', establishes the final solution.

Now to establish the linear relation between $e_{i,2}$ and $e_{i,J}$, it is convenient to introduce additional notation for those two vectors, as shown in Figure 3-17. The final error vector is defined as $F_\ell = e_{i,J}$, where $\ell = (i - 1)$ runs from $\ell = 1$ to $\ell = (I - 2)$. The initial error vector is defined as $E_m = e_{i,2}$, where $m = (i - 1)$ also runs from $m = 1$ to $m = (I - 2)$. The "influence coefficient matrix" $C = [C_{\ell m}]$ is then defined by the equation

$$F_\ell = C_{\ell m} E_m \tag{3-408}$$

Unlike the one-dimensional case, no convenient equation exists for the $C_{\ell m}$. The matrix is established prior to the solution of a particular problem by the following process. Taking a particular value m_1 of m, we set $E_{m_1} = e_{m_1+1,2} = 1$, and all other $E_m = 0$. Then the propagation of the error vector E into e_{ij} is calculated by application of equations (3-405) and (3-406). The resulting final error vector is $F_\ell = e_{i,J}$, where $\ell = (i-1)$ runs from 1 to (I-2). The m_1-column of C is so determined as

$$C_{\ell,m_1} = F_\ell \tag{3-409}$$

Repeating the generation of $E_{m_1} = 1$, all other $E_m = 0$, for m_1 ranging from 1 to (I-2) fills in the influence coefficient matrix, C. Finally, to solve for $e_{i,2}$ in terms of $e_{i,J}$, we invert equation (3-408), using direct Gaussian elimination, and obtain

$$E_m = C^{-1}_{m\ell} F_\ell \tag{3-410}$$

and, finally,

$$e_{m+1,2} = C^{-1}_{m\ell} e_{+1,J} \qquad (3\text{-}411)$$

When applicable, this method is clearly more efficient than direct Gaussian elimination. The original problem of (I - 2) x (J - 2) simultaneous equations with block-tridiagonal matrix form has been reduced to that of solving (I - 2) equations* to determine C^{-1}, and additionally doing the equivalent work of I Richardson iterations: two sweeps of ψ', ψ in equation (3-404) and (I - 2) sweeps of e in equation (3-405). Since the error propagation equation (3-405) is independent of the non-homogeneous term, ζ_{ij}, and since its boundary conditions do not depend on the boundary values of ψ but only on their specification as Dirichlet boundary conditions, then the sweep of e via equations (3-405) and (3-406) and the inversion of C need to be done only once for a family of solutions in the same mesh with the same type of boundary conditions, but with different boundary values of ψ and different ζ. This is precisely the case in fluid dynamics problems.

Compared with iterative methods, we can easily make EVP look 10 to 100 times faster, depending on the mesh size and the convergence criteria used. But EVP does have the disadvantage of requiring considerably more core memory storage, and the method cannot compete with the simplicity of the SOR method. Most important, it is field-size-limited because of its error propagation characteristics.

There is apparently a tendency for C to become ill-conditioned as J becomes large, intuitively seen as a difficulty in discriminating an error at (i,2) from errors at (i±1,2) as the source of any error at J. But in common applications, it is not this behavior which limits the method, but the following. Unlike the recursion relation (3-393) for the one-dimensional error propagation, which is linear in j, the two-dimensional version (3-405) gives a value F_{ic} (ic is at the I-center of the mesh) which increases exponentially in j. For J large, this means that the ability of the method to resolve the error at j = 2 is limited by machine round-off error.

This behavior puts an absolute ceiling on the resolution ability of the method, even if the inversion of C were to be accomplished with perfect accuracy. In fact, there are available Gaussian elimination routines which perform a double-precision iteration to reduce this error to an entirely negligible level. Also, the details of the particular problem (i.e., ψ and ζ values) do not significantly affect the error propagation, provided the boundary values of ψ are reasonably scaled.** Consequently, an order of magnitude estimate of the practical limitation of the method is found by numerically marching out equations (3-405) and (3-406) with unit errors at j = 2.

The error propagation has several fortunate aspects. The largest (and therefore most limiting) error occurs in the center of the mesh, so that we need only consider F_{ic}. Also, the effect of various conditions along the boundaries i = 1 and i = I adjacent to the march have negligible effect on the center value, even for I as small as 7, so we may neglect the I dimension and the adjacent boundary conditions at B3 and B4 as parameters of the error propagation. Finally, there is a strong effect of mesh aspect ratio $\beta = \Delta x/\Delta y$, which may be used to advantage. Small β has an adverse effect on error propagation, while large β has a favorable effect. (In the limit of large β, the error propagation approaches that of the one-dimensional problem, which is merely linear in J. But, in this case, the accuracy is limited by the ill-conditioning of C. This error was found to be dominant in a 101 x 101 mesh problem with β = 10.)

The dimensionless length of practical interest in determining the applicability of the method is $Y/\Delta x$ (that is, the number of x-increments that we can march in the y-direction). For a unit error $E_{ic} = 1$, the value $P = \log_{10}[F_{ic}]$ is plotted in Figure 3-18 with β as the

---

*We have described the method in terms of C^{-1}, but it is well-known that it is more efficient to just solve the system and retain the triangular decomposition of the Gaussian ellimination , rather than to actually compute C^{-1}.

**Errors from these other sources may be somewhat reduced by overall iteration of the entire EVP method, but usually by less than the improvement obtained by reducing J by one. The procedure diverges beyond the second iteration.

parameter. The resolution level (number of significant figures) S of several current U.S. computers is also shown. (The prefixes SP, DP, and HDP, respectively, refer to single precision, double precision, and hardware double precision.)

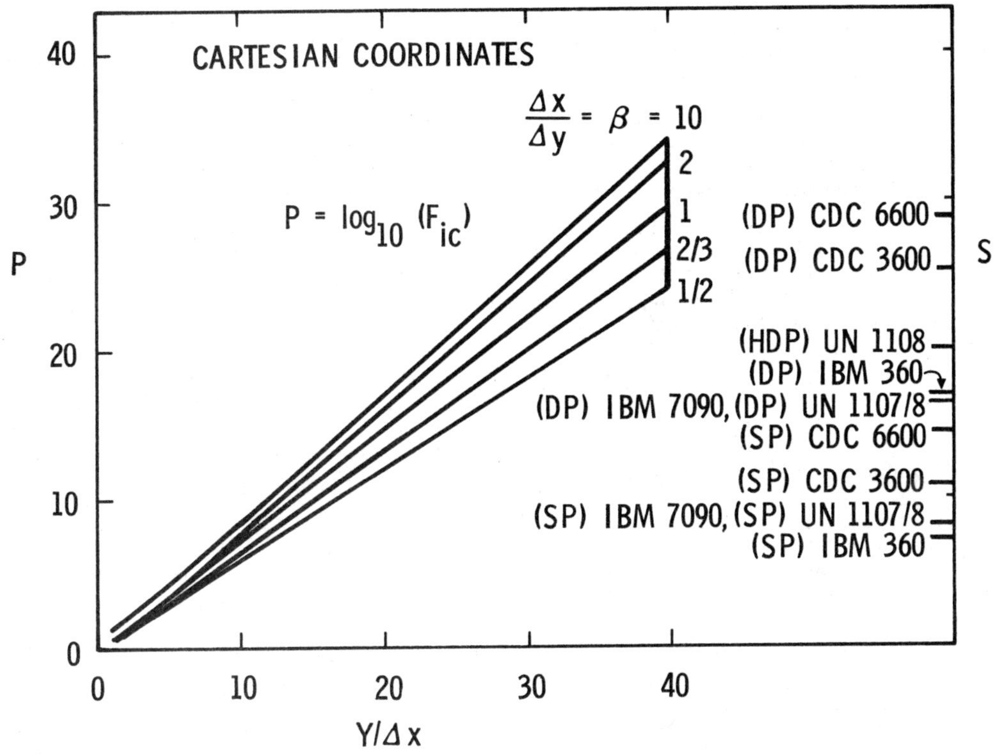

Figure 3-18. Error propagation characteristics for the EVP method applied to the Poisson Equation in Cartesian coordinates (x, y).

As an example of interpreting Figure 3-18, consider a CDC 6600 with single precision, S = 14.45. For $\beta = 1$, $S \simeq P$ at $Y/\Delta x = 20$. At this condition (J = 21), we may expect resolution errors in $\psi_{i,J}$ of order unity, which is unacceptable. But, at $Y/\Delta x = 10$ with $\beta = 2$ (still J = 21), we find $P \simeq 7.5$. The difference $S - P \simeq 14.5 - 7.5 = 7$ indicates that we may expect resolution errors in $\psi_{i,J}$ of order 10^{-7}, which is generally acceptable.

This resolution error of the EVP method differs from the convergence criteria or residual error of iterative methods. The largest resolution errors in the EVP method appear on the single boundary at the end of the march, while the errors at interior points are much smaller. The largest magnitude residual errors of iterative methods appear at internal points, while the specified boundary values remain intact. Thus, a resolution error in ψ of 10^{-6} at the final boundary in EVP is not directly comparable to, but is better than, a residual error in ψ of 10^{-6} in ADI or SOR.

One of the advantages of the formalism of the EVP method over other direct methods is its simple adaptation to irregular boundary geometries and varied combinations of boundary conditions. The only explanation required for the adaption to irregular geometries is the definition of the initial and final error vectors, E and F. Several examples are given in Figure 3-19. The indexing of E and F are not unique.

The EVP characteristics would need to be worked out for each case. But, since the presence of boundaries more than 4 or 5 cells from an interior march path has a relatively

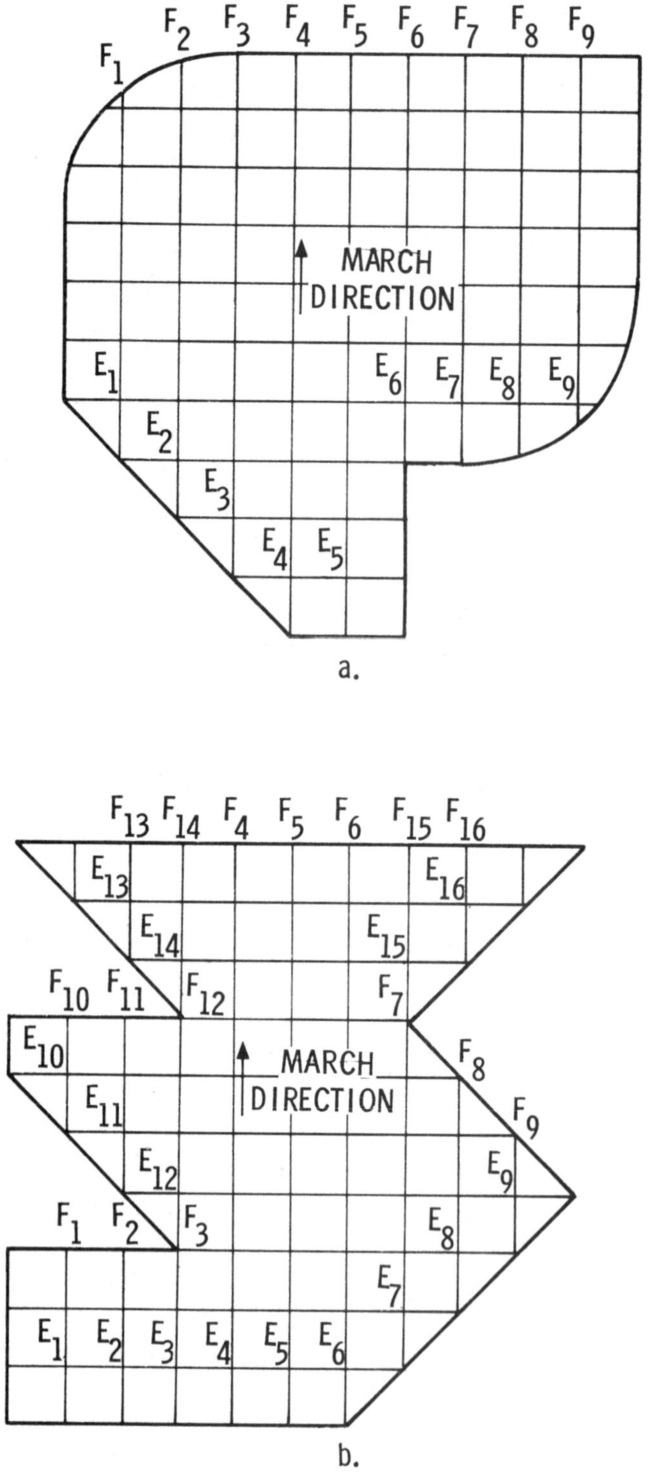

Figure 3-19. EVP method for irregular geometries. E_m = components of the initial error vector, F_ℓ = components of the final error vector.

slight favorable effect on e, it may be expected that the method will frequently be limited by the P of Figure 3-18, based on the longest march path of the problem.

Note also that partial cell treatment of irregular boundaries (see Section VI-A; also, Salvadori and Baron, 1961) may be easily accommodated in equations (3-404) and (3-405).

The Neumann boundary conditions result in zero gradient conditions on e.

Exercise: Show that, for any Neumann boundary condition on ψ, i.e., $\partial\psi/\partial n = c$, where c is not necessarily = 0, results in $\partial e/\partial n = 0$ in the EVP method.

For example, if the condition along B1 is

$$\frac{\psi_{i,2} - \psi_{i-1}}{\Delta y} = c \tag{3-412}$$

then, after the provisional values $\psi'_{i,2}$ are chosen, then $\psi'_{i,1}$ are set from equation (3-412). Similarly, the error vector march (3-405) for each value m_1 is started not with $E_{m_1} = e_{m1+1,2} = 1$ and $e_{m1+1,1} = 0$, but with $e_{m1+1,1} = 1$ also. Along B3, the Neumann condition

$$\frac{\psi_{2,j} - \psi_{1,j}}{\Delta x} = d \tag{3-413}$$

is incorporated into the ψ' and ψ marches of equation (3-404), and the error boundary condition (3-406) is supplanted at B3 with

$$e_{1,j} = e_{2,j} \tag{3-414}$$

Similarly, any Neumann condition along B4 gives $e_{I,j} = e_{I-1,j}$.

The Neumann conditions have a negligible effect on the error characteristics when applied along B3 or B4, a slight favorable effect when applied along B1, and a seriously adverse effect when applied along B2. Robbin's conditions are also easily treated, as are irregular mesh systems. The 3- and n-dimensional problems are readily formulated, although the inversion of C may become difficult. Although the jump from 1 to 2 dimensions greatly deteriorates the error characteristics, higher dimensions have a rapidly diminishing adverse effect.

The diagonal unit square or the nine-point analogs of the Laplacian (see Section III-B-10) can be used in EVP. With these, an implicit march scheme worsens the error characteristics, while an explicit diagonal march (solving for $\psi_{i+1,j+1}$) improves them, for I small. Another worthwhile adaptation is the use of a fourth-order-accurate five-point analog to $\partial^2\psi/\partial x^2$, transverse to the march direction. This produces about a 12% increase in P for $\beta = 1$, but may allow the use of larger β with a consequent decrease in P. The EVP method is also applicable to other linear elliptic equations of fluid dynamics, and can be used in an iterative fashion to solve variable coefficient and nonlinear Poisson equations. For details, see Roache (1971A). It also allows the direct solution of equations of the type $\nabla^2\psi = f(\psi)$, which arise in ICE calculations (see Section V-I), and for problems with internal consistency conditions at material interfaces, which arise in certain electric field problems.

III-B-9. Fourier-Series Methods

The Fourier-series methods are based on the fact that an *exact* solution to the *finite-difference* equation (3-365) can be expressed in terms of finite eigenfunction expansions. For example, on a rectangular region of dimensions X x Y with M x N interior points (N = I-2, M = J-2), with constant Δx and Δy, and $\psi = 0$ on all boundaries, the exact solution of equation (3-365) can be expressed as (Dorr, 1970)

$$\psi_{ij} = \sqrt{\frac{2}{N+1}} \sum_{p=1}^{N} H_{pj} \sin \frac{p\pi x_i}{X} \tag{3-415}$$

where $x_i = (i-1)\Delta x$. The H_{pj} are the solutions, for $1 \leq p \leq N$, of the tridiagonal difference equations,

$$\frac{1}{\Delta y^2}\left(H_{p,j-1} - 2H_{pj} + H_{p,j+1}\right) + \lambda_p H_{pj} = V_{pj} \tag{3-416}$$

with

$$H_{p,1} = H_{p,J} = 0 \tag{3-417}$$

and

$$V_{pj} = \sqrt{\frac{2}{N+1}} \sum_{q=1}^{N} \zeta_{q+1,j+1} \sin \frac{q\pi p\Delta x}{X} \tag{3-418}$$

$$\lambda_p = \frac{2}{\Delta x^2}\left[\cos \frac{p\pi\Delta x}{X} - 1\right] \tag{3-419}$$

Hockney (1965, 1971) has shown how this solution may be used effectively for computations assuming $N = 2^k$ or $N = 3 \cdot 2^k$, where k is an integer. Hockney's method also used "odd-even reduction" (Dorr, 1970; Buzbee, et al., 1970), and its speed depends on the use of the "fast Fourier transforms" (Cooley and Tukey, 1965; Pipes and Hovanessian, 1969; Webb, 1970) which are becoming available in the program libraries of many computer centers. A less complex and less restricted method was used by Williams (1969), who also treats periodic boundary conditions. Related methods are those of Buneman (1969) and Ogura (1969). The reader must refer to these original sources (see also Colony and Reynolds, 1970) for full information, as the methods are beyond the level of this book.

We do note, however, that although these methods in their basic form are quite restricted in the boundary formulation of the problem, they may be applied in a modified procedure to more general problems. Consider first the case of a rectangular region with Dirichlet condition $\psi = f(x,y)$ on the boundary, where $f \neq 0$ everywhere. An auxiliary function, ψ^1, is defined by obtaining the exact solution of $\nabla^2\psi^1 = \zeta$, with boundary conditions of $\psi^1 = 0$ everywhere. Then a second auxiliary function, ψ^{11}, is defined by obtaining the exact solution of the finite-difference Laplace equation, $\nabla^2\psi^{11} = 0$, with the correct boundary conditions $\psi^{11} = f(x,y)$. The *exact* solution is obtained by the separation-of-variables method of PDE's (e.g., Weinberger, 1965) applied to the *finite-difference* equation. (The necessary eigenfunction expansions are already available from the expansion required by the solution of the Poisson equation.) Then the final solution ψ is obtained, because of the linearity of the problem, by superposition. That is, since $\nabla^2\psi^1 = \zeta$ and $\nabla^2\psi^{11} = 0$, then $\nabla^2(\psi^1 + \psi^{11}) = \zeta_1$, and since $\psi^1 = 0$ and $\psi^{11} = f(x,y)$ on boundaries, then $\psi^1 + \psi^{11} = f(x,y)$. So $\psi = \psi^1 + \psi^{11}$ satisfies $\nabla^2\psi = \zeta$ and $\psi = f(x,y)$ on boundaries.

If the zero-gradient Neumann boundary conditions are used, expansion in a cosine series is appropriate. The problem of non-zero boundary specification of the normal gradient $\partial\psi/\partial n = g(x,y)$ can be solved as follows (Williams, 1969). An auxiliary function ψ^1 is written as $\psi^1 = 0$ at all internal points, $\psi^1 = +\, g(x,y)\cdot\Delta n$ at the extremes $i = I$ and $j = J$, and $\psi^1 = -\, g(x,y)\cdot\Delta n$ at the extremes $i = 1$ and $j = 1$. This ψ^1 is a solution of the auxiliary discretized Poisson equation $\nabla^2\psi^1 = \zeta^1$, with $\delta\psi^1/\delta n = g(x,y)$ on boundaries, and with

$\zeta^{1} = 0$ everywhere except at points adjacent to boundaries, where $\zeta^{1} = \nabla^2\psi^{1} \neq 0$. (At a node two positions in from a boundary, $\nabla^2\psi^{1} = 0$, since $\psi^{1} = 0$ at all the neighboring points.) Defining $\psi^{11} = \psi - \psi^{1}$ and $\zeta^{11} = \zeta - \zeta^{1}$, the original problem is then converted to finding the FDE solution of $\nabla^2\psi^{11} = \zeta^{11}$, with conditions of $\delta\psi^{11}/\delta n = 0$ on all boundaries, which can be handled by a cosine expansion. The desired solution is then $\psi = \psi^{11} + \psi^{1}$.

In a similar manner, non-rectangular regions may be solved by using a rectangular mesh which overlaps the desired region. Consider the region shown in Figure 3-20a, formed by taking a small corner off of a rectangular region. The boundary point (2,2) is not on the

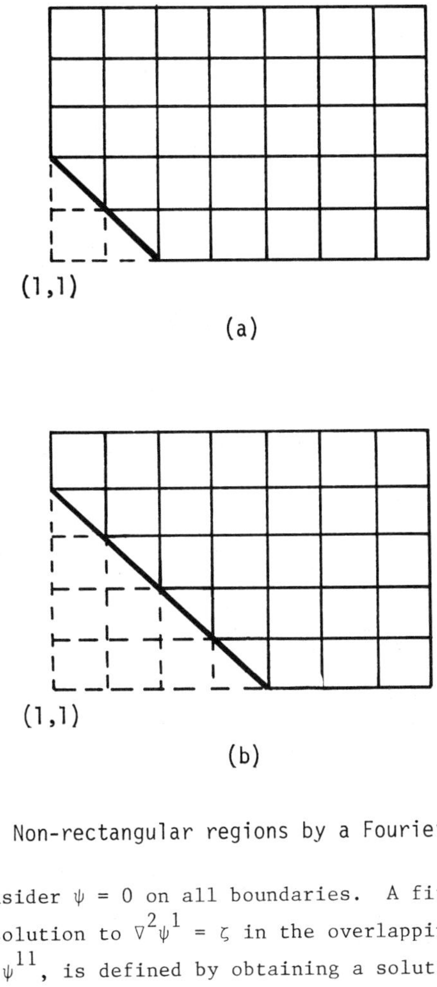

Figure 3-20. Non-rectangular regions by a Fourier series method.

overlapping rectangle. Consider $\psi = 0$ on all boundaries. A first auxiliary function, ψ^{1}, is defined by obtaining a solution to $\nabla^2\psi^{1} = \zeta$ in the overlapping mesh with $\zeta_{2,2} = 0$. A second auxiliary function, ψ^{11}, is defined by obtaining a solution to the equation $\nabla^2\psi^{11} = \zeta^{11}$, where ζ^{11} is defined as $\zeta_{2,2} = 1$, all other $\zeta_{ij} = 0$.* Then a linear combination is found which gives the desired $\psi_{22} = 0$ value without altering ζ at internal points. That is,

$$\psi_{2,2} = 0 = 1\cdot\psi^{1}_{2,2} + a\cdot\psi^{11}_{2,2} \tag{3-420}$$

*In electrical applications where ψ is the potential and ζ is the charge, this technique is known as the "unit charge" method. The present author is indebted to O. Buneman and J. Boris for discussions on this technique.

or

$$a = -\psi^{1}_{2,2}/\psi^{11}_{2,2} \tag{3-421}$$

(The denominator will never equal zero.) The final solution is, by superposition,

$$\psi = \psi^{1} + a^{11}\psi^{11} \tag{3-422}$$

Although the ζ at (2,2) of the composite solution is a^{11} + (true)$\zeta_{2,2}$, the solution is not affected, since (2,2) is a boundary point of the composite problem and therefore ζ_{22} does not enter into that problem.

When more than one boundary point is not on the overlapping rectangle, each point requires an additional Poisson solution. In Figure 3-20b, auxiliary solutions ψ^{11}, ψ^{111}, ψ^{1111}, are associated with $\zeta = 1$ solutions at (2,4), (3,3), (4,2), respectively. Corresponding to equation (3-420), the linear system defined by

$$\begin{aligned}
\psi_{2,4} &= 0 = \psi^{1}_{2,4} + a\psi^{11}_{2,4} + b\psi^{111}_{2,4} + c\psi^{1111}_{2,4} \\
\psi_{3,3} &= 0 = \psi^{1}_{3,3} + a\psi^{11}_{3,3} + b\psi^{111}_{3,3} + c\psi^{1111}_{3,3} \\
\psi_{4,2} &= 0 = \psi^{1}_{4,2} + a\psi^{11}_{4,2} + b\psi^{111}_{4,2} + c\psi^{1111}_{4,2}
\end{aligned} \tag{3-423}$$

must be solved for a, b, c, establishing an "influence coefficient matrix" to zero the boundary values for later solutions with new ζ. For w boundary points, w auxiliary Poisson equations must be solved, and a w-order linear system like (3-423) must be solved by Gaussian elimination; however, this need be done only for the first solution in a family of problems with different ζ, all in the same mesh.

Problems with combinations of non-zero Dirichlet and the Neumann boundary conditions on a non-rectangular region are possible to formulate, but are obviously somewhat clumsy. For techniques of applying these methods to irregular regions, see Buzbee, et al., (1970) and George (1970). For similar methods which are applicable to the 9-point Laplacian analogs of Section III-B-10, see Le Bail (1969), Birdsell and Fuss (1969).* All of these methods are more complex than the EVP method. However, they are not field-size limited, they give almost an exact FDE solution, and they are equally adaptable to three-dimensional and cylindrical coordinate problems (Williams, 1969). Because of their great speed and accuracy, they will undoubtedly see extensive future applications to simple geometry problems.

III.B.10. Higher Order Approximations

Thus far we have considered only the usual 5-point analog of the Laplacian operator in the Poisson equation, but there are others. Three of these are restricted to the case of a square mesh with $\Delta x = \Delta y = \Delta$. The terminology, schematic representation, and historical credits given here are taken from Thom and Apelt (1961). The 5-point analog which we have been considering was first used by Runge in 1908, and is sometimes called the "basic unit square." For a square mesh, it may be schematically represented as in Figure 3-21a, which represents the equation

$$S_1 - 4\psi_{ij} = \Delta^2\zeta_{ij} \tag{3-424}$$

---

*The most complete exposition of the application of Fourier methods is given in a recent article by R.C. Le Bail: "Use of Fast Fourier Transforms for Solving Partial Differential Equations in Physics", Journal of Computational Physics, Vol.9, No. 3, June 1972, pp. 440-465.

where

$$S_1 = \psi_{i+1,j} + \psi_{i-1,j} + \psi_{i,j+1} + \psi_{i,j-1} \tag{3-425}$$

which is second-order accurate in Δ.

$$\begin{bmatrix} & 1 & \\ 1 & -4 & 1 \\ & 1 & \end{bmatrix}$$

5a. BASIC UNIT SQUARE

$$\frac{1}{2}\begin{bmatrix} 1 & & 1 \\ & -4 & \\ 1 & & 1 \end{bmatrix}$$

5b. DIAGONAL UNIT SQUARE

$$\frac{1}{4}\begin{bmatrix} 1 & 2 & 1 \\ 2 & -12 & 2 \\ 1 & 2 & 1 \end{bmatrix}$$

5c. THE "12" FORMULA

$$\frac{1}{6}\begin{bmatrix} 1 & 4 & 1 \\ 4 & -20 & 4 \\ 1 & 4 & 1 \end{bmatrix}$$

5d. THE "20" FORMULA

Figure 3-21. Schematic of finite difference analogs of the Laplacian in a square mesh.

Since the Laplacian operator is invariant to a coordinate rotation, it may be expressed in coordinates rotated 45° with respect to the mesh, in which case the spacing between points becomes $\sqrt{2}\Delta$. The resultant "diagonal unit square" operator is depicted in Figure 3-21b, which represents

$$S_2 - 4\psi_{ij} = 2\Delta^2\zeta_{ij} \tag{3-426}$$

where

$$S_2 = \psi_{i+1,j+1} + \psi_{i+1,j-1} + \psi_{i-1,j+1} + \psi_{i-1,j-1} \tag{3-427}$$

which is second-order accurate in $\sqrt{2}\Delta$.

The two best known 9-point formulas are Thom's "12" formula, Figure 3-21c, and Bickley's "20" formula, Figure 3-21d. The "12" formula is

$$2S_1 + S_2 - 12\psi_{ij} + 4\Delta^2\zeta_{ij} \qquad (3\text{-}428)$$

and the "20" formula is

$$4S_1 + S_2 - 20\psi_{ij} = 6\Delta^2\zeta_{ij} \qquad (3\text{-}429)$$

The "12" formula is fourth-order accurate only if $1/12(\nabla^4 + \partial^4/\partial x^2\partial y^2) = O(\Delta^2)$ or smaller, and the "20" formula, only if $1/12(\nabla^4\psi) = O(\Delta^2)$. These conditions are not ordinarily met in fluid dynamics problems. An exception is a steady Stokes flow (Re = 0), for which case the vorticity transport equation reduces to $\nabla^2\zeta = 0$, which implies $\nabla^4\psi = 0$. Generally, these forms are even less accurate than the basic unit square form (Cyrus and Fulton, 1967; Jenssen and Straede, 1968).

True higher order local accuracy may be obtained with equations which use ψ values from points farther than the immediately adjacent points, e.g., $\psi_{i+2,j}$. Jenssen and Straede (1968) consider several of these, such as

$$\left.\frac{\delta^2\psi}{\delta x^2}\right|_{ij} = \frac{1}{12\Delta x^2}\left[-\psi_{i+2,j} + 16\psi_{i+1,j} - 30\psi_{ij} + 16\psi_{i-1,j} - \psi_{i-2,j}\right] + O(\Delta x^4) \qquad (3\text{-}430)$$

and a corresponding equation for $\delta^2\psi/\delta y^2$. (See also Fairweather, 1969.) Pereyra (1969) considers asymptotically high-order methods in his method of "iterated deferred corrections".

It is important to realize here, as with the vorticity transport equation in Section III-A-23, that *local* fourth-order accuracy does not imply *global* accuracy. The latter may be strongly influenced by boundary-value accuracy, which is often far from fourth order (see Section III-C). Further, methods like equation (3-430) cannot be used at points adjacent to boundaries, since information is required from outside the mesh. Authors usually revert to second-order forms at points adjacent to boundaries.* A more nearly consistent problem formulation would then require locally low-order forms in a fine mesh near boundaries, expanding to a higher order form in a coarse mesh away from boundaries. Such methods are considered by Bahvalov (1968). In any case, the accuracy of the $\nabla^2\psi = \zeta$ solution will certainly be limited by the accuracy of the vorticity transport equation and its boundary conditions, so a uniformly high-order formulation of the entire problem is recommended.

The iteration methods described earlier are applicable to these forms for the Laplacian, although convergence rates may be affected. Some of the direct methods (LeBail, 1969; Roache, 1971A,B) are adaptable, but generally these are less flexible in this regard than iterative methods.

Finally, we note the question of consistency in the discretized form of the Poisson equation and in the determination of velocities. The only dynamic use of the Poisson solution for ψ is in the determination of advection velocities in the vorticity transport equation. The Poisson equation $\nabla^2\psi = \zeta$ is nothing more than a definition of (discretized) vorticity,

$$\zeta = \frac{\delta u}{\delta y} - \frac{\delta v}{\delta x} \qquad (3\text{-}431)$$

---

*Many workers use the fourth-order form near boundaries by defining false points outside the mesh, the values there being set in accordance with a boundary condition. The result is still loss of fourth-order accuracy, except at a true symmetry boundary. Uncentered high-order forms exist (e.g. Southwell, 1946) but complicate stability and rarely are used.

with $u = \delta\psi/\delta y$ and $v = -\ \delta\psi/\delta x$. The solution of $\nabla^2\psi = \zeta$ should be regarded not as a solution for ψ, but as a solution for $\delta\psi/\delta x$ and $\delta\psi/\delta y$. Only the 5-point form of the Laplacian operator, as in equation (3-365), is compatible with the evaluations of u and v by second-order centered differences,* as in

$$u_{ij} = (\psi_{i,j+1} - \psi_{i,j-1})/2\Delta y \qquad (3\text{-}432A)$$

$$v_{ij} = -\ (\psi_{i+1,j} - \psi_{i-1,j})/2\Delta x \qquad (3\text{-}432B)$$

If higher order forms of the Laplacian are used, it would seem that accuracy of the total problem might actually be decreased, unless u and v were evaluated in a manner consistent with the higher order Laplacian form.

Similar remarks apply to the "least squares" methods for elliptic equations (e.g., Davis and Rabinowitz, 1960). In these methods, the ψ-values are "curve-fit" with high-order polynomials or some other m-parameter functional form. The functional form is usually selected to satisfy the differential equation at internal points (which is not difficult for the Laplace equation) and the m parameters (or polynomial coefficients) are solved by least squares minimization of the boundary errors.** These methods are not expected to be generally successful, especially when ψ varies non-monotonically, and are not recommended for use in computational fluid dynamics. But, if they are used, the evaluation of u and v ought to be consistent with the fitted functional form.

III-B-11. Remarks on Evaluating Methods

The remarks made in Section III-A-23 on evaluating methods for the vorticity transport equation also apply to evaluating solution methods and finite-difference forms for the Poisson equation. The remarks made in the previous section (III-B-10) about consistency between the Poisson stream function equation and the vorticity transport equation, for both the order of truncation error and for the velocity evaluations, should also be considered.

Finally, in weighing speed of production-run computation versus complexity and development time, it is necessary to consider *both* equations *at once*. If iterative methods are used for the Poisson stream function equation, $\nabla^2\psi = \zeta$, and an elementary one-step explicit method is used for the vorticity transport equation for $\partial\zeta/\partial t$, it typically occurs that about 90% of the computer time for a transient solution is taken up with the $\nabla^2\psi = \zeta$ solution. So if one changed to a two-step explicit method (such as the Cheng-Allen method of Section III-A-15) for $\partial\zeta/\partial t$, the computer time for the entire ψ - ζ computation is not doubled, but is only increased by about 10%. So the factor of 45 between the speed of calculation

---

*More precisely, the 5-point Laplacian is compatible with a definition of velocities at *cell boundaries* between nodes (see Section III-A-2), as in

$$u_{i,j+1/2} = (\psi_{i,j+1} - \psi_{i,j})/\Delta x \qquad (3\text{-}433A)$$

$$u_{i,j-1/2} = (\psi_{i,j} - \psi_{i,j-1})/\Delta x \qquad (3\text{-}433B)$$

with the node values then defined as

$$u_{i,j} = \frac{1}{2}\,(u_{i,j+1/2} + u_{i,j-1/2}) \qquad (3\text{-}433C)$$

**Alternately, the functional forms are selected to satisfy the boundary conditions exactly, and the residual errors from the PDE are used as the basis of minimization. Or, neither the boundary values nor the PDE is satisfied, and a combination of boundary errors and residual errors is minimized (A. J. Russo, private communication).

for the fourth-order Roberts and Weiss method (Section III-A-19) and the upwind differencing method is greatly reduced, (although it is still significant) to about a factor of 6, for the entire $\psi - \zeta$ solution. Similarly, the number of iterations to converge the $\nabla^2\psi = \zeta$ iteration depends somewhat on the initial estimate for ψ_{ij}. It is natural to use the previous converged $\nabla^2\psi^n = \zeta^n$ solution for the initial estimate of ψ^{n+1} in $\nabla^2\psi^{n+1} = \zeta^{n+1}$. For larger Δt, ζ^{n+1} will differ more from ζ^n, and ψ^n will be a poorer estimate of ψ^{n+1}. In the limit of $\Delta t \to 0$, $\zeta^{n+1} \to \zeta^n$ and the initial estimate $\psi^n \to \psi^{n+1}$. (This also occurs as a steady state is approached.) Thus, increases in the allowable Δt achieved by the use of implicit methods for $\partial\zeta/\partial t$ will be at least partially offset by increases in the computation time for the $\nabla^2\psi = \zeta$ iteration, as well as by the additional computer time required by the implicit methods per se. For this reason, Fromm (1964) reported a set of computations where, within limits, the computer time was practically independent of Δt!

The opposite situation can occur if direct methods are used for the $\nabla^2\psi = \zeta$ solution. The EVP method (Section III-B-8) has typically reduced the computation time for that equation by a factor of 100, meaning that the $\nabla^2\psi = \zeta$ solution now only takes up about 10% of the computer time, and the one-step explicit method for $\partial\zeta/\partial t$ requires about 90%. Changing to a two-step Cheng-Allen method for $\partial\zeta/\partial t$ now will nearly double the total computer time, and the fourth-order Roberts and Weiss will be slower than upwind differencing by about a factor of 40. On the other hand, increases in allowable Δt, except for the added complexity of the $\partial\zeta/\partial t$ equation itself, will directly decrease the computer time because the computation time for $\nabla^2\psi = \zeta$ by a direct method does not depend on any initial estimate of ψ.

The choice between iterative and direct methods will also be determined by the boundary formulation. When the Neumann conditions are used on all boundaries, as with the pressure equation (Section III-E), iteration convergence is greatly slowed down. If a pressure solution is desired at every computational step, the direct methods may be more attractive.

In addition to the stream function and pressure equations, the methods presented here are, of course, applicable to other occurrences of the Poisson equation in fluid dynamics. See, e.g., Rosenbaum (1968), for inviscid potential flow calculations. Iterative methods for nonlinear elliptic equations are treated by Ames (1965, 1969) and Lick (1969).

III-C. BOUNDARY CONDITIONS FOR THE VORTICITY AND STREAM FUNCTION EQUATIONS.

It is easy to conjure up some kind of plausible boundary conditions for ψ and ζ, but attempts to determine realistic, accurate, and stable methods can be highly frustrating. It has been found that the adequacy of any boundary condition, as determined by computational experiments, can depend on the Reynolds number, the interior-point differencing methods, the other boundary conditions, and sometimes on initial conditions. The large number of contingencies make analytical studies difficult and limited in applicability. See, however, Eddy (1949), Keast and Mitchell (1967), Campbell and Keast (1968), P. J. Taylor (1968, 1969, 1970), and Cheng (1968, 1970).

No mathematically rigorous solutions are available for many of the problems. We will draw mainly upon intuition, wind tunnel experience, and computational experimentation. Most of the computational experiments on boundary conditions have been carried out using simple two-level explicit methods for the vorticity transport equations. We will note the few known cases where the same boundary conditions caused instability when used with other methods. ("Instability" is used here in the sense of preventing iteration convergence, rather than necessarily exponential error growth.) These examples will serve as cautions in the applications of these essentially *ad hoc* methods. In this regard, we suggest that, in the initial development phase of a computer program, the most restrictive and lowest order boundary conditions be used to debug the program and to establish stability of the interior-point methods. The less restrictive boundary conditions can then be attempted.

Most of the boundary conditions given are either Dirichlet conditions (specified function value) or the Neumann conditions (specified normal gradient). No fluid dynamics computations have been performed to date with mixed (Robbin's) type, where a weighted combination of function value and normal gradient are specified. (This condition does arise naturally in free-convection heat transfer calculations using Newton's "law" of cooling.) Indeed, Campbell and Keast (1968) have shown that such conditions can cause instabilities in computations with the diffusion equation due to the presence of unbounded solutions to the original PDE, and Keast and Mitchell (1967) have shown that solutions of elliptic equations with such boundary conditions cannot always be obtained by the asymptotic time-limit approach. Neither have any nonlinear boundary conditions, such as specifying a gradient of the squared function, been used.

III-C-1. Remarks on the Dominant Importance of Computational Boundary Conditions

A first-order *ordinary* differential equation such as $df/dx = 0$ specifies the solution of a problem up to an additive constant; the boundary condition determines the value of the *constant*. A first-order *partial* differential equation such as $\partial f(x,y)/\partial x = 0$ specifies very little of the solution; any function $g(y)$ satisfies the PDE, and the boundary conditions must specify the *function*. A PDE such as $\nabla^2\psi = \zeta$ really contains very little information on ψ. All the fantastic flow patterns of common gases and liquids are solutions of the *same* PDE's, the Navier-Stokes equations. The flows (solutions) are distinguished only by boundary and initial conditions, and by the flow parameters such as Re.

It is therefore not surprising that the specification of computational boundary conditions, besides affecting numerical stability, greatly affects the accuracy of the FDE solution. What is surprising is that this importance was not widely recognized for many years, and perhaps is still not. Richardson (1910) certainly gave importance to boundary conditions but, in later years, most of the work was done on the interior-point difference methods. Perhaps a reason was the predominant interest in heat conduction problems where the boundary conditions are often simple and unequivocal. Another reason was that experimentation was not practical until electronic computers became common.* Pioneering work on boundary conditions (pre-computer) was done by Southwell (1946), Allen and Southwell (1955), and Thom and Apelt (1961). The modern pioneering work of Thoman and Szewczyk (1966) was the first to include really extensive experimentation with boundary treatments in a finite computational mesh. The papers of Cheng (1968, 1970) and Moretti (1968B), although often displaying contrasting views, served to emphasize the importance of boundary conditions. Especially, Cheng (1970) showed that, in his numerical experiments on Burger's

---

*Kawaguti (1953) obtained a single solution for Re = 40 flow over a circular cylinder by using a mechanical desk calculator, working 20 hours/week for 1-1/2 years. He could hardly be expected to repeat the calculations to test three different outflow boundary conditions, first- and second-order wall vorticities, etc. etc.

equation, higher order schemes can be less accurate than first-order schemes for cell Reynolds numbers $R_c > 1$ because of boundary treatments, and that boundary errors can be twice as large as interior point truncation errors. He noted that the wide discrepancy between calculations of the same cylinder flow problem by Jenson (1959) and by Hamielec, et al. (1967) were due to boundary errors, and that such errors are persistent, even as $\Delta x \to 0$. Thus, in several senses, boundary conditions are seen to be of dominant importance in computational fluid dynamics.

It is especially this boundary treatment aspect of computational fluid dynamics which has drawn accolades of "artistry" and/or denunciations of "voodoo" from the scientific community.

Boundary conditions will be discussed in relation to the planar 2D backstep problem shown in Figure 3-22, which affords examples of all types of boundaries encountered in a block-rectangular region. (Curved boundaries and mesh mappings will be discussed in Chapter VI.)

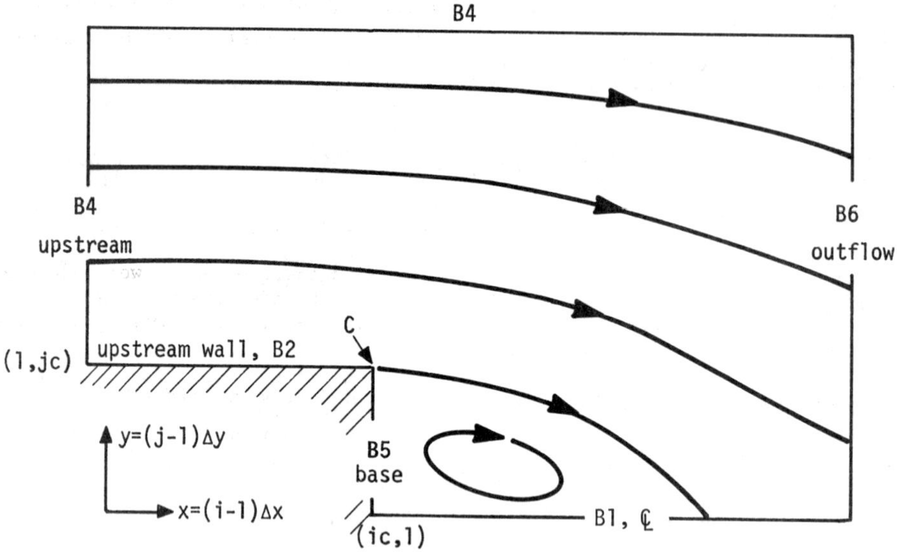

Figure 3-22. Boundary labeling.

This Figure will be cited constantly in this section and again in Chapter V. The reader would do well to mark the page with a folded corner or a paper clip for easy reference.

III-C-2. Walls in the First Mesh System

Solid walls appear in Figure 3-22 at boundary B2, the upstream wall, and at B5, the base. In the first mesh system, node values of ψ and ζ lie along the walls.

Since the line of B2-B5-B1 is a streamline,* any constant value of ψ may be selected; the conventional choice is $\psi = 0$.

The wall vorticity is an extremely important evaluation. The vorticity transport equation (2-12) for $\partial\zeta/\partial t$ determines how ζ is advected and diffused, but the total ζ is conserved at interior points. At no-slip boundaries, however, ζ is *produced*. It is the diffusion and subsequent advection of this wall-produced vorticity which actually drives

*If a permeable wall is to be modeled, say, over the base, then the velocity u along the base may be specified $u = f(y)$. Then ψ may be determined by integration down from the corner to satisfy $\delta\psi/\delta y = f(y)$. See remarks in Section III-C-6.

the problem. Some early geophysical calculations were erroneous, because wall ζ-values were determined by extrapolation procedures which had nothing to do with the physics driving the problem.*

The vorticity is obtained from the no-slip conditions. Using boundary B2 as an example, we expand to $\psi_{i,jc+1}$ by a Taylor series out from the wall values, $\psi_{i,jc}$.

$$\psi_{i,jc+1} = \psi_{i,jc} + \left.\frac{\partial\psi}{\partial y}\right|_{i,jc} \Delta y + \frac{1}{2}\left.\frac{\partial^2\psi}{\partial y^2}\right|_{i,jc} \Delta y^2 + \frac{1}{6}\left.\frac{\partial^3\psi}{\partial y^3}\right|_{i,jc} \Delta y^3 + O(\Delta y^4) \qquad (3\text{-}434)$$

But $\partial\psi/\partial y|_{i,jc} = u_{i,jc} = 0$ by the no-slip condition, and $\partial^2\psi/\partial y^2|_{i,jc} = \partial u/\partial y|_{i,jc}$. Now $\zeta = \partial u/\partial y - \partial v/\partial x$. Along the wall, $\partial v/\partial x|_{i,jc} = 0$ because v = constant (i.e., $v = 0$) along x. Thus, $\partial u/\partial y|_{i,jc} = \zeta_{i,jc}$. Substituting this into equation (3-434) and solving for $\zeta_{i,jc}$ with $\psi_{i,jc} = 0$ gives $\zeta_{i,jc} = 2\psi_{i,jc+1}/\Delta y^2 + O(\Delta y)$. Regardless of the wall orientation or boundary value of ψ, we can write

$$\zeta_w = \frac{2(\psi_{w+1} - \psi_w)}{\Delta n^2} + O(\Delta n) \qquad (3\text{-}435A)$$

where Δn is the distance from (w+1) to (w), normal to the wall.

This first-order form was given as early as 1928 in the pioneering work of Thom (1928, 1933) and has been used extensively since then. It is the safest form to use, and it often gives results essentially equal to higher order forms.** See, for example, the experiments by Esch, reported by Pearson (1965A).

As mentioned in Chapter II, the alternate definition of vorticity as $\zeta' = -\zeta = -\frac{\partial u}{\partial y} + \frac{\partial v}{\partial x}$ does not change the form of the vorticity transport equation. But this sign convention "gets into" the vorticity transport through the boundary conditions, with the counterpart of equation (3-435A) being

$$\zeta'_w = -\frac{2(\psi_{w+1} - \psi_w)}{\Delta n^2} + O(\Delta n) \qquad (3\text{-}435B)$$

Woods (1954) suggested the following second-order form, obtainable from equation (3-434) by retaining the Δy^3 term. By differentiating the definition of vorticity, we obtain

$$\frac{\partial\zeta}{\partial y} = \frac{\partial^2 u}{\partial y^2} - \frac{\partial^2 v}{\partial y\partial x} = \frac{\partial^3\psi}{\partial y^3} - \frac{\partial}{\partial x}\left(\frac{\partial v}{\partial y}\right) \qquad (3\text{-}436)$$

From the continuity equation (2-3), $\partial v/\partial y = -(\partial u/\partial x)$. Substituting this above and evaluating at the wall, we obtain

$$\left.\frac{\partial\zeta}{\partial y}\right|_w = \left.\frac{\partial^3\psi}{\partial y^3}\right|_w + \left.\frac{\partial^2 u}{\partial x^2}\right|_w \qquad (3\text{-}437)$$

The second term, $\partial^2 u/\partial x^2|_w$, is zero by the no-slip condition (u = constant = 0). The term $\partial\zeta/\partial y|_w$ is evaluated by a first-order backward difference from (w+1) as

---

*S. A. Piacsek (private communication).

**In fact, for a Blasius boundary-layer profile, $\partial^3\psi/\partial y^3 = \partial^2 u/\partial y^2 = 0$ at the wall; numerical tests verify that this first-order form is more accurate than second-order forms in this case.

$$\left.\frac{\partial \zeta}{\partial y}\right|_w = \frac{\zeta_{w+1} - \zeta_w}{\Delta n} + O(\Delta n) = \left.\frac{\partial^3 \psi}{\partial y^3}\right|_w \tag{3-438}$$

Substituting this equation for $\left.\frac{\partial^3 \psi}{\partial y^3}\right|_w$ into equation (3-434) and solving for ζ_w gives Woods' method for wall vorticity.

$$\zeta_w = \frac{3(\psi_{w+1} - \psi_w)}{\Delta n^2} - \frac{1}{2}\zeta_{w+1} + O(\Delta n^2) \tag{3-439}$$

Wood's method was used successfully by Russell (1963); see also Lester (1961) and Michael (1966). Hung and Macagno (1966) independently developed equation (3-439) and applied it to moderate Re flows (Re ≃ 300) with success; so also has the present author. But P. J. Taylor's (1968) theoretical calculations indicate possible instability at high cell Reynolds numbers. This was the computational experience of Runchal et al., (1969), who again independently derived equation (3-439). Using steady-flow iteration methods, Runchal, et al., (1969) found that this second-order form was more likely to cause divergence, especially at high Re and with variable Δy.

Second-order formulations for ζ_w in axisymmetric coordinates were evaluated by Lugt and Rimon (1970) by comparison with an exact solution for flow over an oblate spheroid, unfortunately applicable only at Re = 0. They considered an expression which involved $\partial\zeta_w/\partial t$. In terms of the local radius of curvature r_c, their expression was

$$\zeta_w = \frac{\dfrac{3\psi_{w+1}}{\Delta n^2} - \dfrac{1}{2}\zeta_{w+1} + \Delta n^2 \, Re \, \dfrac{\partial \zeta_w}{\partial t} / 24}{1 - \dfrac{\Delta n}{2r_c} + \dfrac{3\Delta n^2}{8r_c^2}} + O(\Delta n^2) \tag{3-440}$$

For $r_c \to \infty$, the equation appropriate for a flat plate is obtained, and is seen to be just Woods' expression (3-439) plus the time-dependent term. Dawson and Marcus (1970) found that the $\partial\zeta_w/\partial t$ term in equation (3-440) "made the equations extremely unstable, and had to be dropped."

A more frequently used second-order form was first used by Jensen (1959) and later by Pearson (1965A), who fit the values of ψ near the wall to a third-order polynomial, with the condition of $\partial\psi/\partial y|_w = u_w = 0$. The result (see also Briley, 1970) is

$$\zeta_w = \frac{-7\psi_w + 8\psi_{w+1} - \psi_{w+2}}{2\Delta n^2} + O(\Delta n^2) \tag{3-441}$$

This form was later used by Wilkes and Churchill (1966) for low Grashoff number free-convection solutions, but when the calculations were extended by Samuels and Churchill (1967), they found it unstable and reverted to the first-order form (3-435). Apparently, Jensen's (1959) calculations also would have become unstable at higher Reynolds numbers. The calculations of Torrance (1968) and of Pao and Daugherty (1969) using equation (3-441) were also limited to moderate Reynolds numbers; see also Southwell (1946) and Paris and Whitaker (1965). Further, Beardsley (1969) found, in comparisons with the one-dimensional exact solution for the axisymmetric cylinder spin-down problem, that Jensen's form (3-441) is actually less accurate than the first-order form,* even when stable.

---

*In axisymmetric coordinates with the maximum radius at $r = J \cdot \Delta r$, Beardsley (1969) found that the best form, compatible with global conservation in the axisymmetric coordinates, was $\zeta_J = (2J - 1)\psi_{J-1}/(J - 1/4)\Delta r^2$. For large $J(r \to \infty)$, this approaches to equation (3-435)

Until quite recently, it was a widely held belief (superstition?) that the instability of equation (3-441) was somehow connected with the fact that information from (w + 2) was being used, and that such methods were doomed. The illuminating work of Briley (1970) has proved this belief incorrect. Briley noted that the polynomial form of ψ assumed in the derivation of equation (3-441) was inconsistent with the evaluation of $u = \partial\psi/\partial y$ at (w + 1) using centered differences, and that a consistent formulation would require the use of the following special form, only at (w + 1).

$$u_{w+1} = \left.\frac{\delta\psi}{\delta y}\right|_{w+1} = \frac{-5\psi_w + 4\psi_{w+1} + \psi_{w+2}}{4\Delta n} + O(\Delta n^2) \qquad (3\text{-}442)$$

When this form was used along with equation (3-441), using ADI differencing methods, the computations were stable, even at high Reynolds numbers. (Of course, there is no really rational connection between stability and "consistency," and we have not really explained the instability, any more than the superstition cited above explained it. From Briley's work, perhaps we are merely substituting a consistent superstition for an inconsistent superstition.) Briley even derived another equation based on a fourth-order Lagrange interpolating polynomial.

$$\zeta_w = \frac{-85\psi_w + 108\psi_{w+1} - 274\psi_{w+2} + 4\psi_{w+3}}{18\Delta n^2} + O(\Delta n^2) \qquad (3\text{-}443)$$

A consistent formulation requires that the velocities at two points near the wall be specially evaluated as

$$u_{w+1} = \frac{-17\psi_1 + 9\psi_2 + 9\psi_3 - \psi_4}{18\Delta n^2} + O(\Delta n^2) \qquad (3\text{-}444A)$$

$$u_{w+2} = \frac{14\psi_1 - 36\psi_2 + 18\psi_3 + 4\psi_4}{18\Delta y^2} \qquad (3\text{-}444B)$$

and furthermore that the term $\partial^2\psi/\partial y^2$ in the Poisson equation must be evaluated as

$$\left.\frac{\delta^2\psi}{\delta y^2}\right|_{w+1} = \frac{29\psi_w - 54\psi_{w+1} + 27\psi_{w+2} - 2\psi_{w+3}}{18\Delta y^2} \qquad (3\text{-}445A)$$

$$\left.\frac{\delta^2\psi}{\delta y^2}\right|_{w+2} = \frac{11\psi_w - 27\psi_{w+2} + 16\psi_{w+3}}{18\Delta y^2} \qquad (3\text{-}445B)$$

These forms were also stable at high Re. The form of equation (3-443) gives very little change in ζ_w compared with Jensen's equation (3-441), since the overall truncation error is unchanged (Briley, 1970). But the ADI iteration for ζ_w^{n+1} (see Section III-A-16) converged faster, allowing larger time steps and a halving of overall computer time. However, additional programming effort is required because the use of equation (3-445) in the ADI solution for the Poisson equation introduces terms from farther than the neighboring nodes; thus, the time-split ADI implicit equations along y are not tridiagonal. Briley (1970) used a simple Gaussian modification to remove the added implicitness at just these two points, (w + 1) and (w + 2). It would appear that the equations (3-445) would hinder SOR convergence of the Poisson equation, and could prohibit the use of many direct methods. (The EVP method of Section III-B-8 could be used with a march in the x-direction.)

For the case of a sloping wall lying diagonally along the mesh, as shown in Figure (3-23), the wall vorticity may be evaluated from the first-order form (3-435) as follows. The wall values at w_a, w_b, w_c which are not on mesh nodes are evaluated as

$$\zeta_{wa} = \frac{2(\psi_a - \psi_w)}{\Delta n^2} \qquad (3\text{-}446A)$$

$$\Delta n^2 = [\Delta y \cdot \cos\theta_w]^2 = \frac{\Delta x^2}{1 + \beta^2} \qquad (3\text{-}446B)$$

where $\beta = \Delta x/\Delta y$ is the mesh aspect ratio. Similarly for ζ_{wb} and ζ_{wc}. Then the wall values

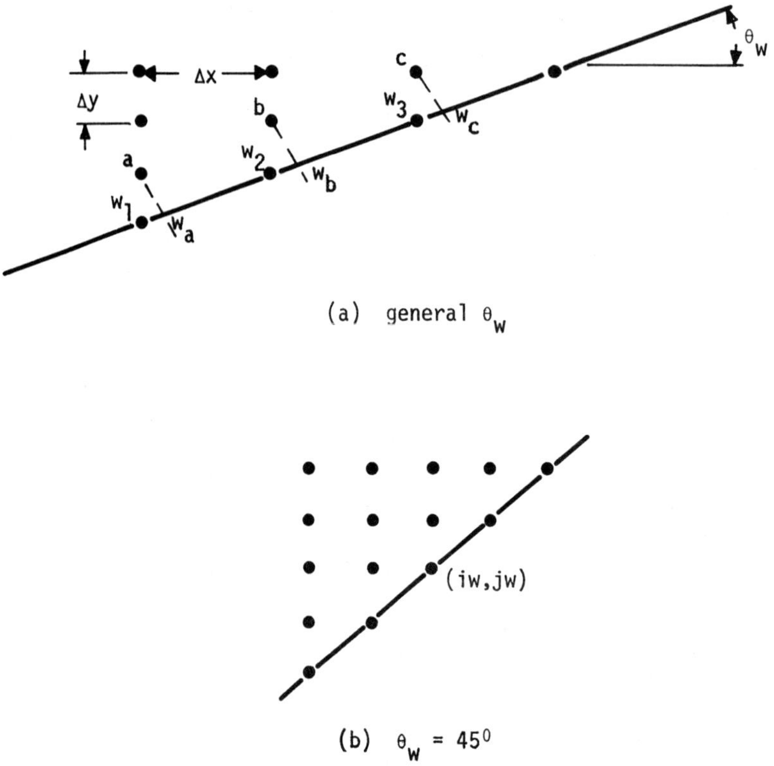

Figure 3-23. Evaluation of wall vorticity on a sloping wall.

at nodes w_1, w_2, w_3 are determined by interpolation along the wall. The method is stable and apparently accurate (Campbell and Mueller, 1968). For the particular case of a 45° wall inclination ($\beta = 1$) this procedure gives, for a point (i_w, j_w) as shown in Figure 3-23b,

$$\zeta_{iw,jw} = \frac{2(\psi_{iw-1,jw} + \psi_{iw,jw+1} - 2\psi_{iw,jw})}{\Delta^2} \qquad (3\text{-}447)$$

where $\Delta = \Delta x = \Delta y$ (O'Leary and Mueller, 1969).* This method appears to be more accurate than using

$$\zeta_{iw,jw} = \frac{(\psi_{iw-1,jw+1} - \psi_{iw,jw})}{\Delta^2} \qquad (3\text{-}448)$$

---

*For a similar treatment using the diagonal-unit square operator, see Thom and Apelt (1961), page 126.

since the ψ values used in equation (3-448) are twice as far from the wall as those in equation (3-447).

It may also be useful to model a "slip" wall condition with the viscous equations. This implies that the boundary layer is less than Δy thick. This condition may be heuristically simulated by using a "slip" wall condition, as in section III-C-5. The ψ value is specified, and the vorticity is evaluated from a Neumann condition,

$$\zeta_w = \zeta_{w+1} \tag{3-449}$$

The validity of this approach is unproved. It cannot provide a mathematically "consistent" system of equations, since the boundary-layer thickness cannot remain less than Δy as $\Delta y \to 0$. Yet, it does seem to provide a meaningful physical approximation.

It should be noted that, if *inviscid flow* is being calculated, it is not enough to set $1/Re = 0$ in the vorticity transport equation; it is also necessary to use a *slip boundary condition*. In fact, it is more important than setting $1/Re = 0$. It is known that inviscid flows can be modelled quite well with Re as low as 300 provided that slip boundary conditions are used (Kentzner, 1970A). For inviscid flow, it is easily seen from equation (2-12) that ζ, like ψ, is constant along any steady streamline, including a slip wall, since $D\zeta/Dt = 0$. Thus ζ_w = constant (set at inflow) is the correct boundary condition for inviscid flow at a wall.

A remark is needed about the possible overspecification of boundary conditions. To simplify, we consider some kind of flow in an enclosure in which *all* boundaries are stationary walls. For an impermeable no-slip wall aligned along x, we know that both $u = 0$ and $v = 0$. In terms of ψ, these give two relations: $\partial\psi/\partial x = -v = 0$, which implies ψ_w = constant (say 0) along a wall, and $\partial\psi/\partial y = u = 0$ normal to the wall. Considering the Poisson equation $\nabla^2\psi = \zeta$ *in isolation*, either condition provides a sufficient boundary condition. Obviously, one cannot use both conditions in the Poisson equation, since that would overspecify the problem. But $\psi_w = 0$ is *not* sufficient to determine the wall *vorticity*, ζ_w; the information that $\partial\psi/\partial y|_w = 0$ is also needed, as in the derivation of equation (3-435A) or (3-439). So the gradient condition, $\partial\psi/\partial y|_w = 0$, is required for the ζ boundary condition and, by default, the condition $\psi_w = 0$ goes to the Poisson equation for ψ. This is the *only* correct distribution of these conditions. (See also Problem 3-27.)

It may be confusing to note that the numerical solution for ψ will not* show $\delta\psi/\delta y = (\psi_{w+1} - \psi_w)/\Delta y = 0$. This paradox arises because this form for $\delta\psi/\delta y|_w$ is only first-order accurate, whereas the entire equation is solved to second-order accuracy. If, instead of splitting the system into two second-order systems for ψ and ζ, we use a single fourth-order equation for ψ, then both conditions for ψ_w and $\partial\psi/\partial y|_w$ would be required and satisfied in the FDE. It might thus appear that the single fourth-order equation would be more accurate, at least in regard to the boundary conditions. But, in order to evaluate fourth-order spatial derivatives, it is necessary to use non-centered differences *near* the wall. These forms are not compatible with the evaluation of $\delta\psi/\delta y|_w = (\psi_{w+1} - \psi_w)/\Delta y$, and we would again find that $\psi_{w+1} \neq \psi_w$, even in the fourth-order form.

A remark is appropriate in regard to determining separation and reattachment points. It may easily be shown that, in the continuum equations, $\zeta_w = 0$ at both separation and reattachment. With solution values of ζ_w known, the locations of $\zeta_w = 0$ may be found by interpolation. But the interpolated values are no better than the wall values, and the separation and reattachment points cannot be located to great precision. The method of determining these points from a solution should be clearly stated in any research work (Lavan, et al., 1969; Roache and Mueller, 1970; and Chavez and Richards, 1970).

Finally, we note that the wall vorticity evaluation results in a kind of boundary conservation error which may be used as a truncation-error convergence check. (See Problem 3-32.)

---

*Israeli (Studies in Applied Mathematics, Vol. 51, March 1972, pp. 67-71) determines ζ_w^{n+1} by iterating (with under-relaxation) to enforce $\delta\psi/\delta n|_w = 0$ at the new time level.

Exercise: Derive* first- and second-order wall vorticity equations for the wall aligned with the x-axis, for the cases of (a) a moving belt wall with $u_w = U_w$ (specified) and $v_w = 0$, and (b) a permeable wall with $u_w = 0$ and a specified blowing rate $v_w = V_w$.

The work of Taylor (1970) indicates that wall blowing may be numerically destabilizing.

III-C-3. Walls in Alternate Mesh Systems

A *time-staggered* mesh system is used in the method of Roberts and Weiss (Section III-A-19). Another type is the *space*-staggered mesh, in which some flow variables are defined at one set of nodes, and other variables in a mesh dislocated diagonally from the first (Harlow and Fromm, 1954; Fromm, 1963; Williams, 1969; see also Section III-G-3.) Still another type is the *shifted* mesh, in which one mesh is shifted half a node-space with respect to the other, but along a coordinate line rather than diagonally. These methods all result in a definition of vorticity ζ at a node located $\Delta n/2$ off the wall, as shown in Figure 3-24a.

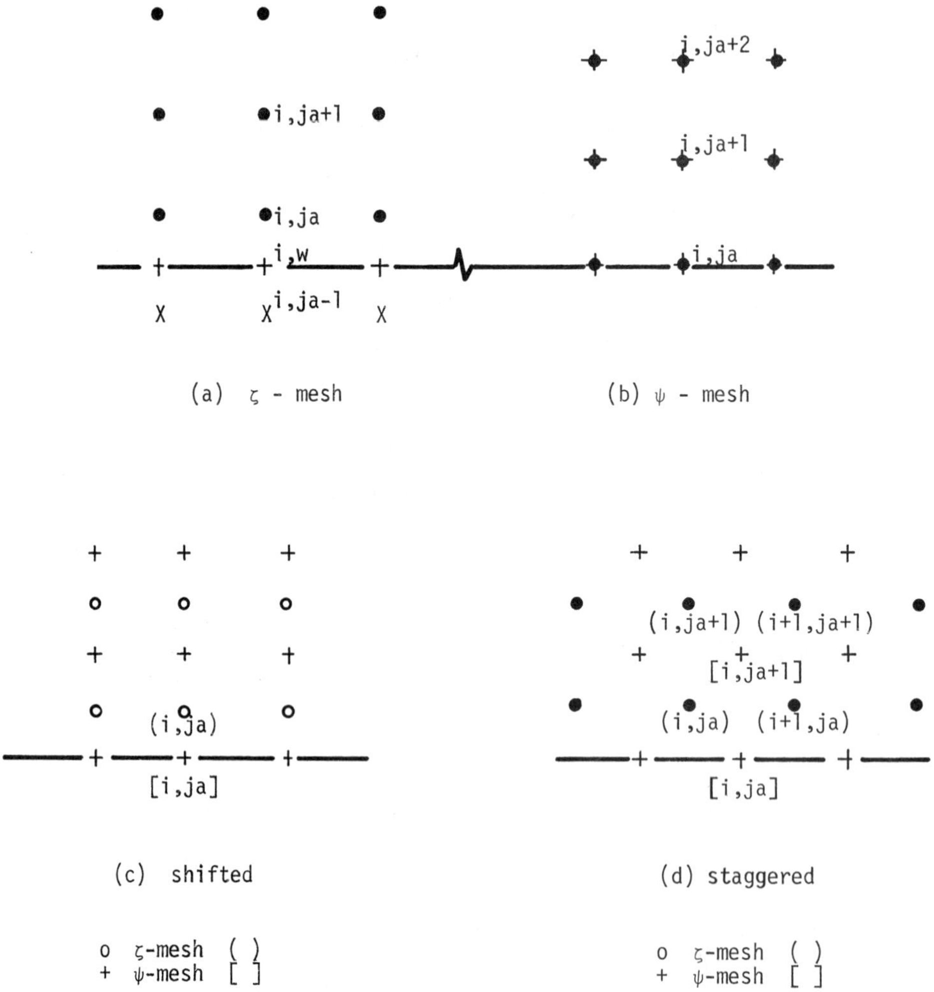

Figure 3-24. Walls in the alternate mesh systems.

---

*Note that the moving wall adds the term $2U_w/\Delta n$ to the first-order form for ζ_w, but adds $3U_w/\Delta n$ to the second-order forms.

The use of this second mesh system for ζ is not recommended. Some care is required in the formulation, lest an inconsistent evaluation of ζ_w make the wall to be effectively offset $\Delta n/2$ (Fromm, 1967). Even when done properly, the use of this second mesh system for ζ decreases accuracy, as we now demonstrate.

The no-slip condition is to be applied *at* the wall, not near it. The vorticity at the (i,ja) location in Figure 3-24, adjacent to the wall, is evaluated from the vorticity transport equation. The wall vorticity, ζ_w, is evaluated to first-order accuracy at the off-node position, w, using equation (3-435) with $\Delta n/2$ replacing Δn.

$$\zeta_{i,ja-1/2} \equiv \zeta_{iw} = \frac{2(\psi_{i,ja} - \psi_w)}{(\Delta n/w)^2} + 0\left(\frac{\Delta n}{2}\right) \qquad (3\text{-}450)$$

(The second-order equations might also be used.)

Then, to perform the differencing at the point (i,ja) adjacent to the wall, it is necessary to use some form of uncentered difference equation. It is this necessity that reduces the accuracy. One approach is to define mid-node values at (i,ja+1/2) by

$$\zeta_{i,ja+1/2} = \frac{1}{2}\left[\zeta_{i,ja+1} + \zeta_{i,ja}\right] \qquad (3\text{-}451A)$$

$$(v\zeta)_{i,ja+1/2} = \frac{1}{2}\left[(v\zeta)_{i,ja+1} + (v\zeta)_{i,ja}\right] \qquad (3\text{-}451B)$$

Then the usual spatial y-differences in the vorticity transport equation at (i,ja) are replaced by differences over these half-cell values, as in

$$\left.\frac{\delta(v\zeta)}{\delta y}\right|_{i,ja} = \frac{(v\zeta)_{i,ja+1/2} - (v\zeta)_{i,w}}{\Delta y} = \frac{(v\zeta)_{i,ja+1/2}}{\Delta y} \qquad (3\text{-}452)$$

$$\left.\frac{\delta^2\zeta}{\delta y^2}\right|_{i,ja} = \frac{\zeta_{i,ja+1/2} - 2\zeta_{i,ja} + \zeta_{i,w}}{(\Delta y/2)^2} \qquad (3\text{-}453)$$

It is readily shown (Problems 3-24, 25) that the definition of the mid-node value of $(v\zeta)$ and $\zeta_{i,ja+1/2}$ as in equation (3-451) is not consistent with the use of centered differencing over the full node spaces; no error is introduced for the first derivatives, but the second-derivative form (3-451B) is reduced to first-order accuracy. In fact, for the simple 1D problem with only the second-derivative term present, it is readily shown (Problem 3-25) that the use of the second mesh system at the wall results in a *boundedness* error; in a fluid dynamics problem, this behavioral error could conceivably result in a false indication of separation.

An illusion of second-order accuracy, as well as some real programming convenience, can be obtained by defining artificial values of v and ζ at the "x" nodes, shown in Figure (3-24a), which are *inside* the wall. These node values of ζ are defined as part of the doubly subscripted array for ζ, call it Z(I,J). For example, for a wall at the lower extremity of the mesh, the elements Z(I,1) would be the "x" values inside the wall, and the elements Z(I,2) would be the $\zeta_{i,ja}$ values. The "x" node values are set after each internal-point calculation of ζ so as to produce the desired boundary condition at (i,w) when the usual interior-point equations are used at (i,ja). Although more convenient, this method still is only first-order accurate. In no case is this second mesh system recommended for vorticity.

Even if the second mesh system is used for vorticity, the stream function definitely should not be defined in this second mesh system. The condition $u_w = (\delta\psi/\delta y)_w = 0$ could be met by defining a zero gradient condition $\psi_{i,ja-1} = \psi_{i,ja}$. The Poisson equation would then be solved with this Neumann boundary condition, which slows iterative convergence. More importantly, this technique does not give consistent ψ-values at the wall. If the technique gives $\psi_w = 0$, then it also sets $\psi_{i,ja} = 0$ at $\Delta n/2$ above the wall. This gives a zero-velocity slug of fluid, of thickness $\Delta y/2$, above the wall. That is, the wall is effectively displaced upward $\Delta n/2$ in the ψ formulation but not in the ζ equations, which is clearly inconsistent.

The fundamental difficulty is that the boundary conditions $\psi_w = 0$ and $u_w = \partial\psi/\partial y|_w = 0$ cannot *both* be used in the Poisson equation along the same boundary, since this would over-specify the problem, either Dirichlet or the Neumann conditions being sufficient. The $\psi = 0$ condition is to be used in the Poisson equation. (See also the previous section, III-C-2, and Problem 3-27.)

It is necessary that ψ be defined in the first mesh system, which is shifted $\Delta y/2$ from the ζ mesh, as shown in Figure 3-24. The ζ-advection velocities are defined at nodes in the ζ-mesh, and are consistently evaluated from ψ-mesh values, using the indexing shown in Figure 3-24, as

$$u_{i,j} = \left(\frac{\psi_{i,j+1} - \psi_{i,j}}{\Delta y}\right) \tag{3-454}$$

The ψ-mesh could also be staggered with respect to the ζ-mesh in the x-direction. This formulation might even appear to decrease the truncation error of the velocity calculations, but additional error is introduced into the solution of the Poisson equation, $\nabla^2\psi = \zeta$. The solution of the Poisson equation requires ζ-values in the ψ-mesh, which we denote by ζ^+. For the (non-unique) indexing shown in Figure 3-24, these could plausibly be evaluated by averaging the ζ values in the ζ-mesh, denoted as $\zeta^{\cdot}$.

$$\zeta^+_{i,j} = \frac{1}{4}\left(\zeta^{\cdot}_{i,j} + \zeta^{i}_{i+1,j} + \zeta^{\cdot}_{i,j+1} + \zeta^{\cdot}_{i+1,j+1}\right) \tag{3-455}$$

The velocities in the ζ-mesh, denoted by $u^{\cdot}$ and $v^{\cdot}$, are evaluated by differencing ψ^+ values in the ψ-mesh, as follows:

$$u^{\cdot}_{ij} = \frac{1}{2}\left[\frac{\psi^+_{i,j+1} - \psi^+_{i,j}}{\Delta y} + \frac{\psi^+_{i-1,j+1} - \psi^+_{i-1,j}}{\Delta y}\right] \tag{3-456a}$$

$$v^{\cdot}_{ij} = -\frac{1}{2}\left[\frac{\psi^+_{i,j+1} - \psi^+_{i-1j+1}}{\Delta x} + \frac{\psi^+_{i,j} - \psi^+_{i-1,j}}{\Delta x}\right] \tag{3-456b}$$

Inflow and outflow conditions may also be somewhat complicated. For example, using the staggered mesh system, the inflow ψ would be specified along a line $\Delta x/2$ away from the inflow ζ line, which is inconsistent.

The first mesh system is evidently much more straightforward to use, as well as more accurate, and is therefore generally recommended. The use of the second mesh system for ζ, and the shifted or staggered first mesh system for ψ, does have the advantage of requiring no special treatment for the sharp convex corner c in Figure 3-22 (see Section III-C-II-a). But this advantage is minimal, and the exclusive use of the first mesh system at walls is recommended.

III-C-4. Symmetry Boundaries

The centerline boundary B1 in Figure (3-22) would have the same boundary conditions as a solid no-slip wall if it represented a splitter plate. If the figure represents a symmetric half of a base flow problem, it is still required that $\psi = 0$, meaning that only subcritical wake solutions will be meaningful.* In this case, B1 represents a slip splitter plate, and the boundary condition on vorticity is very simple. Because v = 0 everywhere on the centerline, $\partial v/\partial x = 0$. Because u is symmetric above and below the centerline, $\partial u/\partial v = 0$. Thus,

$$\zeta_{\mathbb{C}} = 0 \tag{3-457}$$

We cannot use symmetry conditions on ψ derivatives in addition to the condition $\psi = 0$; this would overspecify the Poisson solution.

In axisymmetric flow in cylindrical coordinates, we have $\zeta = \nabla \times \vec{V}$ and equation (3-457) still applies. But the equations are more conveniently written in terms of a $\zeta' = \zeta/r$, where the local radius r = 0 at $\mathbb{C}$. Torrance (1968) points out that $\zeta'_{\mathbb{C}}$, though bounded, is not zero.

A symmetry condition of $\zeta = 0$ was used by Runchal and Wolfshtein (1969) along a 45° line between a wall and the centerline of an impinging normal jet. Symmetry in this case applies strictly only to the potential flow equation, but was used in this work as a high-Re approximation.

III-C-5. Upper Boundary

The upper boundary, B3 in Figure 3-22, can pose an interesting problem. Of course, we can choose to model a physical problem whose upper boundary condition is obvious; for example, if we choose to model an unsymmetric channel expansion, then B3 is a solid no-slip wall and those equations (section III-C-2) for the vorticity apply. The value ψ(B3) is constant and may be evaluated by integrating the input u-velocity profile at B4. (See the next section, III-C-6.) Kawaguti (1965) studied this problem. Or, if we choose the geometry to represent the lower half of a symmetrical channel expansion, then symmetry conditions (as in the slip splitter-plate centerline solution of the previous section III-C-4) give $\zeta = 0$. The ψ evaluation is still obtained by integration at boundary B4. If we choose to combine these same boundary conditions for B3 with a slip splitter plate for the centerline B1, we model the smallest symmetric part of the flow field over an infinite series of rectangular bodies. (These may practically represent an array of heat exchanger fins, for example.)

But if we want to model the condition of *no* boundary at B3, the "free flight case" where a fluid of infinite extent in the y-direction is assumed, the choice is not so clear.

One's first impulse might be to model a wind tunnel situation with a solid no-slip wall at B3. Drawing from wind tunnel experience, we know that, as the tunnel wall moves farther away, the "blockage" problem is reduced and the flow adjacent to the body will model the free flight case. However, the computing time and computer memory limit the number of mesh points, and accuracy requirements limit the mesh spacing Δy, so there is a limitation on the "test section" size of the wind tunnel. (The use of an expanding mesh or a coordinate transformation for this type of problem will be covered in Chapter VI. Even if these devices are used, the treatments given here will still be applicable.)

---

*The supercritical wake, with oscillations and asymmetries, is solved by doubling the mesh and the picture of Figure 3-22, giving an upper and lower boundary like B3. The centerline mesh points are then regular interior points.

It may be of interest, however, to calculate a supercritical condition for the purposes of hydrodynamic stability analysis; in that case symmetry is imposed even for supercritical Re.

**Note that ζ is defined by the vector operation $\zeta = \pm \nabla \times \vec{V}$, so it is invariant to a coordinate rotation. Thus, $\zeta = 0$ along *any* line of symmetry, regardless of its orientation.

A considerable improvement accrues if a *moving* wind tunnel wall is modeled, as done by Fromm (1963) and Fromm and Harlow (1963). They specified v = 0 and $u = U_o$, where U_o may be interpreted as a "freestream" velocity.* The boundary conditions in the (ψ,ζ) system would then be those for a no-slip moving wall, ψ = constant, and (see previous exercise) ζ evaluated from

$$\zeta_{iJ} = \frac{2(\psi_{i,J-1} - \psi_{iJ} - U_o \cdot \Delta y)}{\Delta y^2} + O(\Delta y) \tag{3-458}$$

or

$$\zeta_{iJ} = \frac{3(\psi_{i,J-1} - \psi_{iJ} - U_o \cdot \Delta y)}{\Delta y^2} - \frac{1}{2}\zeta_{i,J-1} + O(\Delta y^2) \tag{3-459}$$

Equation (3-458) was successfully used by Campbell and Mueller (1968).

Another improvement over modeling a physical wind tunnel wall is to model a *frictionless* wind tunnel wall. The value of ψ(B3) is still constant and the vorticity is evaluated using the concept of "image systems" used to evaluate wind tunnel blockage corrections. In this case, the symmetry of the "image" system gives ζ(B3) = 0. This boundary condition was used by Jensen (1959) in an axisymmetric problem, the "tunnel wall" being a circular tube. Thoman and Szewczyk (1966) used a treatment which was similar but less restrictive because of the treatment used at an *adjacent* boundary. They specified the far-field condition of $\zeta = 0$ with $\partial\psi/\partial y \equiv u = U_0$ and $\partial\psi/\partial x \equiv v = 0$. The U_o condition was set by writing the Neumann condition along B3 as

$$\psi_{i,J} = \psi_{i,J-1} + U_o \cdot \Delta y \tag{3-460}$$

If the boundary line B3 is taken *at* j = J, equation (3-460) is only first-order accurate. But by simply taking B3 to lie between j = J and j = J-1, which we denote by (i,J - 1/2), equation (3-460) becomes "second-order" accurate.** This point is frequently misunderstood. The real question is not "what is the truncation error of equation (3-460) in setting $u = U_o$ along B3?" Rather, the question is "how good is the approximation $u = U_o$ along B3 to the "free flight" case which we wish to model?" The condition v = 0 along B3 implies $\partial\psi/\partial x$ along B3. If B3 is at (i,J), the combination with equation (3-460) just gives

$$\psi(B3) = \psi_{i,J} = \psi_{1,J} \tag{3-461}$$

or, if B3 is at (i,J-1/2), it gives

$$\psi_{i,J} = \psi_{1,J-1} + U_o \cdot \Delta y \tag{3-462}$$

In these equations, $\psi_{1,J}$ or $\psi_{1,J-1}$ are values at the top of the *inflow* boundary, B4 in Figure 3-22. If the inflow ψ is completely specified, this treatment is equivalent to the "frictionless wind-tunnel wall" at B3. But if ψ at inflow is not specified but develops as part of the computation, as in Thoman and Szewczyk (1966)(see next section, III-C-6), then this treatment is less restrictive. The mass flux through the "wind tunnel test section"

---

*Continuing the wind tunnel analogy, we note that the physical wall moving at constant speed is actually achieved by a moving belt on rollers, and has been used to simulate the ground-plane effect in aerodynamic wind tunnel testing of automobiles.

**Another programming device is to set B3 at j = J-1, and define the "fictitious" points at j = J so as to obtain U_o along B3. That is, $\psi_{i,J} = \psi_{i,J-2} + 2U_o\Delta y$.

is not pre-determined. Although the "lid" B3 is still a streamline with constant ψ, the value of that ψ is not specified, but is free to develop.

Roache and Mueller (1970) modeled the wind tunnel wall by fixing ζ at inflow, thus making B3 a streamline, but modeled the frictionless "lid" condition on ζ in a slightly less restrictive way. We want the lid to be "frictionless" in the sense of allowing slip, but we want friction to act in the fluid near the lid. Note that, when v = 0 on the lid, $\partial v/\partial x = 0$ on the lid and $\zeta = \partial u/\partial y$. Thus, the condition $\zeta = 0$ specifies $\partial u/\partial y = 0$. We go to the next step in "freeing" the upper boundary, and set

$$\zeta(B3) = \zeta_{i,J} = \zeta_{i,J-1} \tag{3-463}$$

For an interpretation of this condition in terms of velocities, we note that equation (3-461) implies $\partial v/\partial x|_J = 0$. If B3 is far away from B2 so that u_{B3} varies at most linearly in x [i.e., if $\partial^2 u/\partial x^2|_J = 0 + O(\Delta y)$] then it may be shown that $\partial v/\partial x|_{J-1} = 0 + O(\Delta y^2)$, and that equation (3-463) is approximately equivalent to a linear extrapolation of the u-velocity component up to the "lid."*

Higher order extrapolations for ζ cause catastrophic instabilities or drifting of the solution to develop. Compared with the "moving wind tunnel wall" approach of equation (3-458) or (3-459), the latter treatments may better simulate the "free flight" condition, although this would be difficult to substantiate in general. But the "moving wall" approach does have the virtue of correctly modeling *some* physical problem. Except for truncation error, the only question then remaining is how well this physical problem approximates the one of interest, i.e., the "free flight" case. The latter methods are less restrictive, however.

There are other methods of modeling the free-flight case which actually allow for inflow through the upper boundary B3. For fairly high Re flow, one might use an analytical potential flow solution to fix ψ along B3. Thom (1933) used graphical solutions of the potential flow over a cylinder to construct his boundary conditions.** For a low Re flow, one could use a Stokes solution to evaluate ψ and ζ along B3. At higher Re, an Oseen solution may be appropriate. But it is preferable to use the gradient conditions from these solutions; the gradient conditions are not as restrictive, and the errors are not so persistent (Cheng, 1970). These methods do not seem promising to model a problem, like the backstep, in which the flow perturbation from rectilinear flow is dominated by the separated flow region and which has no convenient solution in either Reynolds-number limit.

Briley (1970) computed the boundary-layer separation bubble on a flat plate by specifying the u-velocity at B3 according to the Howarth linearly retarded flow,

$$u = U_o(x) = a + bx \tag{3-464}$$

up to some x_1 such that x(separation) $< x_1 < -a/b$, and u = constant beyond x_1. Along with the condition $\zeta = 0$ along B3, this specification causes the boundary layer along B1 to separate and to subsequently reattach. It allows inflow through boundary B3, and is computationally stable. It has the further virtue that an exact non-similar boundary layer solution is available for comparison for equation (3-464) up to separation.

Pao and Daugherty (1969) used the conditions $\partial\zeta/\partial y = 0$ and $\partial^2\psi/\partial y^2 = 0$ along B3. The sufficiency of this condition depends on the outflow and inflow boundaries, as will be dis- in Section III-C-7.

---

*In Thoman and Szewczyk's (1966) solutions of flow over a circular cylinder, the slightly more restrictive lid condition $\zeta = 0$ was necessitated only when their cylinder was spinning. When it was not, the specification of $\partial u^2/\partial^2 y \simeq 0$ was also successful in their work. (Private communication.)

**Graphical or numerical solutions of the potential flow are probably preferable to the simple analytical solution for flow over a cylinder, which is not precisely compatible with the finite-difference solutions.

In some meteorological model problems, the effect of wind blowing over a liquid surface is represented by a non-deformable surface shape, with $\psi(B3)$ = constant and with an applied constant *wind stress*, $\zeta(B3)$ = specified constant (Festa, 1970).

III-C-6. Upstream Boundary

The upstream boundary B4 in Figure 3-22 cannot have a unique solution because its characteristics will change depending on the physical flow upstream of the mesh and upon the flow solution in the mesh. All authors previous to Thoman and Szewczyk (1966) had completely specified the inflow boundary conditions. For example, in his study of sudden expansion in a channel flow, Kawaguti (1965) used a solution for fully developed Poiseuille flow to fix both ψ and ζ. Thom (1933) used the potential flow solution upstream of a cylinder. Brennen (1968) applied a potential flow solution for the gradient of ψ at inflow, rather than the value of ψ. This is the less restrictive and preferred method. Fromm (1963, 1967), Harlow and Fromm (1963), and Katsanis (1967) specified uniform inflow $u(1,j)$ and set $v(1,j) = 0$. Pao and Daugherty (1969) specified $\zeta = 0$ and $\partial\psi/\partial y = U_o$, fixing ψ. Greenspan (1969B) fixed ψ and assumed $\partial v/\partial x = 0$, giving $\zeta = \partial^2\psi/\partial y^2$. It is not clear that one should completely specify the inflow, lest the elliptic nature of the equations be restricted, yet *something* must be specified. Even von Neumann (Charney et al., 1950) could give only heuristic arguments to assert that the specification of ζ at inflow boundaries is sufficient.

Thoman and Szewczyk (1966) used a less restrictive input for the mesh boundary well ahead of the cylinder. They required $v(1,j) = 0$, which gives the condition

$$\psi_{1,j} = \psi_{2,j} \tag{3-465}$$

This allows $u(1,j)$ to develop as part of the computation. With their upper boundary treatment, $u = U_o$ is specified at B3, but $\psi(B3)$ also develops. With equation (3-465), the upstream effect is felt even at the inflow.

In the study of problems like the backstep flow of Figure 3-22, viscous effects are important at the input and it is desirable to fix $u_{1,j}$ and to let $v_{1,j}$ develop freely. In the study of Roache and Mueller (1970), $\psi_{1,j}$ was fixed by a boundary-layer solution, thus fixing $\partial\psi/\partial y|_{1,j} = u_{1,j}$. This also fixed $\partial^2\psi/\partial y^2|_{1,j} = \partial u/\partial y|_{1,j}$, which is the first of the two terms in $\zeta_{1,j} = \partial u/\partial y|_{1,j} - \partial v/\partial x|_{1,j}$. The second term could also have been fixed from a boundary-layer solution, but instead, less restrictive evaluations were attempted. The most successful method found was to approximate

$$\frac{\partial}{\partial x}\left(\frac{\partial v}{\partial x}\right)_{1,j} = 0 \tag{3-466}$$

by

$$\left.\frac{\partial v}{\partial x}\right|_{1,j} = \left.\frac{\partial v}{\partial x}\right|_{2,j} = \frac{\partial}{\partial x}\left(-\frac{\partial\psi}{\partial x}\right)_{2,j} = -\left.\frac{\partial^2\psi}{\partial x^2}\right|_{2,j} \tag{3-467}$$

or finally

$$\left.\frac{\partial v}{\partial x}\right|_{1,j} = -\frac{\psi_{1,j} + \psi_{3,j} - 2\psi_{2,j}}{\Delta x^2} \tag{3-468}$$

Thus, the input profile fixes ψ and $\partial u/\partial y$, while $\zeta = \partial u/\partial y - \partial v/\partial x$ is set from the input $\partial u/\partial x$ and equation (3-468).

Even less restrictive methods were attempted; these included a linear extrapolation of $\partial v/\partial x|_{1,j}$ back upstream from interior points, and extrapolation by Adams-Bashforth weighting. At high Re, these were acceptable but of no consequence; at lower Re, they caused the solution to wander.

One convenient form of u(y) at the inflow boundary, used by Roache and Mueller (1970), is the one-parameter family of profiles due to Pohlhausen (see Schlichting, 1968). This fourth-order polynomial profile results from an integral solution of the boundary-layer equations. The Pohlhausen parameter Λ is a dimensionless pressure gradient, and $\eta = y/\delta$ where δ is the boundary layer thickness. The u-profile is given by

$$u(B4) = (2\eta - 2\eta^3 + \eta^4) + \frac{1}{6}\Lambda(\eta - 3\eta^2 + 3\eta^3 - \eta^4) \tag{3-469}$$

A Blasius profile or Falkner-Span profiles (Schlichting, 1968) may also be used.

Two methods of specifying the discretized input are possible. We may choose to match discrete values of ψ to the input continuum solution. These can be obtained by accurately integrating equation (3-469) with (say) a Simpson's rule integration to get

$$\psi_{1,j} = \int_0^{y_j} u(B4)\, dy \tag{3-470}$$

However, when we now numerically differentiate $\psi_{1,j}$ to get $u_{1,j}$, the discretized u(1,j) are not the input continuum solution for u(B3). The alternative method is to discretize the u(B3) solution and then evaluate $\psi_{1,j}$ in a manner compatible with the velocity differencing, as in

$$u_{1,j} = \frac{\psi_{1,j+1} - \psi_{1,j-1}}{2\Delta y} \tag{3-471}$$

This method would give

$$\psi_{1,jc} = 0 \tag{3-472A}$$

$$\psi_{1,jc+1} = \frac{1}{2} u_{1,jc+1} \cdot \Delta y \tag{3-472B}$$

$$\psi_{1,j} = u_{1,j-1} \cdot 2\Delta y + \psi_{1,j-2} \tag{3-472C}$$

This second method gives the correct velocities by equation (3-471), but results in a discretization error for $\psi_{1,j}$. Of course, both methods converge for $\Delta y \to 0$. Since the dynamics are affected by u rather than ψ, the second method of matching the desired u and accepting errors in ψ seems preferable.

Note that this discretization error in ψ will cause an error in the calculation of the displacement thickness δ^* (Schlichting, 1968) if it is evaluated as

$$\delta^* = \lim_{y\to\infty} \left[y - \frac{\psi(y)}{u(y)} \right] \tag{3-473}$$

which is usually applied in an approximate manner at some vaguely defined "edge" of the boundary layer, denoted by "e".

$$\delta^* \cong y_e - \frac{\psi_e}{u_e} \tag{3-474}$$

To achieve results compatible with the discretized inflow, δ^* should be evaluated by numerical quadrature (integration of a known function) using the trapezoidal rule. Since u is the normalized velocity, $u = \bar{u}/\bar{U}_o$, this gives

$$\delta^* = \int_0^{y_e} (1 - u)dy \tag{3-475}$$

It is not surprising that close agreement between numerical and experimental inflow boundary conditions is often necessary for overall solution agreement (Mueller and O'Leary, 1970; Fanning and Mueller, 1971; and Shavit and Lavan, 1971).

III-C-7. Outflow Boundary

The evaluation of ψ and ζ at the outflow boundary B6 of Figure 3-22 is one of the most interesting computational boundary problems. We must somehow neglect the details of farther downstream flow and obtain realistic answers upstream. We again fall back on wind tunnel experience; if the "test section" is long enough, the downstream flow is not important. But numerical experience shows that catastrophic instabilities may be propagated upstream from the outflow boundary and destroy the solution. Our aim is to allow the most free-flow adjustment at B6 which still gives a solution.

The safest method from the view of stability is to completely specify the outflow conditions. Thom (1933) used a potential flow solution, as did Allen and Southwell (1955), Michael (1966), Son and Hanratty (1969), and Hamielec and Raal (1969). Katsanis (1967) specified uniform outflow (and inflow) with u = constant and $v = 0$. These are clearly not suitable for separated flows or any flows with a viscous wake. One might also use a Stokes solution for low-Re flows, or an Oseen solution for the far wake, as did Varapáev (1968).

Richardsonn (1910) suggested the general concept of using infinity boundary conditions at the extremes of the mesh. Kawaguti (1965), Friedman (1970), and Lee and Fung (1970) used the Poiseuille flow solution at outflow, for example. Note that the continuum asymptotic solution should be used *in the variables of the problem*, e.g. the Poiseuille solution for ψ and ζ should be used if the FDE are cast in these variables. If the continuum solution for u is used and then ψ is found by quadrature, a discretization error results (see previous page for the same problem at inflow). For more general flows, e.g. asymptotic boundary layer flows, the continuum solution will differ from the asymptotic FDE solution in *all* variables. An FDE solution to the asymptotic *ODE* will then be preferable at outflow (Kawaguti, 1965).

Instead of using the "infinity" downstream condition, one can use asymptotic solutions applicable at large but finite distances from the region of interest. Plotkin (1968) and Yoshizawa (1970) applied a boundary-layer solution for function values at the end of the computational mesh. Varzhanskaia (1969, 1970) changed over to the boundary-layer equations at a large distance from the region of interest (the flat-plate leading-edge region), and recommends matching the boundary-layer equations by specifying continuity of ψ, $v = -\partial\psi/\partial x$, and $\partial v/\partial x = -\partial^2\psi/\partial x^2$. This technique allows further downstream marching with the boundary-layer equations, but does not solve the outflow vorticity problem for the Navier-Stokes equations used upstream.

Fromm (1963, 1964) and Gawain and Pritchett (1970) used periodic inflow and outflow boundary conditions, which have the sole virtue of being mathematically explicit. They represent no meaningful physical situation, with the exception of the free homogeneous turbulence problem, and were part of the instability problem which Fromm encountered at high Re.

Apparently, the first really successful computational outflow condition of the less restrictive type was used by Paris and Whitaker (1965). In a two-dimensional channel flow, they specified that $v = -\partial\psi/\partial x = 0$ and $\partial\zeta/\partial x = 0$ at outflow. They found this less restrictive condition to be an improvement over the parabolic flow specification, as expected, allowing a considerably shorter computational mesh for comparable accuracy in the region of interest. With the present notation of max $i = I$, and dropping the j-index, their conditions are

$$\psi_I = \psi_{I-1} \tag{3-476A}$$

$$\zeta_I = \zeta_{I-1} \tag{3-476B}$$

These conditions were later used by Pao and Daugherty (1969) on a flat-plate problem. Computational experiments by the present author on the one-dimensional inviscid model transport equation, using the three-time level midpoint leapfrog method of Section III-A-6, showed that (3-476B) is destabilizing when used with this method. More restrictive conditions of

$$\psi_I = \psi_{I-1} \tag{3-477A}$$

$$\zeta_I = 0 \tag{3-477B}$$

were applied in the wake of an object by Rimon and Cheng (1969), who used midpoint leapfrog differencing with a DuFort-Frankel (Section III-A-7) treatment of diffusion terms. The computational experiments of Allen (1968) and Cheng (1970) on Burgers' equation confirm the intuitive supposition that gradient conditions on ζ as in equation (3-476B) will produce smaller boundary errors than function specifications on ζ as in (3-477B).*

Thoman and Szewczyk (1966) developed even less restrictive outflow boundary conditions. They approximated $\partial\zeta/\partial x = 0$ and $\partial v/\partial x = 0$. Since $v = -\partial\psi/\partial x$, this second condition implies $\partial^2\psi/\partial x^2 = 0$, which was approximated by linear extrapolation out to $i = I$. For constant Δx, this gives

$$\zeta_I = \zeta_{I-1} \tag{3-478A}$$

$$\psi_I = 2\psi_{I-1} - \psi_{I-2} \tag{3-478B}$$

To second-order accuracy, these really set $\partial\zeta/\partial x = 0$ at $(I - 1/2)$ and $\partial^2\psi/\partial x^2 = 0$ at $(I - 1)$. The extrapolation for $\psi_{I,j}$ was performed after each Liebman iteration for the Poisson equation. Fromm (1967) later used equation (3-478B) and significantly raised his limiting computational Re from previous work (Fromm, 1963) using periodic boundary conditions. Fromm also tried linear extrapolation for ζ outflow but found it destabilizing, using explicit methods for the vorticity transport equation. The present author has had the same experience, using explicit methods.

The Thoman and Szewczyk conditions (3-478) were later used successfully by Roache and Mueller (1970) and Campbell and Mueller (1968) in a variety of problems, using SOR iteration for the Poisson equation.

Hung and Macagno (1966) and Macagno and Hung (1967) used the following expressions at outflow:

$$\psi_I = \psi_{I-4} - 2\psi_{I-3} + 2\psi_{I-1} \tag{3-479A}$$

$$\zeta_I = \zeta_{I-4} - 2\zeta_{I-3} + 2\zeta_{I-1} \tag{3-479B}$$

---

*See also the article "Viscous Fluid Flow in Initial Segment of Two-Dimensional Channel with Porous Walls," by V. N. Varapaev, Izv. AN SSSR, Mekanhika Zhidkosti i Gaza (Fluid Dynamics), Vol. 4, No. 4, pp. 178-181, 1969.

These are obtainable from a second-order Taylor-series expansion about (I - 2) to I, using a $(2\Delta x)$ increment. This method has been used up to a base height Re = 333/2 with success. Hung and Macagno (1966) also experimented with a simple application of a parabolic profile at outflow, similar to Kawaguti's method (1965) already mentioned. They found it reasonably satisfactory, although equation (3-479) is certainly an improvement. Giaquinta and Hung (1968) used equation (3-479) in non-Newtonian (Reiner-Rivlin) flow.

Greenspan (1969A) used methods formulated by R. E. Meyer, approximating the following conditions at outflow:

$$\frac{\partial \psi}{\partial x} = 0 \tag{3-480A}$$

$$\frac{\partial \zeta}{\partial x} + u \times Re\left(\zeta + \frac{\partial u}{\partial y}\right) = 0 \tag{3-480B}$$

The first equation gives v = 0; with this and the assumption of a steady state, the second equation gives $\partial P/\partial y = 0$, approximating the usual boundary-layer condition.*

The extrapolation of equation (3-478B) for ψ_I at outflow can be erroneous. This extrapolation must be done at each iteration of the Poisson equation. Roache (1970) considered the sufficiency of this condition to determine a solution. Consider inflow conditions at i = 1 fixed. For a one-dimensional ODE problem, $d^2\psi/dx^2 = \zeta$, the extrapolation condition which sets $d^2\psi/dx^2 = 0$ at outflow either contradicts the ODE if $\zeta \neq 0$ at outflow, or simply restates the ODE if $\zeta = 0$. That is, it is no boundary condition at all, in one dimension. But, in the two-dimensional problem, the condition $\partial^2\psi/\partial x^2 = 0$ reduces the Poisson equation to the ODE

$$\frac{d^2\psi}{dy^2} = \zeta \tag{3-481}$$

If the conditions at B1 and B3 in Figure 3-22 are such as to provide equation (3-481) along B6 with sufficient boundary conditions, then the extrapolation procedure may be sufficient. Particularly, if $\psi = 0$ along B1 and either ψ or $\partial\psi/\partial y$ is fixed along B3 (as considered earlier) then extrapolation along B6 is sufficient. But if the condition at B3 is also set by extrapolation, then extrapolation along B6 is *not* sufficient. Again, the setting of $\partial^2\psi/\partial y^2 = 0$ at B3 either contradicts (3-481) if $\zeta(B3) \neq 0$, or simply restates (3-481) if $\zeta(B3) = 0$. The sufficiency of the downstream condition $\partial^2\psi/\partial x^2 = 0$ is thus seen to depend on the conditions used at adjacent boundaries, indicating the significance of *dimensionality* to the problem.

---

*The normalized form of the v-momentum equation (2-2) is

$$\frac{\partial v}{\partial t} + u\frac{\partial v}{\partial x} + v\frac{\partial v}{\partial y} = -\frac{\partial P}{\partial y} + \frac{1}{Re}\left(\frac{\partial^2 v}{\partial x^2} + \frac{\partial^2 v}{\partial y^2}\right)$$

For steady flow with v(x,y) = 0 for all y and $\partial P/\partial y = 0$, this reduces to

$$Re\cdot u\frac{\partial v}{\partial x} = \frac{\partial^2 v}{\partial x^2}$$

From $\zeta = \partial v/\partial x - \partial u/\partial y$, we have $\partial v/\partial x = \zeta + \partial u/\partial y$, $\partial^2 v/\partial x^2 = \partial\zeta/\partial x + \partial/\partial y(\partial u/\partial x)$. From continuity, $\partial u/\partial x = -\partial v/\partial y$, so $\partial/\partial y(\partial u/\partial x) = -\partial^2 v/\partial y^2 = 0$. So $Re\cdot u(\zeta + \partial u/\partial y) = \partial\zeta/\partial x$. Although not detailed in Greenspan (1969A), the finite-difference form could be

$$\zeta_I = u_{I-1}\left(Re\ \zeta_{I-1} + \left.\frac{\delta u}{\delta y}\right|_{I-1}\right)\cdot 2\Delta x + \zeta_{I-2}$$

Note that the SOR iteration problem, using extrapolation at both B3 and B6, might "converge" to within some specified tolerance, or that discretization could conceivably make the solution unique, i.e., independent of the initial estimate. But the unique solution so obtained will depend on Δx and Δy and, as Δx and $\Delta y \to 0$, the problem would become indeterminate.

The above arguments also suggest an efficient method of implementing the condition $\delta^2\psi/\delta x^2 = 0$ near the downstream boundary (Roache, 1970; Briley, 1970). Instead of applying it at $i = I - 1$ by linear extrapolation, it may be applied at $i = I$ directly, reducing equation (3-481) to the discretized ordinary differential equation.

$$\frac{\delta^2\psi}{\delta y^2} = \zeta \tag{3-482}$$

with the two-point boundary conditions of $\psi(I,1) = 0$ at B1 and either Dirichlet or the Neumann conditions at B2. The value of $\zeta(I,j)$ at B6 can be determined by any of the several methods already discussed. This ordinary difference equation can then be quickly solved non-iteratively by the tridiagonal matrix algorithm (see Appendix A). With the downstream boundary values of ψ so determined, one can proceed confidently with the solution of the partial difference Poisson equation, with Dirichlet conditions at B6. Convergence is also accelerated (unpublished work by the present author). In fact, this device of using $\partial^2\psi/\partial x^2 = 0$ to reduce the Poisson equation to an ODE is absolutely necessary if ADI methods are used for the Poisson equation, since the ADI splitting effectively reduces the half-step iteration to the tridiagonal solution of an ODE in the x-direction, for which the condition $\partial^2\psi/\partial x^2 = 0$ is *not* sufficient.

Because of the special form of the ODE (3-482), the simple one-dimensional method described in Section III-B-8 may be used instead of the more general tridiagonal algorithm (see Appendix A).

As previously mentioned, Fromm (1967) and the present author tried extrapolation for both ψ and ζ, and found it destabilizing, using explicit methods for the vorticity transport equation. Using ADI methods, Briley (1970) and Fanning and Mueller (1971) were able to achieve stability. With an additional "fictitious" node point at $i = I + 1$ for ζ only, Briley set

$$\zeta_{I+1} = 2\zeta_I - \zeta_{I-1} \tag{3-483}$$

after each computation for $\partial\zeta/\partial t$ up to $i = I$. The condition of $\partial^2\psi/\partial x^2 = 0$ was set by solving the ODE $d^2\psi/dy^2 = \zeta$ at $i = I$, as discussed above.

Even the relatively unrestrictive method of Thoman and Szewzyk (1966), equation (3-478), may give an unrealistically abrupt variation of ζ near B6 for low Re flows, say Re = 0(10). For such low Re flows, Roache and Mueller (1970) used a minimally restrictive method to determine ζ at outflow from the vorticity transport equation. Assuming $u_{I,j} \geq 0$ (i.e., B6 really is an *outflow* boundary), the u-advection term can be evaluated by upwind differencing at $i = I$, with no further approximation. The v-advection term can be evaluated by upwind differencing (depending on the sign of $v_{I,j}$) or by whatever method is being used at internal points; similarly, the y-diffusion term at $i = I$ requires no approximation. The x-diffusion term could be evaluated at $i = I - 1$. But, by itself, this evaluation is statically destabilizing (see section III-A-4) in the term most important in low-Re flows, $(\partial^2\zeta/\partial x^2)/\text{Re}$. This is easily seen by referring back to Figure 3-6 on page 34; the corrective displacement due to the $\partial^2\zeta/\partial x^2$ term for a point at $i = I + 1$ is actually applied at point i = I, thus causing monotonic error growth or a static instability.*

The discrete perturbation stability analysis also suggests the solution to this problem. Stability of the $\partial^2\zeta/\partial x^2$ term is achieved by an additional phase shift, this one in time. The result is

$$\zeta_{I,j}^{n+1} = \zeta_{I,j}^{n} + \Delta t \left\{ - \frac{(u\zeta)_{I,j}^{n} - (u\zeta)_{I-1,j}^{n}}{\Delta x} - \left.\frac{\delta(u\zeta)^n}{\delta y}\right|_{I,j} + \frac{1}{Re}\left.\frac{\delta^2\zeta^n}{\delta y^2}\right|_{I-1,j} + \frac{1}{Re}\left.\frac{\delta^2\zeta^{n-1}}{\delta x^2}\right|_{I-1,j} \right\} \tag{3-484}$$

The appearance of the additional time level (n-1) in equation (3-484) does not require that another complete matrix of $\zeta_{i,j}^{n-1}$ be stored in memory. Since the equation is applied only at the boundary, only the vector $V(j) = \delta^2\zeta/\delta x^2\big|_{I-1,j}^{n-1}$ need be stored. This derivative is calculated at the end of each new vorticity calculation, before updating as $V(j) = \delta^2\zeta/\delta x^2\big|_{I-1,j}^{n-1}$. Updating then indexes the time level down and the vector becomes V(j) at (n-1).

This method has been used successfully by Roache and Mueller (1970), and by O'Leary and Mueller (1969), using upwind differencing at all interior points. Regardless of what method is being used for the internal point calculations, it is recommended that at outflow, one switch to upwind differencing at least for the u-advection term. In the limit Re $\to \infty$, this means that *no* boundary condition on ζ is necessary at outflow; this agrees with the one-dimensional continuum equation $\partial\zeta/\partial t = -(u\zeta)/\partial x$, which needs only inflow conditions to specify the problem. If, for example, the midpoint leapfrog method of Section III-A-6 is being used for the advection terms with forward-time, centered-space differencing for the diffusion terms as in equation (3-165), one can use the following equation at i = I.

$$\zeta_{I,j}^{n+1} = \zeta_{I,j}^{n-1} + 2\Delta t \left\{ - \frac{(u\zeta)_{I,j}^{n-1} - (u\zeta)_{I-1,j}^{n-1}}{\Delta x} - \frac{\delta(u\zeta)^n}{\delta y} + \frac{1}{Re}\left.\frac{\delta^2\zeta^n}{\delta y^2}\right|_{I,j} + \frac{1}{Re}\left.\frac{\delta^2\zeta^{n-1}}{\delta x^2}\right|_{I-1,j} \right\} \tag{3-485}$$

These methods have been tested by the present author and were successful in freeing the downstream boundary and giving a smooth variation in ζ, as long as no vortex shedding occurred. When vortex shedding occurs, $u < 0$ may occur at i=I, so the above forms become *downwind* differencing forms, with resultant instability. In that case, the method of Thoman and Szewczyk (1966), equation (3-478), was again required for stability. Since vortex shedding occurs only at higher Re, the expected improvements of this upwind method would be insignificant in this case.

The stability of the total method may now be limited by the Courant-number restrictions of the upwind differencing at B6. For example, note that, in combination with the midpoint

---

*Following the discrete perturbation methods of Section III-A-5-a and neglecting advection and y-diffusion terms, we have $\zeta_{I,j}^{n+1} + \zeta_{I,j}^{n} + \Delta t/Re\left[\zeta_{I-2}^{n} + \zeta_{I,j}^{n} - 2\zeta_{I-1,j}^{n}\right]/\Delta x^2$. If a steady solution is assumed and a perturbation δ in $\zeta_{I,j}$ is introduced, we get $\left[\zeta_{I,j}^{n+1} = \zeta_{I,j}^{n}\right]/\delta = \Delta t/\left[Re\Delta x^2\right]$, which is positive definite. Thus, $\Delta\zeta$ is in the same direction as δ, and this viscous term is statically destabilizing.

leapfrog method as in equation (3-485), the effective time step for the u-advection term is $2 \cdot \Delta t$. The one-dimensional stability criterion would be $C_x = u\Delta t/\Delta x \leq 1/2$. However, the upwind differencing form at B6 apparently can be used in an ADI method with no stability restriction, as in Pao and Duagherty (1969). Other combinations would have to be tested individually.

The sufficiency of this outflow boundary condition has not been rigourously established, even for the linear PDE, but some inferences can be drawn from the related 1D steady-state model problem,

$$- u \frac{\partial \zeta}{\partial x} + \alpha \frac{\partial^2 \zeta}{\partial x^2} = 0 \tag{3-486}$$

The upwind differencing of the advection term imposes no outflow boundary condition, and the x-diffusion treatment of equation (3-484) gives, as $\Delta x \to 0$,

$$\frac{\partial^3 \zeta}{\partial x^3} = 0 \tag{3-487}$$

For $u > 0$ and $\alpha > 0$, this boundary condition, in conjunction with a fixed value ζ at inflow, is sufficient to specify the solution (see Problem 3-29). If $\alpha = 0$ (that is, $Re = \infty$) this outflow condition *as applied in a continuum solution* overspecifies the problem and causes the singular behavior of the solution as $\alpha \to 0$. (See Section III-C-8). However, in the computational scheme, it is applied only to the diffusion term and there is no such singular behavior as $\alpha \to 0$. There is no counterpart in an analytical treatment of even the linear ODE model equation. In the other limit of $u = 0$ ($Re = 0$) equation (3-487) becomes no boundary condition at all (see Problem 3-29). It would then appear that the outflow condition (3-484), which is appropriate to low-Re flows, is still limited to $Re > 0$ and may possibly be limited to some *minimum* cell Reynolds number.

The idea of using upwind differencing at B6 was also used by Fromm (1967), and the evaluation of x-diffusion terms at (I-1) was used by Eaton and Zumwalt (1967) in a non-steady supersonic problem, without the time shift of the present method. Note that the frequently used artifice of extrapolating $(u\zeta)$ to a fictitious point outside the mesh, say $(u\zeta)_{I+1,j}$, and then evaluating $\partial(u\zeta)/\partial x|_{I,j} = [(u\zeta)_{I+1,j} - (u\zeta)_{I-1,j}]/2\Delta x$ by centered differences, only obscures the problem. It is algebraically equivalent to using the one-sided difference form $\partial(u\zeta)/\partial x|_{I,j} = [(u\zeta)_{I,j} - (u\zeta)_{I-1,j}]/\Delta x$. It certainly does not result in second-order accuracy, as some authors have claimed, and will be unstable when vortex shedding occurs through B6.

A few systematic experiments have been run to test the validity of the computational outflow boundary conditions in multidimensional problems. Briley (1970) ran one calculation and then repeated another with B6 moved an additional five nodes downstream. Using equations (3-481) and (3-483), he found changes in wall vorticity of less than 0.2% ahead of the first boundary. Cheng (1970) has run parallel calculations in compressible flow, and found that the less restrictive methods are generally more accurate, as expected. It should also be noted that, even if higher order extrapolations on ψ and ζ can be made stable, they cannot be expected generally to produce more accurate results, since they are based on computed values at internal points and not on exact values (Cheng, 1968, 1970). Further multi-dimensional parallel computations would be highly desirable. Not only different outflow conditions should be tested, but different combinations with different upper-boundary treatments.

For inviscid flows, Shapiro and O'Brien (1970)(see also Charney, 1962) have used an excellent method which is accurate, stable, and fairly easy to program. The method involves simply following the Lagrangian trajectory of a particle to the outflow boundary, using linear extrapolation. With no diffusion present, the value of vorticity is fixed to a particle; from Figure 3-25a, the outflow value $\zeta^{n+1}_{I,j}$ is given by

$$\zeta^{n+1}_{I,j} = \zeta^{*} \doteq \zeta(X_I - \tilde{u}\Delta t,\ y_j - \tilde{v}\Delta t) \tag{3-488}$$

(See derivation of Leith's method, section III-A-13.) The average velocities $\tilde{u}$ and $\tilde{v}$ are estimated from neighboring points (say by interpolation, or by $\tilde{u} = u^n_{I-1,j}$). The value of

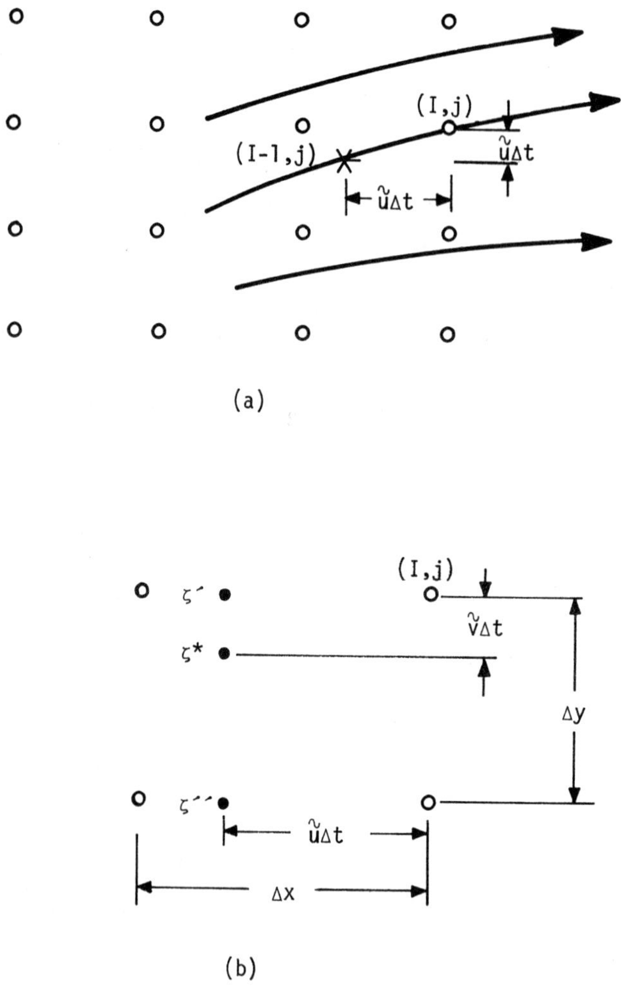

Figure 3-25. The method of Shapiro and O'Brien for outflow vorticity.

$\zeta^n(X^*, y^*)$ is determined by interpolation between the neighboring points. As an example, consider $\tilde{u} > 0$ and $\tilde{v} > 0$, as in Figure 3-25b. We have

$$\zeta' = \zeta^n_{I,j} + \frac{\tilde{u}\Delta t}{\Delta x}(\zeta^n_{I-1,j} - \zeta^n_{I,j}) \tag{3-489A}$$

$$\zeta'' = \zeta^n_{I,j-1} + \frac{\tilde{u}\Delta t}{\Delta x}(\zeta^n_{I-1,j-1} - \zeta^n_{I,j-1}) \tag{3-489B}$$

$$\zeta^{*} = \zeta' + \frac{\tilde{v}\Delta t}{\Delta x}(\zeta'' - \zeta') \tag{3-489C}$$

If $\tilde{u}$ and $\tilde{v}$ are also found by interpolation, the arithmetic can get messy; the procedure is probably best solved by iteration in that case.

This method would also be appropriate to high-Re flows, with the only approximation being that diffusion is neglected over the last two node columns.

Shapiro and O'Brien (1970) compared the results of this method in a 2D meteorological calculation with results obtained using fixed (Dirichlet) ζ values at outflow, obtained from a larger field calculation. Although one might expect the latter method to be more accurate, in fact it was not, and it caused oscillations or "wiggles" in the late-time ζ calculations. (See also Varapaev, op cit.) These "wiggles" are a common occurrence which are explained in the next section.

III-C-8. "Wiggles"

"Wiggles", or spatial oscillations in a flow solution, have been encountered in many works. In supersonic flow, they are usually associated with post-shock oscillations of methods using centered-space derivatives (see Section V-C). But wiggles also arise in long-term incompressible flow calculations. Many authors have associated the behavior with nonlinearities, or with linear instabilities in the transient calculation. (Wiggles can indeed prevent iteration convergence.) We now demonstrate that the actual source of the phenomenon is more elementary.

Figure 3-26 shows the *exact* solutions, obtained by a non-iterative or direct tridiagonal solution (see Appendix A) of the FDE corresponding to the steady-state *linear*, constant coefficient model advection-diffusion equation

$$0 = - u\partial\zeta/\partial x + \alpha\partial^2\zeta/\partial x^2 \qquad \text{(3-490A)}$$

with boundary conditions of

$$\zeta(0) = 0 \quad , \quad \zeta(1) = 1 \qquad \text{(3-490B)}$$

The corresponding centered-difference FDE is

$$0 = - u\frac{\zeta_{i+1} - \zeta_{i-1}}{2\Delta x} + \alpha\frac{\zeta_{i+1} - 2\zeta_i + \zeta_{i-1}}{\Delta x^2} \qquad \text{(3-491A)}$$

With $\Delta x = 1/10$, the discretized boundary conditions are

$$\zeta_1 = 0 \quad , \quad \zeta_{11} = 1 \qquad \text{(3-491B)}$$

The solution of Figure 3-26a was obtained with $\alpha/u = 1$, corresponding to Re = 1, and is smooth. The solution of Figure 3-26b was obtained with $\alpha/u = 0.01$, corresponding to Re = 100, and demonstrates wiggles. We emphasize that Figure 3-26 presents the *exact, steady-state* solution of the *linear, constant coefficient* FDE (3-491). The wiggles are not caused by iterative instability, nonlinearties, or spatially varying coefficients; they simply are the solution of the FDE (3-491).

That such wiggles must occur in a FDE solution is readily shown. Consider first the continuum problem (3-490), with solutions as shown in Figure 3-27a. For u = 0 (zero Re or creeping flow), the continuum diffusion equation solution is simply a straight line, $\zeta = x$. For $u > 0$, this solution is blown downstream.* For u/α large (high Re flow), the solution approaches the horizontal line $\zeta = \zeta(0) = 0$ for most of the region, but then shoots up for $x \to 1$ to meet the second boundary condition of (3-490B), that $\zeta(1) = 1$. For $\alpha = 0$ (Re = ∞) the solution is everywhere $\zeta = \zeta(0) = 0$ and the second boundary condition,

---

*The solution is $\zeta = (1 - e^{xu/\alpha})/(1 - e^{u/\alpha})$.

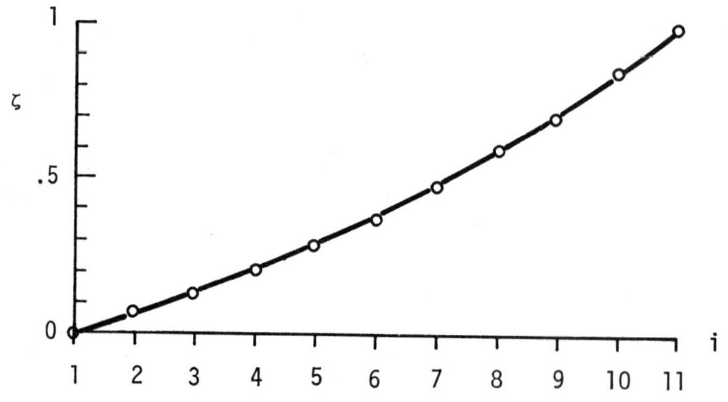

(a) $\alpha/u = 1$, $Re = 1$, $R_c = 0.1$

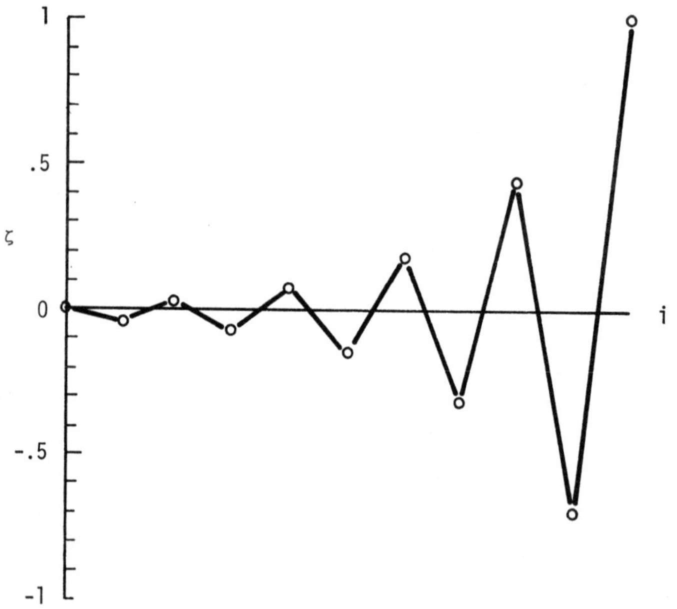

(b) $\alpha/u = 0.01$, $Re = 100$, $R_c = 10$

Figure 3-26. Exact solutions of $0 = -u(\delta\zeta/\delta x) + \alpha(\delta^2\zeta/\delta x^2)$. Centered differences, $\Delta x = 1/10$, $\zeta_1 = 0$, $\zeta_{11} = 1$, constant u.

$\zeta(1) = 1$, cannot be used because it overspecifies the now first-order equation. This change in the order of the differential equation and in the number of admissible boundary conditions is the classic *singular* problem in the small parameter α/u.

Now we consider the FDE solution of equation (3-491). For $u = 0$, the FDE gives the exact continuum solution, as shown in Figure 3-27b. As the wind blows harder, the ζ profile again is blown downstream, as in the continuum solution, everywhere satisfying the conditions from equation (3-491) of

$$u \frac{\delta\zeta}{\delta x} = \alpha \frac{\delta^2\zeta}{\delta x^2} \tag{3-492}$$

But what happens once the "knee" of the profile is blown past the second last node point, at $i = 10$? In the continuum solution, the value of $\partial\zeta/\partial x$ *at* $x = 1$ increases without bound as u/α increases, so as to balance the advection and diffusion terms as in equation (3-492); but, in the FDE solution, the value of $\delta\zeta/\delta x$ is limited. Once the "knee" is blown inside the last cell, we have

$$\left.\frac{\delta\zeta}{\delta x}\right|_{I-1} = \frac{\zeta_I - \zeta_{I-2}}{2\Delta x} = \frac{1-0}{2\Delta x} = \frac{1}{2\Delta x} \tag{3-493}$$

If the value of $\zeta_{I-1} = 0$, we have

$$\left.\frac{\delta^2\zeta}{\delta x^2}\right|_{I-1} = \frac{\zeta_I - 2\zeta_{I-1} + \zeta_{I-2}}{\Delta x^2} = \frac{1}{\Delta x^2} \tag{3-494}$$

Thus the FDE (3-491) will be satisfied at $i = I - 1$ if

$$\frac{u}{\alpha} = \frac{\left.\delta^2\zeta/\delta x^2\right|_{I-1}}{\left.\delta\zeta/\delta x\right|_{I-1}} = \frac{1/\Delta x^2}{1/2\Delta x} = \frac{2}{\Delta x} \tag{3-495}$$

The entire FDE solution is $\zeta_i = 0$ for $i = 1$ to 10, and $\zeta_{11} = 1$. Equation (3-495) again gives the ubiquitous condition on the cell Reynolds number,

$$R_c \equiv u\Delta x/\alpha = 2 \tag{3-496}$$

When this condition is exceeded (with $R_c > 2$), the term $\delta\zeta/\delta x$ is still limited, and in order to achieve the balance of equation (3-492), the term $\delta^2\zeta/\delta x^2$ increases by way of ζ_{I-1} decreasing below zero, as shown in Figure 3-27c. Note that this FDE solution violates the monotonicity and boundedness conditions of the continuum equation, thus constituting a "behavioral error" (see Section III-A-23). The value of $\left.\delta^2\zeta/\delta x^2\right|_{I-2}$ is somewhat decreased when $\zeta_{I-1} < 0$, and the effect feeds forward, causing the wiggle pattern.

The occurrence of the wiggles in the discretized equation is then seen to be analogous to the singularity in the continuum equation; the latter occurs as $Re \to \infty$, and the former occurs when $R_c > 2$. The FDE is "singular" at $R_c = 2$ in the sense that, as the parameter α becomes small enough that $R_c > 2$, the FDE loses the monotonicity and boundedness properties of the continuum solution.

The phenomenon can also arise in regions far from boundaries for the nonlinear Burgers equation. As Re increases, the frequency content of the solution increases until the minimal representation of the solution is the Nyquist frequency, or the $\Lambda = 2\Delta x$ wavelength. This occurs exactly at $R_c = 2$ for the Burgers equation. For higher Re, the structure of the solution cannot even be qualitatively resolved in the mesh. (See page 81 on *aliasing error*.)

There are two ways to eliminate the behavioral error of wiggles in the constant coefficient equation, besides reducing $R_c < 2$ by reducing Δx. These are given in the exercises.

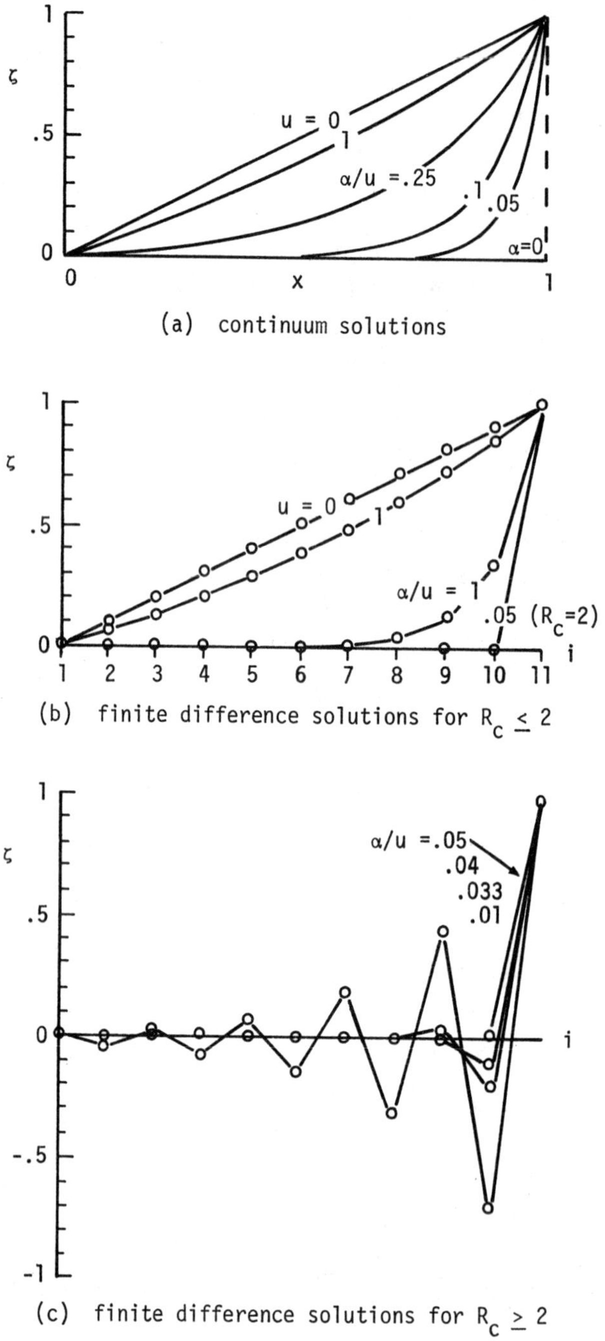

Figure 3-27. Continuum and finite difference solutions to the steady-state linear model advection-diffusion equation: $0 = -u(\partial\zeta/\partial x) + \alpha(\partial^2\zeta/\partial x^2)$, $\zeta(0) = 0$, $\zeta(1) = 1$, constant u. Centered differences, $\Delta x = 1/10$. The cell Reynolds number $R_c = u\Delta x/\alpha$.

Exercise: Show that the upwind differencing method for equation (3-490) does not produce wiggles.

Exercise: Show that the use of the gradient condition $\partial\zeta/\partial x = 0$ at $x = 1$ does not produce wiggles.

In regard to the first remedy (upwind differencing), we note that the condition for wiggles to appear using centered differences ($R_c \geq 2$) is just the condition for the formal accuracy of the upwind-difference method (see Section III-A-8). That is, for $R_c > 2$, the artificial viscosity $\alpha_e > \alpha$, so that, in a sense, this remedy is fictitious.

In regard to the second remedy, we note that we have *fixed* the problem by *changing* the problem, reducing the solution to the trivial value $\zeta(x) = \zeta(0) = 0$. (If $\partial\zeta/\partial x \neq 0$ is used at outflow, the 1D solution is non-trivial, but an R_c limit still exists. See Problem 3-30.) However, the second remedy *is* applicable to outflow problems in 2D and 3D, without reducing the problem to a triviality, and it frequently may be used to eliminate wiggles in multi-dimensional fluid dynamics calculations. Also, the outflow condition of Shapiro and O'Brien (see previous section III-C-7) eliminates wiggles. (For the 1D steady-state problem, the Shapiro-O'Brien method reduces to the gradient condition, $\delta\zeta/\delta x = 0$.

Of course, the effect of spatially varying u and higher dimensions may quantitatively change the behavior just described. Particularly, a variable u may allow $R_c > 2$ away from the boundary without causing wiggles, provided that $R_c < 2$ near the boundary. But the analysis presented here has proved to be surprisingly appropriate for 2D fluid-flow problems. In 2D calculations of multifluid flow in a boundary-layer using the full Navier-Stokes equations, A. J. Russo (private communication) has encountered 1D wiggles in either of the directions parallel to, or transverse to, the wall. In either direction, the wiggles were cured by changing to a Neumann boundary condition or to upwind differencing in that one direction. Another effective means used by Russo is to locally refine the mesh near the wall (see Section VI-A), thus locally reducing $R_c < 2$. Polger (1971) eliminated wiggles near a wall by switching to a diffusive method only at the first node away from the wall. He used Lax's method (Section V-E-4), but upwind differencing (Section III-A-8) would also work. (In the linear 1D problem of Figure 3-26, the use of upwind differencing at i = 10 almost completely eliminates wiggles.)

III-C-9. The Downstream Paradox

It seems at first glance that the free outflow condition at a downstream location is most serious in incompressible flow, since then conditions at outflow affect all other points in the flow. For supersonic flow, it would appear that the downstream condition would be of consequence only to the viscous terms, since the supersonic flow limits the upstream effect. Several authors have stated or implied these points.

We first note that the upstream effect is felt in supersonic flow in the finite-difference method, even without viscous terms present. This is especially true if space-centered differences are used for the advection terms; but even if upwind differencing is used, at least the pressure gradient is felt upstream. This is, in fact, proper and *necessary* if a shock is ever to develop, for example, or if an indraft wind tunnel is ever to be turned off!

But the complete paradox is more emphatic than this; we assert that the downstream outflow problem is *more* important in supersonic flow than in subsonic flow.

Consider two cases of quasi-one-dimensional inviscid flow in a channel, as depicted in Figure 3-28. We consider the simplest case, where the inflow conditions at one are fixed. For the subsonic case (Figure 3-28a) elementary relations show that the outflow at two has a unique solution. For example, if $M \rightarrow 0$, $u_2 = u_1(A_1/A_2)$, and $P_2 = P_1 + 1/2\ \rho(u_1^2 - u_2^2)$, etc. Conceptually at least, all we need do in a computational problem is to allow conditions at two to develop "freely." What we ordinarily think of in regard to upstream effects is the *physical* problem, wherein a change in conditions, even downstream of two, will affect conditions at one. For example, if the back pressure in the duct is raised, the pressure at (1) and (2) will increase and the velocities will decrease. This is precisely the point, however. If P_2 were changed, P_1 and u_1 would change. But, since we *specify* the input flow at (1), all flow properties are everywhere uniquely determinable.

Now the same applies to the case of completely supersonic duct flow (Figure 3-28b).

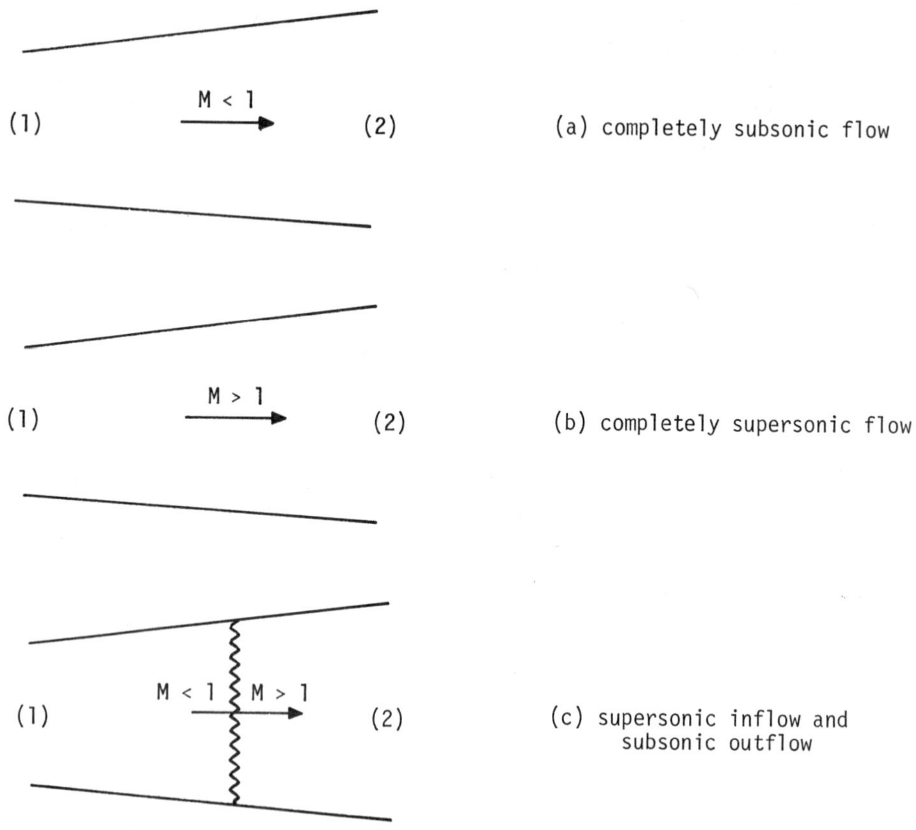

Figure 3-28. The Downstream Paradox; quasi-one-dimensional inviscid flow.

There is also the choking effect, which means that if the back pressure is lowered beyond a limiting value, no effect is felt upstream at (1).

But, if the back pressure is raised, the situation shown in Figure 3-28c develops. A shock wave moves into the duct, its final position depending on the back pressure. There will be a *range* of back pressure, with corresponding *range* in properties at (2) for which the flow at (1) is still supersonic. Thus, for fixed input conditions with supersonic flow, the unique solution depends on the outflow problem.

This effect showed up in Crocco's (1965) calculations of quasi-one-dimensional flow in a duct. To approach a steady solution, Crocco had to specify two downstream properties, pressure and temperature. When the solution was reached, he could free the temperature condition by setting $T_{I,j} = T_{I-1,j}$. But when this was attempted with the density also, the shock wave drifted. Benison and Rubin (1969) also had to fix ρ at outflow in their quasi-one-dimensional calculations, since this determines the shock structure upstream. The present author has experienced the same phenomenon in two-dimensional duct flow calculations.

Of course, if conditions in a two-dimensional problem are such as to assure completely supersonic outflow, we may expect less sensitivity to the outflow conditions, especially if upwind differencing is used for the u-advection terms. (See Section V-G-6.)

III-C-10. Computational versus Analytic Boundary Conditions

We have considered several "computational boundary conditions" in the preceding sections. For lines of symmetry and for no-slip walls, these are no different from, and are actually based on, the analytic conditions. But the computational conditions at the upper, inflow, and outflow boundaries are different. There are a variety of computational boundary conditions from which to choose, and these are not equivalent to the commonly used analytic boundary conditions, so the computational solution will not globally converge to any analytic solution (if one were available).

But we do not use the words "computational" and "analytical" with any suggestion of "approximate" and "exact." A mature view of physics demands that both the computational and analytical approaches be recognized as approximate.

After all, the analytical boundary conditions are chosen by convenience and simplicity to the analytical formulation of the problem. No true physical problem has "freestream" conditions at "infinity." If we wanted to be picayune, we could demand that the upper-boundary analytical condition for the prediction of lift on an airplane include the relative velocity of air-to-airplane caused by earth rotation; and density ρ would not approach the "freestream" value ρ_∞ as $y \to \infty$, but would perhaps approach the exponential atmosphere variation $\rho = \rho_o e^{-c(y+h)}$, where h is the airplane's altitude above sea level. For the lower-boundary condition, we would require a spherical no-slip wall (the earth's surface) at $y = -h$.

This is unnecessary, of course. We know, from comparisons with physical data and from intuition, that the analytical "freestream" conditions give results in the region of interest (near the airplane) that are "accurate," i.e., that agree with the observed physics to within some tolerance. But the analytical boundary conditions are just a *convenient approximation* to the physical problem of interest.

So also are our computational boundary conditions and, as the upper boundary (and the others) are moved far away from the region of interest, we should find that, in the region of interest, the computations agree with the physics to within some tolerance. Both the analytical and computational solutions will agree to within some tolerance locally, in the region of interest, but globally (in particular, near the upper boundary) they need not agree.

And there is no good reason for the analytical boundary conditions to be taken as the standard of comparison, which is the common inclination. It is merely an historical accident that the analytical boundary conditions came first. If electronic computers had become commonplace in the 1660's instead of the 1960's, the common inclination might be to demand that the analyst come up with closed-form solutions to fit the evolved standard computational boundary conditions, which would be very inconvenient for the analyst!

III-C-11. Conditions at "Infinity"

In spite of the preceding comments, it would be *nice* if the computational and analytical boundary conditions were equivalent. This can sometimes be accomplished by the use of coordinate transformations (see Section VI-B).

Richardson (1910) suggested the general idea of applying the analytical "infinity" conditions at the boundaries of the computational mesh, at a finite distance from the region of interest. Indeed, many authors have applied these to at least one of the "infinity" boundaries considered above: the upper, inflow, and outflow boundaries. Boswell and Werle

(1971) have studied the effect of infinity conditions being set at a finite distance for flow past a parabola. Masliyah and Epstein (1970) gave a low-Re perturbation solution for C_D of a sphere, with "freestream" conditions on velocity applied at a spherical envelope located at $1/\gamma$ sphere radii. Their expression is

$$C_D = \frac{120}{Re}\left[\frac{1}{5 - 9\gamma + 5\gamma^3 - \gamma^6}\right] \tag{3-497}$$

With this equation, they compare the drag for true "infinity" boundaries ($\gamma = 0$) with the drag for the computational "infinity" boundary located at 100 sphere radii ($\gamma = 0.01$). The result is

$$\frac{C_D|_{\gamma=0.01}}{C_D|_{\gamma=0}} = 1.018 \tag{3-498}$$

This means that, even for boundaries at 100 sphere radii, which is much more than would ordinarily be used in a computation, the use of "infinity" conditions on velocity gives a 2% error in C_D at low Re. For $\gamma = 0.1$, which is still very demanding on a computational mesh, the ratio is 0.821, or an 18% error, even without the truncation error of the FDE solution.

Leal (1969), in a study of flow over a flat plate using elliptical coordinates, used an asymptotic far-field solution (due to Imai) to evaluate ψ and ζ at the outer boundaries. The solution gives a first-order correction (to the potential flow) which depends on C_D, the plate drag coefficient. C_D was evaluated by quadrature of the skin friction at the plate surface (Problem 2-2) at each iterative step. Thus the far-field computational boundary conditions were iteratively coupled to the wall vorticity evaluation by way of a far-field analytical solution. (The solution applies only to steady state and, when coupled early in the iterative solution, it prevented iterative convergence.)

As stated in the previous sections, we do prefer the less restrictive computational boundary conditions. Still, the application of any one of these "infinity" conditions, taken singly, might be a legitimate approximation. But we note a danger not commonly recognized.* Consider the computation of viscous flow over a cylinder, for example. Without imposing centerline symmetry, the computations are made with four outer boundaries: the upper, lower, inflow, and outflow boundaries. Suppose that "infinity" conditions are applied at all of these, with $u = U_\infty$, $v = 0$, and $P = P_\infty$. Then a finite-difference momentum integral (see, e.g., Schlichting, 1968) around the outer boundaries will show that the cylinder has zero drag!

The results of Leal (1969) do show that satisfactory results can be obtained at interior points, even though such boundaries are incompatible with the overall momentum integral for non-zero drag, but the required distance to the far-field boundary becomes much larger than if "free" computational boundary conditions are used.

III-C-12. The Sharp Corners

The boundary conditions at the sharp concave corner (ic,1) on the base in Figure 3-22 (page 140) are no problem; $\psi = 0$ and $\zeta = 0$, regardless of whether B1 is a line of symmetry or a no-slip wall. (ζ at this point does not even enter into the calculations when the usual 5-point equations are used at interior points, but its value is useful for plotting purposes and for use with 9-point equations.)

But the computational conditions at the sharp convex corner c, at (ic,jc) in Figure 3-22 (page 140) require special considerations of boundary values and of accuracy.

---

*This point was made by Professor M. D. Van Dyke during a paper discussion session at a symposium in August of 1968.

III-C-12-a Boundary Conditions at the Sharp Convex Corner

The stream function at the sharp corner presents no problem. Like the rest of the solid wall, $\psi_c = 0$ or some other constant. But there are several alternatives for the evaluation of ζ_c. The no-slip wall conditions for ζ_w can be applied in a number of ways. We will consider only first-order formulations for ζ_c, consistent with the wall equation (3-435), with $\psi_c = 0$.

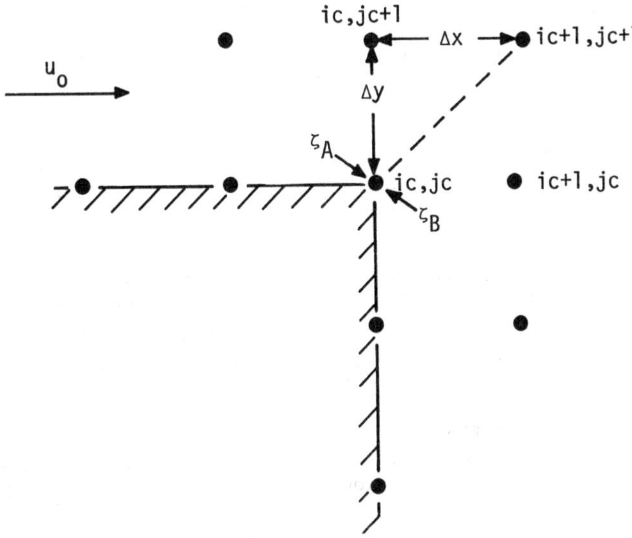

Figure 3-29. Notation for the vorticity at a sharp corner.

Referring to Figure 3-29, if the wall equation is applied to the upstream wall, we obtain a value $\zeta_c = \zeta_A$, where ζ_A is defined by $\zeta_A = 2\psi_{ic,jc+1}/\Delta y^2$. If applied to the downstream side, it yields $\zeta_c = \zeta_B = 2\psi_{ic+1,jc}/\Delta x^2$. The method which we suggest is to do both, using a discontinuous value of ζ_c. (There is, in fact, no reason to assume continuity or single-valuedness of ζ at the corner geometric singularity.) When ζ_c is used in a difference equation about node (ic,jc+1) just above c, then $\zeta_c = \zeta_A$ is used. When ζ_c is used in a difference equation about node (ic+1,jc) just downstream of c, then $\zeta_c = \zeta_B$ is used. Second-order wall equations could also be used in this multivalued approach.

However, there are other possibilities. Seven different methods of handling the corner vorticity in rectangular coordinates were examined by Roache and Mueller (1970). They are listed in Figure 3-30. The first four methods were tested with both the mid-point leapfrog method and upwind differencing, and the last three methods were tested only with upwind differencing. The test case was backstep flow with Re = 10 and an inflow Pohlhausen boundary-layer profile with $\delta/h = 1$, $\Lambda = 0$, and a no-slip wall. (At higher Re, the results are not very sensitive to the method used.)

Method 1 has been described. Discontinuous values of ζ_c are used, each one evaluated by the wall equation (3-435). Such discontinuous treatment of boundary conditions was suggested in Richardoon's (1910) original paper, and was suggested specifically for the vorticity evaluation by Thom and Apelt (1961), Roache and Mueller (1970), Kacker and Whitelaw (1970). Method 2 was derived by Kawaguti (1965) by treating ψ as if it were symmetric about c. That is, fictitious values of ψ^* are defined along the walls by $\psi^*_{ic-1,jc} = \psi^*_{ic+1,jc}$ and $\psi^*_{ic,jc-1} = \psi_{ic,jc+1}$, and the Poisson equation is evaluated at c, based on these ψ^*. Effectively, this method evaluates ζ_c as though the corner were an interior point, with $\nabla \cdot \psi = 0$ set by the no-slip wall conditions $u = v = 0$. For the case of $\Delta x = \Delta y = \Delta$, the method is equivalent to another of Richardson's (1910) suggestions for rounding off the corner, in which ψ_p and Δ_p are evaluated by interpolation on the line from (ic,jc+1) to (ic+1,jc).

Methods 3 and 4 are also attempts to somehow round off the corner. In Method 3, a single corner vorticity equal to the average of the two wall values is used. This gives a value of ζ_c = 1/2 of the ζ_c of Method 2. The same equation can be obtained by evaluating the Poisson equation (3-365) at the corner point, with wall values of $\psi = 0$ at (ic-1,jc) and (ic,jc-1). This interpretation has been given by several authors (see, for example, Greenspan, 1969A) and it would imply second-order accuracy for ζ_c. This interpretation is not valid, since the finite-difference form of the Laplacian used in equation (3-365) is not valid at the corner; there is no reason to assume continuity of $\partial^2\psi/\partial x^2$ along x at c, or of $\partial^2\psi/\partial y^2$ along y at c, so the Taylor-series derivation of equation (3-365) is not valid at c. Note that the 5-point equation for $\nabla^2\psi$ is valid at c only if $\partial^2\psi/\partial x^2$ and $\partial^2\psi/\partial y^2$ are continuous across c, or, equivalently, if $\partial v/\partial x$ and $\partial u/\partial y$ are continuous. But, along B2, the one-sided limit approaching c with $x < x_c$ gives $\partial v/\partial x = 0$ because of the no-slip condition along B2; similarly, $\partial u/\partial y = 0$ along B5.* Since $\zeta = \partial u/\partial y - \partial v/\partial x$, we see that continuity of $\partial^2\psi/\partial x^2$ and $\partial^2\psi/\partial y^2$ at c implies that $\zeta_c = 0$. We conclude that, if it is valid to evaluate $\nabla^2\psi$ at c, then we need not do so, but we can instead write a single-valued $\zeta_c = 0$.

1. discontinuous values

   $\zeta_A = 2\psi(ic,jc+1)/(\Delta y)^2, \quad \zeta_B = 2\psi(ic+1,jc)/(\Delta x)^2$
2. ψ-symmetry about C

   $\zeta_C = 2\psi(ic,jc+1)/(\Delta y)^2 + 2\psi(ic+1,jc)/(\Delta x)^2$
3. average of wall values

   $\zeta_C = \psi(ic,jc+1)/(\Delta y)^2 + \psi(ic+1,jc)/(\Delta x)^2$
4. wall at 45^0

   $\zeta_C = 2\psi(p)/(\Delta p)^2$
5. separation value of ζ_C

   $\zeta_C = 0$
6. separation value of ζ_B

   $\zeta_B = 0, \quad \zeta_A = 2\psi(ic,jc+1)/(\Delta y)^2$
7. upstream wall value

   $\zeta_C = 2\psi(ic,jc+1)/(\Delta y)^2$

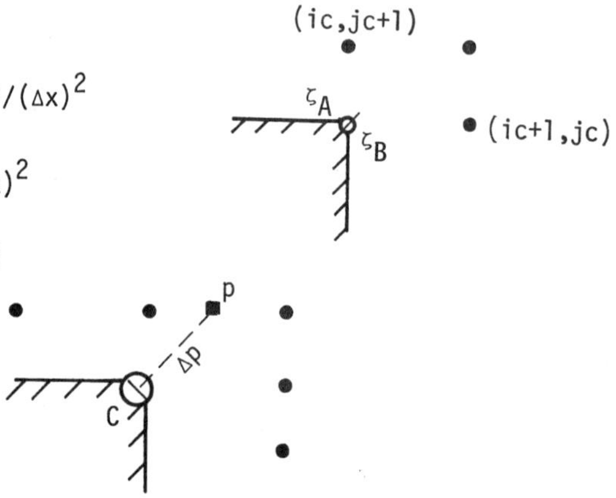

Figure 3-30. Seven methods for the corner vorticity.

In Method 4, the wall angle at node (ic,jc) is set at 45^0 and the wall equation (3-435) is applied to the point $\psi(P)$ found by interpolation between ψ(ic,jc+1) and ψ(ic-1,jc-1).** For $\beta = \Delta x/\Delta y = 1$, this reduces to $\zeta_c = 2\psi(ic+1,jc+1)\Delta P^2$ or $\zeta_c = \psi_{ic+1,jc+1}/\Delta^2$, where $\Delta = \Delta x = \Delta y$.

Methods 5 and 6 are attempts to *force* separation to occur where expected, at the sharp corner. At a separation (or reattachment) point in the continuum flow, the vorticity is zero, as is easily seen. In Method 5, a single-valued ζ_c was set equal to zero; in Method

*A simple sketch of the velocity profile in the neighborhood of c for $\partial u/\partial y = 0$ will assure the reader that this condition is absurd.

**For $\beta > 1$. If $\beta < 1$, ψ(P) is found by interpolation between (ic-1,jc-1) and (ic-1,jc).

6, a double-valued corner vorticity was used, with the downstream value $\zeta_B = 0$. It is generally better not to force such behavior, but to let it develop as part of the solution, so neither of these methods is considered valid. (Nor, surprisingly, are they even successful at forcing corner separation; see Roache and Mueller, 1970.)

Method 7 is based on the treatment of Hung and Macagno (1966). The idea is that, since the dividing streamline (presumed to separate from the sharp corner) is nearly parallel to the upstream wall, then the upstream wall evaluation should be used for ζ_c (single-valued). Hung and Macagno actually used their second-order equation (3-439) for the wall value, rather than the presently used first-order method (3-435). Method 7 gives essentially the same results as Methods 2 and 4.

If the second mesh system is used (Section III-C-3), no special treatment of the convex corner c is required, provided 5-point equations for $\nabla^2\zeta$ are used in the vorticity transport equation. If a 9-point analog is used for $\nabla^2\zeta$, then methods analogous to Methods 2, 3, or 4 in the second mesh system could be used to evaluate ζ_c.

O'Leary and Mueller (1969) also used a discontinuous formulation of ζ_c at sharp corners like c of Figure 3-22 (page 140) and also at the V-notch corner in a square mesh, as shown in Figure 3-31a. The upstream value,

$$\zeta_A = \frac{2\psi_{ic,jc+1}}{\Delta^2} \tag{3-499}$$

was used for interior point differencing at (ic,jc+1), and the value

$$\zeta_B = \frac{2\psi_{ic+1,jc+1}}{\Delta n^2} = \frac{\psi_{ic+1,jc+1}}{\Delta^2} \tag{3-500}$$

was used for interior point differencing at (ic+1,jc).

We would also recommend discontinuous treatment of ζ for the infinitely thin flat plate leading- and trailing-edge problems, using three values ζ_A, ζ_B, or ζ_D, as shown in Figure 3-31b. (It is certainly clear that different ζ values are required on opposite sides of the plate if symmetry is not assumed.) In connection with this problem, we note that Yoshizawa (1970) (and others; see Section VI-D) numerically solved the flat plate leading-edge problem using the Navier-Stokes equations by first transforming to parabolic coordinates (ξ,η), which are optimum for the solution of the ultimate boundary-layer growth. He further transformed the dependent variable to $\theta = -\zeta(\xi^2 + \eta^2)$, which removes the singularity at the leading edge.

While the relative quality of each of the above seven methods is perhaps debatable, it is clear that procedures which evaluate ζ_c by artibrary extrapolation from interior points values are not correct, and may be destabilizing.

III-C-12-b Convergence and Accuracy at the Sharp Convex Corner

The special numerical problems associated with the solution of elliptic equations at the sharp corner have been treated by Woods (1953), Wasow (1957), Laasonen (1958A,B), and others. Laasonen (1958B) used the Green's Function integral formulation of the solution to Poisson's equation (see Weinberger, 1965) to demonstrate truncation convergence for a finite number of discontinuities in ζ. His theorem requires that the continuum solution be piecewise continuous in ζ, and requires that the discontinuities occur inside the mesh points. The second of these conditions is not met in our discontinuous treatment of ζ_c. The necessity of the condition was not proved. Wasow (1957) used asymptotic expansion methods to prove truncation convergence of a finite-difference scheme when the boundary values are piecewise analytic, but required that the boundary be an analytic curve without corners. He also proved the existence (and gave the form) of an asymptotic expansion for the case of a sharp corner caused by the intersection of two analytic arcs. Woods (1953) assumed various forms of singularities in boundary values of ψ, including the case where ψ is finite but has infinite derivatives, and showed how to formally subtract out the singularities and solve the remaining finite-difference equations by Southwell's method. He also quotes Southwell as saying that iteration convergence is inhibited by rounding off the corner. In another paper, Laasonen (1958A) used both theory and numerical experiments to study truncation convergence for the sharp corner. This paper is of fundamental importance to the computational fluid dynamics problems. Laasonen showed deleterious effects on truncation convergence rates for both the corner shape and also the discontinuity. If, for a straight-line boundary with discontinuous boundary values, $E = 0(\Delta^2)$, where Δ is the mesh spacing, then, for a straight-line boundary with discontinuous boundary values, $E = 0(\Delta)$; for a right-angle corner with continuous boundary values, $E = 0(\Delta^{4/3})$; for a right-angle corner with discontinuous boundary values, $E = 0(\Delta^{2/3})$. The situation at the corner c in Figure 3-22 (page 140) is the last case, so such solutions appear to be *less* than first-order accurate in the neighborhood of the sharp corner. Forsythe and Wasow (1960) suggest that corner effects may have global, rather than merely local, effects on accuracy, although the experiments of Cheng (1970B) in compressible flow indicate that corner errors die out quickly. (See also Mehta and Lavan, 1968, for incompressible flow.)

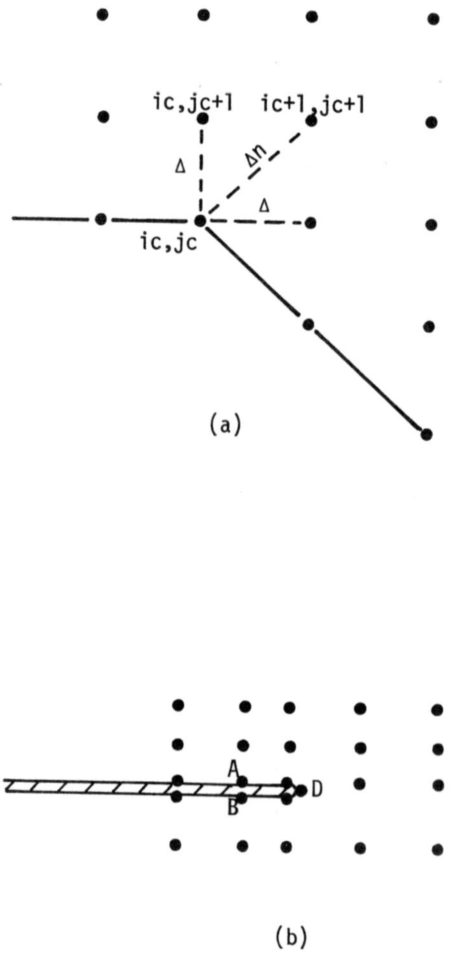

Figure 3-31. Vorticity at a 45° sharp corner and at the trailing edge of a flat plate.

Other papers dealing with the sharp-convex-corner singularities are the following. Whiteman (1967) treated the geometric singularity for the Laplace equation, $\nabla^2\psi = \zeta = 0$, by conformal transformation methods, which are, however, not applicable to the Poisson equation. He did state that, even for $\zeta = 0$, the 9-point methods are still less accurate than the 5-point methods near the corner. Whiting (1968) used the Motz method (see also Woods, 1953); during the relaxation process, node equations for the Laplacian near the corner are not expressed in terms of 5- or 9-point analogs but in terms of a truncated series of circular harmonics. Thuraisamy (1967) considered the problem of the effect of boundary data singularities on the rate of truncation convergence for the Poisson equation. Sinnott (1960) and Wigley (1969) also treated the effect of the corner on the Poisson equation, while Jamet (1968) considered an existence theorem for the effect of singularities in ζ on the parabolic vorticity transport equation. Howell and Spong (1969) considered related problems of geometric singularity at a wedge corner in inviscid, compressible potential flow.

It would appear that good accuracy near a sharp convex corner, and a fully satisfactory resolution of the question of separation in the vicinity of the sharp corner, will be achieved only by a local solution in polar coordinates centered on the sharp corner.

III-D. CONVERGENCE CRITERIA AND INITIAL CONDITIONS

The word "convergence" is used in two different senses. *Iteration convergence* refers to finally arriving at a solution of an FDE via iteration. It includes the attainment, to within some tolerance, of an acceptable solution to the discretized Poisson equation which requires $\psi^{k+1} \simeq \psi^{k}$, in some sense. Also, it includes the solution of implicit ζ boundary conditions by iterations, as in $(\zeta_w^{n+1})^{k+1} \simeq (\zeta_w^{n+1})^{k}$. Both of these iterations are nested in the iteration convergence to a steady state (if it exists), as in $\zeta^{n+1} \simeq \zeta^{n}$. The other "convergence," what mathematicians are usually talking about when they use the word, is what we here refer to as *truncation convergence* in order to distinguish it from iteration convergence. Truncation convergence has to do with the convergence of the solution of the FDE to the solution of the PDE as Δx, $\Delta t \to 0$.

There are no definitive criteria for either iteration or truncation error convergence.

The usual stated criterion of *iteration convergence,* say for steady-state ζ, is one where the entire array is checked for

$$\max_{ij} |\zeta_{ij}^{n+1} - \zeta_{ij}^{n}| \leq \varepsilon \tag{3-501A}$$

and similarly for ψ^{k+1} and $(\zeta_w^{n+1})^{k+1}$ Often, a relative error criterion is used, such as

$$\max_{ij} \left| \frac{\zeta_{ij}^{n+1} - \zeta_{ij}^{n}}{\zeta_r} \right| \leq \varepsilon \tag{3-501B}$$

where ζ_r might be some representative ζ of the problem, or $\zeta_r = \max_{ij} |\zeta_{ij}^{n}|$, or just ζ_{ij}^{n}. This latter criterion is often more meaningful, but is obviously dangerous, since near-zero values of ζ_r^n can appear locally and cause "divide overflows" in equation (3-501B). Values of ε used in equation (3-501), stated in the open literature, have varied from $\varepsilon = 10^{-3}$ to 10^{-8}, from which one might conclude that the whole idea is not very rational. In fact, this is the case. The ε in equation (3-501) can be made arbitrarily small and possibly still stop the iteration of a solution prematurely, if the solution behaves as depicted in Figure 3-32. This behavior is not uncommon for the total ψ-ζ solution; see, for example, Ingham (1968). (The elliptic solution is usually better behaved.) We can attempt to avoid this premature halt in the computations by also taking a "second derivative," as in

$$\Delta^n\zeta = \max_{ij} |\zeta_{ij}^{n+1} - \zeta_{ij}^{n}| < \varepsilon_1 \tag{3-502A}$$

$$\Delta^n\Delta^n\zeta = |\Delta^{n+1}\zeta - \Delta^n\zeta| < \varepsilon_2 \tag{3-502B}$$

and/or by testing over more iterative steps, as in

$$\max_{ij} |\zeta_{ij}^{n+10} - \zeta_{ij}^{n}| < \varepsilon \tag{3-503}$$

But none of these can substitute for actually examining the iterative behavior as in Figure 3-32 and *subjectively* judging convergence. Most workers then select a numerical value of ε which coincides with their subjective judgment of convergence, and then publish only the ε. In fact, people usually satisfy themselves that their *procedure* is convergent, by examining plots like Figure 3-32 for a representative problem, and then set ε according to the computer time available!

In order to make a point, we introduce additional notation. Let ζ^P be the (exact) solution of the PDE, ζ^F be the exact solution of the FDE, ζ^∞ be $\lim_{k\to\infty} \zeta^k$ in the iteration procedure, and ζ^{k+1} be the iterative value at which we satisfy a convergence criterion. We note (Paris and Whitaker, 1965) that ζ^∞ does not necessarily equal ζ^F because the ζ^∞ is arrived at by an iteration procedure in which round-off error may systematically accumulate. Then our convergence criterion, giving a bound on $|\zeta^{k+1} - \zeta^k|$, does not imply any bound on $|\zeta^{k+1} - \zeta^\infty|$, which does not imply any bound on $|\zeta^{k+1} - \zeta^F|$. Further, we are really interested in a bound on $|\zeta^P - \zeta^F|$, which involves the truncation convergence criterion and is just as uncertain as the other steps in the sequence.

There are seven firm suggestions which we can make about attempting to establish iteration convergence. (1) We note that, although we prefer to examine plots of the iterative behavior like Figure 3-32, the internal iteration on the Poisson equation for ψ^{k+1} and the

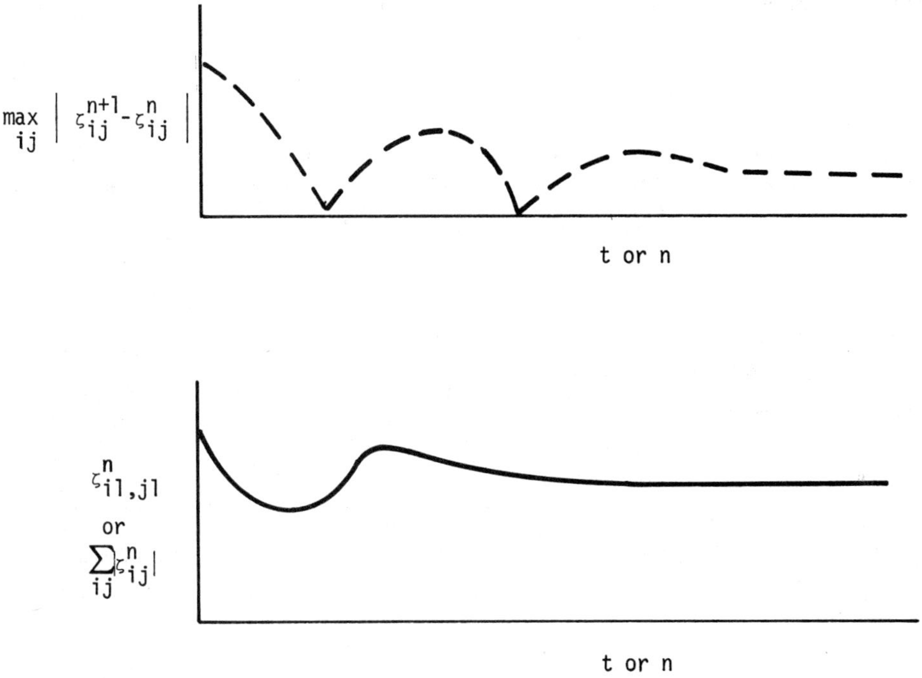

Figure 3 - 32. Representative time behavior for a viscous problem with inflow and outflow boundaries.

iteration for implicit wall values $(\zeta_w^{n+1})^{k+1}$ must be done with quantitative criteria like (3-501) through (3-503). The adequacy of these criteria can be subjectively judged beforehand by examining the iterative behavior of these subproblems, but in the total problem it is not practical to monitor anything but the time convergence. (A real attraction of direct methods for the Poisson equation is that no agonizing over convergence criteria is required.) (2) We note that it requires some computer time to perform the convergence tests like (3-501), so it may save time to only test every 10 or so iterations for ψ, as in (3-503). (3) If quantitative tests are made for ζ^n convergence, allowance should be made for possible time-splitting instability; a time-splitting instability will not be discovered by any test of the form $|\zeta^{n+p} - \zeta^n| < \varepsilon$ if p is an even number. (4) Different flow variables, and variables at different locations, will converge at different rates. If the slowest converging variable is known, it can be tested; otherwise, all variables must be tested. (Usually, ζ converges slower than velocity components which converge slower than ψ, because of the differention involved.) It is also dangerous to test a variable at only one location; flow properties in different regions of the problem can converge at vastly different rates. For example, using steady-state methods with "combined iteration" (see Section III-A-22), Textor (1968) and Textor, et al., (1969) found that ψ^{n+1} converges more slowly than ζ^{n+1}; the opposite is true in unsteady methods. Also, in the compressible flow detached shock problem, stagnation pressure converges much slower than stagnation density, making the latter a poor test variable. (5) The ε should be a function of the mesh size; if Δ is refined, ε should be reduced accordingly. (6) Any statement about a steady-state criterion, either how long the solution was run or what "ε" criterion was satisfied, would be better stated in terms of a physically relevant elapsed time, τ, rather than an iteration index, n. For example, a criterion such as (3-501) can be met at *any* stage for *any* ε, simply by selecting a small enough Δt. A more meaningful criterion is

$$\max_{ij}|\zeta_{ij}^{t+1} - \zeta_{ij}^{t}| < \varepsilon \tag{3-504A}$$

where $t = \bar{t}(\bar{U}_o/\bar{L})$ for an advection-dominated (Re >> 1) problem, or

$$\max_{ij}|\zeta_{ij}^{t'+1} - \zeta_{ij}^{t'}| < \varepsilon \tag{3-504B}$$

where $t' = \bar{t}(\bar{\nu}/\bar{L}^2) = t/Re$ for a diffusion dominated (Re << 1) problem (see section II-D). Similarly, a statement that solutions were carried out to $n = 1000$ is not as meaningful as giving the maximum t or t′ of the solution, which, of course, depends on the Δt of the method. Son and Hanratty (1969) estimated steady-state drag coefficients of spheres by plotting C_D versus $1/t_{max}$ and extrapolating to $1/t_{max} = 0$. No details of the extrapolation were given and, in fact, the use of a "french curve" or "eyeballing" is probably as valid as any other procedure, although neither procedure is repeatable. (7) Finally, it is strongly recommended that the convergence capability of the entire computational method, including boundary treatments, be tested as stringently as the machine word length allows, in an extremely coarse mesh. The use of a coarse mesh with just a few regular internal points will usually allow convergence to be met to within the last decimal-place accuracy of the machine, in a reasonable time. The test can then be repeated with greatly different initial conditions, preferably requiring an iterative approach path from the opposite "side" of the first solution. If the two solutions agree, one can state, not rigorously but at least in good conscience, that the method is convergent.

There are a few other aspects of convergence testing which may be worth considering in any problem. Thom (1933) and Thom and Apelt (1961) suggest convergence criteria based on the size of the "residual" (see Section III-B-3,4). Generally, even for linear equations, this type of test may be inadequate; see Forsythe (1970). Brown (1967) noted, for a thermal convection problem, that the temperatures and heat transfer rates (which were of primary interest) converged long before the flow velocities; after verifying that the method was convergent in velocities, he could stop the iteration early, when just the temperatures had converged. Briley (1970) tested iteration convergence of the Poisson equation for $(\psi^{n+1})^{k+1}$ by evaluating wall values of ζ_w at each ψ iteration and testing for $\Delta^k \zeta_w < \varepsilon_1$. This is a meaningful and sensitive test for ψ, and has the further advantage of requiring the same kind of calculations as the test for implicit wall boundary conditions, $\Delta^n \zeta_w < \varepsilon_2$. In order for this second test to be meaningful, it is clear that $\varepsilon_1 < \varepsilon_2$ is required; Briley (1970) used $\varepsilon_1 = 1/2\ \varepsilon_2$. Roache and Mueller (1970) and others have also found that considerable computer time can be saved by using a coarse criterion for the Poisson solution, say $\varepsilon = 10^{-4}$, early in the transient solution, and then refining ε to 10^{-6} as a steady state is approached. Another commonly used technique is to specify that the Poisson iteration be terminated with either $\max_{ij}|\psi_{ij}^{k+1} - \psi_{ij}^{k}| < \varepsilon$ or $k > k_{max}$, and perhaps setting $k_{max} = 50$. Torrance (1968) used this technique, and further noted that, as steady state was approached, the Poisson iteration criterion was easily met for $k < k_{max}$. He then updated the Poisson solution only for ψ^{n+p}, where $p \simeq 1 + 1/4(k_{max} - k_{conv})$ and where k_{conv} was the k required for convergence in the previous Poisson iteration.* Thus the velocity field for the vorticity transport equation was "frozen" for p steps, resulting in a considerable savings in computer time.

Many of the above remarks on iteration convergence also apply to *truncation convergence*. After obtaining an FDE solution with Δ_1, we obtain another solution with, say, $\Delta_2 = 1/2\times\Delta_1$, where Δ can be any or all of Δx, Δy, or Δt. Then a test for convergence of the form

$$\zeta(\Delta_2) = \zeta(\Delta_1) + \varepsilon \tag{3-505}$$

etc., can be used. Of course, the real test would come from plotting solution functionals versus Δ or $1/\Delta$ and judging convergence from the plot. The difference between $\zeta(\Delta_2)$ and $\zeta(\Delta_1)$ may be stated and it is of some value to the reader in judging convergence, but the judging remains subjective. The ε is usually set by available computer time and core storage. It is usually impractical to continue halving the mesh size, even to a value of $\Delta_3 = 1/4\times\Delta_1$ for PDE's; in fact, only one mesh solution is commonly computed and that one is presented, for better or for worse. It would be nice if we could confidently draw at least qualitative ideas about flow-parameter behavior from coarse mesh solutions, and sometimes we can, but the practice is dangerous. For example, Burggraf (1966) computed driven cavity flows and found opposite trends in the movement of the driven vortex center with Re for the coarse-mesh and fine-mesh solutions.

It is worth noting that methods with $O(\Delta^2)$ accuracy converge quadratically while $O(\Delta)$ methods converge linearly, and the adequacy of convergence is much easier to judge in the $O(\Delta^2)$ case.

We feel obliged to mention a well-known technique, called Richardson extrapolation ** (Richardson, 1910; Shortley and Weller, 1938; Salvadori and Baron, 1961) for estimating the final truncation-converged solution for the $O(\Delta^2)$ methods. Let ζ = true PDE solution with the computational boundary conditions, that is, $\zeta = \lim_{\Delta\to 0} \zeta(\Delta)$. Then the error in the solution value $\zeta(\Delta_1)$ at some particular x_1, y_1, t_1 can be written in a Taylor series as

$$E_1 = \zeta - \zeta(\Delta_1) = a\Delta_1{}^2 + b\Delta_1{}^4 + \ldots \tag{3-506}$$

If $\Delta_2 = (1/p)\Delta_1$, where p is an integer, then a node-point $\zeta(\Delta_2)$ solution is available at the same point x_1, y_1, t_1, and the error can be written as

$$E_2 = \zeta - \zeta(\Delta_2) = a\Delta_2{}^2 + b\Delta_2{}^4 + \ldots \tag{3-507}$$

Eliminating a from these two equations, we obtain a higher-order approximation.

$$\zeta = \zeta(\Delta_2) - \frac{\zeta(\Delta_2) - \zeta(\Delta_1)}{1 - (\Delta_1/\Delta_2)^2} + O(\Delta_1^2\Delta_2^2, \Delta_2^4) \tag{3-508A}$$

For $\Delta_2 = \frac{1}{2}\Delta_1$,

$$\zeta = \frac{4}{3}\zeta(\Delta_2) - \frac{1}{3}\zeta(\Delta_1) + O(\Delta^4) \tag{3-508B}$$

*For example, suppose $k_{max} = 50$. Also, suppose that the Poisson solution at time level n = 1000 converged in $k_{conv} = 18$ iterations. Then $p = 1 + 1/4(50-18) = 18$, so the Poisson equation for ψ was not updated again until time level n = 1018.

**These techniques also go by the names "extrapolation to the limit", "deferred approach to the limit", and "iterated extrapolation."

The trouble with the method is that, practically, we have no way of knowing the size of the coefficients of Δ^4 and higher terms, nor of knowing if the convergence is monotonic in Δ.* If it is not, Richardson extrapolation could give a worse estimate of ζ than $\zeta(\Delta_1)$. Also, extrapolation methods magnify round-off error (Burgess, 1971) and iteration errors.

An aid to estimating mesh convergence without changing the mesh size is to recompute the problem with a different order method. For $\Delta x = \Delta y$, Thom and Apelt (1961) suggest re-computing the Laplacian operators ($\nabla^2\psi$ in the Poisson equation and $\nabla^2\zeta/Re$ in the vorticity transport equation) using the diagonal unit square operator (see section III-B-10), which is accurate to $O(\sqrt{2}\Delta)$, or the other Laplacian analogs. Recomputation with the first-, second-, and fourth-order methods of section III-A also suggests itself. Note that the accuracy order of boundary conditions should also be consistently changed. This method has not been used in published computational fluid dynamics solutions.

An important point about truncation convergence was made forcefully by Cheng (1968, 1970). Even if we could take $\lim_{\Delta\to 0} \zeta(\Delta)$, or if we obtain $\zeta(\Delta_2) = \zeta(\Delta_1) + \varepsilon$ with ε arbitrarily small, we still have an effect of computational boundary conditions for "free flight" problems. Strictly speaking, we should also check for convergence in the region of interest as the "infinity" boundaries (upper, lower, inflow, and outflow) are moved farther away from the region of interest. Checks on mesh-size convergence alone, or checks with higher-order methods, give *no clue* as to convergence in the sense of diminishing boundary effect. The method used by Hamielec and Raal (1969) is worth mentioning. They computed flow over a cylinder in polar coordinates with the computational mesh extending to $r = r_b$; all the "infinity" conditions were applied at this coordinate. They computed with two values of r_b; judging the convergence to be quadratic, they estimated the "free flight" drag coefficient by Richardson extrapolation to $1/r_b = 0$.

Other indicators of truncation error are obtained by (1) evaluating the conservation error (see section III-A-3) for a non-conservative method, (2) evaluating the line integral of $\partial\zeta/\partial n$ around a body (see problem 3-32), and (3) comparing the two sharp-corner values of pressure (see section III-E-2).

The *initial conditions* used for the entire incompressible fluid dynamics problem are not important to the steady solution if unsteady two-level methods are used and if iterations for ψ^{k+1} and $(\zeta_w^{n+1})^{k+1}$ using implicit methods are sufficiently converged.** [The initial conditions obviously determine the initial transient solution, but even this effect often decays exponentially in time. See Beardsley (1969), for example.] Further, the initial conditions usually make little difference on the computer time required to obtain a completely converged solution, although many people assume that the opposite is obviously true. The initial conditions make little difference because the error of our initial guess is usually bounded and many orders of magnitude larger than our convergence criteria. For example, if the flow rate through a channel is normalized to $\psi_{max} = O(1)$, then a guess of $\psi^0 = 0$ at all internal points would only give errors of $O(1)$. A very good guess for a channel expansion separated flow field might give errors in ψ^0 of $O(0.1)$. However, the improvement is insignificant when convergence criteria like $\varepsilon = 10^{-6}$ are being used. It also helps in understanding the convergence process to recognize that for high-Re flows, an error in ζ at a single node value near the inflow boundary and outside the boundary layer must essentially be *advected* out of the mesh. The required number of time steps is then bounded from below by $n = \tau/\Delta t$, where τ is the transit time for a particle at that location to traverse the computational mesh. This transit time τ is independent of the size of the error and so, roughly, is the convergence time.

---

*Truncation convergence is dependably monotone for the Laplace equation $\nabla^2\psi = 0$ and for the simple diffusion equation $\partial\zeta/\partial t = a\nabla^2\zeta$, but not necessarily for the Poisson equation or for the nonlinear advection-diffusion equation. Other references for extrapolation methods applied to PDE's can be found in Burgess (1971) and in section V-H. Schoenerr and Churchill (1970) consider extrapolation of transient solutions to the steady-state limit for the diffusion equation. Stetter (1970) considers Richardson extrapolation for stiff equations.

**If not, over-all stability often may be enhanced by decreasing Δt during the initial transient stage (Pao and Daugherty, 1969).

As a specific example of the unimportance of initial conditions, consider the backstep problem shown in Figure 3-22 (page 140). The present author has computed this problem with an initial estimate consisting of nothing more than setting $\psi = 0$ at all interior points and along B1-B5-B2, fixing the inflow boundary layer ψ and the "lid" $\psi(B3) = \psi(1,J)$ at inflow, with $\zeta = 0$ everywhere. This is a completely non-sensical initial estimate. Yet, after the first Poisson iteration, with the boundary conditions at outflow described by equation (3-478), non-zero advection velocities appear everywhere. By $n = 30$, a realistic-looking recirculation region had developed, which was a better initial estimate than one could be expected to achieve with an analysis. This crude method also gave essentially the same computer time for ultimate convergence as the commonly used method of starting with the results of some previously converged computation (obtained with different Re or different inflow parameters) for $Re > 1$.

The initial conditions can make a difference to steady-flow methods using "continuous substitution," to unsteady-flow methods if the Poisson equation convergence is inadequate, and in implicit methods if wall values $(\zeta_w^{n+1})^{k+1}$ are not sufficiently iterated. In these cases, poor initial conditions can lead to a true nonlinear instability. (The latter two cases can be cured by reducing Δt during early transients.) Initial conditions can also cause meaningless oscillations when multi-level methods are used, even on the simplest equations.

The accuracy of the initial condition can also make an important difference to the iterative solution of the Poisson equation, simply because very good initial estimates are sometimes available from the previous time step. The new time level converged solution ψ^{n+1} differs from ψ^n only in the source term ζ and in the boundary conditions on ψ. If Δt is small enough, or if time convergence of the entire problem is near, we will have $\zeta^{n+1} \simeq \zeta^n$ and $\psi^{n+1} \simeq \psi^n$ on boundaries. (For problems such as the driven cavity, we have ψ fixed on boundaries for all time.) Then the use of the ψ^n solution as the initial estimate of $(\psi^{n+1})^{k=0}$ is very efficient. In fact, if ζ convergence is near, the convergence criteria on $|\psi^{k+1} - \psi^k| < \varepsilon$ is often satisfied in one iteration, so that the method reverts automatically to a continuous-substitution, steady-flow iteration.

Lynch and Rice (1968) have shown that, with ADI methods, convergence proceeds faster with smooth initial errors. This is usually met in practice unconsciously, since the final solution is smooth and the initial estimates (even $\psi = 0$) are also.

The accuracy of initial conditions can also be more important for supersonic flows, where bad initial data can cause persistent wave propagations.

III-E. PRESSURE SOLUTION

One of the advantages of working with the incompressible equations is that the number of variables is reduced. Pressure was eliminated from the primitive momentum equations by cross-differentiation in Section II-B. We now wish to extract the pressure solution from our numerical solution in terms of ψ and ζ.

The pressure equation to be derived is the Poisson equation, identical to the equation for stream function. But a major difficulty arises in the use of SOR iteration because of the different type of boundary condition used.

III-E-1. Numerical Cubature

The normalized forms of the primitive equations (2-1) through (2-3) are (see first Exercise in Section II-D)

$$\frac{\partial u}{\partial t} + u\frac{\partial u}{\partial x} + v\frac{\partial u}{\partial y} = -\frac{\partial P}{\partial x} + \frac{1}{Re}\nabla^2 u \tag{3-509A}$$

$$\frac{\partial v}{\partial t} + u\frac{\partial v}{\partial x} + v\frac{\partial v}{\partial y} = -\frac{\partial P}{\partial y} + \frac{1}{Re}\nabla^2 v \tag{3-509B}$$

$$\frac{\partial u}{\partial x} + \frac{\partial v}{\partial y} = 0 \tag{3-509C}$$

where pressure P has been normalized by

$$P = \bar{P}/\bar{\rho}\ \bar{U}_o^{\ 2} \tag{3-510}$$

To obtain a pressure solution, one might start at an arbitrary point with an arbitrary pressure level (constant of integration) and numerically integrate the discretized equations (3-509) for $\partial P/\partial x$ and $\partial P/\partial y$, the velocity components being obtained from the stream-function solution. [This procedure has been used by Kawaguti (1965), Pearson (1965A), Burggraf (1966), Hung and Macagno (1966), Thoman and Szewczyk (1966), Rimon and Cheng (1969), Lavan, et al., (1969), Son and Hanratty (1969), Masliyah and Epstein (1970), and Shavit and Lavan (1971).] If one's concern is only with surface pressure, as in the calculation of a drag coefficient, the method may be acceptable. Surface pressure integration is facilitated by using the easily derived relation (Problem 32),

$$\frac{\partial P}{\partial s} = \frac{1}{Re}\frac{\partial \zeta}{\partial n} \tag{3-511}$$

where (n,s) are normal and tangential to the surface (see Pearson, 1965A). (A thorough exposition of the numerical difficulties may be found in Masliyah and Epstein, 1970.) But in general, the method will give different answers when different paths are used to get to the same point. The discrepancy will be partly due to solution errors for ψ. But even if the exact solution values for ψ were known at all mesh points, different paths of integration would yield different answers due to quadrature* errors. Further, note that velocity *gradients* are integrated in equation (3-509), requiring double differentiation of the numerically determined stream function, which is usually inaccurate. This method is especially susceptible to error in problems like that depicted in Figure 3-22 (page 140) when the path of integration is close to the sharp convex corner.

III-E-2. Poisson Equation for Pressure

A more accurate solution can be determined from the Poisson form of the pressure equation, obtained as follows.

---

*Buckingham (1962) has termed such integration of a known function f(x,y) over two independent variables as "cubature."

Differentiating equation (3-509A) with respect to x gives

$$\frac{\partial}{\partial t}\left(\frac{\partial u}{\partial x}\right) + \left(\frac{\partial u}{\partial x}\right)^2 + u\left(\frac{\partial^2 u}{\partial x^2}\right) + \left(\frac{\partial v}{\partial x}\right)\left(\frac{\partial u}{\partial y}\right) + v\,\frac{\partial^2 u}{\partial x \partial y} = -\frac{\partial^2 P}{\partial x^2} + \frac{1}{Re}\frac{\partial}{\partial x}\left(\nabla^2 u\right) \tag{3-512}$$

Differentiating equation (3-509B) with respect to y gives

$$\frac{\partial}{\partial t}\left(\frac{\partial v}{\partial y}\right) + \left(\frac{\partial v}{\partial y}\right)^2 + v\left(\frac{\partial^2 v}{\partial y^2}\right) + \left(\frac{\partial u}{\partial y}\right)\left(\frac{\partial v}{\partial x}\right) + u\,\frac{\partial^2 v}{\partial x \partial y} = -\frac{\partial^2 P}{\partial y^2} + \frac{1}{Re}\frac{\partial}{\partial y}\left(\nabla^2 v\right) \tag{3-513}$$

Adding equation (3-512) and (3-513) gives

$$\frac{\partial}{\partial t}\left(\frac{\partial u}{\partial x} + \frac{\partial v}{\partial y}\right) + \left(\frac{\partial u}{\partial x}\right)^2 + \left(\frac{\partial v}{\partial y}\right)^2 + 2\left(\frac{\partial v}{\partial x}\right)\left(\frac{\partial u}{\partial y}\right) + u\left(\frac{\partial^2 u}{\partial x^2} + \frac{\partial^2 v}{\partial x \partial y}\right) + v\left(\frac{\partial^2 u}{\partial x \partial y} + \frac{\partial^2 v}{\partial y^2}\right)$$

$$= -\nabla^2 P + \frac{1}{Re}\left[\frac{\partial}{\partial x}\left(\nabla^2 u\right) + \frac{\partial}{\partial y}\left(\nabla^2 v\right)\right] \tag{3-514}$$

From the continuity equation (3-510),

$$\frac{\partial}{\partial t}\left(\frac{\partial u}{\partial x} + \frac{\partial v}{\partial y}\right) = \frac{\partial}{\partial t}(0) = 0 \tag{3-515}$$

Also,

$$u\left(\frac{\partial^2 u}{\partial x^2} + \frac{\partial^2 v}{\partial x \partial y}\right) = u\,\frac{\partial}{\partial x}\left(\overbrace{\frac{\partial u}{\partial x} + \frac{\partial v}{\partial y}}^{0}\right) = 0 \tag{3-516}$$

$$v\left(\frac{\partial^2 u}{\partial x \partial y} + \frac{\partial^2 v}{\partial y^2}\right) = v\,\frac{\partial}{\partial y}\left(\overbrace{\frac{\partial u}{\partial x} + \frac{\partial v}{\partial y}}^{0}\right) = 0 \tag{3-517}$$

Finally,

$$\frac{\partial}{\partial x}\left(\nabla^2 u\right) + \frac{\partial}{\partial y}\left(\nabla^2 v\right) = \frac{\partial^3 u}{\partial x^3} + \frac{\partial^3 u}{\partial x \partial y^2} + \frac{\partial^3 v}{\partial y \partial x^2} + \frac{\partial^3 v}{\partial y^3}$$

$$= \frac{\partial^2}{\partial x^2}\left(\overbrace{\frac{\partial u}{\partial x} + \frac{\partial v}{\partial y}}^{0}\right) + \frac{\partial^2}{\partial y^2}\left(\overbrace{\frac{\partial u}{\partial x} + \frac{\partial v}{\partial y}}^{0}\right) = 0 \tag{3-518}$$

Substituting equations (3-515) through (3-518) into (3-514) gives

$$\left(\frac{\partial u}{\partial x}\right)^2 + \left(\frac{\partial v}{\partial y}\right)^2 + 2\left(\frac{\partial v}{\partial x}\right)\left(\frac{\partial u}{\partial y}\right) = -\nabla^2 P \tag{3-519}$$

Consider

$$0 = \left(\frac{\partial u}{\partial x} + \frac{\partial v}{\partial y}\right)^2 = \left(\frac{\partial u}{\partial x}\right)^2 + \left(\frac{\partial v}{\partial y}\right)^2 + 2\left(\frac{\partial u}{\partial x}\right)\left(\frac{\partial v}{\partial y}\right) \tag{3-520}$$

or

$$\left(\frac{\partial u}{\partial x}\right)^2 + \left(\frac{\partial v}{\partial y}\right)^2 = -2\left(\frac{\partial u}{\partial x}\right)\left(\frac{\partial v}{\partial y}\right) \tag{3-521}$$

Substituting this into equation (3-519) gives

$$-\frac{1}{2}\nabla^2 P = \left(\frac{\partial v}{\partial x}\right)\left(\frac{\partial u}{\partial y}\right) - \left(\frac{\partial u}{\partial x}\right)\left(\frac{\partial v}{\partial y}\right) \tag{3-522}$$

(Note from this equation that $\nabla^2 P = 0$ along a no-slip wall.)

To cast equation (3-522) in terms of the stream function ψ, we substitute

$$u = \partial\psi/\partial y \quad , \quad v = -\partial\psi/\partial x \tag{3-523}$$

giving

$$-\frac{1}{2}\nabla^2 P = \left(\frac{\partial^2\psi}{\partial x^2}\right)\left(-\frac{\partial^2\psi}{\partial y^2}\right) - \left(\frac{\partial^2\psi}{\partial x\partial y}\right)\left(-\frac{\partial^2\psi}{\partial x\partial y}\right) \tag{3-524}$$

or

$$\nabla^2 P = S = 2\left[\left(\frac{\partial^2\psi}{\partial x^2}\right)\left(\frac{\partial^2\psi}{\partial y^2}\right) - \left(\frac{\partial^2\psi}{\partial x\partial y}\right)^2\right] \tag{3-525\_}$$

This is the Poisson form of the pressure equation. The usual centered finite-difference forms (Section III-A-1) give, to $O(\Delta x^2, \Delta y^2)$,

$$S_{i,j} = 2\left[\left(\frac{\psi_{i+1,j} + \psi_{i-1,j} - 2\psi_{i,j}}{\Delta x^2}\right)\left(\frac{\psi_{i,j+1} + \psi_{i,j-1} - 2\psi_{i,j}}{\Delta y^2}\right)\right.$$
$$\left. - \left(\frac{\psi_{i+1,j+1} - \psi_{i+1,j-1} - \psi_{i-1,j+1} + \psi_{i-1,j-1}}{4\Delta x\Delta y}\right)^2\right] \tag{3-526}$$

Equation (3-526) is analogous to the stream-function equation, $\nabla^2\psi = \zeta$, with the source term S analogous to ζ. All the methods of solving this Poisson equation which are given in Section III-B are applicable, but the boundary conditions are different, as we discuss in the next section.

III-E-3. Boundary Conditions of the Second Kind on Pressure

In the solution of equation $\nabla^2\psi = \zeta$, at least some of the boundary conditions are of the first kind, or Dirichlet boundary conditions: $\psi(x,y)$ is a specified function along the boundaries. In the solution of the Poisson form of the pressure equation, the boundary conditions are of the second kind (the Neumann boundary conditions) specifying $\partial P/\partial n\ (x,y)$, where n is normal to the boundary. The values of the pressure gradients are found from equation (3-509).

Exercise: Express the finite-difference pressure gradients $\delta P/\delta x$ and $\delta P/\delta y$ in terms of ψ.

Exercise: Show that, on a no-slip wall (Pearson, 1965A),

$$\frac{\delta P}{\delta n} = -\frac{1}{Re}\frac{\delta\zeta}{\delta x} \tag{3-527}$$

where (n,s) are normal and tangential to the wall.

The surface tangential gradient $\delta\zeta/\delta s$ in equation (3-527) is readily evaluated by the use of centered differences. Note that *any* procedure can be used, since the gradient is not re-evaluated during the Poisson solution, and the lack of feedback eliminates the possibility of instabilities. Centered second-order differences appear to be preferable, consistent with $O(\Delta^2)$ accuracy of the entire method. For transient solutions, the terms $\partial u/\partial t$ and $\partial v/\partial t$ in equation (3-509) are often neglected, and the simpler approximation $\partial P/\partial n \simeq 0$ is also frequently used.* These approximations are not necessary, and the more accurate methods based on equation (3-509) are recommended. Note that the appearance of the coefficient $1/Re$ does not mean that these terms are negligible in high-Re flows. For a boundary layer, we know that $\partial P/\partial n = 0$ (boundary-layer thickness), which is small, but for more general cases, $\partial P/\partial n$ may be large. (Of course, across a line of symmetry, $\delta P/\delta n = 0$ holds.) Near a sharp convex corner, use of Neumann boundary conditions can result in a double-valued pressure at corner c in Figure 3-22 (page 140). Although this is non-physical, it is recommended that the double value be retained, as in the Method 1 of Figure 3-30, for double-valued ζ_c. The error in single-valuedness, $P_A - P_B$, can be used as an index of truncation error near the corner.

III-E-4. Iterative Solution Methods

Neumann conditions require two special formulations: first, in the incorporation of the gradient condition in the SOR method. An obvious method of solution is to sweep the mesh for the new (k+1) iterative estimates at all interior points, and then set new iterative (k+1) estimates for the boundary values from the known slope $\delta P/\delta n$ and the newly calculated adjacent interior points. For a point (i,jc) on B2 of Figure 3-22 (page 140) and using SOR according to equation (3-380), we would have the following.

During interior sweep:

$$P^{k+1}_{i,jc+1} = \frac{\omega}{2(1+\beta^2)}\Big[P^k_{i+1,jc+1} + P^{k+1}_{i-1,jc+1} + \beta^2 P^k_{i,jc+2} + \underbrace{\beta^2 P^k_{i,jc}}_{\text{previous boundary value}}$$

$$- \Delta x^2 S_{i,jc+1} - 2(1+\beta^2)P^k_{i,jc+1}\Big] + P^k_{i,jc+1} \tag{3-528A}$$

Boundary-value determination:

$$P^{k+1}_{i,jc} = P^{k+1}_{i,jc+1} - \frac{\delta P}{\delta n}\cdot\Delta y \tag{3-528B}$$

This plausible method does not converge. The solution drifts, slowly but endlessly. This is the reason many papers published even as recently as the late 1960's did not contain pressure solutions, and others contained pressure distributions which were erroneous, although they satisfied some kind of "convergence criterion" like equation (3-501).

The meteorologist Miyakoda (1962) recommends that the derivative boundary conditions be incorporated directly into the SOR difference scheme at interior points adjacent to the boundaries. Thus, an equation of the form of (3-528A) is used only at interior points more than one node removed from the boundary. At points adjacent to the boundary, equation (3-528A) is replaced by

---

*This approximation is familiar from boundary-layer theory (e.g., Schlichting, 1968). It is actually a more mild approximation here, since constant P would not be applied throughout the boundary layer, but only over the lowest Δn in the boundary layer.

**The present author is especially indebted to Dr. S. Piacsek for providing this valuable reference.

$$P^{k+1}_{i,jc+1} = \frac{\omega}{2(1 + \beta^2)} \left[P^k_{i+1,jc+1} + P^{k+1}_{i-1,jc+1} + \beta^2 P^k_{i,jc+2} + \beta^2 \left(P^{k+1}_{i,jc+1} - \frac{\delta P}{\delta n}\Big|_{i,jc} \Delta y \right) - \Delta x^2 S_{i,jc+1} - 2(1 + \beta^2) P^k_{i,jc+1} \right] + P^k_{i,jc+1} \qquad (3\text{-}529)$$

which is solved algebraically for the $P^{k+1}_{i,jc+1}$ appearing on both sides. After convergence is completed, the final boundary values may be computed from equation (3-528B).

Equation (3-529) differs from the cyclic use of (3-528A) and (3-528B) only in the time level of the term $\beta^2 \left(P^{k+1}_{i,jc+1} - \delta P/\delta n|_{i,jc} \times \Delta y \right)$.

The second special formulation for Neumann conditions is that the boundary gradients should be compatible with the source term. By Green's theorem, a continuum solution to (3-525) over the area R exists only if $E \equiv \int_R S dR - \int_{\partial R} (\partial P/\partial n) d\ell = 0$. (See page 348 for the 1D version.) Because of truncation error, the boundary values usually will fail to meet this constraint, causing a slow divergence of SOR iteration. Miyakoda (1962) recommends that $\delta P/\delta n$ be determined to meet this constraint. A remedy used by Briley (J.Comp. Phys., Vol. 14, Jan. 1974, p.20) and by Ghia and Ghia (private communication) is to compute the discretized value of E, and then solve a modified equation $\nabla^2 P = S - E/R$. When the quadrature for E is done in a consistent manner (Ghia and Ghia) as in (3-533A,B), a considerable improvement is made in the iterative convergence for SOR and ADI. Note that the second gradient boundary condition may require special treatment to avoid an indeterminacy in the tridiagonal solution with the ADI method (see page 348). Provided that $\delta P/\delta n$ are determined to second-order accuracy, the overall method appears to be second-order accurate also. Experiments (Miyakoda,1962) indicate that the optimum ω_o is increased by the Neumann conditions.

If P_{ij} is a solution, then $(P_{ij} + C)$, where C is a constant, is also a solution. The particular solution is chosen by specifying P at one point. If some boundary point can be identified as a "freestream" location, it is recommended that P = 0 be set there, making P easily relatable to the common engineering pressure coefficient term, C_p, by $P = 2C_p$. This one point may be omitted from the iteration scheme.

III-E-5. Pressure Level

An important point related to the above discussion has been neglected in published solutions. For a "freestream" flow problem, we may arbitrarily set P = 0 at the "freestream" location. Any other value is legitimate; for example, P = 1 has often been used. For a confined flow problem, like the popular driven-cavity problem (Kawaguti, 1961; Burggraf, 1966; Pan and Acrivos, 1967; Donovan, 1970) the initial pressure of the time-dependent problem may also be set arbitrarily. For example, Donovan (1970) sets P = 1 at a reference location, chosen as the center of the wall opposite the "driving" lid, say reference pressure $P_r = 1$. The point is this: once $P_r = 1$ is set initially, its later values are determined by the physics.

In order to evaluate P_r, and thus the level of pressures at all field points at later times, we must resort to a thermodynamic calculation using an equation of state. For any gas or liquid, we have some equation of state for the dimensional thermodynamic variables, expressed as

$$\bar{P} = \text{fcn}\,(\bar{\rho},\bar{T}) \qquad (3\text{-}530)$$

(For example, if we assume an ideal gas, we have $\bar{P} = \bar{\rho}\bar{R}_g\bar{T}$, where $\bar{R}_g$ is the gas constant.) Suppose the initial conditions are those of no flow. Then, at n = 0, the initial condition on dimensional pressure is $\bar{P}_{ij}$ = constant everywhere; we call this constant our reference pressure $\bar{P}_r^o$. As the time solution proceeds, the temperature solution is also calculated (e.g., Section III-F) and through it, the pressure level. These calculations may or may not be coupled.* The temperature can be raised by either non-adiabatic walls or by dissipation (see next Section III-F). The latter will eventually be important for a long-time solution; however, for illustrative purposes, we consider the case of adiabatic walls and negligible dissipation. Taking the global (space) integral of equation (3-530) gives

$$\int_R \bar{P}\, dR = \int_R \text{fcn}\ (\bar{\rho},\bar{T})\, dR \tag{3-531}$$

where R is the interior volume of the cavity. With the assumption of incompressibility and of $\bar{T}$ = constant, equation (3-531) becomes

$$\int_R \bar{P}\, dR = R\cdot\bar{P}_r^o \tag{3-532}$$

If we define a normalized pressure by $P = \bar{P}/\bar{P}_r^o$, this equation gives

$$\int_R P\, dR = R \tag{3-533A}$$

The finite-difference form of this integration is easily expressed by a summation, weighting P over its cell area, consistent with the finite-difference method. For second-order methods, the appropriate area weighting is $\Delta x\cdot\Delta y$. For a point such as (1,j) lying on the wall boundary (in the first mesh system), the appropriate cell area is $1/2\ \Delta x\cdot\Delta y$. For corner points such as (1,1) the area is $1/4\ \Delta x\cdot\Delta y$. With $X = (I-1)\Delta x$, $Y = (J-1)\Delta y$, the summation becomes

$$\sum_{i=2}^{I-1}\sum_{j=2}^{J-1} P_{ij}\Delta x\cdot\Delta y + \sum_{j=2}^{J-1} P_{1j}\frac{1}{2}\Delta x\cdot\Delta y + \sum_{j=2}^{J-1} P_{Ij}\frac{1}{2}\Delta x\cdot\Delta y + \sum_{i=2}^{I-1} P_{i1}\frac{1}{2}\Delta x\cdot\Delta y$$

$$+ \sum_{i=2}^{I-1} P_{iJ}\frac{1}{2}\Delta x\cdot\Delta y + \frac{1}{4}\Delta x\cdot\Delta y(P_{11} + P_{1J} + P_{I1} + P_{IJ}) = X\cdot Y \tag{3-533B}$$

Dividing by $\Delta x\cdot\Delta y$ and noting that $X\cdot Y/(\Delta x\cdot\Delta y) = (I-1)(J-1)$, we obtain

$$\sum_{i=2}^{I-1}\sum_{j=2}^{J-1} P_{ij} + \sum_{j=2}^{J-1}\frac{1}{2}(P_{1j} + P_{Ij}) + \sum_{i=2}^{I-1}\frac{1}{2}(P_{i1} + P_{iJ})$$

$$+ \frac{1}{4}(P_{11} + P_{1J} + P_{I1} + P_{IJ}) = (I-1)(J-1) \tag{3-534}$$

After the Poisson solution for P_{ij} is obtained with $P_r^n = 1$ (or anything else), equation (3-534) will not be satisfied, generally. The summation on the left-hand side of (3-534), call it S, is then evaluated, and the pressures are corrected as follows.

Exercise: Show that the pressure P_{ij} from the Poisson solution should be replaced by $(P_{ij}-S/L)$ where S is the summation on the left side of (3-534) and $L = (I-1)(J-1)$.

Since the physical pressure is undetermined without specification of the thermal problem, the published solutions with $P_r^n = P_r^o = 1$ are not really incorrect, but are perhaps incomplete. We suggest that the adiabatic, dissipation-free solution, consistent with the use of equation (3-534), is a more meaningful reference solution than $P_r^n = 1$.

*As the bulk temperature increases, Re will increase as μ increases for gases.

III-F. TEMPERATURE SOLUTIONS AND CONCENTRATION SOLUTIONS

In incompressible flow with constant properties and no body force, the dynamics are independent of the thermodynamics. Once the kinematic flow field is described by ψ and ζ solutions, we can solve for any number of temperature distributions with different thermal boundary conditions. This decoupling is also possible for many concentration problems.

III-F-1. Basic Equations

The unsteady energy equation reduces to the following (Schlichting, 1968) for incompressible flow with no source terms and with constant properties. All variables are dimensional.

$$\rho C_p \frac{D\bar{T}}{D\bar{t}} = k\bar{\nabla}^2\bar{T} + \mu\bar{\Phi} \tag{3-536}$$

where the dissipation function $\bar{\Phi}$ is given by

$$\bar{\Phi} = 2\left[\left(\frac{\partial\bar{u}}{\partial\bar{x}}\right)^2 + \left(\frac{\partial\bar{v}}{\partial\bar{y}}\right)^2\right] + \left(\frac{\partial\bar{v}}{\partial\bar{x}} + \frac{\partial\bar{u}}{\partial\bar{y}}\right)^2 \tag{3-537}$$

As in section II-B, we normalize these equations with respect to a characteristic velocity, $\bar{u}_o$, a characteristic length, L, and a characteristic advection time, $(L/\bar{u}_o)$. In addition, a characteristic positive temperature difference is written as $(\bar{T}_1 - \bar{T}_o)$. (For example, if the backstep problem depicted in Figure 3-22 on page 140 were computed for a base heating solution, $\bar{T}_1$ might be the base temperature and $\bar{T}_o$ the freestream inflow temperature.) Using these, equation (3-536) becomes

$$\rho C_p\left(\frac{\bar{u}_o}{L}\right)(\bar{T}_1 - \bar{T}_o)\frac{DT}{Dt} = k\frac{1}{L^2}(\bar{T}_1 - \bar{T}_o)\nabla^2 T + \mu\left(\frac{\bar{u}_o}{L}\right)^2\Phi \tag{3-538}$$

Division by $\rho C_p(\bar{u}_o/L)(\bar{T}_1 - \bar{T}_o)$ and expanding gives

$$\frac{\partial T}{\partial t} = -\vec{V}\cdot(\nabla T) + \frac{k}{\rho C_p\,\bar{u}_o\,L}\nabla^2 T + \frac{\mu\bar{u}_o}{\rho C_p L(\bar{T}_1 - \bar{T}_o)}\Phi \tag{3-539}$$

Now

$$\frac{k}{\rho C_p\,\bar{u}_o\,L} = \left(\frac{k}{C_p\mu}\right)\left(\frac{\mu}{\rho\bar{u}_o\,L}\right) = \frac{1}{Pr\ Re} = \frac{1}{Pe} \tag{3-540}$$

where Pr is the Prandtl number and Pe is the Peclét number. Also,

$$\frac{\mu\bar{u}_o}{\rho C_p L(\bar{T}_1 - \bar{T}_o)} = \frac{(\bar{u}_o^2/C_p)}{(\bar{T}_1 - \bar{T}_o)}\left(\frac{\mu}{\rho\bar{u}_o L}\right) = E/Re \tag{3-541}$$

where E is the Eckert number. Using the continuity equation, $\nabla\cdot\vec{V} = 0$, we have

$$\nabla\cdot(\vec{V}T) = \vec{V}\cdot\nabla T + T\,\nabla\cdot\vec{V} = \vec{V}\cdot\nabla T \tag{3-542}$$

Using these relations in equation (3-539) we obtain the normalized energy equation as

$$\frac{\partial T}{\partial t} = -\nabla\cdot(\vec{V}T) + \frac{1}{Pe}\nabla^2 T + \frac{E}{Re}\Phi \tag{3-543}$$

where

$$\Phi = 2\left[\left(\frac{\partial u}{\partial x}\right)^2 + \left(\frac{\partial v}{\partial y}\right)^2\right] + \left[\frac{\partial v}{\partial x} + \frac{\partial u}{\partial y}\right]^2 \tag{3-544}$$

Except for the dissipation term, equation (3-543) is the same as the vorticity equation (2-12), with T replacing ζ and Pe replacing Re. Accordingly, all the finite-difference methods of this chapter are applicable to this energy equation. Since the dissipation function Φ does not depend on the independent variable T (i.e., there is no feedback), it does not affect the stability analyses. In fact, since the velocity field is now fixed, the linearized stability analysis is now more appropriate than when applied to the vorticity transport equation.

For gases, Pr = 0(1), so Pe = 0(Re), and the same methods are appropriate to both equations. For oils, Pr >> 1 and for liquid metals, Pr << 1. In these cases, the Peclét and Reynolds numbers differ greatly, and different computational methods may be appropriate to each equation. Further, the transient behavior of each equation can proceed on different time scales. For these reasons, Brown (1966) suggests the consideration of using different Δt for the two equations.

Note that the decoupling of the flow from the energy equations allows one flow solution to be used for many temperature solutions, with varying Pe, E, and boundary conditions (see, e.g., Roache and Mueller, 1970). Note also that, since the T equation is linear, its solution does not require additional Poisson solutions, so that the computation of many temperature solutions is economically feasible. Also, direct solution of the steady-state temperature field without iteration is possible with the EVP method (section III-B-8).

Equation (3-543), without the Φ term, can also be used to represent the simplest two-specie diffusion problems governed by the elementary Fick's law. See, e.g., Bird, et al., (1960). Then T is replaced by a specie concentration, and Pe is replaced by a Schmidt number.

III-F-2. Retention of Dissipation

The dissipation term is frequently dropped from incompressible temperature solutions, since E → 0 as the Mach number M → 0 for a fixed characteristic temperature difference. However, even at low Mach numbers, Φ can be important if $(\bar{T}_1 - \bar{T}_0)$ is small.

By definition,

$$E = \frac{\bar{u}_0^2}{C_P(\bar{T}_1 - \bar{T}_0)} \tag{3-545}$$

$$\frac{1}{E} = \left(\frac{\bar{T}_0}{\bar{T}_0}\right)\frac{C_P\,\bar{T}_1}{\bar{u}_0^2} - \frac{C_P\,\bar{T}_0}{\bar{u}_0^2} = \frac{C_P\,\bar{T}_0}{\bar{u}_0^2}\left(\frac{\bar{T}_1}{\bar{T}_0} - 1\right) \tag{3-546}$$

Now,

$$\frac{C_P\,\bar{T}_0}{\bar{u}_0^2} = \frac{C_P(\gamma R_g \bar{T}_0)}{\bar{u}_0^2\,\gamma R_g} = \frac{C_P}{\gamma R_g}\left(\frac{\bar{a}}{\bar{u}_0}\right)^2 \tag{3-547}$$

where $\bar{a}$ is the speed of sound, R_g is the gas constant, $R_g = C_P - C_V$, and $\gamma = C_P/C_V$.

Continuing,

$$\frac{C_P\,\bar{T}_0}{\bar{u}_0^2} = \frac{C_P}{\gamma R_g}\left(\frac{\bar{a}}{\bar{u}_0}\right)^2 = \frac{C_P}{\gamma R_g}\left(\frac{1}{M_0^2}\right) = \frac{C_P}{\gamma(C_P - C_V)}\,\frac{1}{M_0^2} = \frac{1}{(\gamma - 1)M_0^2} \tag{3-548}$$

$$\frac{1}{E} = \frac{1}{(\gamma - 1)M_o^2}\left(\frac{\bar{T}_1}{\bar{T}_o} - 1\right) \tag{3-549}$$

or

$$E = \frac{(\gamma - 1)M_o^2}{(\bar{T}_1/\bar{T}_o - 1)} \tag{3-550}$$

Thus $E \to 0$ as $M_o \to 0$ for $\bar{T}_1/\bar{T}_o > 1$. But for any small but non-zero M_o and for $\gamma > 1$, there will be a small enough temperature difference, expressed as $(\bar{T}_1/\bar{T}_o - 1)$, to insure large E and the importance of the dissipation term.

Note that, for $E = 0$, one can replace T by $-T$ in equation (3-543), meaning that solutions for temperature *defect* are identical to solutions for temperature *excess*. However, the solutions will differ for $E > 0$ because of the positive definite dissipation function.

III-F-3. Finite-Difference Representation of Dissipation

With $u = \partial\psi/\partial y$ and $v = -\,\partial\psi/\partial x$, equation (3-544) may be written as

$$\Phi = 2\left[\left(\frac{\partial^2\psi}{\partial x\partial y}\right)^2 + \left(\frac{\partial^2\psi}{\partial x\partial y}\right)^2\right] + \left[\frac{\partial^2\psi}{\partial y^2} - \frac{\partial^2\psi}{\partial x^2}\right]^2 \tag{3-551}$$

$$\Phi = 4\left[\frac{\partial^2\psi}{\partial x\partial y}\right]^2 + \left[\frac{\partial^2\psi}{\partial y^2} - \frac{\partial^2\psi}{\partial x^2}\right]^2 \tag{3-552}$$

Using the usual centered finite-difference forms of Section III-A-1, we evaluate Φ to $O(\Delta x^2, \Delta y^2)$ as

$$\Phi_{i,j} = \frac{1}{4}\left[\frac{\psi_{i+1,j+1} - \psi_{i+1,j-1} - \psi_{i-1,j+1} + \psi_{i-1,j-1}}{\Delta x\Delta y}\right]^2 + \left[\frac{\psi_{i,j+1} + \psi_{i,j-1} - 2\psi_{i,j}}{\Delta y^2} - \frac{\psi_{i+1,j} + \psi_{i-1,j} - 2\psi_{i,j}}{\Delta x^2}\right]^2 \tag{3-553}$$

There is never any need for one-sided difference evaluations, since Φ is calculated only at internal points and never on boundaries. The form (3-553) retains the positive definite nature of the dissipation function.

An interesting relation which is readily derived is

$$\Phi = \zeta^2 + S \tag{3-554}$$

where S is the source function in the Poisson form of the pressure equation (3-525). This form would be convenient and accurate to use, since ζ and S may be already calculated, but it is not positive definite. Truncation error may cause slight negative values of $(\zeta^2 + S)$. The error will not usually be appreciable, but the appearance of negative Φ can be confusing and misleading. However, this equation may be useful in calculating Φ on no-slip boundaries for the sake of a plotting routine. Since $S = 0$ on a no-slip wall, then

$$\Phi_w = \zeta_w^2 \tag{3-555}$$

which is again positive definite.

III-F-4. Boundary Conditions for Temperature and Concentration

The boundary conditions for temperature are analogous to those for vorticity, except at walls, where they are simpler. The boundary labeling has been given in Figure 3-22 (page 140). The wall temperatures may be fixed, for example, at the normalizing temperatures. For the base heating problem, this would give

$$T(B5) = 1 \quad , \quad T(B2) = 0 \tag{3-556}$$

For a base cooling problem, we would have

$$T(B5) = 0 \quad , \quad T(B2) = 1 \tag{3-557}$$

Any distributions are possible, of course, as, for example, a base specification along B5 of $T(ic,j) = f(y)$.

The wall temperature boundary condition may also be specified by the Neumann condition on the gradient $\partial T/\partial n$. In normalized variables, $\partial T/\partial n$ = Nu, the Nusselt number, which is a dimensionless heat transfer rate. The most common boundary condition is the adiabatic wall case, or Nu = 0. Even if Nu is not specified, it is of interest to calculate it as part of the solution. In the first mesh system, a Taylor-series expansion gives

$$T_{w+1} = T_w + \left.\frac{\partial T}{\partial n}\right|_w \Delta n + \frac{1}{2}\left.\frac{\partial^2 T}{\partial n^2}\right|_w \Delta n^2 + O(\Delta n^3) \tag{3-558}$$

or

$$\left.\frac{\partial T}{\partial n}\right|_w \equiv Nu = (T_{w+1} - T_w)/\Delta n + O(\Delta n) \tag{3-559}$$

To fix an adiabatic condition at a wall, we set Nu = 0, or

$$T_w = T_{w+1} \tag{3-560}$$

If explicit methods are used for equation (3-543), this condition is simply set after the interior-point calculation for T_{w+1}^{n+1}, by setting $T_w^{n+1} = T_{w+1}^{n+1}$. If implicit methods are used, equation (3-560) should be incorporated into the interior-point scheme, as in Miyakoda's method (3-529) for the pressure solution.

In evaluating these gradients, it is particularly important to be consistent with the finite-difference scheme used at interior points. If we treated interior calculated temperatures merely as data points, we might try a higher order extrapolation for $\partial T/\partial n = 0$ at an adiabatic surface. For example, one such evaluation would give

$$T_w = 4T_{w+1} - 3T_{w+2} \tag{3-561}$$

for an adiabatic wall temperature. However, such an evaluation is not *computationally* adiabatic. To be specific, consider a wall along x, at j = 1. Then, at (i, w+1), the heat-flux gradient is given by $\nabla^2 T|_{i.1}$ which is evaluated as

$$\nabla^2 T\Big|_{i,2} = \frac{\partial^2 T}{\partial x^2}\Big|_{i,2} + \frac{\partial^2 T}{\partial y^2}\Big|_{i,2} \tag{3-562}$$

and

$$\frac{\delta^2 T}{\delta t^2}\Big|_{i,2} = \frac{T_{i,3} + T_{i,1} - 2T_{i,2}}{\Delta y^2} = \frac{\dfrac{(T_{i,3} - T_{i,2})}{\Delta y} - \dfrac{(T_{i,2} - T_{i,1})}{\Delta y}}{\Delta y} = \frac{\dot{q}_{23} - \dot{q}_{12}}{\Delta y} \tag{3-563}$$

This formulation shows the essence of the heat-flux gradient term $\partial^2 T/\partial y^2$ at (i,2). In the control volume terminology of section III-A-2, $\dot{q}_{12}$ is the heat flux out of node 1 (on the wall) and into node 2. Thus, for the wall to be *computationally* adiabatic, $\dot{q}_{12}$ must equal zero, requiring $T_{i,1} = T_{i,2}$. That is, energy transfer between the wall and the fluid will occur, computationally, unless $T_w = T_{w+1}$. Higher order extrapolations for $\partial T/\partial y|_w = 0$ will be computationally adiabatic only if they are consistent with some higher order analog for $\partial^2 T/\partial y^2$ at (i,w+1).

Widespread confusion exists on the interpretation of the formal order of the truncation error of these expressions. Equation (3-559) would seem to imply that the evaluation of the Nusselt number is only first-order accurate. This is misleading. When consistent interior-point differencing is used, as shown above, equation (3-559) gives the *algebraically correct* dimensionless heat transfer rate which actually occurs, computationally, for the specified wall temperature. Or, when an adiabatic wall condition is set via equation (3-560), the wall *is* computationally adiabatic; no energy is conducted from the wall to the fluid. The Nusselt number so determined has no error at all, but is specified as a parameter of the problem, just as the Reynolds number is specified. The question of the formal truncation error is then not how accurately is Nu evaluated; rather the question is, how accurately is the wall *temperature* evaluated, for the specified Nusselt number? Writing equation (3-558) as

$$T_{w+1} = T_w + Nu\cdot\Delta n + O(\Delta n^2) \tag{3-564}$$

we see that the wall temperature evaluation is indeed second-order accurate.

The second mesh system, shown back in Figure 3-24, could be used for temperature solutions with a gradient boundary condition. A specified wall Nusselt number determines T_{ja-1}, inside the wall, from

$$T_{ja-1} = T_{ja} - Nu\cdot\Delta n \tag{3-565}$$

The computationally adiabatic condition is then met by $T_{ja-1} = T_{ja}$. But if a desired wall temperature T_w is specified by linear interpolation, setting

$$T_{ja-1} = 2T_w - T_{ja} \tag{3-566}$$

the result is first-order accuracy, and possible boundedness errors. See the discussion of the vorticity wall condition in section III-C-2.

The inflow temperature distribution across B4 in Figure 3-22 (page 140) is to be specified. The form is arbitrary, of course, but a particularly simple and significant form to consider is a similarity solution in the inflow boundary layer (Schlichting, 1968).

For a hot or cold upstream wall, a convenient boundary-layer temperature similarity solution can exist only for E = 0 and Pr = 1. In these cases, the temperature (difference) profile and the velocity profile are identical.

If the normalizing temperature difference for equation (3-543) is taken as the difference between the inflow "freestream" temperature at B4 in Figure 3-22 (page 140) outside the boundary layer and the wall temperature at B2, then the similarity temperature solution is

$$T_w = 0 \quad , \quad T(B4) = u(B4) \quad \text{for a cold wall at B2} \tag{3-567A}$$

$$T_w = 1 \quad , \quad T(B4) = 1 - u(B4) \text{ for a hot wall at B2} \tag{3-567B}$$

In this form, the external temperature range (with E = 0) is from 0 to +1 in both cases.

Note that, if a staggered mesh (Figure 3-24) is used for temperature, the inflow temperature profile must be set at $\Delta x/2$ away from the inflow velocity profile, which is inconsistent.

The temperature boundary conditions at the other boundaries may be handled as were the vorticity conditions in section III-C. The temperature and vorticity conditions should be consistent; particularly, a slip wall condition at B3 should be adiabatic. If the dissipation function Φ is needed at outflow B6 in the counterparts of equation (3-484) or (3-485), it may be evaluated by an extrapolation of any order or by one-sided differencing, since it is fixed for all time and therefore has no destabilizing feedback effect. Second-order extrapolation is recommended, as in

$$\Phi_{I,j} = \frac{3}{2}\,\Phi_{I-1,j} - \frac{1}{2}\,\Phi_{I-2,j} \tag{3-568}$$

The same remarks apply to the simple concentration diffusion equation. The gradient boundary condition is commonly expected on walls, if the wall is a permeable membrane.

The considerations given to convergence criteria and initial conditions in section III-D also apply to the temperature and concentration solutions.

III-F-5. Source Terms and Stiff Equations

Temperature and concentration equations like (3-543) can contain an additional term like +aT. This source term (a > 0) or sink term (a < 0) can represent, for example, the release or absorption of internal energy from a chemical reaction with a temperature-dependent reaction rate, or the reduction of dissolved oxygen in blood as the oxygen molecules are captured by red blood cells. Other forms for the term are possible, such as $a\cdot(T - T^*)$ or aT^p.

The addition of such a term to equation (3-543) looks innocent enough, since no additional spatial discretization is required. In fact, the stability considerations of this term, which makes the equation "stiff" (Curtiss and Hirschfelder,* 1952), can overshadow all others.

For conceptual clarity, we consider the source term alone. In this case, the problem is just the ODE

$$\frac{\partial T}{\partial t} = aT \tag{3-569}$$

---

*Curtiss and Hirschfelder (1952) actually define a "stiff" equation in terms of the FDE, with $|1/a\Delta t| << 1$ implying a stiff equation. They also consider a more general ODE than equation (3-569).

which has the exact solution

$$T(t) = T_1 e^{at} \tag{3-570}$$

where T_1 is the initial temperature. For $a > 0$, an exponentially increasing temperature is the true solution. Obviously, a "stable" finite-difference solution would be erroneous, since the correct solution is "unstable." The von Neumann stability condition is readily modified to

$$|G| \leq 1 + O(\Delta t) \tag{3-571}$$

in order to allow for such solutions. But a more meaningful criterion is the relative stability criterion given by Hamming (1962). The idea is that the exponential growth of the error E(t) should not swamp the true exponential error growth; with $E(t) = E_1 e^{bt}$, relative stability requires $b < a$.

But even for the exponential decay solution ($a < 0$), this term is important to stability. Consider leapfrog differencing for equation (3-569).

$$\frac{T^{n+1} - T^{n-1}}{2\Delta t} = aT^n \tag{3-572}$$

$$T^{n+1} = (2a\Delta t)T^n + (1)T^{n-1} \tag{3-573A}$$

$$T^n = (1)T^n + (0)T^{n-1} \tag{3-573B}$$

$$G = \begin{bmatrix} 2a\Delta t & 1 \\ 1 & 0 \end{bmatrix} \tag{3-574}$$

The eigenvalues, λ, are determined by

$$\begin{vmatrix} 2a\Delta t - \lambda & 1 \\ 1 & -\lambda \end{vmatrix} = 0 \tag{3-575}$$

$$\lambda^2 - (2a\Delta t)\lambda - 1 = 0 \tag{3-576}$$

$$\lambda_\pm = \frac{(2a\Delta t) \pm \sqrt{4a^2\Delta t^2 + 4}}{2} = a\Delta t \pm \sqrt{1 + a^2\Delta t^2} \tag{3-577}$$

Consider small Δt so that $a^2\Delta t^2 << 1$; then

$$\sqrt{1 + a^2\Delta t^2} \simeq 1 + \frac{1}{2} a^2\Delta t^2 \tag{3-578}$$

This gives

$$\lambda_+ = 1 + a\Delta t + \frac{1}{2} a^2\Delta t^2 \tag{3-579A}$$

$$\lambda_- = - 1 + a\Delta t - \frac{1}{2} a^2\Delta t^2 \tag{3-579B}$$

For $a > 0$, $|\lambda_+| > 1$ which is the counterpart of the exact, exponentially growing solution. But for $a < 0$, $|\lambda_-| > 1$. Even though the true solution decays exponentially and has a finite-difference counterpart in λ_+, the discretization has introduced a second solution mode which grows exponentially and will swamp the true solution.

Exercise: Show the FTCS differencing of equation (3-569) with $a < 0$ leads to the condition $\Delta t \leq 2/|a|$ for stability, and $\Delta t \leq 1/|a|$ for the zero overshoot.

A review of numerical methods for stiff equations is given by Seinfeld et al., (1970); their use in fluid dynamics specifically is treated by Blottner (1970). Generally, it is recommended that the stiff terms in the equation be evaluated at (n+1); this "implicit" formulation does not require the "implicit" solution of simultaneous equations at (n+1), however, since only ζ^{n+1} at (i) is involved. The fully "implicit" formulation is unconditionally stable for $a < 0$, and also gives zero overshoot (see problem 3-33).

III-G. METHODS FOR SOLVING THE PRIMITIVE EQUATIONS

The normalized Navier-Stokes equations written in terms of the primitive variables (u,v,p) have been given in equations (3-509) and (3-510). The general approach to such a problem formulation is quite similar to that for the (ψ,ζ) systems, with several interesting additional aspects. The relative advantages and disadvantages of the (u,v,p) system compared to the (ψ,ζ) system will be discussed.

III-G-1. General Considerations

The pressure in the (u,v,p) system should be solved from a Poisson equation, similar to that of section III-E, in order to model the elliptic nature of the continuum equations. The stability analyses for the (ψ,ζ) system may be directly applied to the (u,v,p) system. The pressure gradient terms are linearized out of equation (3-509), and terms like $u(\partial u/\partial x)$ are linearized to $\tilde{u}(\partial u/\partial x)$ where $\tilde{u}$ is the linear coefficient. The forms of the momentum equations (3-509) are then identical to the vorticity transport equation; the same methods and the same stability restrictions apply. Any method in section III-A can be applied to the Poisson pressure equation, at least from the point of linear stability analysis. The momentum equations also have a simple conservation form, obtainable by replacing $\vec{V}\cdot\nabla u$ by $\nabla\cdot(u\vec{V})$ as in the vorticity transport equation. But the conservation concept is more difficult in regard to the mass conservation. A change of form is required in the Poisson equations, which we now consider.

III-G-2. Basic Equations

The conservation form of the momentum equations is easily obtained as

$$\frac{\partial u}{\partial t} + \frac{\partial(u^2)}{\partial x} + \frac{\partial(uv)}{\partial y} = -\frac{\partial P}{\partial x} + \frac{1}{Re}\left(\frac{\partial^2 u}{\partial x^2} + \frac{\partial^2 u}{\partial y^2}\right) \qquad (3\text{-}580A)$$

$$\frac{\partial v}{\partial t} + \frac{\partial(uv)}{\partial x} + \frac{\partial(v^2)}{\partial y} = -\frac{\partial P}{\partial y} + \frac{1}{Re}\left(\frac{\partial^2 v}{\partial x^2} + \frac{\partial^2 v}{\partial y^2}\right) \qquad (3\text{-}580B)$$

These equations "conserve" momentum, as equation (2-10) "conserves" vorticity. A Poisson equation for pressure, similar to the previously obtained equation (3-525), is obtained by differentiation and addition of (3-580A,B).

$$\nabla^2 P = -\frac{\partial^2(u^2)}{\partial x^2} - 2\frac{\partial^2(uv)}{\partial x\partial y} - \frac{\partial^2(v^2)}{\partial y^2} - \frac{\partial D}{\partial t} + \frac{1}{Re}\left(\frac{\partial^2 D}{\partial x^2} + \frac{\partial^2 D}{\partial y^2}\right) \equiv S_P \qquad (3\text{-}581A)$$

where the "dilation" term D is

$$D = \frac{\partial u}{\partial x} + \frac{\partial v}{\partial y} \qquad (3\text{-}581B)$$

Obviously, S_P in equation (3-581) is equal to S in (3-525) in the continuum.

This remarkable equation and its curious features was first presented by the Los Alamos group (Harlow and Welch, 1965; Welch et al., 1966) in conjunction with their famous MAC method (section III-G-4). The remarkable aspect is that the continuity equation (3-509C) just states that the continuum D = 0, yet it is important to evaluate the terms containing D in the Poisson equation (3-581). Because of incompatible initial conditions or because of incomplete iterative solution of the Poisson equation, the finite-difference $D_{ij} \neq 0$. The term could still be set = 0 in equation (3-581) without changing the order of the truncation error, but, because the Poisson equation is solved iteratively, the error accumulates. The result is not only inaccuracy, but also a nonlinear instability in the momentum equations. The inclusion of D can eliminate the nonlinear instability. The time derivative, $\partial D/\partial t$, is to be evaluated by whatever time differencing is used for $\partial u/\partial t$ and $\partial v/\partial t$, by forcing $D^{n+1} = 0$. (For forward time differences, this gives $\partial D/\partial t = -D^n_{ij}/\Delta t$.)

This instability and its cure were first presented by Harlow and Welch (1965). The same behavior has been noted since then by other workers using the MAC method, such as Pagnani (1968), Gawain and Pritchett (1966), Slotta et al., (1969), and by Chan et al., (1969) using a modified MAC method. The concept has been generalized to other time-like solutions by Hirt and Harlow (1967); their paper is recommended reading. Donovan (1968, 1970) and Putre (1970) found the same behavior using slightly different equations and the same cell structure as MAC (see Section III-G-4). Their equations are obtained by using the continuity equation (3-509C) to evaluate

$$\frac{\partial^2 u}{\partial x^2} + \frac{\partial}{\partial x}\left(-\frac{\partial v}{\partial y}\right) = -\frac{\partial^2 v}{\partial x \partial y} \tag{3-582A}$$

$$\frac{\partial^2 v}{\partial y^2} = \frac{\partial}{\partial y}\left(-\frac{\partial u}{\partial x}\right) = -\frac{\partial^2 u}{\partial x \partial y} \tag{3-582B}$$

in equation (3-580) and setting D = 0 in the spatial terms of equation (3-581). The result is

$$\frac{\partial u}{\partial t} = \frac{\partial (u^2)}{\partial x} + \frac{\partial (uv)}{\partial y} = -\frac{\partial P}{\partial x} + \frac{1}{Re}\left(\frac{\partial^2 u}{\partial y^2} - \frac{\partial^2 v}{\partial x \partial y}\right) \tag{3-583A}$$

$$\frac{\partial v}{\partial t} + \frac{\partial (uv)}{\partial x} + \frac{\partial (v^2)}{\partial y} = -\frac{\partial P}{\partial y} + \frac{1}{Re}\left(\frac{\partial^2 v}{\partial x^2} - \frac{\partial^2 u}{\partial x \partial y}\right) \tag{3-583B}$$

$$\nabla^2 P = -\frac{\partial^2 (u^2)}{\partial x^2} - 2\frac{\partial^2 (uv)}{\partial x \partial y} - \frac{\partial^2 (v^2)}{\partial y^2} - \frac{\partial D}{\partial t} \tag{3-584}$$

From these works, it would appear that the $\nabla^2 D$ term in equation (3-581A) is not as important as the $\partial D/\partial t$ term. But, since the continuity equation is not satisfied exactly, then equation (3-582) is not satisfied exactly in finite-difference form. Thus, the x-diffusion of u-momentum in equation (3-483A) is not conservative, nor is the y-diffusion of u-momentum in equation (3-583B). Williams (1969) also retained the $\partial D/\partial t$ term, using leapfrog time differencing in conjunction with Arakawa's method (section III-A-21). Since he used a direct method for the Poisson equation, the iteration error in the cylindrical coordinate version of equation (3-581) was zero.

It is also possible to devise methods which do identically have D = 0, that is, which conserve mass (volume); see Piacsek and Williams (1970).

III-G-3. Boundary Conditions in Primitive Variables

The boundary conditions along a no-slip wall are just $u_w = 0$ and $v_w = 0$ for all time. This is evidently a great advantage for the use of implicit methods, since no iteration is required for the advance boundary condition. However, there is an impediment to the successful application of implicit methods to the primitive equations, due to a nonlinear instability of the pressure term (Aziz, 1966; Aziz and Hellums, 1967), which may be curable by retaining $\partial D/\partial t$ as in equation (3-581) or (3-584). Note that sharp convex-corner velocities are single-valued. The slip-wall condition may be used along the upper boundary or some other slip wall. For the slip wall along x, the conditions are $v_w = 0$, and (plausibly) $\partial u/\partial y|_w = 0$. For a node on the wall, the latter gives $u_w = u_{w+1}$ using second-order spatial differences. At a sharp convex corner, the slip-wall velocities would be multi-valued.

At inflow (see Figure 3-22 on page 140) u is specified, often as $u(y) = U_o$. The v-component can also be set, often as v = 0, or can be allowed to develop by using $\partial v/\partial x = 0$ or $v_{1,j} = v_{2,j}$ as done by Slotta, et al., (1969). The same v-condition at outflow (Slotta, et al., 1969) gives

$$v_I = v_{I-1} \qquad (3\text{-}585)$$

This results in

$$\left.\frac{\delta v}{\delta y}\right|_I = \left.\frac{\delta v}{\delta y}\right|_{I-1} = \left.\frac{\delta v}{\delta y}\right|_{I-1/2} \qquad (3\text{-}586)$$

This term can be evaluated from known internal values at (I-1). Applying the continuity equation at (I-1/2) gives

$$\left.\frac{\delta u}{\delta x}\right|_{I-1/2,j} = -\left.\frac{\delta v}{\delta y}\right|_{I-1/2,j} = \left.\frac{\delta v}{\delta y}\right|_{I-1,j} \qquad (3\text{-}587)$$

Then the outflow value of u is set from equation (3-587) from known internal values. To second-order accuracy, it gives

$$u_{I,j} = U_{I-1,j} - \frac{\Delta x}{2\Delta y}\,(v_{I-1,j+1} - v_{I-1,j-1}) \qquad (3\text{-}588)$$

The indexing in equation (3-588) often appears different due to the cell definitions of the MAC method (section III-G-4). In a more general notation, applicable to both the mesh systems, equation (3-588) can be written as

$$u_{I,j} = u_{I-1,j} - \frac{\Delta x}{\Delta y}\,(v_{I-1,j+1/2} - v_{I-1,j-1/2}) \qquad (3\text{-}589)$$

The pressure conditions are still the Neumann gradient conditions, as in section III-E.

Putre (1970) specified P constant along inflow and outflow and specified the total pressure drop $P_{in} - P_{out}$. This was appropriate to his study of hydrodynamic step bearings. He allowed the u-component to develop at inflow and outflow by using $\delta u/\delta x = 0$. With the continuity equation, this requires $\partial v/\partial y = 0$; and since $v_w = 0$, it sets $v_{in} = v_{out} = 0$. It is not known if this evaluation of velocities would provide convergence if a less restrictive pressure evaluation were used.

III-G-4. The MAC Method

The MAC method (for Marker And Cell) of Harlow and Welch (1965) and Welch et al., (1966), is distinguished by four features: the form of the primitive-variable equations used, the differencing scheme, the cell structure, and the use of "marker particles."

The form of the equations and the rationale have already been given in equations (3-580) and (3-581). The differencing scheme for the momentum equations is basically just the forward-time, centered-space (FTCS) method described in sections III-A-1, et seq. The stability has been discussed in section III-A-5 et seq. It is unconditionally unstable for inviscid flows. The MAC equations are a little different, however, because of the cell structure.

The MAC cell structure is shown in Figure 3-33. Pressure is defined at the node locations in the cell centers, as in $P_{i,j}$ and $P_{i+1,j}$. The u-velocities are defined along the right- and left-hand boundaries, as in $u_{i+1,2,j}$ and $u_{i-1/2,j}$. The v-velocities are defined along the top and bottom boundaries as in $v_{i,j+1/2}$ and $v_{i,j-1/2}$. The finite-difference spatial derivatives are then evaluated by centered differences over a single mesh spacing, where possible. For example, in evaluating

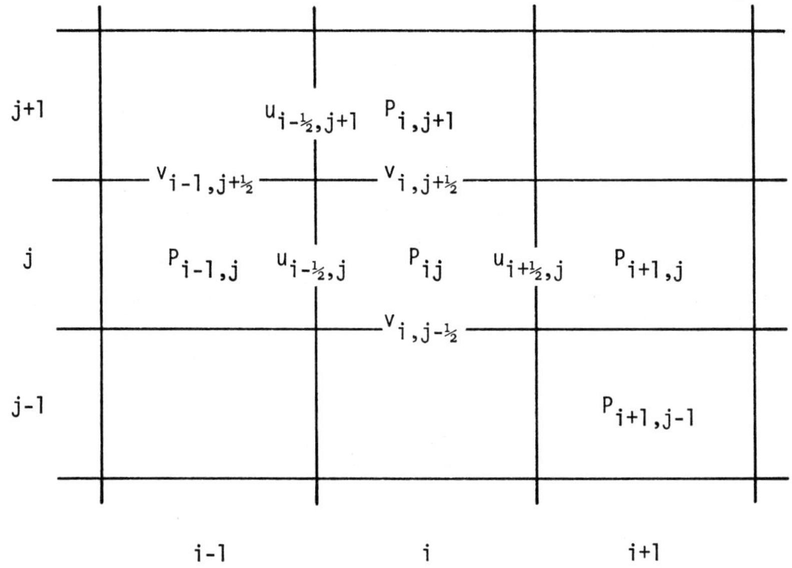

Figure 3-33. MAC cell structure.

$$\left.\frac{\partial u}{\partial t}\right|_{i+\frac{1}{2},j} \simeq \left.\frac{\delta u}{\delta t}\right|_{i+\frac{1}{2},j} = \frac{u^{n+1}_{i+\frac{1}{2},j} - u^{n}_{i+\frac{1}{2},j}}{\Delta t} \tag{3-590}$$

the term $\left.\frac{\partial P^n}{\partial x}\right|_{i+\frac{1}{2},j}$ is evaluated as

$$\left.\frac{\delta P^n}{\delta x}\right|_{i+\frac{1}{2},j} = \frac{P^n_{i+1,j} - P^n_{i,j}}{\Delta x} \tag{3-591}$$

But terms such as $\left.\frac{\partial^2 u}{\partial x^2}\right|^n_{i+\frac{1}{2},j}$ are evaluated as

$$\left.\frac{\delta^2 u^n}{\delta x^2}\right|_{i+\frac{1}{2},j} = \frac{u^2_{i+1\frac{1}{2},j} - 2u_{i+\frac{1}{2},j} + u_{i-\frac{1}{2},j}}{\Delta x^2} \tag{3-592}$$

where the superscript n is understood on the right-hand side. The evaluation of

$$\left.\frac{\delta(u^2)}{\delta x^2}\right|^n_{i+\frac{1}{2},j} = \frac{u^2_{i+1,j} - u^2_{i,j}}{\Delta x} \tag{3-593}$$

requires the definition of u at the node values. These are obtained by averaging, as in

$$u_{i+1,j} = \frac{1}{2}\left(u_{i+1\frac{1}{2},j} + u_{i+\frac{1}{2},j}\right) \tag{3-594}$$

The product terms like (uv) are evaluated as the product of the averages, not as the average of the products.

$$(uv)_{i+\frac{1}{2},j+\frac{1}{2}} = \frac{1}{2}\left(u_{i+\frac{1}{2},j} + u_{i+\frac{1}{2},j+1}\right) \times \frac{1}{2}\left(v_{i+1,j+\frac{1}{2}} + v_{i,j+\frac{1}{2}}\right) \tag{3-595}$$

The final equations corresponding to equation (3-580) are as follows. The superscript n is understood on the right-hand side.

$$u^{n+1}_{i+\frac{1}{2},j} = u_{i+\frac{1}{2},j} + \Delta t\left\{-\frac{u^2_{i+1,j} - u^2_{ij}}{\Delta x} - \frac{(uv)_{i+\frac{1}{2},j+\frac{1}{2}} - (uv)_{i+\frac{1}{2},j-\frac{1}{2}}}{\Delta y} - \frac{P_{i+1,j} - P_{ij}}{\Delta x}\right.$$
$$\left. + \frac{1}{Re}\left(\frac{u_{i+3/2,j} - 2u_{i+\frac{1}{2},j} + u_{i-\frac{1}{2},j}}{\Delta x^2} + \frac{u_{i+\frac{1}{2},j+1} - 2u_{i+\frac{1}{2},j} + u_{i+\frac{1}{2},j-1}}{\Delta y^2}\right)\right\} \tag{3-596A}$$

$$v^{n+1}_{i,j+\frac{1}{2}} = v_{i,j+\frac{1}{2}} + \Delta t\left\{-\frac{v^2_{i,j+1} - v^2_{ij}}{\Delta y} - \frac{(uv)_{i+\frac{1}{2},j+\frac{1}{2}} - (uv)_{i-\frac{1}{2},j+\frac{1}{2}}}{\Delta x} - \frac{P_{i,j+1} - P_{ij}}{\Delta y}\right.$$
$$\left. + \frac{1}{Re}\left(\frac{v_{i,j+3/2} - 2v_{i,j+\frac{1}{2}} + v_{i,j-\frac{1}{2}}}{\Delta y^2} + \frac{v_{i+1,j+\frac{1}{2}} - 2v_{i,j+\frac{1}{2}} + v_{i-1,j+\frac{1}{2}}}{\Delta y^2}\right)\right\} \tag{3-596B}$$

The S_P term in equation (3-581) is evaluated with the same type of differencing and by using

$$\left.\frac{\partial D}{\partial t}\right|^n_{ij} = \left.\frac{D^{n+1} - D^n}{\Delta t}\right|_{ij} = -\frac{D_{ij}}{\Delta t} \tag{3-597}$$

(That is, the equation *sets* $D^{n+1} = 0$.) The final result is

$$-S_P\Big|_{ij} = \frac{u^2_{i+1,j} - 2u^2_{ij} + u^2_{i-1,j}}{\Delta x^2} + \frac{2}{\Delta x\Delta y}\left[(uv)_{i+\frac{1}{2},j+\frac{1}{2}} - (uv)_{i+\frac{1}{2},j-\frac{1}{2}} - (uv)_{i-\frac{1}{2},j+\frac{1}{2}}\right.$$
$$\left. + (uv)_{i-\frac{1}{2},j-\frac{1}{2}}\right] + \frac{v^2_{i,j+1} - 2v^2_{ij} + v^2_{i,j-1}}{\Delta y^2} - \frac{D_{ij}}{\Delta t}$$
$$- \frac{1}{Re}\left(\frac{D_{i+1,j} - 2D_{ij} + D_{i-1,j}}{\Delta x^2} + \frac{D_{i,j+1} - 2D_{ij} + D_{i,j-1}}{\Delta y^2}\right) \tag{3-598A}$$

wherein

$$D_{ij} = \frac{u_{i+\frac{1}{2},j} - u_{i-\frac{1}{2},j}}{\Delta x} + \frac{v_{i,j+\frac{1}{2}} - v_{i,j-\frac{1}{2}}}{\Delta y} \tag{3-598B}$$

The boundary conditions on u and v are evaluated in accordance with the averaging definitions of node points. For example, a no-slip wall along j = w in Figure 3-33 gives

$$u_w = 0 \quad , \quad u_{i+\frac{1}{2},w} = 0 \quad , \quad u_{i-\frac{1}{2},w} = 0 \tag{3-599}$$

$$v_w = 0 \quad , \quad v_{i,w} = \frac{1}{2}\left(v_{i,w+\frac{1}{2}} + v_{i,w-\frac{1}{2}}\right) \tag{3-600}$$

or

$$v_{i,w-\frac{1}{2}} = - v_{i,w+\frac{1}{2}} \tag{3-601}$$

The use of this equation to determine a fictitious value of $v_{i,w-1/2}$ inside the wall gives an effective $v_{i,w} = 0$ when regular interior-point differencing of $O(\Delta y^2)$ is used at the first internal points (i,w+1) and (i,w+1/2). In section III-C-2 (see also Problems 3-24, 25, 26), we noted that the use of equations similar to (3-600) to define wall values leads to errors in the vorticity diffusion terms. However, in the MAC system, the variables so defined only affect advection terms, which are correctly treated by this procedure. Since the diffusion of v does not enter into the u-momentum equation, no error occurs.

A plausible formulation for a slip wall would be

$$u_{i+\frac{1}{2},w} = u_{i+\frac{1}{2},w+1} \tag{3-602A}$$

$$v_{i,w-\frac{1}{2}} = - v_{i,w+\frac{1}{2}} \tag{3-602B}$$

The boundary conditions at other boundaries are also obvious analogs of those previously given in the first mesh system.

The marker particles used in MAC are "massless particles" which move with the advection field. They do not participate directly in the computation. At internal points, there is no feedback from the markers, and therefore no stability considerations arise. By tracing and plotting the position of the marker particles, one obtains a "streakline" picture, analogous to a smoke tunnel or dye visualization photograph.

The positions, or Lagrangian coordinates, of each particle (x_p^n, y_p^n) are obtained by numerical integration from some initial position (x_p^o, y_p^o) at time = 0;

$$x_p^n = x_p^o + \int_o^t u_p \, dt \tag{3-603A}$$

$$y_p^n = y_p^o + \int_o^t v_p \, dt \tag{3-603B}$$

where u_p and v_p are the velocities in the Eulerian mesh at the time-dependent location of the particle. Consistent with the forward time integration of MAC, equation (3-603) is evaluated sequentially as

$$x_p^{n+1} = x_p^n + u_p \cdot \Delta t \tag{3-604A}$$

$$y_p^{n+1} = y_p^n + v_p \cdot \Delta t \tag{3-604B}$$

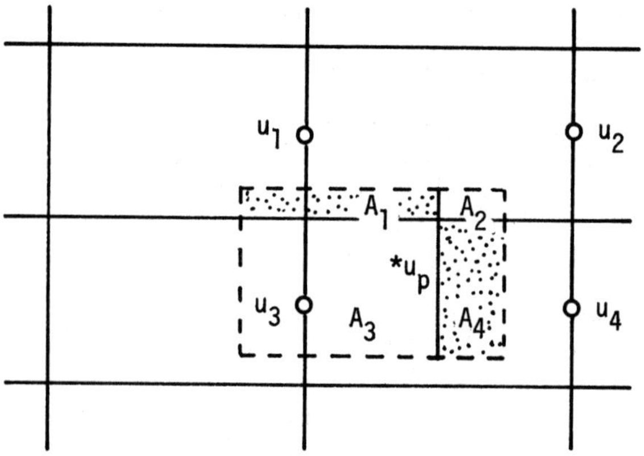

$$u_p = \frac{A_1u_1 + A_2u_2 + A_3u_3 + A_4u_4}{\Delta x \cdot \Delta y}$$

(a) bivariate linear interpolation for u_p

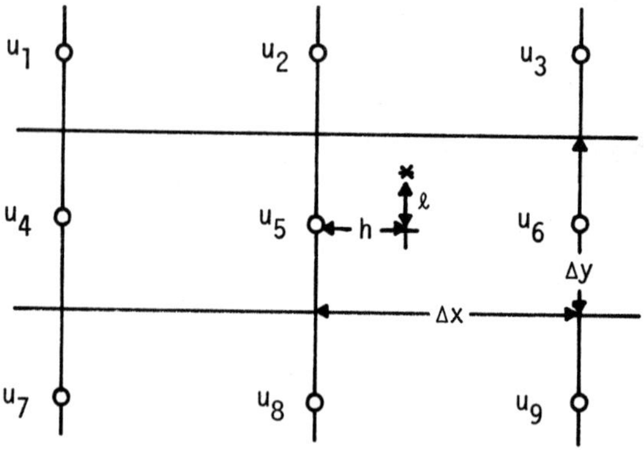

$a = h/\Delta x, \quad b = \ell/\Delta y,$

$$u_p = \frac{a}{2}(u_6 - u_4) + (u_2 - u_8)$$
$$+ \frac{1}{2}[a^2(u_6 + u_4 - 2u_5) + b^2(u_2 + u_8 - 2u_5) + \tfrac{1}{2}ab(u_3 + u_7 - u_1 - u_9)]$$

(b) second-order interpolation for u_p (after Chan et al., 1969)

Figure 3-34. Particle velocity weighting for MAC.

In MAC, the particle velocities are evaluated by two-variable linear interpolation. The areas A_1 through A_4 are defined by the location (x_p^n, y_p^n) as shown in Figure 3-34a.

$$u_p = \frac{A_1 u_1 + A_2 u_2 + A_3 u_3 + A_4 u_4}{\Delta x \cdot \Delta y} \tag{3-605}$$

Chan et al., (1969) have used the second-order evaluation shown in Figure 3-34b, resulting in increased accuracy near velocity extrema. Most computations with MAC-like codes have been done with a global average of 4 or 5 "marker particles" per cell.

This streakline computation is not restricted to MAC, nor even to primitive variable equations (see, e.g., Thoman and Szewczyk, 1966). It is considered an essential part of the MAC method only because of the extensive use of MAC in *free surface* problems, such as surface wave formation. The shape of the free surface is not known *a priori*; it is defined, in the solution, by the position of the marker particles. (Additional references on free surface problems will be given in Chapter VI.) We note here that the boundary conditions at the free surface require zero tangential stress and a normal stress which balances any externally applied normal stress. The application of these conditions requires a knowledge of not only the location of the free surface in a cell but also its slope and curvature. Then the shape of the surface does affect the dynamics, through surface tension effects and through the "partial cell" formulation required of the difference equations. Through this route, the particle position calculation *does* feed back to the dynamics. Possible anomalous behavior includes nonlinear instability and a numerical *surface tension* effect.

Important improvements in the MAC partial cell treatment and in the computation of marker positions were made by Chan et al., (1969, 1970) and Nichols (1970). Pagnani (1968) used MAC calculations of natural convection. Daly (1969A,B) presented two-fluid MAC computation in cylindrical coordinates, including the details of the surface tension calculation. Daly and Pracht (1969) used the full two-fluid approach and the solute transport approximation to study density current surges. Viecelli (1969) presented an ingenious method of treating curved surfaces with MAC in a regular mesh, using the techniques developed for free-surface calculations. At each time step, he specifies an external pressure which forces the "free surface" to align with the desired boundary shape. Other applications of MAC include the work of Hwang (1966), Gawain and Pritchett (1968), Slotta et al., (1969), Donovan (1968, 1970), Crowley (1970A), Putre (1970), Mitchell (1970), and Easton (1969).

The SMAC method (Amsden and Harlow, 1970A,B) is a somewhat simplified MAC method in which pressure is not solved, but the divergence (continuity) condition on velocities is satisfied directly. Some techniques of the (ψ,ζ) system are used and some boundary treatments are simplified. The MACRL method of Pracht (1971) uses an iteratively implicit formulation for the diffusion and pressure-gradient terms. The implicit equations are solved by an iteration process which includes the Poisson equation and the boundary conditions. The SUMMAC method (Chan and Street, 1970; Chan et al., 1971) improves the free-surface boundary treatment.

III-G-5. Other Methods Using Primitive Variables

An imaginative method for obtaining a steady-state solution in primitive variables was introduced by Chorin (1967)(see also Chu, 1968) and later used by Plows (1968). The incompressible problem is solved by an asymptotic time solution of equations containing an artificial *compressibility*, which is designed to vanish as steady state is approached. A similar concept was used with a fractional-time-step method by Vladimirova et al., (1966). We briefly note here that the true compressible equations should not generally be used for an incompressible flow problem just by using low Mach number (see the more complete discussion in section V-I). Likewise, the use of the incompressible equation for pressure,

$$\frac{DP}{Dt} \equiv \frac{\partial P}{\partial t} + u\frac{\partial P}{\partial y} + v\frac{\partial P}{\partial y} = \frac{\partial P}{\partial t} + \frac{\partial (uP)}{\partial x} + \frac{\partial (vP)}{\partial y} = 0 \tag{3-606}$$

(see Schlichting, 1968) is not recommended. This equation is based on the energy equation, with low-Mach-number approximations. This energy-equation coupling is responsible for the hyperbolic* form of this equation contrasted to the elliptic form of the Poisson pressure equation. While valid, the hyperbolic equation allows for wave propagations which dominate the stability considerations. Implicit methods would have to be used, and the accuracy would be limited by these unwanted wave solutions. It has been well-known since Charney et al., (1950) that the Poisson form, which suppresses these unwanted waves, is preferable.

The method of Callens (1970) gives solutions valid only in the steady state. Its truncation error is $O(\Delta x, \Delta y)$. Although all the terms of the full Navier-Stokes equations are used, the method is not applicable unless the conditions required by boundary-layer assumptions (see section VI-D) exist.

The upwind differencing methods of section III-A-8 through III-A-11 can also be used with the primitive equations, e.g. Jamet et al. (1970). Considerations of artificial viscosity still apply, as shown in the next exercise.

Exercise: Linearize the left-hand-side of equation (3-580A) to

$$\frac{Du}{Dt} = \frac{\partial u}{\partial t} + \tilde{u}\frac{\partial u}{\partial x} + \tilde{v}\frac{\partial u}{\partial y}$$

Show that the use of upwind differencing on the primitive equations gives the same artificial viscosity effect as it does on the vorticity transport equation. See equations (3-176) to (3-179), section III-A-8.

III-G-6. *Relative Merits of the (ψ,ζ) and (u,v,P) Systems*

The relative merits of using the (ψ,ζ) system or the (u,v,P) system depend on the problem. Although previous experience is always a major factor, it will be seen that, for most cases excluding free surfaces (or other fluid interface problems), the (ψ,ζ) system is recommended.

For a basis of comparison, we first consider two-dimensional flows *using iterative methods* for the Poisson equations, with no free surfaces.

If the transient pressure solution is *not* required, the (ψ,ζ) system involves one parabolic vorticity transport equation and one elliptic equation $\nabla^2\psi = \zeta$ with some (perhaps all) Dirichlet boundaries. (The steady-state pressure solution involves an elliptic equation at just the last time step, which should not greatly influence the choice of a method.) The (u,v,P) system involves two parabolic momentum transport equations, and one elliptic equation, $\nabla^2 P = S_P$, with all Neumann boundaries. The ζ transport equation involves two additional differentiations of ψ to obtain the advection velocities, but the u and v momentum transport equations are somewhat more complicated, either because of the retained dilation terms D_{ij} (in the MAC mesh system, they are considerably more complicated) or because of special forms required for mass (volume) conservation. Implicit methods may be used for the ζ transport equation, although iteration may be required because of implicit values of ζ_w^{n+1} at no-slip walls. In the (u,v,P) system, the implicit values of u_w^{n+1} and v_w^{n+1} are known explicitly for all times, but a difficulty exists with a nonlinear instability (see page 195). The elliptic equation, $\nabla^2 P = S_P$, takes much longer than $\nabla^2\psi = \zeta$ to achieve iteration convergence, because of the difference in boundary conditions.

If the transient pressure solution *is* required, a Poisson equation for $\nabla^2 P = S$ with the Neumann boundary conditions is also required in the (ψ,ζ) system. If iterative methods are used and a pressure solution is demanded at every Δt, the (u,v,P) system is faster. But note that, with the use of the explicit methods which have historically been required in (u,v,P), the time step Δt is so small that a pressure solution may not be required at *every* step in the (ψ,ζ) solution, but only periodically. (It is usually difficult to make

*Its influence spreads only forward, in time and in space. In one spatial dimension, the linearized form is $\partial P/\partial t = -\tilde{u}\,\partial P/\partial x$, and the influence of a point (x,t) spreads along the characteristic line through (x,t) of slope $1/u$.

intelligent use of all those numbers, anyhow.) If the sampling rate for the pressure solution is about one every 10 steps or smaller, the (ψ,ζ) system is again recommended. Also, we compare the two systems if transient streamline patterns are required. In the (ψ,ζ) system, streamline values at constant ψ need only be found by interpolation. In the (u,v,P) solution, they must be found by integration, the most accurate way being by solution of the *Poisson* equation, $\nabla^2\psi = \delta u/\delta y - \delta v/\delta x$.

Now we change the basic case by considering the use of *direct methods for the Poisson equations*. These take a great deal of program development time (depending on previous experience, of course) and the development time will be longer in the (ψ,ζ) system if pressure solutions are required, since the direct solution of two Poisson equations with different boundary conditions is required. If the pressure solution is not required, the development time may be somewhat shorter in the (ψ,ζ) system, since Dirichlet conditions are a little easier to set up than the Neumann conditions in most direct methods. (The EVP method of section III-B-8 is an exception.) In these cases, the Poisson solutions take less time than the parabolic (transport) solutions. In running time, the single ζ transport equation, simpler than each of the two momentum transport equations, again makes the (ψ,ζ) system faster.

These relative merits are unaffected by the addition of body-force terms; by the specification of cylindrical, spherical, or other orthogonal coordinates; or by the use of variable property equations. The definitions of ζ are altered (or terms merely analogous to ζ may be used) but similar equations may be derived. For example, a variable viscosity momentum equation may be written as

$$\frac{\partial u}{\partial t} + u\frac{\partial u}{\partial x} + v\frac{\partial u}{\partial y} = -\frac{\partial P}{\partial x} + \frac{1}{Re}\left\{2\frac{\partial}{\partial x}\left[\mu\frac{\partial u}{\partial x}\right] + \frac{\partial}{\partial y}\left[\mu(u_y + v_x)\right]\right\} \tag{3-607}$$

where

$$Re = \bar{\rho}\bar{U}_o\bar{L}/\bar{\mu}_o \quad , \quad \mu = \bar{\mu}/\bar{\mu}_o = \mu(x,y) \tag{3-608}$$

and $\bar{\mu}_o$ is the reference viscosity (perhaps a freestream value). A corresponding vorticity transport equation can be derived* (see Problem 3-34) as

$$\frac{\partial\zeta}{\partial t} = -\nabla\cdot(\vec{V}\zeta) + \frac{1}{Re}\left[\nabla^2\mu\zeta - 2(\mu_{xx}\psi_{yy} - 2\mu_{xy}\psi_{xy} + \mu_{yy}\psi_{xx})\right] \tag{3-609}$$

Much more complicated problems are possible in the (ψ,ζ) system; Shavit and Lavan (1971) use a two-fluid, variable-property vorticity transport equation in cylindrical coordinates.

We also repeat that marker particle positions are no more difficult to calculate in the (ψ,ζ) system than in the MAC (u,v,P) system, so that the availability of streakline plots do not constitute an advantage of MAC over a (ψ,ζ) system. The streamline information is, of course, easier in the (ψ,ζ) system.

However, if we consider problems with free surfaces or fluid interfaces, the use of the primitive variable system (u,v,P) becomes recommended by history, since it has often been used successfully with these problems. But in the (ψ,ζ) system, there exists the difficulty of formulating the boundary conditions at the free surface, especially for *transient* free surface flows such as fuel sloshing. (See references in Section VI-D.)

We also note that, although many authors have stated otherwise, the extension of the (ψ,ζ) system to three-dimensional flows *is* possible, and the system still retains some advantages over the primitive variable system (see the next section).

---

*Private communication from D. Reed and Prof. W. Oberkampf, University of Texas at Austin.

III-H. THREE-DIMENSIONAL FLOWS

It is practical to calculate some three-dimensional flows on the best of currently available computers. The 3-D Navier-Stokes equations which correspond to equations (3-509) and (3-510) are

$$\frac{\partial u}{\partial t} + \frac{\partial (u^2)}{\partial x} + \frac{\partial (uv)}{\partial y} + \frac{\partial (uw)}{\partial z} = -\frac{\partial P}{\partial x} + \frac{1}{Re} \nabla^2 u \qquad (3\text{-}610A)$$

$$\frac{\partial v}{\partial t} + \frac{\partial (uv)}{\partial x} + \frac{\partial (v^2)}{\partial y} + \frac{\partial (vw)}{\partial z} = -\frac{\partial P}{\partial y} + \frac{1}{Re} \nabla^2 v \qquad (3\text{-}610B)$$

$$\frac{\partial w}{\partial t} + \frac{\partial (uw)}{\partial x} + \frac{\partial (vw)}{\partial y} + \frac{\partial (w^2)}{\partial z} = -\frac{\partial P}{\partial z} + \frac{1}{Re} \nabla^2 w \qquad (3\text{-}610C)$$

$$D \equiv \nabla \cdot \vec{V} \equiv \frac{\partial u}{\partial x} + \frac{\partial v}{\partial y} + \frac{\partial w}{\partial z} = 0 \qquad (3\text{-}611)$$

where

$$\nabla^2 f = \frac{\partial^2 f}{\partial x^2} + \frac{\partial^2 f}{\partial y^2} + \frac{\partial^2 f}{\partial z^2} \quad ; \quad f = u,v,w \qquad (3\text{-}612)$$

These primitive variable equations may be solved by the use of the same methods as applied to the 2D equations. We have already noted some extensions to 3D in section III-A. For example, the typical Courant-number restriction for explicit methods (without time-splitting) becomes

$$C_x + C_y + C_z \leq 1 \qquad (3\text{-}613)$$

or

$$\frac{u\Delta t}{\Delta x} + \frac{v\Delta t}{\Delta y} + \frac{w\Delta t}{\Delta z} \leq 1 \qquad (3\text{-}614)$$

A Poisson equation of the form

$$\nabla^2 P = \frac{\partial^2 P}{\partial x^2} + \frac{\partial^2 P}{\partial y^2} + \frac{\partial^2 P}{\partial z^2} = S_P \qquad (3\text{-}615)$$

is derivable; the proper form again includes a dilation term, $\partial D/\partial t$, even in cylindrical coordinates (Williams, 1969).

In the MAC mesh, w velocities are defined at $(i,j,w \pm 1/2)$, etc. The 3D Poisson equation still has the Neumann boundary conditions; the added dimension makes iteration convergence very slow, and direct methods become attractive. Even then, the problems may be expected to be very time consuming. Testing and development in a coarse mesh is now extremely important. For example, Williams (1969) used Arakawa's method (section III-A-21) for the advection terms and a direct method which reduced the time for the Poisson solution to 25% of the total computer time. On a Univac 1108, the computer time, per time step Δt, was 2 seconds for a 14 x 14 x 14 grid, and 96 seconds for a 60 x 34 x 34 grid.

3D incompressible flows may also be formulated and computed in a (ψ,ζ)-like system, as shown by Aziz and Hellums (1967); see also Hirasaki (1967). The vector vorticity is now defined conventionally as

$$\vec{\zeta} = \nabla \times \vec{V} \qquad (3\text{-}616)$$

or by components, in terms of unit vectors $(\vec{i}, \vec{j}, \vec{k})$, as

$$\vec{\zeta} = \zeta_x \vec{i} + \zeta_y \vec{j} + \zeta_z \vec{k} \tag{3-617A}$$

$$\zeta_x = + \left(\frac{\partial w}{\partial y} - \frac{\partial v}{\partial z} \right) \tag{3-617B}$$

$$\zeta_y = - \left(\frac{\partial w}{\partial x} - \frac{\partial u}{\partial z} \right) \tag{3-617C}$$

$$\zeta_z = + \left(\frac{\partial v}{\partial x} - \frac{\partial u}{\partial y} \right) \tag{3-617D}$$

If the flow is two-dimensional in (x,y), $\zeta_x = 0 = \zeta_y$ and $\zeta_z = \partial v/\partial x - \partial u/\partial y$, which is the negative of our previous definition of ζ. (The definition of two-dimensional $\zeta = -\zeta_z$ is common, but not universal.) The corresponding vorticity transport equation is a vector equation, for each of the three components. All the 2D methods are again applicable.

$$\frac{\partial \zeta_x}{\partial t} = - \nabla \cdot (\vec{V} \zeta_x) + \frac{1}{Re} \nabla^2 \zeta_x + \vec{\zeta} \cdot \nabla u \tag{3-618A}$$

$$\frac{\partial \zeta_y}{\partial t} = - \nabla \cdot (\vec{V} \zeta_y) + \frac{1}{Re} \nabla^2 \zeta_y + \vec{\zeta} \cdot \nabla v \tag{3-618B}$$

$$\frac{\partial \zeta_z}{\partial t} = - \nabla \cdot (\vec{V} \zeta_z) + \frac{1}{Re} \nabla^2 \zeta_z + \vec{\zeta} \cdot \nabla w \tag{3-618C}$$

(See* e.g. Lamb's classic "Hydrodynamics", Dover Publications, pg. 578.) The terms like $\vec{\zeta} \cdot \nabla u$, which cause vorticity intensificaiton by vortex-filament stretching, do not exist in 2D. The vortex stretching terms may be lagged or iterated in implicit methods, like the nonlinear advection velocities.

The stream function ψ as such does not exist for general 3D flows; i.e., there is no function ψ such that a ψ-isoline is a streamline. But for a "solenoidal" vector field (i.e., one which obeys the 3D continuity equation, $\nabla \cdot \vec{V} = 0$) a so-called vector potential $\vec{\psi} = \psi_x \vec{i} + \psi_y \vec{j} + \psi_z \vec{k}$ does exist such that velocities are obtainable from curl $\vec{\psi}$. (This "vector potential" is not to be confused with the "velocity potential" of 2D inviscid flow, which requires $\zeta = 0$ everywhere.)

$$\vec{V} = \nabla x \vec{\psi} \tag{3-619A}$$

$$u = + \frac{\partial \psi_z}{\partial y} - \frac{\partial \psi_y}{\partial z} \tag{3-619B}$$

$$v = - \frac{\partial \psi_z}{\partial z} + \frac{\partial \psi_x}{\partial z} \tag{3-619C}$$

$$w = + \frac{\partial \psi_y}{\partial x} - \frac{\partial \psi_x}{\partial y} \tag{3-619D}$$

In 2D, $\vec{\psi} = (0,0,\psi_z)$, and $\psi = \psi_z$ gives back the usual 2D relations for u and v.

---

*See also "Convection in the Tanks of a Rotating Spacecraft", NASA TR R-386, June 1972, page 25 et seq. The pertinent section was authored by E.D.Martin and B.S.Baldwin of NASA-Ames Research Center. Dr. Martin advised the present author of the error in the first printing of this book, in which the vortex-stretching terms were omitted from (3-618).

Comparing equations (3-619A) and (3-616) gives

$$\nabla x(\nabla x\vec{\psi}) = \vec{\zeta} \tag{3-620}$$

An added degree of arbitrariness is present in the definitions of $\vec{\psi}$. We can arbitrarily require ψ to be solenoidal, i.e.,

$$\nabla \cdot \vec{\Psi} = 0 \tag{3-621}$$

It can be shown that this allows equation (3-620) to be written as a vector Poisson equation,

$$\nabla^2\vec{\psi} = -\vec{\zeta} \tag{3-622}$$

Thus, at each time step, there are three 3D Poisson equations to be solved.

The boundary conditions at a no-slip boundary are tricky. The $\vec{\psi}$ components are *not* all zero; rather, the components tangential to the surface are zero and the normal derivative of the normal component is zero. For example, for a wall along (y,z) at $x = x_a$, we have

$$\frac{\partial\psi_x}{\partial x} = 0 \quad , \quad \psi_y = \psi_z = 0 \text{ at } x = x_a \tag{3-623}$$

The no-slip wall vorticity components can be expressed in terms of their fundamental velocity-component definitions. With our example of a wall along (y,z) at $x = x_a$, we have

$$\left.\begin{aligned} \zeta_x &= 0 \\ \zeta_y &= -\frac{\partial w}{\partial x} \\ \zeta_z &= +\frac{\partial v}{\partial x} \end{aligned}\right\} \text{ at } x = x_a \tag{3-624}$$

Aziz and Hellums suggest evaluating wall vorticity components directly from equation (3-624), using forms such as the following for a wall at (ia).

$$\zeta_y(ia,j,k) = -\left.\frac{\partial w}{\partial x}\right|_{ia,j,k} = -\left[\frac{4w(ia+1,j,k) - w(ia+2,j,k)}{2\Delta x}\right] + 0(\Delta x) \tag{3-625}$$

Second-order formulations have not been successfully used, nor have any flow-through problems been calculated with this method. It is obvious that the computational boundary conditions for such problems, covered in section III-C, must be cast in the basic (u,v) form before being applied (cautiously) in this $(\vec{\psi},\vec{\zeta})$ system.

The discussion of the relative merits of the primitive variable system $(\vec{V},P)$ compared with the $(\vec{\psi},\vec{\zeta})$ system is now modified. *Both* systems now require three parabolic transport equation solutions; the $(\vec{V},P)$ equations are still more complicated because of the dilatational D_{ijk} terms, and implicit methods have not yet been successfully applied in $(\vec{V},P)$. Aziz and Hellums did successfully apply ADI implicit methods in the $(\vec{\psi},\vec{\zeta})$ system. The $(\vec{V},P)$ system requires the solution of one 3D Poisson equation, $\nabla^2 P = S_P$, with the Neumann conditions on all boundaries whereas the $(\vec{\psi},\vec{\zeta})$ system requires the solution of *three* 3D Poisson equations, $\nabla^2\vec{\psi} = -\vec{\zeta}$. However, in the natural convection problem of Aziz and Hellums (1967), these three 3D Poisson equations each have Dirichlet conditions at boundaries along two coordinates, and the relatively simple zero-gradient Neumann condition along the third. If iterative methods are used for the Poisson equations (as in Aziz and Hellums, 1967) then 2D experience would indicate that it is faster to solve the three equations $\nabla^2\vec{\psi} = -\vec{\zeta}$ with some Dirichlet conditions than to solve the one equation $\nabla^2 P = S_P$ with all Neumann conditions. If direct methods are used (which is more likely) the program development time may be shorter in $(\vec{\psi},\vec{\zeta})$ because of the simpler boundary conditions, but the run time for the Poisson equations probably will be shorter in the $(\vec{V},P)$ system. Also, the $(\vec{V},P)$ system requires storage of only four 3-dimensional arrays, while the $(\vec{\psi},\vec{\zeta})$ system requires six such arrays.

So the use of direct methods for the Poisson equation may make the $(\vec{V},P)$ system preferable in 3D problems. The preference would be enhanced if implicit methods could be developed for the $(\vec{V},P)$ system. In any case, Aziz and Hellums (1967) have demonstrated the feasibility of the $(\vec{\psi},\vec{\zeta})$ system in three dimensions, by calculating a fairly large problem (11 x 11 x 11) on a modest computer.

As an alternative to solving the three Poisson equations for the velocity potential $\vec{\psi}$, we may consider solving three Poisson equations for the velocity components directly. These equations are readily derived from the continuity equation $\nabla\cdot\vec{V} = 0$ and the definition of vorticity $\vec{\zeta} = \nabla\times\vec{V}$. The vector equation is

$$\nabla^2\vec{V} = \nabla \times \vec{\zeta} \qquad (3\text{-}626A)$$

In expanded form,

$$\nabla^2 u = \frac{\partial\zeta_z}{\partial y} - \frac{\partial\zeta_y}{\partial z} \qquad (3\text{-}626B)$$

$$\nabla^2 v = -\frac{\partial\zeta_z}{\partial x} + \frac{\partial\zeta_x}{\partial z} \qquad (3\text{-}626C)$$

$$\nabla^2 w = \frac{\partial\zeta_y}{\partial x} - \frac{\partial\zeta_x}{\partial y} \qquad (3\text{-}626D)$$

Fasel* has used this vorticity-velocity system in 2D calculations of boundary-layer stability, and found that the Poisson equaitons for u and v, which are a higher order differential system than the stream function equaiton, allowed less restrictive computational boundary conditions to be used. In 3D, the boundary conditions on the velocity components are straightforward, as contrasted to (3-623) for the $\vec{\psi}$ components. However, volume conservation may be lost even in 2D, since the solution of $\nabla^2\psi = \zeta$ assures that the discrete continuity equation for velocity components is satisfied identically (see exercise below) whereas the solution of (3-626) does not.

Exercise: Show that the use of the stream function ψ in 2D to determine velocities u and v by centered differences assures that

$$\frac{\delta u}{\delta x} + \frac{\delta v}{\delta y} = 0 \qquad (3\text{-}627)$$

identically, regardless of the accuracy of the solution for ψ.

---

*Fasel, H., "Numerical Solution of the Unsteady Navier-Stokes Equations for the Investigation of Laminar Boundary Layer Stability", pp. 151-160 of Proc. Fourth International Conference on Numerical Methods in Fluid Dynamics, R.D. Richtmyer, ed., Lecture Notes in Physics, Vol. 35, Springer-Verlag, Berlin.

CHAPTER IV

COMPRESSIBLE FLOW EQUATIONS IN RECTANGULAR COORDINATES

This chapter begins with a brief discussion of the computational problems unique to compressible flow. The customary fundamental flow equations are then presented, followed by a derivation of the conservation form. Supplemental relations (equation of state, etc.) are presented, the conservation equations are normalized, and alternate normalizing systems are considered. The commonly used short form (vector) equations are then presented. Finally, the physical and mathematical existence of shocks is discussed.

IV-A. Fundamental Difficulties

Basically, there are four dependent variables in an ideal-gas, two-dimensional flow problem: two velocity components and any two thermodynamic properties. In incompressible flow, the energy equation is not required to solve the momentum and continuity equations, removing one thermodynamic property, temperature. The pressure may be removed by cross differentiation, and vorticity may be introduced. The two velocity components are then removed by introduction of the stream function, leaving the two unknowns of vorticity and stream function to be solved by two equations, one parabolic and one elliptic. In the case of compressible flow, the energy equation is required to solve the others, and the stream function is not defined for unsteady flow. We must work with four coupled partial differential equations.*

The inviscid equations for supersonic flow are hyperbolic. The finite-difference equations must retain something of the continuum domain of dependence, which leads to the famous Courant-Friedrichs-Lewy stability limitation.

The most formidable numerical problem unique to supersonic flow is the existence of *shock waves*. In the high Reynolds number limit, shock waves are *discontinuities* in the flow solution. Most of the efforts of workers in this area have been directed towards smearing out these discontinuities while retaining accuracy away from the discontinuities, but we will also consider some shock-fitting methods in which the discontinuity is maintained.

IV-B. Customary Equations

The continuity equation and the two momentum equations for planar two-dimensional flow are presented here in their dimensional customary form as in Schlichting (1968) and the energy equation** as in Owczarek (1964). [See also standard gas dynamics textbooks, such as Shapiro (1953), Liepman and Roshko (1957), Chapman and Walker (1971).] We assume an ideal gas, i.e., internal energy $\bar{e} = f(\bar{T})$ only, but allow variable properties. The body force is zero; bulk viscosity κ is retained (for a while) with $\bar{\lambda} = \bar{\kappa} - 2/3\ \bar{\mu}$. The overbar indicates a dimensional quantity.

$$\frac{\partial \bar{\rho}}{\partial \bar{t}} + \nabla\cdot(\bar{\rho}\vec{\bar{V}}) = 0 \tag{4-1}$$

$$\bar{\rho}\frac{D\bar{u}}{D\bar{t}} = -\frac{\partial \bar{P}}{\partial \bar{x}} + \frac{\partial}{\partial \bar{x}}\left[2\bar{\mu}\frac{\partial \bar{u}}{\partial \bar{x}} + \bar{\alpha}\bar{D}\right] + \frac{\partial}{\partial \bar{y}}\left[\bar{\mu}\left(\frac{\partial \bar{u}}{\partial \bar{y}} + \frac{\partial \bar{v}}{\partial \bar{x}}\right)\right] \tag{4-2}$$

$$\bar{\rho}\frac{D\bar{v}}{D\bar{t}} = -\frac{\partial \bar{P}}{\partial \bar{y}} + \frac{\partial}{\partial \bar{x}}\left[\bar{\mu}\left(\frac{\partial \bar{v}}{\partial \bar{x}} + \frac{\partial \bar{u}}{\partial \bar{y}}\right)\right] + \frac{\partial}{\partial \bar{y}}\left[2\bar{\mu}\frac{\partial \bar{v}}{\partial \bar{y}} + \bar{\alpha}\bar{D}\right] \tag{4-3}$$

$$\bar{\rho}\frac{D\bar{e}_s}{D\bar{t}} + \bar{\nabla}\cdot\vec{\bar{q}} - \bar{\nabla}\cdot(\overleftrightarrow{\bar{T}}\cdot\vec{\bar{V}}) = 0 \tag{4-4}$$

---

*If constant transport coefficients and specific heats are assumed, the equations may be profitably re-cast in terms of vorticity and entropy, as in Tsien (1958). Then, for a restrictive class of problems with no viscous or heat-transfer effects and no shock waves, constant entropy may be assumed, thus eliminating one variable. This approach has not been widely used in computational fluid dynamics solutions. In another approach, Gol'din et al., (1969) replaced the energy equation, including heat conduction, with an entropy transport equation, and sacrificed conservation of energy for *conservation of entropy*.

**Some computations (Kentzner, 1970B) have been performed with the energy equation (4-4) replaced by the following pressure equation.

$$\frac{\partial \bar{P}}{\partial \bar{t}} = -\left[\bar{u}\frac{\partial \bar{P}}{\partial \bar{x}} + \bar{v}\frac{\partial \bar{P}}{\partial y} + \gamma\bar{P}\bar{D}\right] \tag{4-5}$$

where $\gamma = C_p/C_v$, the ratio of specific heats. The equation of state of an ideal gas,

$$\bar{P} = \bar{\rho}\bar{R}\bar{T} \tag{4-6}$$

has already been used in equation (4-5). Equation (4-4) is the more fundamental and the more commonly used, so only it will be considered in this book.

In these equations, $D/D\bar{t} = \partial/\partial\bar{t} + \bar{u}\,\partial/\partial\bar{x} + \bar{v}\,\partial/\partial\bar{y}$ is the substantive derivative, $\bar{D} = \bar{\nabla}\cdot\vec{\bar{V}}$ is the dilation, $\bar{e}_s = \bar{e} + \bar{V}^2/2$ is the reservoir* internal energy per unit mass, $\vec{\bar{q}}$ is the heat-flux vector, and $\overleftrightarrow{\bar{T}}$ is the total stress tensor. Additional laws are required to evaluate these terms. $\overleftrightarrow{\bar{T}}$ contains pressure as well as viscous terms. The gravitational constant used in Schlichting (1968) is not shown explicitly, but is implicitly included in a compatible system of units.

IV-C. Conservation Form

The conservation form of the inviscid compressible flow equations was given by Courant and Friedrichs (1948), but Lax (1954) was the first to actually use this form to achieve a conservative finite-difference scheme.

When the usual differential equations are rearranged so that the conservation variables ρ, ρu, ρv, and E_s (to be defined) are the independent variables, then the use of conservative finite-difference methods assures identical conservation of mass, momentum, and energy. The Rankine-Hugoniot relations** for a normal shock wave are based only on this gross conservation and are independent of the details within the shock structure. The result is that *all* stable, consistent, and conservative finite-difference methods applied to the conservation equations satisfy the Rankine-Hugoniot relations and therefore produce the correct jump conditions across a shock.† Longley (1960) tested four separate differencing methods and found that all produced the correct shock speeds, because the conservation equations were used. Gary (1964) showed that using the Lax-Wendroff method on non-conservative equations produced significant errors in shock speed (although the calculation of a rarefaction wave was slightly more accurate). Many subsequent calculations have shown that the conservation form is more accurate for calculations across a shock wave, without using "shock patching" methods (to be discussed). This is easily understood by considering a stationary normal shock wave. The truncation error of an FDE depends on the size of higher derivatives in the Taylor series expansions for the differentials. In the variables ρ, u, v, T, the shock causes discontinuities in the continuum solution, but the solution is actually *continuous* in the conservation variables. (For moving and oblique shocks, the conservation variables may also be discontinuous.)

The other attraction of the conservation-law form is that the finite-difference forms can then be interpreted as integral laws over the cell "control volume", as discussed in Chapter III, section III-A-3. In this interpretation, no assumptions about continuity of flow variables need be made. The integral formulation is really to be preferred, and many people hold that *all* physical laws should be expressed in integral form. The FDE's for the Navier-Stokes equations are derived in integral form by Allen (1968) and by Rubin and Preiser (1968, 1970).

The overbar notation becomes clumsy, so at this point we introduce a "d" over the = sign to indicate dimensional quantities for an entire equation.

The continuity equation (4-1) is already in conservation form. The conservation form of the other equations may be derived by way of manipulations like the following. For the u-momentum equation, consider the following terms.

$$\frac{\partial(\rho u)}{\partial t} + \frac{\partial(\rho uu)}{\partial x} + \frac{\partial(\rho vu)}{\partial y} \overset{d}{=} u\frac{\partial\rho}{\partial t} + u\frac{\partial(\rho u)}{\partial x} + u\frac{\partial(\rho u)}{\partial x} + \rho\frac{\partial u}{\partial t} + \rho u\frac{\partial u}{\partial x} + \rho v\frac{\partial u}{\partial y}$$

$$\overset{d}{=} u\left[\frac{\partial\rho}{\partial t} + \frac{\partial(\rho u)}{\partial x} + \frac{\partial(\rho v)}{\partial y}\right] + \rho\left[\frac{\partial u}{\partial t} + u\frac{\partial u}{\partial x} + v\frac{\partial u}{\partial y}\right]$$

$$\overset{d}{=} u\left[\frac{\partial\rho}{\partial t} + \nabla\cdot(\rho\vec{V})\right] + \rho\frac{Du}{Dt} \tag{4-7}$$

---

*"Reservoir" or "total" or "stagnation" value, which would exist if the flow were brought to rest reversibly.

**See Owczarek (1964), for example, or any gas dynamics textbook.

†Note that the use of the conservation form of the differential equations does not in itself guarantee such conservation; the finite-difference method must also be conservative.

The bracketed term is zero from the continuity equation (4-1). The remaining terms state

$$\rho \frac{Du}{Dt} \stackrel{d}{\equiv} \frac{\partial(\rho u)}{\partial t} + \frac{\partial(\rho uu)}{\partial x} + \frac{\partial(\rho vu)}{\partial y} \stackrel{d}{\equiv} \frac{\partial(\rho u)}{\partial t} + \nabla\cdot\left[(\rho u)\cdot\vec{V}\right] \tag{4-8}$$

Equation (4-8) is substituted into (4-2) to obtain the conservation form of the x-momentum equation.

$$\frac{\partial(\rho u)}{\partial t} \stackrel{d}{\equiv} -\frac{\partial P}{\partial x} - \nabla\cdot\left[(\rho u)\cdot\vec{V}\right] + \frac{\partial}{\partial x}\left[2\mu\frac{\partial u}{\partial x} + \lambda D\right] + \frac{\partial}{\partial y}\mu\left[\frac{\partial u}{\partial y} + \frac{\partial v}{\partial x}\right] \tag{4-9A}$$

or, in a shorthand form,

$$\frac{\partial(\rho u)}{\partial t} \stackrel{d}{\equiv} -\frac{\partial P}{\partial x} - \nabla\cdot\left[(\rho u)\vec{V}\right] + D_1 + D_2 \tag{4-9B}$$

The conservation form for y-momentum is derived analogously.

$$\frac{\partial(\rho v)}{\partial t} \stackrel{d}{\equiv} -\frac{\partial P}{\partial y} - \nabla\cdot\left[(\rho v)\vec{V}\right] + \frac{\partial}{\partial x}\left[\mu\left(\frac{\partial v}{\partial x} + \frac{\partial u}{\partial y}\right)\right] + \frac{\partial}{\partial y}\left[2\mu\frac{\partial v}{\partial y} - \lambda D\right] \tag{4-10A}$$

or, in shorthand form,

$$\frac{\partial(\rho v)}{\partial t} \stackrel{d}{\equiv} -\frac{\partial P}{\partial y} - \nabla\cdot\left[(\rho v)\vec{V}\right] + D_3 + D_4 \tag{4-10B}$$

In equations (4-8) and (4-9), the independent variables u and v of equations (4-2) and (4-3) have been replaced with the conservation variables (ρu) and (ρv). The momentum equation is now written with the x-momentum (ρu) as the independent variable. To emphasize the distinction, we write $(\rho u) = f$; then equation (4-9) becomes

$$\frac{\partial f}{\partial t} + \nabla\cdot(f\vec{V}) = S_f \tag{4-11}$$

where S_f is a term consisting of the stress gradient terms (pressure and friction) of equation (4-9). The variable f is now clearly presented as a transport property. This form is similar to that of equation (2-10) for vorticity. The discussion of section III-A-3 on the conservative property then applies.

We have derived the conservation form for $\partial(\rho u)/\partial t$ from the non-conservation form for $\partial u/\partial t$ because the latter is more commonly used in textbooks. In fact, the conservation form is more fundamental, and is the basis for the non-conservation form. The *only* physical law used in the $\partial(\rho u)/\partial t$ conservation equation is Newton's second law of motion. When combined with the law of mass conservation (continuity equation) it reduces to the non-conservation equation for $\partial u/\partial t$. The same remarks apply to the energy equation, which we now consider.

In the conservation form for the energy equation, the conservation property will be

$$E_s \stackrel{d}{\equiv} \rho e_s \stackrel{d}{\equiv} \rho\left[e + \frac{1}{2}(u^2 + v^2)\right] \tag{4-12}$$

E_s is thus the "*reservoir*" or "stagnation" *internal* energy *per unit volume*.

If the reader is not familiar with tensor notation, or does not care for such details as just *when* the Stoke's hypothesis is used, etc., he may skip to equation (4-36) et seq.

In terms of E_s, the energy equation (4-4) is

$$\frac{\partial E_s}{\partial t} \stackrel{d}{=} -\nabla\cdot(\vec{V}E_s) - \nabla\cdot\vec{q} + \nabla\cdot(\overline{\overline{T}}\cdot\vec{V}) \tag{4-13}$$

We now express the complete stress tensor $\overline{\overline{T}}$ as the sum of its hydrostatic tensor and the viscous stress tensor $\overline{\overline{\pi}}$ as

$$\overline{\overline{T}} \stackrel{d}{=} -P\overline{\overline{I}} + \overline{\overline{\pi}} \tag{4-14}$$

where $\overline{\overline{I}}$ is the unit dyadic and P is the hydrostatic pressure, by definition. Then

$$\nabla\cdot(\overline{\overline{T}}\cdot\vec{V}) \stackrel{d}{=} -\nabla\cdot(P\overline{\overline{I}}\cdot\vec{V} + \overline{\overline{\pi}}\cdot\vec{V}) \tag{4-15}$$

Now*

$$\nabla\cdot(P\overline{\overline{I}}\cdot\vec{V}) \stackrel{d}{=} \nabla\cdot(\vec{V}P),$$

so equation (4-15) becomes

$$\nabla\cdot(\overline{\overline{T}}\cdot\vec{V}) \stackrel{d}{=} -\nabla\cdot(\vec{V}P) + \nabla\cdot(\overline{\overline{\pi}}\cdot\vec{V}) \tag{4-16}$$

Using this in equation (4-13) gives

$$\frac{\partial E_s}{\partial t} \stackrel{d}{=} -\nabla\cdot(\vec{V}E_s) - \nabla\cdot\vec{q} - \nabla\cdot(\vec{V}P) + \nabla\cdot(\overline{\overline{\pi}}\cdot\vec{V}) \tag{4-17}$$

$$\frac{\partial E_s}{\partial t} \stackrel{d}{=} -\nabla\cdot[\vec{V}(E_s + P)] - \nabla\cdot\vec{q} + \nabla\cdot(\overline{\overline{\pi}}\cdot\vec{V}) \tag{4-18}$$

This is the *conservation form of the energy equation*. Note that the stagnation internal energy per unit volume, $E_s = \rho(e + V^2/2)$, is the *conservation* property, but the *advected* property is $(E_s + P)$, the stagnation *enthalpy* per unit volume. The term $\vec{V}P$ is the "flow work" term of classical thermodynamics.

Note that the conservation form for the continuity equation is obtainable from the customary form (e.g., Bird et al., 1960) by just replacing the substantial derivative $D\rho/Dt$ by $\partial\rho/\partial t + \nabla\cdot(\rho\vec{V})$. The same is true for the momentum equations, since the viscous terms are not affected. But for the energy equation, this is *not* true; rather, the viscous terms are affected by the conservation form. The introduction of the conservation variable $E_s = \rho(e + 1/2\ V^2)$ gives rise to a term $\partial/\partial t(1/2\ V^2)$. This term can be evaluated from the "mechanical energy equation" (see page 87, Bird et al., 1960) and it does depend on the viscous terms. Thus, the form of the viscous terms in the energy equation is altered by the introduction of the conservation variable E_s.

Equations (4-1), (4-9), (4-10), and (4-18) are the dimensional compressible flow equations in the conservation variables ρ, ρu, ρv, E_s. But additional relations are required before a solution is possible.

*$\nabla\cdot(P\overline{\overline{I}}\cdot\vec{V}) \stackrel{d}{=} \nabla\cdot[P(\vec{i}\vec{i} + \vec{j}\vec{j})\cdot(u\vec{i} + v\vec{j})] \stackrel{d}{=} \nabla\cdot[P(\vec{i}u + \vec{j}v)] \stackrel{d}{=} \frac{\partial(uP)}{\partial x} + \frac{\partial(vP)}{\partial y} \stackrel{d}{=} \nabla\cdot(\vec{V}P)$

IV-D. Supplemental Relations

We need supplemental relations to solve this system of equations. We need an "equation of state" to determine P from the conservation variables, and equations for the heat conduction and viscous stress terms. In this chapter, we consider only the simplest case. The assumptions made are applicable to simple gases at "moderate" temperatures and pressures. (Air can often be assumed to be a simple one-component gas, since both of its principal constituents, N_2 and O_2, are diatomic and have similar thermodynamic properties.)

The "perfect gas" law is written

$$\bar{P} = \bar{\rho}\bar{R}_g\bar{T} \tag{4-19}$$

where $\bar{R}_g$ is a constant of the gas. $\bar{R}_g$ is related to the "specific heats" at constant pressure, $\bar{C}_p$, and at constant volume, $\bar{C}_v$, by*

$$\bar{R}_g = \bar{C}_p - \bar{C}_v \tag{4-20}$$

We further assume a "calorically perfect" gas, for which the internal energy $\bar{e}$ is given by

$$\bar{e} = \bar{C}_v\bar{T} \tag{4-21}$$

where $\bar{C}_v$ is *constant*. Using equations (4-20) and (4-21) in (4-19) gives

$$\bar{P} = \bar{\rho}\,\frac{(\bar{C}_p - \bar{C}_v)}{\bar{C}_v}\,\bar{e} = \bar{\rho}\bar{e}(\gamma - 1) \tag{4-22}$$

where $\gamma = \bar{C}_p/\bar{C}_v$ is the "specific heat ratio." Approximate values of γ range from $5/3 \simeq 1.67$ for monotomic gases such as helium, $7/5 = 1.4$ for diatomic gases such as air, $9/7 \simeq 1.28$ for triatomic molecules such as carbon dioxide, and can be as low as 1.1 for the complex gases found in rocket exhausts. The caloric assumption (4-21) is best for the monatomic gases, and is not of much value even for triatomic gases.** All calorically perfect gases have an equation of state of the form of equation (4-22), and often are referred to as "γ-law gases." Using equation (4-12), $\bar{P}$ may be expressed in conservation variables as

$$\bar{P} = \left[\bar{E}_s - \frac{1}{2}\,\bar{\rho}(\bar{u}^2 + \bar{v}^2)\right](\gamma - 1) \tag{4-23}$$

For a FORTRAN program to calculate the state variables of real gas mixtures, see Dahl (1969).

The "dynamic pressure" term $1/2\ \bar{\rho}(\bar{u}^2 + \bar{v}^2)$ is obviously expressible in terms of the conservation variables, $\bar{\rho}$, $\overline{\rho u}$, $\overline{\rho v}$ as

$$\frac{1}{2}\,\bar{\rho}(\bar{u}^2 + \bar{v}^2) = \frac{1}{2}\,\frac{(\overline{\rho u})^2 + (\overline{\rho v})^2}{\bar{\rho}} \tag{4-24}$$

We now relate the heat flux term $\bar{\nabla}\cdot\bar{q}$ and the viscous stress tensor term $\bar{\nabla}\cdot(\overleftrightarrow{\bar{\tau}}\cdot\bar{V})$ to fundamental quantities, using Fourier's law for the first and Stokes' hypothesis for the second.

---

*Equation (4-20) is often found in earlier references as $R_g = (C_p - C_v)/J$, where J is Joule's constant. The form (4-20) just assumes a consistent set of units.

**The limit of incompressible flow requires $\gamma = 1$, but the application of this limit must be handled delicately to avoid erroneous equations.

Fourier's law of heat conduction for an assumed isotropic medium is

$$\vec{q} = - \bar{k}\bar{\nabla}\bar{T} \tag{4-25}$$

giving

$$- \nabla\cdot\vec{q} \overset{d}{=} \nabla\cdot k\nabla T = \frac{\partial}{\partial x}\left(k\,\frac{\partial T}{\partial x}\right) + \frac{\partial}{\partial y}\left(k\,\frac{\partial T}{\partial y}\right) \tag{4-26}$$

To evaluate the viscous stress tensor term, we write (Owczarek, 1964)

$$\overset{\Leftrightarrow}{\pi} \overset{d}{=} \pi_{11}\vec{i}\vec{i} + \pi_{12}\vec{i}\vec{j} + \pi_{21}\vec{j}\vec{i} + \pi_{22}\vec{j}\vec{j} \tag{4-27}$$

$$\vec{V} \overset{d}{=} u\vec{i} + v\vec{j} \tag{4-28}$$

Then

$$\overset{\Leftrightarrow}{\pi}\cdot\vec{V} \overset{d}{=} (\pi_{11}\vec{i}\vec{i} + \pi_{12}\vec{i}\vec{j} + \pi_{21}\vec{j}\vec{i} + \pi_{22}\vec{j}\vec{j})\cdot(u\vec{i} + v\vec{j}) \overset{d}{=} \pi_{11}u\vec{i} + \pi_{12}\vec{i}v + \pi_{21}u\vec{j} + \pi_{22}v\vec{j}$$

$$\overset{d}{=} \vec{i}(\pi_{11}u + \pi_{12}v) + \vec{j}(\pi_{21}u + \pi_{22}v) \tag{4-29}$$

With

$$\vec{a} \overset{d}{=} a_1\vec{i} + a_2\vec{j} \tag{4-30}$$

and

$$\nabla\cdot\vec{a} \overset{d}{=} \frac{\partial}{\partial x}(a_1) + \frac{\partial}{\partial x}(a_2) \tag{4-31}$$

we get

$$\nabla\cdot(\overset{\Leftrightarrow}{\pi}\cdot\vec{V}) \overset{d}{=} \frac{\partial}{\partial x}(\pi_{11}u + \pi_{12}v) + \frac{\partial}{\partial y}(\pi_{21}u + \pi_{22}v) \tag{4-32}$$

The stress components π are obtained from equation (10-17) of Owczarek (1964); κ is the bulk viscosity and D is the dilatation, $D = \nabla\cdot\vec{V}$.

$$\begin{aligned} \pi_{11} &\overset{d}{=} \lambda D + 2\mu\frac{\partial u}{\partial x} \\ \pi_{12} &\overset{d}{=} \mu\left(\frac{\partial v}{\partial x} + \frac{\partial u}{\partial y}\right) \\ \pi_{21} &\overset{d}{=} \pi_{12} \\ \pi_{22} &\overset{d}{=} \lambda D + 2\mu\frac{\partial v}{\partial y} \end{aligned} \tag{4-33}$$

Using these, equation (4-32) becomes

$$\nabla\cdot(\overleftrightarrow{\pi}\cdot\vec{V}) \stackrel{d}{=} \frac{\partial}{\partial x}\left[u\lambda D + \mu u \frac{\partial u}{\partial x}\right] + \frac{\partial}{\partial x}\left[v\mu\left(\frac{\partial v}{\partial x} + \frac{\partial u}{\partial y}\right)\right] + \frac{\partial}{\partial y}\left[u\mu\left(\frac{\partial v}{\partial x} + \frac{\partial u}{\partial y}\right)\right] + \frac{\partial}{\partial y}\left[v\lambda D + 2\mu v \frac{\partial v}{\partial y}\right] \tag{4-34}$$

For the purposes of finite-difference representation, it is desirable to group terms into the form $\partial/\partial x\ [f\ \partial g/\partial y]$, $\partial/\partial x\ [f\ \partial g/\partial x]$, etc. Define

$$\lambda \stackrel{d}{=} \kappa - \frac{2}{3}\mu$$

$$\eta \stackrel{d}{=} \kappa + \frac{4}{3}\mu$$

Using these in equation (4-34), expanding $D = \partial u/\partial x + \partial v/\partial y$, and grouping gives

$$\nabla\cdot(\overleftrightarrow{\pi}\cdot\vec{V}) \stackrel{d}{=} \frac{\partial}{\partial x}\left[(\eta u)\frac{\partial u}{\partial x} + (\mu v)\frac{\partial v}{\partial x}\right] + \frac{\partial}{\partial x}\left[(\lambda u)\frac{\partial v}{\partial y} + (\mu v)\frac{\partial u}{\partial y}\right]$$

$$+ \frac{\partial}{\partial y}\left[(\eta v)\frac{\partial v}{\partial y} + (\mu u)\frac{\partial u}{\partial y}\right] + \frac{\partial}{\partial y}\left[(\lambda v)\frac{\partial u}{\partial x} + (\mu u)\frac{\partial v}{\partial x}\right] \tag{4-35}$$

Additional relations are also required to determine μ and k. An assumption of constant-property μ and k may simplify the viscous terms and may be useful for testing methods. However, the constant $\bar{\mu}$ assumption is poor for gases at even moderate supersonic Mach numbers, so its use is not recommended except to test out methods. The $\bar{k}$ assumption is more realistic, but we will consider variable $\bar{k}$ for generality. There are several equations which can be used to determine the transport properties μ and k. For convenience, they will be discussed in the next section.

Equations (4-35) and (4-26), along with the equation of state (4-23) and the relations for the transport properties μ and k, then complete the expression of the conservation form of the energy equation (4-18) in terms of fundamental quantities suitable for finite-difference expression. But, before getting down to computing, it is preferable to first normalize the equations.

IV-E. Normalized Conservation Equations

It is of considerable advantage to compute with normalized equations, so that the characteristic parameters of the problem may be independently varied.

Normalizing the compressible flow equations involves many more variables than the incompressible equations, especially when variable properties are allowed as they are here.* Different forms appear from different normalizing conditions. Crocco (1965) normalized with respect to reservoir conditions, for example. Skoglund and Cole (1966) and others used the inflow sonic velocity for a normalizing reference velocity. Moretti (1969A) normalized velocities by $\sqrt{\bar{P}_s/\bar{\rho}_s}$, where s is the reservoir value. In the present work, the reference condition is a "freestream" condition, outside (or "near the edge of") the boundary layer at inflow. We revert to the overbar notation for dimensional quantities, and normalize as follows,

$$u = \bar{u}/\bar{u}_o \quad , \quad v = \bar{v}/\bar{u}_o$$

$$x = \bar{x}/L \quad , \quad y = \bar{y}/L$$

*The complications of variable properties are not nearly as messy in the conservation form as they are in the customary form of the equations.

$$
\begin{aligned}
t &= \bar{t}/(L/\bar{u}_o) \\
\rho &= \bar{\rho}/\bar{\rho}_o \\
T &= \bar{T}/\bar{T}_o \\
P &= \bar{P}/(\bar{\rho}_o \bar{u}_o^{\,2}) \\
\mu = \bar{\mu}/\bar{\mu}_o \quad &, \quad \kappa = \bar{\kappa}/\bar{\mu}_o \\
k &= \bar{k}/\bar{k}_o \\
e &= \bar{e}/\bar{u}_o^{\,2} \\
E_s &= \bar{E}_s/\bar{\rho}_o \bar{u}_o^{\,2}
\end{aligned}
\tag{4-36}
$$

Note that both $\bar{u}$ and $\bar{v}$ are normalized by $\bar{u}_o$, and both $\bar{\mu}$ and $\bar{\kappa}$ by $\bar{\mu}_o$.* The internal energy e is normalized by $\bar{u}_o^{\,2}$ so that the form $(\bar{E}_s + \bar{P})$ in the energy equation is unchanged in normalized form.

We also define a Reynolds number

$$Re = \frac{\bar{\rho}_o \bar{u}_o L}{\bar{\mu}_o} \tag{4-37}$$

a Prandtl number

$$Pr = \frac{\bar{C}_p \bar{\mu}_o}{\bar{k}_o} \tag{4-38}$$

and a reference Mach number

$$M_o = \frac{\bar{u}_o}{\bar{a}_o} \tag{4-39}$$

where $\bar{a}_o$ is the isentropic speed of sound at the reference (freestream) condition. For the perfect gas considered, the elementary gas-dynamics relation (e.g., Owczarek, 1964) is

$$\bar{a}_o = \sqrt{\gamma \bar{R}_g \bar{T}_o} \tag{4-40}$$

Note that Re and M_o are based only on the $\bar{u}_o$ component of the inflow velocity. The true freestream Mach number M_∞ is given by

$$M_\infty = \frac{\sqrt{\bar{u}_o^2 + \bar{v}_o^2}}{\bar{a}_o} = M_o \sqrt{1 + v_\infty^2} \tag{4-41}$$

The normalized forms of the continuity, momentum, and energy equations are then as follows. (The algebra is left to the student in problem 4-1.)

$$\frac{\partial \rho}{\partial t} = - \nabla \cdot (\rho \vec{V}) \tag{4-42A}$$

---

*The bulk viscosity $\bar{\kappa}_o$ is virtually always taken as zero for gases.

$$\frac{\partial(\rho u)}{\partial t} = -\frac{\partial P}{\partial x} - \nabla\cdot[(\rho u)\vec{V}] + (D_1 + D_2)/Re \qquad (4\text{-}42B)$$

$$\frac{\partial(\rho v)}{\partial t} = -\frac{\partial P}{\partial y} - \nabla\cdot[(\rho v)\vec{V}] + (D_3 + D_4)/Re \qquad (4\text{-}42C)$$

$$\frac{\partial E_s}{\partial t} = -\nabla[\vec{V}(E_s + P)] + \frac{1}{N}\nabla\cdot(k\nabla T) + \frac{1}{Re}\nabla\cdot(\bar{\pi}\cdot\vec{V}) \qquad (4\text{-}42D)$$

In these equations,

$$D_1 = \frac{\partial}{\partial x}\left[2\mu\frac{\partial u}{\partial x} + \lambda D\right] = \frac{\partial}{\partial x}\left[\mu\left(\frac{4}{3}\frac{\partial u}{\partial x} - \frac{2}{3}\frac{\partial v}{\partial y}\right)\right] \text{ for } \kappa = 0 \qquad (4\text{-}43A)$$

$$D_2 = \frac{\partial}{\partial y}\left[\mu\left(\frac{\partial u}{\partial y} + \frac{\partial v}{\partial x}\right)\right] \qquad (4\text{-}43B)$$

$$D_3 = \frac{\partial}{\partial x}\left[\mu\left(\frac{\partial v}{\partial x} + \frac{\partial u}{\partial y}\right)\right] \qquad (4\text{-}43C)$$

$$D_4 = \frac{\partial}{\partial y}\left[2\mu\frac{\partial v}{\partial y} + \lambda D\right] = \frac{\partial}{\partial y}\left[\mu\left(\frac{4}{3}\frac{\partial v}{\partial y} - \frac{2}{3}\frac{\partial u}{\partial x}\right)\right] \text{ for } \kappa = 0 \qquad (4\text{-}43D)$$

$$D = \frac{\partial u}{\partial x} + \frac{\partial v}{\partial y} \qquad (4\text{-}44A)$$

$$N = PrReM_o^2(\gamma - 1) \qquad (4\text{-}44B)$$

$$-\nabla\cdot\vec{q} = \nabla\cdot k\nabla T = \frac{\partial}{\partial x}\left(k\frac{\partial T}{\partial x}\right) + \frac{\partial}{\partial y}\left(k\frac{\partial T}{\partial y}\right) \qquad (4\text{-}45)$$

$$\nabla\cdot(\vec{\pi}\cdot\vec{V}) = \frac{\partial \pi_1}{\partial x} + \frac{\partial \pi_2}{\partial y} \qquad (4\text{-}46A)$$

$$\frac{\partial \pi_1}{\partial x} = \frac{\partial}{\partial x}\left[(\eta u)\frac{\partial u}{\partial x} + (\mu v)\frac{\partial v}{\partial x}\right] + \frac{\partial}{\partial x}\left[(\lambda u)\frac{\partial v}{\partial y} + (\mu v)\frac{\partial u}{\partial y}\right] \qquad (4\text{-}46B)$$

$$\frac{\partial \pi_2}{\partial x} = \frac{\partial}{\partial y}\left[(\eta v)\frac{\partial v}{\partial y} + (\mu u)\frac{\partial u}{\partial y}\right] + \frac{\partial}{\partial y}\left[(\lambda v)\frac{\partial u}{\partial x} + (\mu u)\frac{\partial v}{\partial x}\right] \qquad (4\text{-}46C)$$

$$\eta = \kappa + \frac{4}{3}\mu \ , \quad \lambda = \kappa - \frac{2}{3}\mu \qquad (4\text{-}47)$$

From these relations, the normalized local speed of sound a is obtained as

$$a^2 = T/M_o^2 \qquad (4\text{-}48)$$

and the local Mach number is obtained as

$$M = M_o\sqrt{(u^2 + v^2)/T} \qquad (4\text{-}49)$$

The equation of state (4-19) becomes

$$P = \frac{\rho T}{\gamma M_o^2} \qquad (4\text{-}50)$$

or, from the γ-law form (4-22),

$$P = \rho e(\gamma - 1) \tag{4-51}$$

That is, with this normalizing system, the γ-law equation of state does not change form from the dimensional to the normalized form.

The internal energy e is obtained as

$$e = \frac{T}{\gamma(\gamma - 1)M_o^2} \tag{4-52}$$

This equation may be regarded as specifying a normalized specific heat, C_v, from the relations

$$e = C_v T \tag{4-53}$$

$$C_v = \frac{1}{\gamma(\gamma - 1)M_o^2} \tag{4-54}$$

Then T may be obtained from

$$T = \frac{1}{C_v}\left[\frac{E_s}{\rho} - \frac{1}{2}(u^2 + v^2)\right] \tag{4-55}$$

Equations (4-51) through (4-54) may be combined to give an equation of state as

$$P = (\gamma - 1)\left[E_s - \frac{1}{2}\rho(u^2 + v^2)\right] \tag{4-56}$$

Exercise: Show that the equation of state in normalized variables may be written as $P = R_g \rho T$, where the normalized gas constant is $R_g = C_v(\gamma - 1) = 1/\gamma M_o^2$.

Similarly, a dimensionless relation for entropy may be derived. To avoid confusion with other symbols, we use Ey to denote entropy.

Exercise: Show that the dimensional equation for entropy of a perfect gas,

$$\bar{E}y - \bar{E}y_1 = \bar{C}_p \ln \bar{T}/\bar{T}_1 - \ln \bar{P}/\bar{P}_1 \tag{4-57A}$$

where "1" is an arbitrary reference state, may be written in dimensionless form as

$$\Delta Ey = \frac{\gamma}{\gamma - 1} \ln T - \ln P/P_o \tag{4-57B}$$

where P_o is the dimensionless reference pressure $\left(= \bar{P}_o/\bar{\rho}_o\bar{u}_o^{-2}\right)$, and where the dimensionless entropy is

$$Ey = \bar{E}y/\bar{R}_g \tag{4-58}$$

Relations are also necessary to determine μ and k. The Sutherland law of viscosity is quite accurate for air and other gases over the range of temperatures for which the calorically perfect assumption is accurate. It gives (Schlichting, 1968)

$$\frac{\mu}{\mu_o} \doteq \left(\frac{T}{T_o}\right)^{3/2} \left(\frac{T_o + S_1}{T + S_1}\right) \tag{4-59}$$

or in normalized form

$$\mu = T^{3/2}\left(\frac{1 + S_1}{T + S_1}\right) \tag{4-60}$$

where

$$S_1 = \bar{S}_1/\bar{T}_o \tag{4-61}$$

and $\bar{S}_1 = 110^oK$ for air. Note that, if a Sutherland viscosity law is used, another characteristic parameter, $S_1 = \bar{S}_1/\bar{T}_o$, is added to the solution. Practically, this implies selecting a *dimensional* value of the reference temperature $\bar{T}_o$ of the problem. The same is true if $\bar{k}$ is fit as a linear function of $\bar{T}$ as in $\bar{k} = \bar{a} + \bar{b}\bar{T}$ when $\bar{a} \neq 0$.

A commonly used approximation to the Sutherland law is the power law,

$$\mu = T^{\omega} \tag{4-62}$$

where ω varies from 1/2 to 1. In this form, an additional characteristic parameter is introduced in ω, although the dependence may be very weak, and $\omega = 1$ might suffice for a wide range of gases. For high-M flows, the use of different viscosity laws may produce substantially different results (e.g., Butler, 1967).

If the Sutherland law is used, care should be taken in programming to keep from running up excessive computer time (see Chapter VII).

The compressible flow equations (4-42) and their necessary supplementary relations such as equation (4-50) are then characterized by four parameters: M_o, Re, Pr, and γ, or alternately M_o, Re, N, and γ, when the transport properties μ and k are assumed constant. For variable μ and k, an additional reference temperature (dimensional) may be introduced.

IV-F. Short-Form Equations

The compressible flow equations are obviously more complicated than the incompressible equations. Several short forms may be used to illustrate and test methods.

A popular representation of equation (4-42) is the "vector" form, where the "vectors" U,F,G are ordered sets of the independent variables. This short form is

$$\frac{\partial U}{\partial t} + \frac{\partial F}{\partial x} + \frac{\partial G}{\partial y} = 0 \tag{4-63A}$$

wherein

$$U = \begin{Bmatrix} \rho \\ \rho u \\ \rho v \\ E_s \end{Bmatrix} \tag{4-63B}$$

$$F = \begin{Bmatrix} \rho u \\ P + \rho u^2 - \frac{1}{Re}\left[\mu\left(\frac{4}{3}\frac{\partial u}{\partial x} - \frac{2}{3}\frac{\partial v}{\partial y}\right)\right] \\ \rho uv - \frac{1}{Re}\left[\mu\left(\frac{\partial v}{\partial x} + \frac{\partial u}{\partial y}\right)\right] \\ u(E_s + P) - \frac{1}{N} k \frac{\partial T}{\partial x} - \frac{\pi_1}{Re} \end{Bmatrix} \tag{4-63C}$$

$$G = \begin{Bmatrix} \rho v \\ \rho uv = \frac{1}{Re}\left[\mu\left(\frac{\partial u}{\partial y} + \frac{\partial v}{\partial x}\right)\right] \\ P + \rho v^2 - \frac{1}{Re}\left[\left(\mu \frac{4}{3}\frac{\partial v}{\partial y} - \frac{2}{3}\frac{\partial u}{\partial x}\right)\right] \\ v(E_s + P) - \frac{1}{N} k \frac{\partial T}{\partial y} - \frac{\pi_2}{Re} \end{Bmatrix} \tag{4-63D}$$

For example, the continuity equation (4-42A) is recovered from the first elements of U, F, G.

$$\frac{\partial U_1}{\partial t} + \frac{\partial F_1}{\partial x} + \frac{\partial G_1}{\partial y} = 0 \tag{4-64}$$

$$\frac{\partial \rho}{\partial t} + \frac{\partial(\rho u)}{\partial x} + \frac{\partial(\rho v)}{\partial y} = 0 \tag{4-65}$$

Equation (4-63) has been written for bulk viscosity $\kappa = 0$.

The inviscid, non-conducting equations* are obtained with $1/Re = 0$ and $1/N = 0$. The one-dimensional inviscid equations are commonly used, and are given by

$$\frac{\partial U}{\partial t} + \frac{\partial F}{\partial x} = 0 \tag{4-66A}$$

$$U = \begin{Bmatrix} \rho \\ \rho u \\ E_s \end{Bmatrix} \tag{4-66B}$$

$$F = \begin{Bmatrix} \rho u \\ P + \rho u^2 \\ u(E_s + P) \end{Bmatrix} \tag{4-66C}$$

The number of arithmetic operations involved in differencing equation (4-63) can also be considerably reduced by assuming constant properties $\bar{\mu}$ and $\bar{k}$, giving $\mu = 1$, $k = 1$ in equation (4-63). But, as we said in section IV-D, the more general forms are required for accuracy.

---

*From both the molecular viewpoint and the viewpoint of Reynold's analogy (see Schlichting, 1968), the inviscid assumption consistently implies the adiabatic assumption.

IV-G. Existence of Shocks - Physical and Mathematical

Physically, a shock wave is a near-discontinuity* in a flow. It is distinguished from a slip-line near-discontinuity in that the flow passes *through* a shock, but *along* a slip line. It is a most remarkable physical phenomenon. A deceleration of 1000 m/sec can occur in 10^{-4} cm; Lagrangian accelerations are measured in billions of g's. Shocks occur when supersonic flow is turned through a compression corner or otherwise meets a sudden pressure rise. Only compression shocks exist; rarefactions occur through smooth expansions. Hydraulic jumps and bores are also shock-like phenomena.

Mathematically, the shock is *essentially* nonlinear. A one-dimensional mathematical prototype of both turbulence and real-fluid shock waves is Burger's equation (see Chapter II),

$$\frac{\partial u}{\partial t} + u\frac{\partial u}{\partial x} = \alpha\frac{\partial^2 u}{\partial x^2} \qquad (4\text{-}67)$$

If the coefficient α of the dissipation term $\partial^2 u/\partial x^2$ is identically zero, corresponding to the Re = ∞ case, then equation (4-67) reduces to

$$\frac{\partial u}{\partial t} = -u\frac{\partial u}{\partial x} \qquad (4\text{-}68)$$

This equation does not necessarily have continuous solutions for all time. It has an analytical solution which experiences a mathematical shock for certain initial conditions. Following Bellman et al., (1958), the analytical solution to equation (4-68) is

$$u(x,t) = f(Z) \quad , \quad Z = x - u(x,t)t \qquad (4\text{-}69)$$

where the function f is the initial condition at t = 0,

$$u(x,o) = f(x) \qquad (4\text{-}70)$$

From this solution we obtain, by differentiation,

$$\frac{\partial u}{\partial x}(t) = \frac{f'(Z)}{1 + tf'(Z)} \qquad (4\text{-}71)$$

If the initial condition f(x) is such that $1 + tf'(Z) \to 0$ for some $t \to t_1$, then

$$\frac{\partial u}{\partial x}(t_1) \to \infty \qquad (4\text{-}72)$$

This is the shock.**

In a higher dimensional problem, the incipient mathematical shock manifests itself in a method of characteristics solution by a crossing of characteristics of the same family (e.g., see Thomas, 1954).

---

*Shocks are mathematically discontinuous only in the inviscid limit; viscosity and heat conduction smooth out the discontinuity. Practically speaking, at sea-level conditions, shock thicknesses are of the order of molecular mean free paths.

**For an alternate description of shock development in Burger's equation, see Lax (1969). Such PDE solutions which are not differentiable are termed "weak solutions", as opposed to "genuine solutions." The shock joins two regions of genuine solutions.

If α in equation (4-67) is non-zero but still very small, a discontinuity can still develop. For slightly larger α, discontinuities may not appear, but large gradients, $\partial u/\partial x$, can develop. This is what happens in real gases at a high Reynolds number. The derivatives like $\partial u/\partial x$ are so large that the shock thickness becomes of the order of the molecular mean free path. For the practical purposes of calculation, it may be represented by a discontinuity.

That the expected inviscid limit behavior of $\alpha \to 0$ occurs smoothly has been shown by Foy (1964). He proved that "for any two states which can be connected by a sufficiently weak shock for the hyperbolic [$\alpha = 0$] system, there exists a continuous solution to the viscous system. As the viscosity coefficient tends toward zero, the solution tends toward the discontinuous generalized solution of the hyperbolic system." See also the discussions and references in Lax (1957).

Burger's equation has also been used in numerical studies of shock development in the conservation form

$$\frac{\partial u}{\partial t} + \frac{\partial}{\partial x}\left(\frac{1}{2} u^2\right) = 0 \tag{4-73}$$

by van Leer (1969) and Lax (1969). However, there are differences between it and the full system (4-63); see Lax (1969).

CHAPTER V

BASIC COMPUTATIONAL METHODS FOR COMPRESSIBLE FLOW

In this chapter, the basic computational methods for solving planar compressible flow problems in rectangular coordinates will be discussed. Most of the methods are built on methods and explanations already presented in Chapter III for incompressible flow, so that a reading of that chapter is essential to an understanding of this one.

Since the publication of the original printing of this text in 1972, an excellent treatise on compressible flow calculations has been published. AGARDograph No. 212 on "Computation of Viscous Compressible Flows Based on the Navier-Stokes Equations", September 1975, by R. Peyret and H. Viviand, is recomended for all aspects of the computational problems to be discussed in the present chapter.

V-A. Preliminary Considerations

Before getting into the main content of this chapter, we dispense with three topics which will not occupy our attention in the rest of the chapter. These topics are (1) shock-free and shock-patching methods, (2) stability analyses, and (3) implicit methods.

V-A-1. Shock-Free Methods and Shock-Patching Methods

Although shocks may occur in compressible flow, shock-free solutions are also of some importance. We briefly review some numerical methods applicable only to shock-free flows; these will not be the principal concern of this book.

Not all compressible flow problems are supersonic; completely subsonic compressible flow problems obviously will not develop shock waves. Thus, Trulio et al. (1966) have used compressible flow equations applied to a subsonic flow. (But see section V-I.)

For supersonic inviscid shock-free flow, where the equations are purely hyperbolic, the method of characteristics (Courant and Friedricks, 1948; Owczarek, 1964) is the natural choice for numerical computation. In this well-known method, the computational mesh is not rectangular and is not known beforehand, but develops as the solution is explicitly stepped out during the computation. This method gives the most accurate results possible, since the calculation node points lie along "characteristics," across which derivatives may be discontinuous. (Further description and references for 2D and 3D methods of characteristics are given in section VI-D.) The greatest limitation of MOC is that viscous effects cannot be included, except by boundary-layer methods.

Boynton and Thomson (1969) devised a spatial marching method which is the spatial counterpart of time-dependent Lagrangian methods. Diffusion is allowed normal to the streamline coordinate, and shocks are patched as in the method of characteristics.

An unusual graphical method for shock-free supersonic flow has been given by Ringleb (1963) and extended by Chou and Mortimer (1966). It is restricted to two-dimensional inviscid flow between two defined streamlines, as in a nozzle. It was to be extended (Chou and Mortimer, 1966) to include viscous effects.

Several time-dependent methods which are designed for hyperbolic equations, but not for handling shock waves, are considered by Gourlay and Morris (1968A) and by Babenko and Voskresenskii (1961).

Any of these or other shock-free methods may be used in conjunction with *shock-patching* methods of various kinds, in which the shock is maintained as a discontinuity, and the Rankine-Hugoniot relations (Owczarek, 1964) are applied across the discontinuity. This approach may be feasible in a fixed Eulerian mesh for a one-dimensional problem (Richtmyer, 1957) but 2D shock patching in a fixed mesh appears to be unworkable (Skoglund and Cole, 1966). Shock-patching in a curvilinear mesh and shock transformation methods are difficult but very accurate (see section VI-B).

Another concept for treating shocks involves changing the numerical procedure to continue the integration through the shock. Thomas (1954) used high order, one-dimensional polynomial curve-fitting techniques and a fine mesh near the shock. T. D. Taylor (1964) had also proposed a local integration scheme through the shock. Bellman et al. (1958) followed characteristics to integrate through the shock; starting values along (x,t) characteristics were found from six-point Lagrange interpolation between rectangular (x,t) mesh points. They found good shock characteristics and marginal stability only for Burger's equation, while calculations for more general hyperbolic equations were unstable. These older methods do not appear well suited to two-dimensional problems, machine calculation, and/or time dependent flows.

In some instances, a weak shock in inviscid flow may be computed as a nearly linear wave compression by the method of characteristics. Weinbaum's (1966) anisentropic characteristics accurately show the lip shock position and pressure rise, for example; see also Baum and Ohrenberger (1970). The method of characteristics may also be used with a shock-patching procedure, the shock build-up being indicated by a crossing of characteristics of the same family. The development of shocks in a characteristic solution was described by Thomas (1954). Xerikos (1968) describes the details of bow shock and flare shock patching in cylindrical coordinates. D. B. Taylor (1968) describes a method of shock patching in a characteristics solution which allows one to keep track of a large number of weak, oblique shock waves.

Moretti's method (Moretti and Abbett, 1966B; Moretti and Bleich, 1967; Moretti, 1968A, 1968B) is a transient method based on characteristics calculations at the moving shock wave, but other calculations are done in a quadrilateral mesh which is not a characteristics mesh. This patching method is successful on the blunt-body problem, in contrast to other patching methods, because the quadrilateral mesh is not rectangular and is not fixed in space. It is defined by a non-orthogonal transformation which changes at each time step, so that both the detached shock wave and the body are always along coordinate lines. The method has several drawbacks. The description of the method, the equations, and the programming are very complex, especially when viscous effects are included and/or when the extension is made to three dimensions. One must have an idea beforehand of the two-dimensional shock structure in order to design the program; for example, the unanticipated appearance of a Mach reflection (Owczarek, 1964) would likely upset the calculation. Also, at the time of this writing, no method has been suggested for treating the gradual merging in space of weak compression waves to form a shock, such as occurs in a gradual channel contraction or in the formation of a near wake shock. When applicable, the method gives very fast and accurate solutions. See section VI-B for further discussion.

A particular patching technique which does not involve a moving shock is Van Dyke's inverse method for the detached-shock blunt body problem (Van Dyke, 1958; Garabedian and Lieberstein, 1958). Again, the solution is not obtained in a fixed Eulerian mesh, but in a mesh which adjusts with each iteration. The shape of the detached bow shock is assumed, and the subsonic flow aft of the shock is integrated directly until the surface is found; i.e., one specifies the shock shape and solves for the body shape. A desired body shape can be found, in principle, by iteration on the shock shape, but abrupt body curvatures are not easily matched. A weakness of these methods is that the subsonic (elliptic) equations are solved as an initial-value problem rather than as a boundary-value problem, which is spatially unstable (see section III-B-8). The success of the method apparently depends on the 11-point interpolation used to smooth initial data of derivatives along the shock, and might be unstable if more than seven points between shock and body were used. However, the method converges rapidly for smooth bodies and high Mach numbers, and is in common use.

The following are references on other methods and programs which use this inverse approach. Webb et al. (1967) used double-precision computer calculations to minimize sources of round-off error, and still had to resort to data smoothing techniques to maintain some kind of stability. Their method does not appear satisfactory, as it is not clear what is driving the solution -- the physics or the data smoothing. Briggs (1960) solved a three-dimensional problem of an elliptic cone at angle of attack using inverse methods. The report of Powers et al. (1967) is a good source of details of the inverse method. Jones (1968) used a three-dimensional inverse method, but instead of driving the shape iterations by deviations in the body shape (where the normal velocity vanishes), he used the residue velocities at the desired body locations. He worked up from the exact (ODE) Taylor-Macoll solution (Owczarek, 1964) for a conical body at zero angle of attack, making small perturbations in angle of attack or in body shape. Makhin and Syagaev (1966) also solved the blunt body at angle of attack using an inverse method. Kyriss (1970) combined shock-free finite-difference calculation in a mixed Lagrangian-Eulerian mesh with an inverse shock approach. Many inverse-method and method-of-characteristics programs are in use in industry; e.g., see Moreno (1967), which also has references on these subjects. Moreno (1967) notes that some initial shock shapes can cause the appearance of singularities as the computation proceeds, preventing iteration convergence. It should also be noted that, in hypersonic flow, the shock curvature on even simple sphere-cones is not always monotonic (Wilson, 1967) but may have an inflection point. This could conceivably upset the calculation procedure.

Our preoccupation in this book is with transient flow methods in an Eulerian mesh which allows the formation of shocks, if they develop, but which allow the calculation to proceed without requiring any special treatment of the shock, or even requiring the detection of the shock. These are called "shock-smearing" or "shock-capturing" methods in the English literature, or "through" methods in the Russian literature (e.g., Godunov and Semendyaev, 1962). If we are interested only in the steady state, we perform the time-dependent calculation and approach the steady state (if it exists) for large time, as in Chapter III. In the Russian literature, this is called (or at least it has been translated as) the "asymptotic" method (Brailovskaya et al.,1970) or the "build-up" method (Brailovskaya et al., 1968).

Of course, if we have no interest in the transient solution, flexibility in the approach is added, in that no real physical problem need be modeled. For example, methods with truncation errors of $O(\Delta t)$ or less may be used. Crocco (1965) suggests using FDE equations which are really inconsistent with the PDE's except in the steady state, when this technique will speed convergence. Simuni (see Brailovskaya et al., 1968) suggested an asymptotic approach using time-dependent boundary conditions which approach the desired problem conditions only as the steady state is approached. Freudiger et al. (1967) also viewed the time-dependent calculation only as an iteration technique, and suggested the possibility of using a spatially varying Δt in order to speed convergence. None of these concepts have been developed into a systematic method and therefore cannot be treated further in this book, but the ideas can be kept in mind for future developments.

V-A-2. Stability Considerations

The remarks of section III-A-5-d on stability criteria and methods of analysis also apply in this chapter. Additional references specifically concerned with the stability of hyperbolic systems are Courant, Friedrichs, and Lewy (1928)*, Lax (1954, 1957, 1958, 1961), Lax and Richtmyer (1956), Lax and Wendroff (1960), Kreiss (1964), Strang (1964), Lax and Nirenberg (1966), Parlett (1966), and Brailovskaya et al. (1970).

Of course, the definitions and conditions discussed in these references will not necessarily be adequate for the nonlinear problems, especially when shocks are present.

The stability analysis of systems of equations like (4-55) is more difficult than for incompressible flows. In the (ψ,ζ) system, the one parabolic equation was independently examined for stability, with the effects of ψ linearized away. Likewise for the primitive variable equations, the coupling of the equations through the advection terms and through the pressure gradient terms was eliminated by linearizing. But for the compressible flow equations, this is not possible. Although the advection terms like $\partial(\rho uv)/\partial y$ may be linearized to $\tilde{v}\ \partial(\rho u)/\partial y$, the pressure gradient terms cannot. They contribute to the essential behavior of even the linearized system. Thus, the model equation

$$\frac{\partial u}{\partial t} = -\tilde{u}\frac{\partial u}{\partial x} + \alpha\frac{\partial^2 u}{\partial x^2} \tag{5-1}$$

which often modeled the stability of the incompressible equations, does not model the behavior of the compressible flow momentum equation (4-42c) even in one dimension, because of the absence of pressure gradient terms.** The analysis of the entire linearized system of equations then requires matrix stability analysis methods (see Ames, 1969; Mitchell, 1969) which we do not cover in this book.

But we do present a trick which allows us to make inferences about the stability of compressible flow methods from the simple model equation (5-1). As we have seen in Chapter III, the typical stability limitation for equation (5-1) is the Courant number restriction,

$$c = \frac{|\tilde{u}|\Delta t}{\Delta x} \leq 1 \tag{5-2}$$

This term arises just from the inviscid terms in equation (5-1), i.e., with $\alpha = 0$. Now in equation (5-1) with $\alpha = 0$, the information is carried at the continuum advection speed, $\tilde{u}$. In each computational time step, a perturbation at (i) affects the new value at (i+1). That is, the information is transmitted a distance Δx in every time step Δt; thus, the *computational* information speed = $\Delta x/\Delta t$. The restriction (5-2) just says that the continuum information speed must be $\leq$ computational information speed, or that

$$|\tilde{u}| \leq \frac{\Delta x}{\Delta t} \tag{5-3}$$

---

*See also IBM Journal, March 1967, for an English translation.

**It is true, luckily enough, that the cross-derivative viscous terms of the compressible flow equations are not important, and equation (5-1) appears adequate in that respect. See section V-F.

Now in compressible flow, the pressure gradient terms in the momentum equations affect the continuum information speed; it is no longer $\tilde{u}$, but something greater than $\tilde{u}$. Richtmyer and Morton (1967) give a verbal description of how this pressure dependence may be derived. Here, we simply rely on the well-known elementary gas dynamics relations. A small pressure disturbance travels at the local speed of sound, a, with respect to the fluid, which is itself moving at $\tilde{u}$. The pressure disturbance moves in all directions, so only $a > 0$ need be considered. Thus, the magnitude of the continuum information speed is now $|\tilde{u}| + a$, and the applicable Courant number restriction is

$$C = \frac{(|\tilde{u}| + a)\Delta t}{\Delta x} \leq 1 \qquad (5\text{-}4A)$$

This is the Courant, Friedrichs, and Lewy (1928) or CFL stability restriction. Using a different vocabulary, it states that the finite-difference domain of influence must be at least as large as the continuum domain of influence (see section III-A-5-e). Physically, it means that a sound wave cannot travel more than one cell length in one time increment.

Thus, the stability restrictions of compressible flow methods are often found by studying the model equation (4-1) and just replacing the incompressible Courant number $C = |u|\Delta t/\Delta x$ by the compressible Courant number, $C = (|\tilde{u}| + a)\Delta t/\Delta x$. The conversion is even extendable to the case of *large* pressure distrubances, wherein the sound speed is replaced by the *shock speed* (see section V-D-1).

The method often gives the same result as an elaborate matrix stability analysis and dependably gives at least a necessary stability condition. More restrictive conditions are necessary for upwind differencing (see section V-E-1), for example; the failure is easily understood here, because the advection terms and the pressure gradients are not differenced by the same method. The trick also fails for implicit methods (see next section). Also, dimensionality effects are different for compressible flow. For $\Delta x = \Delta y = \Delta$, the condition (5-4A) typically changes to

$$C \leq \frac{1}{\sqrt{2}} \text{ or } \frac{1}{\sqrt{3}} \qquad (5\text{-}4B)$$

in two or three dimensions, respectively, when the one-dimensional method is applied to all higher dimensions simultaneously.* When the Marchuk (1965) time-splitting procedure is used (see section III-A-13) and each dimensional contribution is separately calculated, just the one-dimensional restriction (5-4A) holds typically. See also Gourlay and Mitchell (1969B) and MacCormack (1971).

An additional mesh restriction has appeared in the literature. Goodrich (1969) calculated laminar compressible flows over flat plates and backsteps, using the time-dependent Rusanov method (section V-D-3) and the full viscous equations. He found that a spatial mesh restriction was necessary in order to achieve truncation-error convergence and accuracy for a steady-state solution. For flow approximately parallel to the x-coordinate, he required

$$\frac{\Delta y}{\Delta x} \leq \tan \mu_M \qquad (5\text{-}5A)$$

where μ_M is the Mach angle,

$$\mu_M = \arcsin (1/M) \qquad (5\text{-}5B)$$

---

*More generally,

$$C \leq \frac{\Delta x \Delta y}{\sqrt{\Delta x^2 + \Delta y^2}} \quad \text{or} \quad C' \leq \frac{\Delta x \Delta y \Delta z}{\sqrt{\Delta x^2 + \Delta y^2 + \Delta z^2}} \qquad (5\text{-}4C)$$

or, equivalently,

$$\frac{\Delta x}{\Delta y} \geq \sqrt{M^2 - 1} \tag{5-5C}$$

When this criterion was violated, unrealistically high pressures developed near the inflow boundary (private communication).

The requirement does hold, of course, for an inviscid hyperbolic characteristics-type solution marched out in a fixed mesh, being nothing more than the Courant-Friedrichs-Lewy condition (see Courant et al., 1928). However, stable and apparently accurate solutions which violate this criterion have appeared in the literature. It is also known that such a condition has arisen in simpler equations because of particular boundary conditions (private communication, A. J. Chorin).

The exact status of this criterion has not been ascertained, at the time of this writing. It appears that Goodrich's experience may be related to boundary problems, but that it also may not be unique.

V-A-3. Implicit Methods

Almost all computations of actual multi-dimensional compressible flow problems published to date have been achieved with explicit methods. Some of the multi-step methods (sections V-E-7 and V-F-3) may be interpreted as iterative approximations to implicit methods, but in fact prove to be better when applied with just one iteration rather than many. Early papers which considered implicit methods for hyperbolic systems were Wendroff (1960), Anucina (1964), and Gary (1964). Brailovskaya et al. (1970) point out the reason for discouragement; many implicit methods which are unconditionally stable for the model equatin (5-1) are *not* unconditionally stable for the compressible flow equations. This was the experience of Schroeder and Thomsen (1969); they developed an implicit multi-dimensional method for hyperbolic equations which was still limited by $C \leq 1$. Polezhaev (1966, 1967) obtained compressible flow solutions with an ADI method which removed the diffusion time-step limitation, but still required $C \leq 1$, treating the ∇P terms explicitly.

However, Gourlay and Mitchell (1966A) also considered ADI methods and later (Gourlay and Mitchell, 1966B) presented a two-dimensional ADI method for hyperbolic equations, based on 9-point equations applied to each of two levels, which is unconditionally stable. But the method has not been exercised on a nonlinear problem, and has not been proved in an actual fluid dynamics calculation. Schwartz and Wendroff (1970) solved a 1D shock propagation with a method that is *nonlinearly* implicit, using iteration at each step. It would appear to be difficult to adapt to 2D and 3D.

Although implicit methods for compressible flows may become important in the future, we will consider only the proven explicit methods in this book.

V-B. Methods for the Numerical Treatment of Shocks

Rather than follow the shock discontinuity and try to decide when to merge compression waves into a shock, we would prefer to just "turn on" the equations and let shocks develop naturally, wherever they will. The difficulty is that the thickness δ_s of a normal shock wave in real, viscous gases varies as 1/Re for fixed Pr. For high-Re flows, it then becomes likely that $\delta_s < \Delta x$. When this occurs, oscillations develop aft of the shock, as shown in Figure 5-1a. These oscillations in the finite Eulerian mesh are physically representative of the process whereby the ordered kinetic energy lost in the velocity defect across the shock is degraded into internal energy by collisions of molecules (Richtmyer and Morton, 1967). But in the computational model, the node-point molecules are spread too thin. If $1/Re = 0$ and if the finite-difference method has no artificial viscosity, then there is no dissipative mechanism by which the kinetic energy can be converted to internal energy, so the oscillations will persist. For $1/Re > 0$, they will eventually die out; but for most practical computations, the method is inadequate. Note that the correct values of the shock speed are *not* obtained by averaging (Richtmyer, 1957).

If only low-Re flows are studied, the shocks may not pose any special problem. Crocco (1965) encountered only small oscillations when $\delta_s = 2\Delta x$ for a normal shock. Scala and

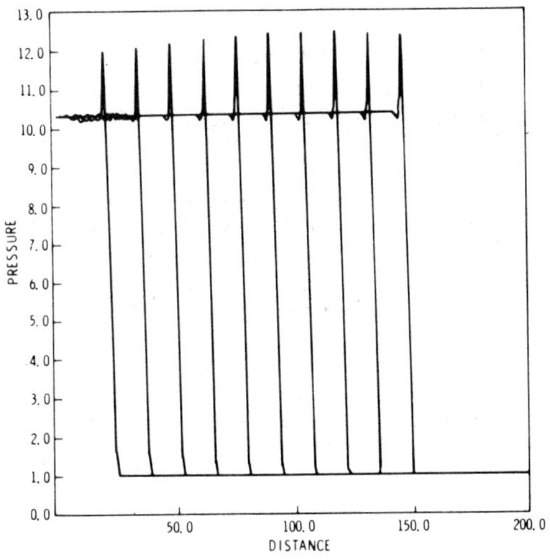

(a) Richtmyer two-step method, $b_1 = 0$.

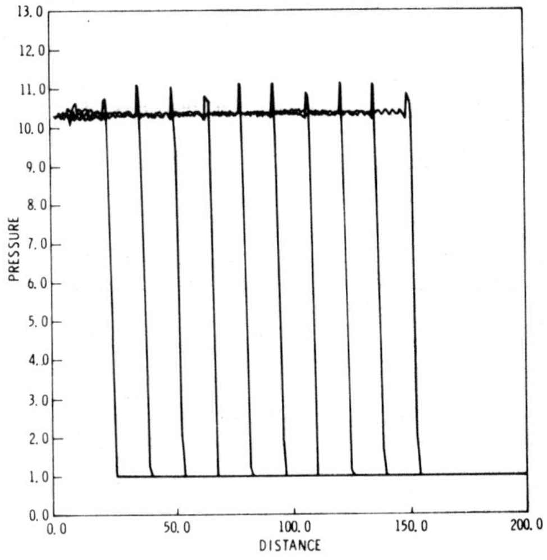

(b) MacCormack permuted method, $b_1 = 0$.

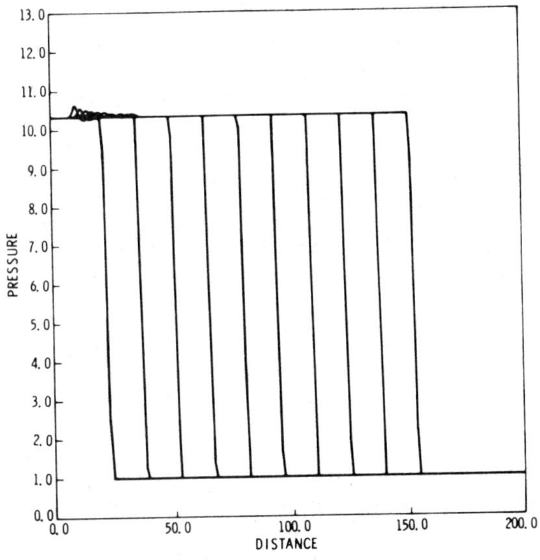

(c) Richtmyer two-step method, $b_1 = 0.15$

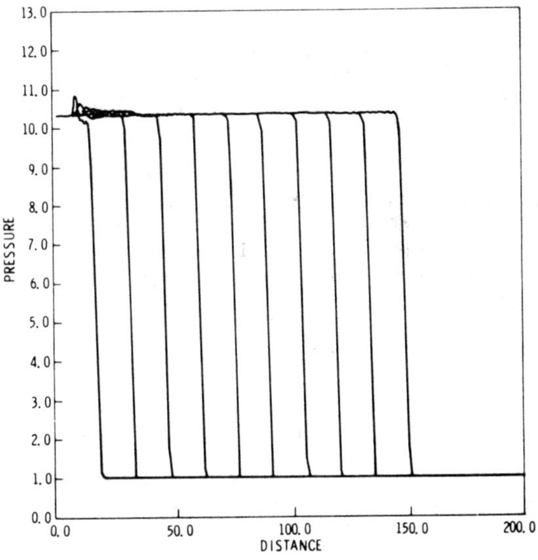

(d) MacCormack permuted method, $b_1 = 0.325$.

Figure 5-1. Calculations of an M = 3 shock wave in an Eulerian mesh, using two-step Lax-Wendroff methods, max Courant number = 0.95. The shock is propagating from left to right. Pressure profiles are plotted at equal time increments. (After Tyler,1970)

Gordon (1967) experienced no difficulty with Re $\leq$ 40 and with a fine mesh which gave 20 points within the shock. The problem will also be alleviated when only oblique shocks (weaker than normal shocks) and when viscous wall boundaries are present.

In an Eulerian system, it is not feasible to use extremely fine mesh spacing at the shock because its position is not known *a priori*. In a Lagrangian system with "rezoning," the shock position can be followed with a fine mesh as it develops. This is successful in a one-dimensional problem (Richtmyer, 1957) such as planar or spherical shock prapagation, but is difficult for higher dimensional problems (Goad, 1960). McNamara (1966, 1967) has developed a method of patching discontinuities into a floating Eulerian mesh which is periodically realigned to follow a contact surface discontinuity and the shocks. Though generally successful, the moving-mesh calculation does cause some error.

The difficulties of shock-patching methods have already been described. These are not successful in a fixed Eulerian mesh, and their use in transformed coordinates will be considered in Chapter VI.

The most generally successful method of handling shocks is to artificially smear out the discontinuity so that δ_s = 3 to 5(Δx), thus losing details in the region of the shock, while maintaining correct properties across the shock.

V-C. *Shock Smearing by Artificial Dissipation*

The concept of introducing an artificial dissipation term to smear out shocks was first introduced by von Neumann (1944), and in the open literature by von Neumann and Richtmyer (1950). In their method, and in later ones patterned after it, an *explicit* artificial viscosity term is added to the equations. It is so devised as to be effective only in regions where high compressions occur, i.e., where shocks are building up. Away from the shock region, the effect is negligible, thus differing from merely using a large μ (low Re) in the viscous equations as done by Ludford, Polachek, and Seeger (1953).

The alternate concept is to devise a finite-differencing method to accomplish the shock smearing automatically, without a viscosity term ever appearing explicitly in the equations. These are the *implicit* artificial viscosity or implicit damping methods. Some of these methods may also require *explicit* artificial viscosities to stabilize strong shocks. An early shock calculation using implicit artificial viscosity was presented by Ludloff and Friedman (1954). In either the explicit or implicit approaches, the method must be "dissipative" in the mathematical sense (Richtmyer and Morton, 1967) of providing more damping for the short wavelength disturbances than for the long-wavelength disturbances. This property is necessary for the differencing method to satisfy the entropy jump conditions across the shock by automatically rejecting rarefaction shocks (e.g., see Owczerek, 1964). Fortunately, this property is easily (even unconsciously) met.

V-D. *Methods Using Explicit Artificial Viscosities*

V-D-1. *von Neumann-Richtmyer Method*

The classic paper introducing the concept of an explicit artificial viscosity was published in 1950 by von Neumann and Richtmyer. To stabilize one-dimensional inviscid calculations of a propagating shock wave using non-conservative Lagrangian equations, they introduced an artificial term into the pressure. But the method is more easily understood if the added term is interpreted as a viscosity term; in fact, an unambiguous extension to higher dimensions follows if we consider the term to be an artificial *bulk* viscosity $\bar{\kappa}_o$.

In an Eulerian system, the concept is explained as follows. Consider the one-dimensional equations obtained from equations (4-42) through (4-47) with $\bar{\mu}$ = 0 but $\kappa > 0$. Note the product

$$\frac{\lambda}{Re} = \frac{\bar{\lambda}/\bar{\mu}_o}{\bar{\rho}_o\bar{u}_o L/\bar{\mu}_o} = \frac{\bar{\kappa}_o}{\bar{\rho}_o\bar{\mu}_o L} = \alpha_B \tag{5-6}$$

The term α_B is a diffusion coefficient, equal to the inverse of a Reynolds number based not on $\bar{\mu}_o$, but $\bar{\kappa}_o$. Then (4-42) through (4-47) become

$$\frac{\partial U}{\partial t} + \frac{\partial F}{\partial x} = 0 \tag{5-7A}$$

$$U = \begin{Bmatrix} \rho \\ \rho u \\ E_s \end{Bmatrix} \tag{5-7B}$$

$$F = \begin{Bmatrix} \rho u \\ P + \rho u^2 - \alpha_B \dfrac{\partial u}{\partial x} \\ u(E_s + P) - \alpha_B u \dfrac{\partial u}{\partial x} \end{Bmatrix} \tag{5-7C}$$

Von Neumann and Richtmyer proposed the following form for α_B.

$$\alpha_B = \rho (b_1 \Delta x)^2 \left|\frac{\partial u}{\partial x}\right| \tag{5-8}$$

where b_1 is an adjustable constant*, always of the order of unity. The actual form they used was the addition of a term,

$$q_1 = -\alpha_B \frac{\partial u}{\partial x} \tag{5-9}$$

which, from equation (5-7), is seen to be equivalent to an artificial pressure.

The idea is this: We make no attempt to calculate the flow inside the shock with any accuracy, but are interested only in the essentially inviscid flow on either side of the shock. If a simple constant value of the artificial diffusion coefficient α_B were used, and if it was large enough to damp out oscillations behind the shock, the "shock" would be spread out, perhaps over 50 or 100 mesh points. But the Rankine-Hugoniot relations (see any gas dynamics text) are valid *across* the shock, no matter what the details of the dissipative process inside the shock. (For example, the Rankine-Hugoniot relations can be written for the complex pattern of the diamond-shock/boundary-layer interaction in a supersonic diffuser, which may be spread over several feet.) The advantage of the form α_B given equation (5-8) is that the diffusion process is now driven by $\frac{\partial}{\partial x}\left[\left(\frac{\partial u}{\partial x}\right)^2\right]$ instead of $\frac{\partial}{\partial x}\left[\frac{\partial u}{\partial x}\right]$. Consequently, the dissipation occurs in a shorter distance. Of course, error arises outside the shock region but, if gradients there are small, the error will be tolerable.

The particular coefficients $(b_1 \Delta x^2)$ in equation (5-8) are chosen so that a constant shock width, in mesh increments, is obtained independent of the shock strength (pressure jump). With this method, shock widths of $\delta_s = 3$ to $5\Delta x$ are obtained; see Figure 5-1b. The definition of δ_s is vague, of course, as is the definition of a boundary-layer thickness. If a slope thickness definition is used, $\delta_s \simeq 3\Delta x$. A mild strengthening of the Courant condition, $C \leq 1$, is required for stability. It was soon realized by Rosenbluth (see Richtmyer and Morton, 1967) that there is no need to smear out rarefaction waves, so most workers used equation (5-8) only when $u(\delta u/\delta x) < 0$ and set $\alpha_B = 0$ when $u(\delta u/\delta x) > 0$.

---

*The constant b_1 plays a role similar to that of the "mixing length" in Prandtl's model of boundary-layer turbulence (see Schlichting, 1968).

The exact value of b_1 is manipulated in numerical experiments to produce a compromise between two desirable properties, minimum shock thickness and minimum amplitude of the oscillations behind the shock, which are never completely eliminated.

The different methods discussed in the following paragraphs are all attempts to obtain these desirable properties, and some methods have certain advantages. But it is important to note five points. (1) If the conservation form of the 1D equations is used, and if some form of dissipation is present, the Rankine-Hugoniot relations will be satisfied across the shock, since they are based only on over-all conservation of mass, momentum, and energy. Thus, the correct shock speed will be attained, regardless of the method used.* This was validated computationally by Longley (1960). (2) This exact satisfaction of the Rankine-Hugoniot relations will *not* necessarily occur in multi-dimensional flows when the shock is oblique to the mesh. The smearing of the shock may allow a gradient of normal momentum flux to develop in the direction tangential to the shock wave. This violates the basic gas dynamics assumption necessary to transform the normal-shock Rankine-Hugoniot relations into oblique-shock equations (see any gas dynamics text.) Thus, the oblique-shock speeds will be in error, and any steady-state oblique shock will not exhibit the exact jump conditions. (3) The post-shock oscillations are *intrinsic* to explicit methods using second-order forms for the spatial derivatives; they occur because of the violation of the characteristic sense of the supersonic flow (see discussion in section V-E-2). (4) In an Eulerian mesh, it would seem that the shock thickness δ_s, defined in terms of the asymptotic approach to upstream and downstream flow variables, *must* be more than one cell. For if the flow properties just changed exactly from $i = w$ to $i = (w+1)$, the position of the shock, x_s, could not be resolved to better than $\pm\Delta x/2$. Then there could be no computational distinction between two shocks, both of which were between (i) and (i+1). For $C < 1$, the shock moves less than Δx, and C could be chosen so that shock a is still between (i) and (i+1). But then it would *never* move. The only computational speeds possible for a shock with $\delta_s < \Delta x$ are either $V_s = 0$ or $V_s = \Delta x/\Delta t$. Then, in order to achieve the correct shock speed, it follows that $\delta_s < \Delta x$ is not possible. (5) Most of the published comparisons of methods have been made on the basis of a forward-time differencing method, which may depend on α_B for linear stability. But the artificial viscosity terms can be used in conjunction with other methods, and the relative merits may then change.

The von Neumann-Richtmyer method is still being used, and frequently compares well with more recently developed methods. Schwartz (1967) used it in spherical coordinates to solve the relativistic fluid dynamics problem of gravitational collapse of a star. Hicks and Pelzl (1968) found it gave better results than the Lax-Wendroff method (sections V-E-5,6) for strong shocks and rarefactions. (See other comparisons in section V-D-4.) Laval (1969) used it to study the starting process in a nozzle. Hofmann and Reaugh (1968) and Wilkins (1969) used it in two-dimensional Lagrangian problems; the latter employed a remapping of the Lagrangian mesh to counteract large mesh distortions. Wilkins (1970) calculated 2D problems ranging from elasticity theory to gas dynamics. A form of varying α_B was employed by van Leer (1969) to compute shock propagation in Lagrangian coordinates with variable mesh spacing. The Eulerian form of α_B (5-8) was used by Tyler and Ellis (1970) in a comparative study. Plooster (1970) has successfully applied it to cylindrical explosion problems.

V-D-2. Landshoff's Method and Longley's Method

Landshoff (1955) experimented in Lagrangian coordinates with a q term that was linear rather than quadratic in velocity gradient, using

$$q_2 = -\frac{1}{2}\rho_o(\Delta x)\left|\frac{\delta u}{\delta x}\right| \tag{5-10}$$

which is equivalent to an artificial diffusion of

$$\alpha_B = \frac{1}{2}\rho_o(\Delta x) \tag{5-11}$$

---

*An idealization. Until the finite shock thickness becomes constant in a steady propagation, the shock speed will be ambiguous, most noticeably for $O(\Delta x)$ methods.

He found that the von Neumann-Richtmyer q_1 gave larger initial overshoot but faster damping of oscillations behind the shock than his q_2. He recommended a linear combination, $q = q_1 + q_2$, with $b_1 = 1/2$ in equation (5-8) to give a good compromise. Emery (1968) says that the method (5-10) gives large density oscillations.

Longley (1960) experimented in an Eulerian system with four artificial viscosities. In addition to the von Neumann-Richtmyer q_1 and Landshoff's q_2, he used

$$q_3 = - b_1 \frac{1}{2} \rho(\Delta x) \; |u| \left|\frac{\partial u}{\partial x}\right| \tag{5-12}$$

which gives an explicit analog of the implicit viscosity of the upwind differencing of the PIC method (see sections V-E-1,3); that is,

$$\alpha_B = b_1 \frac{1}{2} \rho u \cdot \Delta x \tag{5-13}$$

He also experimented with his own form

$$q_4 = - b_1 \frac{1}{2} \Delta x \frac{\rho a}{P} \left|\frac{\partial u}{\partial x}\right| \tag{5-14}$$

which gives

$$\alpha_B = \frac{1}{2} b_1 \frac{\rho a}{P} \cdot \Delta x = \frac{1}{2} b_1 \sqrt{\frac{\gamma \rho}{P}} \Delta x \tag{5-15}$$

where a is the speed of sound. The last two are most effective in stagnation regions of shock reflection, where the other methods tend to fail. All four q methods, and four differencing methods, gave correct shock propagation speeds because of the use of the conservation form of the equations.

V-D-3. Rusanov's Method

An explicit artificial dissipation method that is especially effective for 2D calculations is Rusanov's method (1961). Basically, Rusanov added artificial diffusion terms of the general form $\frac{\partial}{\partial x}\left(\alpha \frac{\partial U}{\partial x}\right)$ in finite difference form to all the frictionless equations for $\partial U/\partial t$, (where $U = \rho$, ρu, ρv, and E_s) using forward-time, centered-space differencing. The method therefore adds not only an artificial viscosity, but also an artificial conductivity and mass diffusion.* The term α_B is an artificial diffusion coefficient proportional to $C = |V| + a$ and to an adjustable parameter ω. The form $\frac{\partial}{\partial x}\left(\alpha \frac{\partial U}{\partial x}\right)$, rather than the simpler form $\alpha \partial^2 f/\partial x^2$, is apparently necessary in order to obtain correct shock solutions (van Leer, 1969).

The actual formulation of Rusanov's method is somewhat more involved, with special emphasis being given to the complication of $\Delta x \neq \Delta y$, which is of considerable practical importance. The method and its stability restrictions may be concisely described by defining the two-dimensional Courant number, C_{2D}, by

$$C_{2D} = \frac{(V + a)\Delta t}{\sqrt{\Delta x^2 + \Delta y^2}} \tag{5-16}$$

---

*In this regard, Rusanov's method is similar to the method of Lax (see Section V-E-4).

where $V = \sqrt{u^2 + v^2}$ is the local velocity magnitude, and a is the local speed of sound. Then, in terms of the equations (4-55), the method changes equation (4-63A) to

$$\frac{\partial U}{\partial t} + \frac{\partial (F + \alpha_x U)}{\partial x} + \frac{\partial (G + \alpha_y U)}{\partial y} = 0 \tag{5-17}$$

$$\beta = \Delta x/\Delta y \tag{5-18}$$

$$\alpha_x = \frac{\omega}{\beta} C_{2D} \quad , \quad \alpha_y = \omega\beta C_{2D} \tag{5-19}$$

Stability requires the usual condition,

$$C_{2D} \leq 1 \tag{5-20}$$

and another condition, arising from the explicit artificial diffusion term,

$$C_{2D} \leq \omega \leq \frac{1}{C_{2D}} \tag{5-21}$$

everywhere in the field. (Van Leer, 1969, has shown that this form of a sufficient stability requirement is remarkably general for this class of methods.)

Skoglund and Cole (1966) tested all combinations of seven different σ (where $\sigma = \max_{ij} C_{2D}$) and six ω on a shock-wave calculation and found that $\sigma < 0.9$, $\omega = 0.6$ gave the best results. They also reduced the modulus (V + a) of the Courant number in equation (5-19) to $\sqrt{V^2 + a^2}$ and thereby considerably improved the accuracy in a boundary-layer calculation.* They applied Rusanov's method to a calculation of shock-wave/boundary-layer interaction, including physical laminar viscosity terms. G. W. Zumwalt and his associates have successfully applied Rusanov's method to a variety of problems, some including a turbulent eddy viscosity; see Tyler (1965), Eaton and Zumwalt (1967), Ruo (1967), Walker et al., (1966), Bauer et al., (1968), and Prentice (1971). Kessler (1968) and Emery (1968) have also used the method in 2D, and Rusanov and Lyubimov (1968) and Rusanov (1969) have extended it to 3D problems. Goodrich (1969) used it in 2D problems including laminar viscous terms. Emery and Ashurst (1971) used it in spherical coordinates.

The method has often been compared with others and usually fares very well, except for problems where transients are difficult to calculate, in which case methods of $O(\Delta t^2)$ are preferred (Emery, 1968). For unsteady flow problems, the explicit artificial viscosity is not as bad as it looks at first glance. Like Leith's method (section III-A-13), the added "diffusion" term in a forward-time method can actually approximate the contribution of a second time derivative in the inviscid equations, for the proper combination of parameters. When applied to the incompressible flow model equation (5-1), the artificial diffusion is zero for $\omega = c$,** and the exact transient solution is obtained for $\omega = 1$ and $c = 1$ (Tyler and Ellis, 1970). For steady-state solutions, the artificial errors persist (see section III-A-8).

It is unfortunately true that the optimum combination of σ and ω, judged in terms of minimum shock thickness and minimum diffusion errors, has proved to be problem dependent. It is also noteworthy that the artificial diffusion terms are *necessary* in this method, not only to smear out shock waves but just to maintain linear stability. In spite of these unfortunate aspects, Rusanov's method is justifiably regarded by many as the *best* of the explicit artificial viscosity methods for multi-dimensional computations in an Eulerian mesh; see, for example, Emery (1968) and van Leer (1969).

---

*Cole's modification may also be interpreted in a transient analysis in the light of van Leer's (1969) analysis; it puts the artificial viscosity somewhere between that of the original Rusanov method and that minimum viscosity needed for linear stability.

**c is the Courant number for equation (5-1), $c = \tilde{u}\Delta t/\Delta x$.

V-D-4. Errors Arising from Artificial Viscosities

The use of artificial viscosities is often unavoidable, and it can be acceptable; but some strange errors can arise from explicit artificial viscosities, aside from the obvious ones common to incompressible flow calculations (section III-A-8). Schulz (1964) pointed out that simple application of the von Neumann-Richtmyer q_1 in cylindrical or spherical coordinates causes a diffusion of radial momentum. He extended q_1 to a tensor form which maintains strict conservation of radial momentum. Cameron (1966) showed that explicit artificial viscosities introduced surprising errors in the calculation of shocks propagating across a material interface or across a change in mesh spacing, Δx. The von Neumann-Richtmyer q_1-term causes spurious fluctuations for the changes in entropy and density as the shock crosses a material interface. Also, when Δx changes, a false shock wave is *reflected off the mesh change,* and the speed of the original propagating shock is altered. He also found that Landshoff's q_2 did not adversely affect shock speed at a mesh change, but was less useful than the von Neumann-Richtmyer q_1 because the shock *thickness* now changed abruptly at the mesh change. Cameron used both errors to partially cancel each other. By changing Δx at the material interface, he obtained the correct speed for the propagating shock. The false reflected shock still appeared, however. Higbie and Plooster (1968) varied the von Neumann-Richtmyer q_1 for a shock propagation problem in Lagrangian coordinates in such a way that the shock thickness in mesh increments stayed constant as the mesh spacing continually changed, thus eliminating oscillations.

V-E. Methods Using Implicit Artificial Damping

Instead of adding explicit artificial viscosity terms like q_1 to the equations, artificial damping may be added implicitly, just from the form of the difference equations. Sometimes these methods add an artificial *viscosity* in the sense of a non-zero coefficient of second space derivatives, and sometimes they just add artificial *damping* in the sense of the eigenvalues of the amplification matrix being less than one in magnitude. In either case, these methods may require *additional explicit* artificial viscosities in order to stabilize strong shock calculations.

V-E-1. Upwind Differencing

The second method of Courant, Isaacson, and Rees (1952) is a one-sided or upwind differencing scheme, as described in section III-A-8. It was also suggested by Lelevier (see Richtmyer, 1957) for Lagrangian equations, and is frequently referred to as Lelevier's method (e.g., Crocco, 1965; Roberts and Weiss, 1966; Kurzrock and Mates, 1966). In equation (4-63), each of the advected properties, U, that appear in F and G is differenced according to the sign of the advection velocity, u or v, respectively. However, the pressure gradients in the momentum equations must *not* be evaluated by upwind differences, as will be discussed in the next section. In terms of the 1D inviscid equations (4-66), the first upwind differencing method is as follows.

$$\frac{\rho_i^{n+1} - \rho_i^n}{\Delta t} = -\frac{(\rho u)_i^n - (\rho u)_{i-1}^n}{\Delta x} \quad \text{for} \quad u_i > 0 \tag{5-22A}$$

$$= -\frac{(\rho u)_{i+1}^n - (\rho u)_i^n}{\Delta x} \quad \text{for} \quad u_i < 0 \tag{5-22B}$$

$$\frac{(\rho u)_i^{n+1} - (\rho u)_i^n}{\Delta t} = -\frac{P_{i+1}^n - P_{i-1}^n}{2\Delta x} - \frac{(\rho u^2)_i^n - (\rho u^2)_{i-1}^n}{\Delta x} \quad \text{for} \quad u_i > 0 \tag{5-23A}$$

$$= -\frac{P_{i+1}^n - P_{i-1}^n}{2\Delta x} - \frac{(\rho u^2)_{i+1}^n - (\rho u^2)_i^n}{\Delta x} \quad \text{for} \quad u_i < 0 \tag{5-23B}$$

$$\frac{E_{s_i}^{n+1} - E_{s_i}^{n}}{\Delta t} = -\frac{\left[u(E_s + P)\right]_i^n - \left[u(E_s + P)\right]_{i-1}^n}{\Delta x} \quad \text{for } u_i > 0 \tag{5-24A}$$

$$= -\frac{\left[u(E_s + P)\right]_{i+1}^n - \left[u(E_s + P)\right]_i^n}{\Delta x} \quad \text{for } u_i < 0 \tag{5-24B}$$

The 2D difference equations follow this form in an obvious way. The analysis of Kurzrock (1966) indicates that stability is limited, in addition to the Courant-number restriction, by

$$\Delta t \leq \frac{|u|/\Delta x + |v|/\Delta y}{\left[|u|/\Delta x + |v|/\Delta y + a/\Delta x \sqrt{1 + \beta^2}\right]^2} \tag{5-25A}$$

or, for $\Delta x = \Delta y = \Delta$(or $\beta = 1$),

$$\Delta t \leq \frac{(|u| + |v|)\Delta}{(|u| + |v| + a\sqrt{2})^2} \tag{5-25B}$$

This limitation will become dominant in stagnation regions and in recirculating flow regions, where $u, v \to 0$. (See also section V-E-3.)

The modifications of this first upwind differencing method, which are necessary to achieve strict conservation near a region of velocity reversal, follow the description in section III-A-10. The more accurate second upwind differencing follows the description in section III-A-11.

These upwind differencing methods introduce effective "viscosity" through the truncation errors of the one-sided differences. The method adds artificial diffusion terms to $U = \rho, \rho u, \rho v, E_s$ in equation (4-63). From the analysis in section III-A-8, the x- and y-diffusion terms for the transient analysis are

$$\alpha_x = \frac{1}{2} u\Delta x(1 - u\Delta t/\Delta x)$$
$$\alpha_y = \frac{1}{2} v\Delta y(1 - v\Delta t/\Delta y) \tag{5-26A}$$

and, for the steady-state analysis,

$$\alpha_x = \frac{1}{2} u\Delta x$$
$$\alpha_y = \frac{1}{2} v\Delta y \tag{5-26B}$$

Note that the viscosity effect is not really equivalent to a physical viscosity, since the coefficients are directional and dependent on the velocity components.

Exercise: In a flow parallel to the x-axis with $\partial U/\partial x = 0$, but with an arbitrary density distribution in the y-direction, contrast the artificial diffusion behavior of the upwind differencing method with that of Rusanov's method.

For strong shocks appearing in inviscid calculations, this implicit viscosity is not usually sufficient to stabilize the calculations (Richtmyer, 1957), but Kurzrock and Mates

(1966), Scala and Gordon (1967), and Roache and Mueller (1970) have applied it to low (cell) Reynolds-number flows with success.* This method is also the basis of the PIC and FLIC codes, to be described shortly.

The upwind difference method possesses the transportive property (sections III-A-9, 10) which is significant for both subsonic and supersonic flow. The associated lack of second-order spatial accuracy is somewhat less significant in supersonic than in subsonic flow, as we now discuss.

V-E-2. The Domain of Influence and Truncation Error

In this section, we will compare and relate the domain of influence in continuum and in finite-difference equations. Our objective is to show how upwind differencing maintains something of the correct characteristic sense of the continuum equations and does not necessarily have a worse spatial truncation error than do centered difference methods.

Consider first the incompressible continuum flow equations,

$$\nabla^2\psi = \zeta \tag{5-27}$$

and

$$\frac{\partial\zeta}{\partial t} = -\nabla\cdot(\vec{V}\zeta) + \frac{1}{Re}\nabla^2\zeta \tag{5-28}$$

The vorticity transport equation (5-28) is parabolic and, by itself, represents an initial-value problem with limited spatial domain of influence in the inviscid limit $1/Re = 0$. But the Poisson equation (5-27) is elliptic and represents a boundary-value problem. Therefore, a disturbance in ζ at any point in the flow immediately affects all other points in the field, even with $1/Re = 0$, through the nonlinear term $\vec{V}$ which depends on ψ, and thus ζ, through equation (5-27). This property is shared by the finite-difference equations. We say that the system (5-27) and (5-28) possesses *infinite signal propagation speed,* and so does the finite-difference equation.

The inviscid compressible flow equations are all transport equations like (5-28) and therefore represent initial-value problems. The signal propagation speed is *finite*; for small linearized disturbances, the signal propagates at the isentropic sound speed (a) relative to the fluid, or at (V + a) relative to an Eulerian mesh. Consequently, for $V > a$, i.e., $M > 1$, no disturbance is propagated upstream. This leads directly to the well-known Mach-cone principle, or the principle of limited upstream influence.

Consider now the signal propagation in a finite-difference equation. If space-centered differences are used, any disturbance at (i) at time (n) is felt at $(i\pm1, j\pm1)$ at $(n+1)$, no matter what the value of Δt. Thus, the propagation distances are always the same, Δx and Δy. The propagation speeds are then $\Delta x/\Delta t$ and $\Delta y/\Delta t$. The Courant-Friedrichs-Lewy (1928) or CFL necessary stability requirement is that the finite-difference domain of influence at least include the continuum domain of influence, i.e., $\Delta x/\Delta t < V + a$, or

$$C = \frac{(V + a)\Delta t}{\Delta x} \leq 1 \tag{5-29}$$

where C is the Courant number. In strong shock problems, where the small-disturbance assumption is not valid, replacement of "a" by the nonlinear shock propagation speed $a_s > a$ leads to the von Neumann-Richtmyer (1950) requirement.

Courant et al. (1928) did not require anything else from the finite-difference equations, since their objective was only to demonstrate the existence of solutions. But it clearly

*Scala and Gordon (1967) used upwind differencing for the advection terms, but with a more complex pattern of operations, as in Sheldon's method for the Poisson equation (see section III-B-7).

would be desirable also to maintain something of the limited upstream influence of the continuum system. Working with a rectangular mesh, the most we can accomplish is to restrict the sense, + or −, of perturbations along u and v. This led Courant, Isaacson, and Rees (1952) to their method for differenicng in a rectangular mesh, upwind differencing.

This leads again to the notion of transportive differencing for the advection terms, as discussed in sections III-A-8, 9, 10. But allowance must be made for the possible nonlinear upstream propagation in the case $a_s > V$. This leads to the space-centered differencing of the pressure gradient terms of the momentum equations, so that pressure gradient effects are felt upstream.* Note that P is *not* an advected quantity in $\partial P/\partial x$ and $\partial P/\partial y$, but *is* an advected quantity in the flow-work term, $\nabla\cdot(\vec{V}P)$, of the energy equation; consequently, upwind differencing *is* used on the flow-work term.

The distinction between the behavior of these equations and the incompressible system is that no elliptic equation like (5-27) appears, so the compressible inviscid system is purely hyperbolic.

The second-order accuracy of space-centered difference methods is still highly desirable, of course, as it was in incompressible flow. But in supersonic flow, we sacrifice less to achieve the transportive property. The accuracy evaluation of centered differences of section III-A-1 is based on Taylor series expansions for the flow properties, assuming continuity of the flow variables and their derivatives. But, in inviscid supersonic flow, the inviscid equations do not necessarily display continuity of derivatives. In fact, characteristic curves may be *defined* (Courant and Friedricks, 1948; Shapiro, 1953) as curves across which flow variables may have discontinuous derivatives.** Therefore, the Taylor-series expansion is not always valid, and the loss of truncation order of the differentials is not as important in supersonic flow.***

For viscous flow, the characteristics do not exist and the above arguments are weakened. It does seem reasonable, however, to base arguments on the differencing methods for the advection terms on only the behavior of the inviscid equations. This approach is conceptually vague, but the known success of method-of-characteristics solutions in computing real flows with small viscosity supports the approach.

Lax (1969) has shown that the upwind difference form gives a very good shock calculation in the inviscid form of Burger's equation, but fails for the full system of compressible flow inviscid equations and also, surprisingly, for the *linearized* inviscid Burger's equation. That is, the calculations of the nonlinear equation are *more* accurate than those of the linear equation.

V-E-3. PIC and FLIC

A well known method originally devised by Evans and Harlow (1957) is the Particle-in-Cell or PIC method. The genesis of this method is different from most, in that the attempt is not made to model the differential equations so much as the fundamental physical process, through a finite-particle approach. PIC may unequivocally be called a "simulation" method. The calculations proceed in several phases at each time level, with several key intermediate

---

*Kurzrock (1966) experimented with forward, backward, and centered pressure differences. His experiments and his stability calculations show that centered pressure differencing is preferable for his boundary-layer calculations.

Note the physical absurdity that would result from using upwind differencing for pressure and all advection terms. Then, in the quasi-1D duct flow problem described in section III-C-9, the effects of flow perturbations at outflow (i = I) could never be felt upstream, and a shock could not propagate upstream. It would therefore not be possible to computationally turn off an indraft supersonic wind tunnel!

**It is precisely this property that gives the method of characteristics its utility, allowing different flow regions to be patched together along characteristics.

***McNamara (1967) credits Trulio (1964) for showing that, for time-marching methods with discontinuous derivatives, the truncation error tends to zero no faster than $(\Delta x)^{3/2}$.

cell properties being calculated on the basis of pressure contributions, followed by advection calculations. The method is too complicated to describe in complete detail here, but the most unique aspect is that continuum flow is not modeled; rather, a finite number of particles is used, their locations and velocities being traced by Lagrangian kinematics as they move through a computational Eulerian mesh. They are not merely marker particles as in the MAC code (see section III-G-4), but they actually participate in the calculation, even when free surfaces and interfaces are not present. Cell-averaged thermodynamic properties are calculated, based on the numbers of particles in the cell. As few as six particles/cell on the average and three particles/cell locally have been used. The results display high frequency oscillations in cell density and pressure, as expected.

A continuum method which evolved out of the PIC code is the Fluid in Cell or FLIC code of Gentry, Martin, and Daly (1966), based on earlier work by Rich (1963). They departed from the finite particle approach of PIC but retained most of the other aspects. It is a two-step method. In the first part of the first step, provisional values, $\overline{u^{n+1}}$ and $\overline{v^{n+1}}$, are calculated using only the contribution of the pressure gradients and the explicit artificial viscosity terms, if present. [A form like (5-10) is used for the explicit artificial viscosity.] Non-conservation forms are used. Then a provisional internal energy, $\overline{e^{n+1}}$, is calculated only from the pressure term of the equation

$$\frac{\partial e}{\partial t} = -\vec{V}\cdot\nabla e - P\nabla\cdot\vec{V} \tag{5-30}$$

plus its artificial viscosity terms. The divergence $\nabla\cdot V$ is based on velocities $\tilde{u}_{ij} = 1/2\ u^n_{ij} + \overline{u^{n+1}_{ij}}$ wherein the provisional values $\overline{u^{n+1}_{ij}}$ have already been calculated; likewise for $\tilde{v}$. In the second step, only the contributions of advection terms are calculated. The mass flux across each cell interface is calculated, using donor cell differencing (second upwind difference method, section III-A-11) based on the provisional values of velocities $\overline{u^{n+1}}$ and $\overline{v^{n+1}}$. This mass flux is used to calculate a new density ρ^{n+1}, and then to calculate only the advective contribution to u, v and $e_s = \overline{E_s}/\rho$. Note that this final advective contribution must be added to the provisional value $\overline{u^{n+1}}$, etc., rather than the original values u^{n+1}, etc.

The PIC calculation is similar, but the mass flux calculation is based on a *finite number of particles* from the donor cell. The particles are not located at the center of the cell, but each particle p has its own Lagrangian coordinates, x_p and y_p. The particles are moved by the same velocity weighting used in the MAC code (see section III-G-4, equation 3-605). If the particle crosses the cell boundary, it contributes its mass, momentum, and internal energy to the averages in the new cell, upon which the pressures for that cell are calculated. As mentioned earlier, momentary crowding or depletion of particles in the cells will occur, producing a random high frequency oscillation of cell properties. This oscillation models the molecular behavior of the gases, but with very few computational molecules.

Both the PIC and FLIC methods use donor cell (second upwind) differencing for the advection terms and therefore have an implicit artificial viscosity (see sections V-E-1,2). Gentry, Martin, and Daly (1966) pointed out that the effect of $q \sim |u|$ in PIC and FLIC means that the artificial diffusion is not Galilean-invariant, i.e., the "wind tunnel transformation" does not apply to these computations.* Also, the method is locally unstable at stagnation points without the additional explicit q terms because the implicit $q \sim |u|$, according to Evans and Harlow (1958, 1959) and Longley (1960). See also equation (5-25) et seq. Both methods are presented in the original papers for both Cartesian and cylindrical coordinate systems.

The PIC method is most advantageously applied to interface problems (free surface or multiple materials), because the discrete particles may be assigned different masses, specific heats, etc., to represent two fluids, a free fluid surface, or even a fluid and a deformable solid. Solutions to the early problems of empty cells, boundary conditions,

---

*Also true of all upwind differencing methods.

and details of the particle weighting procedures have evolved over the years of successful application (Evans and Harlow, 1957, 1958, 1959; Evans et al., 1962; Harlow, 1963, 1964). A review of these techniques was given by Amsden (1966). Mader (1964) has extended the approach to include chemically reactive fluid dynamics in his Explosive-in-Cell or EIC method; Hirt (1965) also presented PIC calculations of shock detonation by explosives. The PIC approach was extended to plasma stability calculations by Dickman et al. (1969) and Morse and Nielson (1971). Armstrong and Nielsen (1970) demonstrated the good agreement of PIC transient computations with transform method calculations of the nonlinear development of a strong two-stream plasma instability. The accuracy has also been demonstrated by several PIC-like multi-material codes at Physics International (Buckingham et al., 1970; Watson and Godfrey, 1967; Watson, 1969). Amsden and Harlow (1965) calculated the gross features of supersonic turbulent flow in a base region. Crane (1968) attempted an accurate calculation of a hypersonic near wake problem using PIC with inviscid equations; the method is not well suited to this problem, and the calculation was unsuccessful. The accuracy of the FLIC method was independently ascertained by Gururaja and Dekker (1970) on several complex 2D shock-propagation problems, and by Satofuka (1970) in calculating 2D planar and cylindrical shock tube problems, Another FLIC-type code is the TOIL code of Johnson (1967); see also Hill and Larsen (1970) and Reynolds (1970). For references of other work on PIC and FLIC codes performed at Los Alamos Scientific Laboratory, see Harlow and Amsden (1970A).

Butler (1967) included viscosity and heat conduction in both PIC and FLIC, and found that the two methods produced comparable results.

V-E-4. Lax's Method

Lax's method* appears in Lax's((1954) fundamental paper on conservation equations. Lax was most concerned with the conservation principles and only secondarily with the finite-difference scheme. To stabilize calculations of the inviscid 1D equations (4-66) using forward-time, centered space differences, as in

$$U_i^{n+1} = U_i^n - \Delta t \left.\frac{\delta F}{\delta x}\right|_i^n \tag{5-31}$$

he replaced the U_i^n in the right-hand member by its space average at time n.

$$U_i^{n+1} = \frac{1}{2}\left(U_{i-1}^n + U_{i+1}^n\right) - \Delta t \left.\frac{\delta F}{\delta x}\right|_i^n \tag{5-32}$$

This simple and historically important method has several instructive properties. The space derivatives are centered and therefore appear to be second-order accurate, but the method is also diffusive. (Richtmyer, 1963, identifies it by the term "diffusing".) Consider the model equation (5-1) with $\alpha = 0$. Lax's method then gives

$$u_i^{n+1} = \frac{1}{2}\left(u_{i+1}^n + u_{i-1}^n\right) - \tilde{u} \left.\frac{\delta u}{\delta x}\right|_i^n \tag{5-33}$$

Expanding in Taylor series, as in Hirt's stability analysis (section III-A-5-c), we obtain

$$\begin{aligned} u_i^n + \frac{\partial u}{\partial t}\Delta t + \frac{1}{2}\frac{\partial^2 u}{\partial x^2}\Delta t^2 + O(\Delta t^3) &= \frac{1}{2}\left[u_i^n + \frac{\partial u}{\partial x}\Delta x + \frac{1}{2}\frac{\partial^2 u}{\partial x^2}\Delta x^2 + \frac{1}{6}\frac{\partial^3 u}{\partial x^3}\Delta x^3 + O(\Delta x^4)\right] \\ &+ \frac{1}{2}\left[u_i^n - \frac{\partial u}{\partial x}\Delta x + \frac{1}{2}\frac{\partial^2 u}{\partial x^2}\Delta x^2 - \frac{1}{6}\frac{\partial^3 u}{\partial x^3}\Delta x^3 + O(\Delta x^4)\right] \\ &- \tilde{u}\left.\frac{\partial u}{\partial x}\right|_i^n + O(\Delta x^2) \end{aligned} \tag{5-34}$$

*Commonly referred to as Lax's method. It first appears in open literature in a footnote of Courant et al. (1952) as the "scheme of J. Keller and P. Lax." Richtmyer (1963) also mentions K. O. Friedrichs in connection with it.

or

$$\frac{\partial u}{\partial t} = - \tilde{u} \frac{\partial u}{\partial x} + \frac{\Delta x^2}{2\Delta t} \frac{\partial^2 u}{\partial x^2} - \frac{\Delta t}{2} \frac{\partial^2 u}{\partial t^2} + 0(\Delta x^2) \tag{5-35}$$

Using the relation

$$\frac{\partial^2 u}{\partial t} = \frac{\partial}{\partial t} \left(- \tilde{u} \frac{\partial u}{\partial x} \right) = - \tilde{u} \frac{\partial}{\partial x} \left(\frac{\partial u}{\partial t} \right) = + \tilde{u}^2 \frac{\partial^2 u}{\partial x^2} \tag{5-36}$$

we obtain

$$\frac{\partial u}{\partial t} = - \tilde{u} \frac{\partial u}{\partial x} + \left(\frac{\Delta x^2}{2\Delta t} - \frac{\Delta t}{2} \tilde{u}^2 \right) \frac{\partial^2 u}{\partial x^2} + 0(\Delta x^2) \tag{5-37}$$

From this transient analysis, Lax's method is seen to introduce an effective artificial diffusion coefficient,

$$\alpha_e = \left(\frac{\Delta x^2}{2\Delta t} - \frac{\Delta t}{2} \tilde{u}^2 \right) = \frac{\Delta x^2}{2\Delta t} \left[1 - \frac{u^2 \Delta t^2}{\Delta x^2} \right] \tag{5-38}$$

or

$$\alpha_e = \frac{\Delta x^2}{2\Delta x} \left[1 - c^2 \right] \tag{5-39}$$

Stability in the model equation requires $\alpha_e \geq 0$ or $c \leq 1$, as usual. For $c = 1$, the exact solution of the model equation is obtained. Since the method is applied to all variables $U = \rho, \rho u, E_s$, the artificial diffusion represents not only an artificial viscosity, but also artificial mass diffusion and heat conduction.*

The order of the truncation error is determined from equation (5-37) to be

$$E = 0\left(\Delta x^2, \Delta t, \frac{\Delta x^2}{\Delta t} \right) \tag{5-40}$$

This equation indicates that, as $\Delta t \to 0$ for fixed Δx, the truncation error becomes unbounded. This indication *is* meaningful. It is disconcerting in the extreme to accidentally run a shock propagation code with $\Delta t = 0$, as the present author has done, and find that the shock still propagates! [Consider equation (5-32) with $\Delta t = 0$.] The disturbance does not actually propagate with a wave front, as a shock does, but diffuses out from the initial jump condition for $\Delta t = 0$.

For small enough Δt, the method obviously provides sufficient α_B to stabilize a strong shock calculation. For $c = 1$, the damping vanishes and the method cannot be used with shocks.

Lax's method is very easily extended to two and three dimensions, as

$$U_{ij}^{n+1} = \frac{1}{4}\left[U_{i+1,j}^n + U_{i-1,j}^n + U_{i,j+1}^n + U_{i,j-1} \right] + \Delta t \frac{\delta U^n}{\delta t} \tag{5-41}$$

$$U_{ijk}^{n+1} = \frac{1}{6}\left[U_{i+1,j,k}^n + U_{i-1,j,k}^n + U_{i,j+1,k}^n + U_{i,j-1,k}^n + U_{i,j,k+1}^n + U_{i,j,k-1}^n \right] + \Delta t \frac{\delta U^n}{\delta t} \tag{5-42}$$

*A diffusive scheme doubly violates the transportive property. Whereas the leapfrog methods (section III-A-6), for example, advect the effect of a perturbation upstream, against the velocity, a diffusive scheme also advects it at right angles to the velocity.

The corresponding stability requirements are

$$C_{2D} \equiv \frac{(V + a)\Delta t}{\Delta x \Delta y} \sqrt{\Delta x^2 + \Delta y^2} \leq 1 \tag{5-43}$$

$$C_{3D} \equiv \frac{(V + a)\Delta t}{\Delta x \cdot \Delta y \cdot \Delta z} (\Delta x^2 + \Delta y^2 + \Delta z^2)^{3/2} \leq 1 \tag{5-44}$$

Thus, for $\Delta x = \Delta y = \Delta z$, the largest possible Δt is reduced by a factor of $1/\sqrt{3} \simeq 0.58$.

Exercise: Derive expressions for α_e of Lax's method in two and three dimensions.

Exercise: Determine the conditions for which Rusanov's method reduces to Lax's method.

Moretti and Abbett (1966A) used the two-dimensional version of Lax's method in conjunction with a patched characteristics solution in an attempt to calculate base flow. They noted a phenomenon which they called "stalling". That is, with a spatial gradient of properties such that

$$U_i^n \neq \hat{U}_i^n \equiv \frac{1}{4}\left(U_{i+1,j}^n + U_{i-1,j}^n + U_{i,j+1}^n + U_{i,j-1}^n\right) \tag{5-45}$$

the time solution adjusted to a condition where

$$U_i^n - \hat{U}_i^n = \frac{\delta U^n}{\delta t} \Delta t \tag{5-46}$$

so that $U_i^{n+1} = U_i^n$ for all i. The situation could be changed by changing Δt. Of course, the method was not intended to be used on this subsonic shock-free problem, but the example shows up another shortcoming of the method.

In spite of its shortcomings, the method has an important asset: simplicity. It is also easily adapted to cylindrical, spherical, and 3D problems. This appears to be the major reason for its use by Bohachevsky and Rubin (1966), Bohachevsky and Mates (1966), Bohachevsky and Kostoff (1971), Barnwell (1967), Xerikos (1968), and Emery and Ashurst (1971). Kentzner (1970B) experimented with using the Lax method and the midpoint leapfrog method (section III-A-6) at different time steps and in different weighted combinations, in a two-dimensional problem in which the shock discontinuity was treated as a boundary.

Because it is easily programmed and is dependable, Lax's method can be used to advantage in the early stages of program development. The program can be converted to more complex methods afterwards.

Exercise: Show that the use of Lax's method on the advection terms and FTCS differencing on the diffusion term of the model equation results in an unconditionally unstable method.

HINT: Use the analysis for the FTCS method, replacing α by $(\alpha + \alpha_e)$.

Exercise: Show that α_e of Lax's method by the steady-state analysis is $\alpha_e = \Delta x^2/2\Delta t$.

V-E-5. Lax-Wendroff Method

Lax and Wendroff (1960, 1964) investigated a class of methods which has attained considerable stature in theoretical studies of difference methods, and which led to a class of two-step methods (next section V-E-6) which are currently the most popular methods for solving compressible flow problems. Like Leith's method (section III-A-13), all these are based on a second-order Taylor series expansion in time, and all are identical to Leith's method for the constant-coefficient model equation.

Compared with Leith's method for incompressible flow, the application of the time expansion to compressible flow is greatly complicated because a *system* of equations is involved, rather than a single equation. For the inviscid model equation (5-1),

$$\frac{\partial u}{\partial t} + \tilde{u}\,\frac{\partial u}{\partial x} = 0 \tag{5-47}$$

the Taylor series expansion in time gives

$$u^{n+1} = u^n + \Delta t\,\frac{\partial u}{\partial t} + \frac{1}{2}\,\Delta t^2\,\frac{\partial^2 u}{\partial t^2} + O(\Delta t^3) \tag{5-48}$$

The term $\partial u/\partial t$ in equation (5-48) is given by (5-47), and the second derivative, $\partial^2 u/\partial t^2$, in (5-48) is obtained from (5-47) as

$$\frac{\partial^2 u}{\partial t^2} = -\tilde{u}\,\frac{\partial^2 u}{\partial x \partial t} = -\tilde{u}\,\frac{\partial}{\partial x}\,(-\tilde{u}\,\frac{\partial u}{\partial x}) = \tilde{u}^2\,\frac{\partial^2 u}{\partial x^2} \tag{5-49}$$

But consider the inviscid compressible flow equation for 1D, equation (4-66).

$$\frac{\partial U}{\partial t} + \frac{\partial F}{\partial x} = 0 \tag{5-50}$$

U and F are vectors, as in equations (4-66B,C). Now the same expansion in time gives

$$U^{n+1} = U^n + \Delta t\,\frac{\partial U}{\partial t} + \frac{1}{2}\,\Delta t^2\,\frac{\partial^2 U}{\partial t^2} + O(\Delta t^3) \tag{5-51}$$

and $\partial U/\partial t$ is given by equation (5-50). But the second derivative term, from equation (5-50), is

$$\frac{\partial^2 U}{\partial t^2} = -\,\frac{\partial^2 F}{\partial x \partial t} = -\,\frac{\partial}{\partial x}\left[\frac{\partial F}{\partial t}\right] \tag{5-52}$$

In this case, the original equation (5-50) does not directly give the required second time derivative.

The time expansion applied to the system (5-50) is accomplished by re-writing it as

$$\frac{\partial U}{\partial t} + A\,\frac{\partial U}{\partial x} = 0 \tag{5-53}$$

where A is a 3 x 3 matrix. This commonly used equation (5-53) is so misleading in its simplicity as to be almost incorrect; whereas the left member refers to just one element at a time of the vector U, the right member must index over all three elements each time. That is,

$$\frac{\partial U_1}{\partial t} + A_{11}\,\frac{\partial U_1}{\partial x} + A_{12}\,\frac{\partial U_2}{\partial x} + A_{13}\,\frac{\partial U_3}{\partial x} = 0 \tag{5-54A}$$

$$\frac{\partial U_2}{\partial t} + A_{21}\,\frac{\partial U_1}{\partial x} + A_{22}\,\frac{\partial U_2}{\partial x} + A_{23}\,\frac{\partial U_3}{\partial x} = 0 \tag{5-54B}$$

$$\frac{\partial U_3}{\partial t} + A_{31}\,\frac{\partial U_1}{\partial x} + A_{32}\,\frac{\partial U_2}{\partial x} + A_{33}\,\frac{\partial U_3}{\partial x} = 0 \tag{5-54C}$$

The matrix elements $A_{\ell m}$ are Jacobians, defined by

$$A_{\ell m} = \frac{\partial F_\ell}{\partial U_m} \tag{5-55}$$

For example, from equation (4-66B,C), we have

$$\begin{Bmatrix} U_1 \\ U_2 \\ U_3 \end{Bmatrix} = \begin{Bmatrix} \rho \\ \rho u \\ E_s \end{Bmatrix} \tag{5-56}$$

$$\begin{Bmatrix} F_1 \\ F_2 \\ F_3 \end{Bmatrix} = \begin{Bmatrix} \rho u \\ P + \rho u^2 \\ u(E_s + P) \end{Bmatrix} \tag{5-57}$$

Then the continuity equation

$$\frac{\partial \rho}{\partial t} + \frac{\partial(\rho u)}{\partial x} = 0 \tag{5-58}$$

is retrieved from equation (5-54A) by

$$A_{11} = 0 \quad , \quad A_{12} = 1 \quad , \quad A_{13} = 0 \tag{5-59}$$

This row of $A_{\ell m}$ is then very simply determined. The other two are more complicated; because of the P terms in F_2 and F_3, their form depends on the gas law used. For the perfect gas law in the form of equation (4-55B),

$$P = (\gamma - 1)\left[E_s - \frac{1}{2}\rho u^2\right] \tag{5-60}$$

the equations are obtained as follows.

We define $m = \rho u$, and write the equation of state as

$$P = (\gamma - 1)\left[E_s - \frac{1}{2} m^2/\rho\right] \tag{5-61}$$

Then we re-write equations (5-56) and (5-57) in terms of the conservation variables ρ, m, E_s.

$$\begin{Bmatrix} U_1 \\ U_2 \\ U_3 \end{Bmatrix} = \begin{Bmatrix} \rho \\ m \\ E_s \end{Bmatrix} \tag{5-62}$$

$$\begin{Bmatrix} F_1 \\ F_2 \\ F_3 \end{Bmatrix} = \begin{Bmatrix} m \\ (\gamma - 1)E_s + \left(\frac{3 - \gamma}{2}\right) m^2/\rho \\ \gamma m E_s/\rho - \left(\frac{\gamma - 1}{2}\right) m^3/\rho^2 \end{Bmatrix} \tag{5-63}$$

Using the forms of equations (5-62) and (5-63), the Jacobian matrix elements, $A_{\ell m} = \partial F_\ell / \partial U_m$, are readily evaluated as

$$A_{11} = \frac{\partial F_1}{\partial U_1} = \frac{\partial m}{\partial \rho} = 0 \tag{5-64A}$$

$$A_{12} = \frac{\partial F_1}{\partial U_2} = \frac{\partial m}{\partial m} = +1 \tag{5-64B}$$

$$A_{13} = \frac{\partial F_1}{\partial U_3} = \frac{\partial m}{\partial E_s} = 0 \tag{5-64C}$$

$$A_{21} = \frac{\partial F_2}{\partial U_1} = \frac{\partial \left[(\gamma - 1)E_s + \left(\frac{3-\gamma}{2}\right) m^2/\rho\right]}{\partial \rho} = -\left(\frac{3-\gamma}{2}\right) m^2/\rho^2 \tag{5-65A}$$

$$A_{22} = \frac{\partial F_2}{\partial U_2} = \frac{\partial \left[(\gamma - 1)E_s + \left(\frac{3-\gamma}{2}\right) m^2/\rho\right]}{\partial m} = (3 - \gamma)m/\rho \tag{5-65B}$$

$$A_{23} = \frac{\partial F_2}{\partial U_3} = \frac{\partial \left[(\gamma - 1)E_s + \left(\frac{3-\gamma}{2}\right) m^2/\rho\right]}{\partial E_s} = (\gamma - 1) \tag{5-65C}$$

$$A_{31} = \frac{\partial F_3}{\partial U_1} = \frac{\partial \left[\gamma m E_s/\rho - \left(\frac{\gamma-1}{2}\right) m^3/\rho^2\right]}{\partial \rho} = -\gamma m E_s/\rho^2 - (\gamma - 1)m^3/\rho^3 \tag{5-66A}$$

$$A_{32} = \frac{\partial F_3}{\partial U_2} = \frac{\partial \left[\gamma m E_s/\rho - \left(\frac{\gamma-1}{2}\right) m^3/\rho^2\right]}{\partial m} = \gamma E_s/\rho + \frac{3}{2}(\gamma - 1)m^2/\rho^2 \tag{5-66B}$$

$$A_{33} = \frac{\partial F_3}{U_3} = \frac{\partial \left[\gamma m E_s/\rho - \left(\frac{\gamma-1}{2}\right) m^3/\rho^2\right]}{\partial E_s} = \gamma m/\rho\ , \tag{5-66C}$$

Summarizing,

$$A = \begin{bmatrix} 0 & +1 & 0 \\ -\left(\frac{3-\gamma}{2}\right) m^2/\rho & (3-\gamma)m/\rho & (\gamma - 1) \\ -\gamma m E_s/\rho^2 - (\gamma - 1)m^3/\rho^3 & \gamma E_s/\rho + \frac{3}{2}(\gamma - 1)m^2/\rho^2 & \gamma m/\rho \end{bmatrix} \tag{5-67}$$

Now, the time expansion (5-51) can be used, with the second time derivatives evaluated as follows. For the ℓ-th element of U,

$$-\frac{\partial U_\ell}{\partial t} = A_{\ell 1}\frac{\partial U_1}{\partial x} + A_{\ell 2}\frac{\partial U_2}{\partial x} + A_{\ell 3}\frac{\partial U_3}{\partial x} = \frac{\partial F_\ell}{\partial U_1}\frac{\partial U_1}{\partial x} + \frac{\partial F_\ell}{\partial U_2}\frac{\partial U_2}{\partial x} + \frac{\partial F_\ell}{\partial U_3}\frac{\partial U_3}{\partial x} = \frac{\partial F_\ell}{\partial x} \tag{5-68}$$

$$-\frac{\partial^2 U_\ell}{\partial t^2} = \frac{\partial^2 F_\ell}{\partial t \partial x} = \frac{\partial}{\partial x}\left[\frac{\partial F_\ell}{\partial t}\right] \tag{5-69}$$

But

$$\frac{\partial F_\ell}{\partial t} = \frac{\partial F_\ell}{\partial U_1}\frac{\partial U_1}{\partial t} + \frac{\partial F_\ell}{\partial U_2}\frac{\partial U_2}{\partial t} + \frac{\partial F_\ell}{\partial U_3}\frac{\partial U_3}{\partial t} = - A_{\ell 1}\frac{\partial F_1}{\partial x} - A_{\ell 2}\frac{\partial F_2}{\partial x} - A_{\ell 3}\frac{\partial F_3}{\partial x} \tag{5-70}$$

So

$$\frac{\partial^2 U_\ell}{\partial t^2} = \frac{\partial}{\partial x}\left[A_{\ell 1}\frac{\partial F_1}{\partial x} + A_{\ell 2}\frac{\partial F_2}{\partial x} + A_{\ell 3}\frac{\partial F_3}{\partial x}\right] \tag{5-71}$$

The time expansion (5-51) can now be written for the ℓ-th component of U as

$$U_\ell^{n+1} = U_\ell^n - \Delta t\frac{\partial F_\ell}{\partial x} + \frac{1}{2}\Delta t^2\frac{\partial}{\partial x}\left[A_{\ell 1}\frac{\partial F_1}{\partial x} + A_{\ell 2}\frac{\partial F_2}{\partial x} + A_{\ell 3}\frac{\partial F_3}{\partial x}\right] + O(\Delta t^3) \quad ,$$

$$\text{for } \ell = 1,2,3 \tag{5-72A}$$

or, less explicitly,

$$U^{n+1} = U^n - \Delta t\frac{\partial F}{\partial x} + \frac{1}{2}\Delta t^2\frac{\partial}{\partial x}\left[A\frac{\partial F}{\partial x}\right] \tag{5-72B}$$

where the presence of the matrix A indicates the interpretation of equation (5-72A). The Lax-Wendroff method then consists of using centered-space differences for the time expansion (5-72). Dropping the ℓ-index for clarity, we have at the i-th location,

$$\left.\frac{\partial F}{\partial x}\right|_i = \frac{F_{i+1} - F_{i-1}}{2\Delta x} + O(\Delta x^2) \tag{5-73}$$

and for each term like $\frac{\partial}{\partial x}\left[A_{\ell 1}\frac{\partial F_1}{\partial x}\right]$ we have

$$\begin{aligned}\frac{\partial}{\partial x}\left[A\frac{\partial F}{\partial x}\right]_i &= \frac{A_{i+\frac{1}{2}}(F_{i+1} - F_i)/\Delta x - A_{i-\frac{1}{2}}(F_i - F_{i-1})/\Delta x}{\Delta x} + O(\Delta x^2)\\ &= \frac{1}{2\Delta x^2}\Big[(A_{i+1} + A_i)(F_{i+1} - F_i) - (A_i + A_{i-1})(F_i - F_{i-1})\Big] + O(\Delta x^2)\end{aligned} \tag{5-74}$$

The additional "averaging" used in defining*

$$A_{i\pm\frac{1}{2}} = A \text{ of } \left(U^n_{i\pm\frac{1}{2}}\right) = A \text{ of } \left[\frac{1}{2}\left(U^n_i + U^n_{i\pm 1}\right)\right] \tag{5-75}$$

*Alternately, $A_{i\pm\frac{1}{2}}$ may be defined by $A_{i\pm\frac{1}{2}} = \frac{1}{2}[A_i + A_{i\pm 1}] = \frac{1}{2}[A \text{ of } (U_i^n) + A \text{ of } (U^n_{i\pm 1})]$. Both expressions are legitimate approximations, but equation (5-75) is preferred by Abarbanel and Zwas (1969) by conservation considerations; also, it involves less arithmetic work.

does not reduce the order of the truncation error, which is $O(\Delta t^2, \Delta x^2)$. The second-order time truncation error is important to some transient solutions (see Emery, 1968).

Exercise: Verify that the Lax-Wendroff method [equations (5-72) to (5-74)] reduces to Leith's method (section III-A-13) for A constant and for the single equation (5-47).

Like Leith's method (see section III-A-13), the Lax-Wendroff method does not introduce an artificial diffusion in the transient analysis, but does give fourth-order damping errors through a non-zero coefficient of $\partial^4 U/\partial x^4$, and gives third-order dispersion errors through a non-zero coefficient of $\partial^3 U/\partial x^3$ (Richtmyer and Morton, 1967). For a steady-flow solution, the analysis of section III-A-13 indicates artificial diffusion $\alpha_e = 1/2\, \tilde{u}^2 \Delta t$, so that the steady solution does depend on Δt (see problem 5-7). For $c \simeq 1$, the steady-state solutions are only first-order accurate (Roache, 1971C; see Appendix B).

This method gives a much sharper shock (i.e., smaller shock width) than other methods, but higher overshoot. Lax and Wendroff (1964) argued that all higher order (in time) methods will produce oscillations behind a shock; see also Vreugdenhil (1969) for results on the linear model equation (5-47). (This appears not to hold for multi-step approximations to implicit methods; see section V-E-7.) To reduce the overshoot and give satisfactory computations for strong shocks, it is necessary to add some form of explicit artificial viscosity (Lax and Wendroff, 1960, 1964; Richtmyer and Morton, 1967).

The method is extendable to higher dimensions, as in

$$U^{n+1} = U^n - \Delta t \frac{\partial F}{\partial x} - \Delta t \frac{\partial G}{\partial y} + \frac{1}{2} \Delta t^2 \frac{\partial}{\partial x} \left[A \frac{\partial F}{\partial x} \right] + \frac{1}{2} \Delta t^2 \frac{\partial}{\partial y} \left[B \frac{\partial G}{\partial y} \right] \qquad (5\text{-}76)$$

where A and B are suitably defined 4 x 4 matrices. However, the method becomes very cumbersome and time consuming; Emery (1968) says that it requires almost 4 times the computing time per step as Lax's method. Also, it is subject to a nonlinear instability. The linear analysis (based on constant A matrices; see, e.g., Parlett, 1966; Burstein, 1965, 1967) shows that the method is neutrally stable at stagnation points ($u = v = 0$) and at sonic points ($V = a$, or $M = 1$). Burstein (1965, 1967) and Thommen (1966, 1967) found that nonlinear effects caused this neutral stability to go over to instability. In Burstein's calculations of steady-state detached shock problems, the instability started in the stagnation region and near the shoulder, where approximate sonic conditions exist. He found that the addition of the explicit fourth-order damping suggested by Lax and Wendroff (1964) had no effect, but the addition of a Rusanov-like second-order explicit artificial viscosity (see section V-D-3) did stabilize the computation; it will also decrease or eliminate the shock overshoot. The penalty is reduced order of truncation error. Burstein found analytically that a sufficient stability condition in two-dimensional flow is

$$C \leq \frac{1}{2\sqrt{2}} \qquad (5\text{-}77)$$

which is 1/2 the usual condition of $C < 1/\sqrt{2}$. But, in computational experiments, the usual condition proved sufficient and even could be exceeded by about 10%. See also Emery (1968).

Gary (1964) first applied the Lax-Wendroff time differencing to the non-conservation equations and found the computation of expansion waves was actually more accurate. Likewise, Moretti used the time expansion with non-conservation variables in two-dimensions (see section VI-B). In this case, the Jacobian matrix A need not be computed, but rather the second time derivatives are evaluated by cross-differentiation, as in the linear model equations. Moretti used shock-fitting techniques (see section VI-B) with this time differencing; Watkins (1970) demonstrated that the shock-fitting may be accomplished just as easily with the conservation variables, at least in 1D. Moretti's method has been applied in 3D and its accuracy appears to be good (Eaton, 1970). The same type of time differencing was used by Bastianon (1969) and for quasi-1D flows by Anderson (1969A,B). Armitage (1967) did not obtain a stable calculation using the Lax-Wendroff method to compute axisymmetric swirl flows utilizing a curvilinear coordinate transformation to fit the general nozzle wall contour; the difficulty was likely due to the instability described by Burstein (1965, 1967). Note however, that Saunders (1966) did calculate a transonic nozzle flow using the Lax-Wendroff method, in spite of theoretical indications (Parlett, 1966) that the method would be inadequate for transonic flows. Ciment (1968) considered the effects of changes in mesh spacing Δx on the stability of Lax-Wendroff methods. The Lax-Wendroff method may also be applied in Lagrangian coordinates, in which case it is the only method which does not smear the shock (Lax and Wendroff, 1964; Richtmyer and Morton, 1967; van Leer, 1969).

Skoglund and Gay (1969) modified the two-dimensional Lax-Wendroff method by expanding terms in equation (5-76) to

$$\frac{\partial}{\partial x}\left[A\,\frac{\partial F}{\partial x}\right] = A\,\frac{\partial^2 F}{\partial x^2} + \frac{\partial A}{\partial x}\,\frac{\partial F}{\partial x} \tag{5-78A}$$

$$\frac{\partial}{\partial y}\left[B\,\frac{\partial G}{\partial y}\right] = B\,\frac{\partial^2 G}{\partial y^2} + \frac{\partial B}{\partial y}\,\frac{\partial G}{\partial x} \tag{5-78B}$$

The terms $\partial F/\partial x$ and $\partial G/\partial y$ are already evaluated for the first time derivative methods, and the other terms were factored and grouped. The result was a considerable decrease in arithmetic operations at the sacrifice of conceptual simplicity and strict conservation. They successfully computed the difficult problem of shock-wave/boundary-layer interaction with this method.

The original Lax-Wendroff method is still of theoretical interest and has served as a spur to the development of other methods (Fischer, 1965A,B; Kasahara, 1965; Kasahara et al., 1965; Zwas and Abarbanel, 1970) and it has been applied in 2D by Burstein (1965,1967) and Skoglund and Gay (1969). But, for multi-dimensional problems, it has generally been supplanted by two-step methods, which we now consider.

V-E-6. Two-Step Lax-Wendroff Methods

Richtmyer (1963) proposed a two-step version of the Lax-Wendroff method, which is much simpler than the original, especially in multi-dimensional problems.* The first step uses Lax's method (see section V-E-4) and the second step is a midpoint leapfrog calculation (see section III-A-6). For the vector equation (4-58), $\partial U/\partial t = -\,\partial F/\partial x$, the method is as follows.

$$U_i^{n+1} = \frac{1}{2}\left[U_{i+1}^n + U_{i-1}^n\right] - \Delta t\left[\frac{F_{i+1}^n - F_{i-1}^n}{2\Delta x}\right] \tag{5-79A}$$

$$U_i^{n+2} = U_i^n - (2\Delta t)\left[\frac{F_{i+1}^{n+1} - F_{i-1}^{n+1}}{2\Delta x}\right] \tag{5-79B}$$

The values $F_{i\pm1}^{n+1}$ in the second step are based on the $U_{i\pm1}^{n+1}$ results on the first step. The first step is to be considered a provisional step, with significance attached only to the results of the second step in each sequence. This method does not look anything like the original Lax-Wendroff method of equations [(5-72) through (5-74)] ,but substitution of (5-79A) into (5-79B) shows that the methods are equivalent for the linearized constant coefficient case.

Exercise: Show the equivalence of Richtmyer's two-step method and the Lax-Wendroff method for the constant-coefficient equations.

The extension to multiple dimensions is obvious and neat.

$$\frac{\partial U}{\partial t} = -\left(\frac{\partial F}{\partial x} + \frac{\partial G}{\partial y}\right) \tag{5-80}$$

---

*Burstein (1967) says that the one-dimensional version of this method was originally suggested by Wendroff; see also the method of Chudov, reported by Brailovskaya et al. (1968).

$$U_{ij}^{n+1} = \frac{1}{4}\left[U_{i+1,j}^{n} + U_{i-1,j}^{n} + U_{i,j+1}^{n} + U_{i,j-1}^{n}\right] - \Delta t\left[\frac{F_{i+1,j}^{n} - F_{i-1,j}^{n}}{2\Delta x} + \frac{G_{i,j+1}^{n} - G_{i,j-1}^{n}}{2\Delta y}\right] \quad (5\text{-}81A)$$

$$U_{ij}^{n+2} = U_{ij}^{n+1} - (2\Delta t)\left[\frac{F_{i+1,j}^{n+1} - F_{i-1,j}^{n+1}}{2\Delta x} + \frac{G_{i,j+1}^{n+1} - G_{i,j-1}^{n+1}}{2\Delta y}\right] \quad (5\text{-}81B)$$

This two-dimensional method requires about 1/4 the computer time of the original Lax-Wendroff method (Emery, 1968) and produces less shock overshoot than the original (e.g., see Rubin and Burstein, 1967). The method has also been applied in general transformed coordinates by Lapidus (1967) and in a geophysical problem with coriolis terms included by Houghton et al., (1966), who introduced additional artificial diffusion through "Fickian smoothing." Sinha, et al., (1970) calculated the plumes of underexpanded jets, including the "Mach disk" shock. Although the first step is diffusive, the overall method is not, at least in the transient analysis. From a steady-state analysis, the method does introduce an artificial viscosity dependent on Δt; see section III-A-13.

Following Richtmyer (1963), it has become customary to refer to any method which can be interpreted as a second-order Taylor series expansion in time as a "two-step Lax-Wendroff method," or "a method of the Lax-Wendroff type," etc. This seems to be a very broad and somewhat inadequate classification; for example, the Adams-Bashforth method (section III-A-12) and Heun's method (section III-A-15), which predate Lax-Wendroff, would qualify, as would Leith's method (section III-A-13) and MacCormack's method (to follow). Although we feel that the various methods should be identified more specifically in references, we follow custom in presenting all the methods in this section.

Burstein (1965, 1967) and later Rubin and Burstein (1967) modified the Richtmyer method (5-79) by applying it over half-mesh increments, $\Delta x/2$ and $\Delta t/2$, as follows.

The first step uses equations such as

$$U_{i+\frac{1}{2}}^{n+\frac{1}{2}} = \frac{1}{2}\left[U_{i+1}^{n} + U_{i}^{n}\right] - \frac{\Delta t}{2}\left[\frac{F_{i+1}^{n} - F_{i}^{n}}{\Delta x}\right] \quad (5\text{-}82A)$$

and

$$U_{i-\frac{1}{2}}^{n+\frac{1}{2}} = \frac{1}{2}\left[U_{i}^{n} + U_{i-1}^{n}\right] - \frac{\Delta t}{2}\left[\frac{F_{i}^{n} - F_{i-1}^{n}}{\Delta x}\right] \quad (5\text{-}82B)$$

The second step is then

$$U_{i}^{n+1} = U_{i}^{n} - \Delta t\left[\frac{F_{i+\frac{1}{2}}^{n+\frac{1}{2}} - F_{i-\frac{1}{2}}^{n+\frac{1}{2}}}{\Delta x}\right] \quad (5\text{-}82C)$$

Burstein (1965,1967) has also applied the method in cylindrical coordinates.

This method is not equivalent to Richtmyer's method, which is to be preferred because of boundary considerations. Boundary conditions must be applied after each of the two steps, and the evaluation of wall boundary conditions after equations (5-82A and B) at the "staggered" mesh locations like (i+1/2) is not satisfactory* (see section V-G-1,2). Similarly, Gourlay and Morris (1968B) presented the two-step Lax-Wendroff method as

$$U_{i}^{n+\frac{1}{2}} = \frac{1}{2}\left(U_{i+\frac{1}{2}}^{n} + U_{i-\frac{1}{2}}^{n}\right) - \frac{\Delta t}{2}\left[\frac{F_{i+\frac{1}{2}}^{n} - F_{i-\frac{1}{2}}^{n}}{\Delta x}\right] \quad (5\text{-}83A)$$

---

*For the usual test case of one-dimensional shock propagation, a satisfactory boundary treatment is possible.

$$U_i^{n+1} = U_i^n - \Delta t \left[\frac{F_{i+\frac{1}{2}}^{n+\frac{1}{2}} - F_{i-\frac{1}{2}}^{n+\frac{1}{2}}}{\Delta x} \right] \tag{5-83B}$$

Although conceptually similar, this method is not identical to Richtmyer's (5-79), nor even to Burstein's (5-82). Note that, in the Lax replacement of the first step (5-83A), we have

$$\frac{1}{2}\left[U_{i+\frac{1}{2}}^n + U_{i-\frac{1}{2}}^n\right] = \frac{1}{4} U_{i+1}^n + \frac{1}{2} U_i^n + \frac{1}{4} U_{i-1}^n \tag{5-84}$$

which is different from the usual Lax method applied in (i), (i±1). In fact, the method is not entirely defined, since the F-terms in the second step (5-83B) could be ambiguously interpreted as either

$$F_{i\pm\frac{1}{2}}^{n+\frac{1}{2}} = F \text{ of } \left(U_{i\pm\frac{1}{2}}^{n+\frac{1}{2}}\right) = F \text{ of } \left[\frac{1}{2}\left(U_i^{n+\frac{1}{2}} + U_{i\pm 1}^{n+\frac{1}{2}}\right)\right] \tag{5-85}$$

or as

$$F_{i\pm\frac{1}{2}}^{n+\frac{1}{2}} = \frac{1}{2}\left[F \text{ of } \left(U_i^{n+\frac{1}{2}}\right) + F \text{ of } \left(U_{i+1}^{n+\frac{1}{2}}\right)\right] \tag{5-86}$$

which are not equivalent in the case of non-constant U. This method still has boundary problems, as does Burstein's (5-82).

Another method of Rubin and Burstein (1967) involves a different approach to the time centering. Provisional values at half-spaces (i±1/2) are calculated at the full Δt advance time step, and the centered $F^{n+\frac{1}{2}}$ values by averaging

$$\overline{U_{i+\frac{1}{2}}^{n+1}} = \frac{1}{2}\left(U_i^n + U_{i+1}^n\right) - \Delta t \left[\frac{F_{i+1}^n - F_i^n}{\Delta x}\right] \tag{5-87A}$$

$$\overline{U_{i-\frac{1}{2}}^{n+1}} = \frac{1}{2}\left(U_i^n + U_{i-1}^n\right) - \Delta t \left[\frac{F_i^n - F_{i-1}^n}{\Delta x}\right] \tag{5-87B}$$

The second step is then

$$U_i^{n+1} = U_i^n - \Delta t \left\{\frac{1}{2}\left[\frac{F_{i+1}^n - F_{i-1}^n}{2\Delta x} + \frac{\overline{F_{i+\frac{1}{2}}^{n+1}} - \overline{F_{i-\frac{1}{2}}^{n+1}}}{\Delta x}\right]\right\} \tag{5-87C}$$

where $\overline{F^{n+1}}$ values are unambiguously defined by

$$\overline{F_{i\pm\frac{1}{2}}^{n+1}} = F \text{ of } \left(\overline{U_{i\pm\frac{1}{2}}^{n+1}}\right) \tag{5-88}$$

This method produces even less overshoot at the shock than equation (5-82) for high Courant numbers, but the wall boundary conditions still pose a problem. Rubin (1970) has applied the method to viscous flow with chemical reaction and radiation in 1D.

For all the above methods, the stability analysis of Rubin and Preiser (1968) holds, giving the usual Courant number restrictions (5-5) as both necessary and sufficient for stability.*

Singleton (1968) used a kind of time-splitting approach in a two-step Lax-Wendroff scheme, wherein the provisional values at (i±1/2,j) are calculated by a Lax method that is

---

*This work supplants the early, sufficient condition of Burstein (1965, 1967), which was twice as restrictive.

one-dimensional in x, and the provisional values at $(i,j\pm1/2)$ used a Lax method that is one dimensional in y. For equation (5-80), the first step uses

$$U^{n+\frac{1}{2}}_{i+\frac{1}{2},j} = \frac{1}{2}\left(U^n_{ij} + U^n_{i+1,j}\right) - \frac{\Delta t}{2}\left\{\frac{F^n_{i+1,j} - F^n_{ij}}{\Delta x} + \frac{G^n_{i+\frac{1}{2},j+1} - G^n_{i+\frac{1}{2},j-1}}{2\Delta y}\right\} \tag{5-89A}$$

and

$$U^{n+\frac{1}{2}}_{i,j+\frac{1}{2}} = \frac{1}{2}\left(U^n_{ij} + U^n_{i,j+1}\right) - \frac{\Delta t}{2}\left\{\frac{F^n_{i+1,j+\frac{1}{2}} - F^n_{i-1,j+\frac{1}{2}}}{2\Delta x} + \frac{G^n_{i,j+1} - G^n_{i,j}}{\Delta y}\right\} \tag{5-89B}$$

The second step is the usual leapfrog calculation.

$$U^{n+1}_{ij} = U^n_{ij} - \Delta t\left\{\frac{F^{n+\frac{1}{2}}_{i+\frac{1}{2},j} - F^{n+\frac{1}{2}}_{i-\frac{1}{2},j}}{\Delta x} + \frac{G^{n+\frac{1}{2}}_{i,j+\frac{1}{2}} - G^{n+\frac{1}{2}}_{i,j-\frac{1}{2}}}{\Delta y}\right\} \tag{5-89C}$$

Singleton chose to define half-space values at $(i\pm1/2)$ and $j\pm1/2)$ in equations (5-89A and B) in accordance with (5-86) rather than (5-85). In the second step (5-89C), the half-space values are unambiguously defined as in equation (5-88). The stability limits were not given for this method.

A very interesting two-step method is that of MacCormack (1969, 1970). The method alternately uses forward and backward differences for the two steps. In one dimension, the basic method is as follows:

$$\overline{U^{n+1}_i} = U^n_i - \Delta t\left[\frac{F^n_{i+1} - F^n_i}{\Delta x}\right] \tag{5-90A}$$

$$U^{n+1}_i = \frac{1}{2}\left\{U^n_i + \overline{U^{n+1}_i} - \Delta t\left[\frac{\overline{F^{n+1}_i} - \overline{F^{n+1}_{i-1}}}{\Delta x}\right]\right\} \tag{5-90B}$$

A variation of the method is obtained by the forward-backward differencing at alternate (complete) time steps. In two dimensions, the forward-backward differencing can be applied differently in x and y, and cyclically over two or four complete time steps. Also, the method can be applied with the Marchuk time-splitting approach (section III-A-13). Details may be found in MacCormack (1971). A description of the derivation can be found in Kutler and Lomax (1971). More aspects of the time-splitting are given by MacCormack and Paulley (1972); approximations are required for the cross-derivative viscous terms.

It is not at all obvious that the method is a Lax-Wendroff type, nor even that it is consistent with the PDE, but the excellent results obtained (MacCormack, 1969, 1970; Kutler, 1969; Lomax, et al., 1970; Kutler and Lomax, 1971) bolster confidence. Because half-space values such as $F_{i\pm\frac{1}{2}}$ are not required, the boundary conditions are not complicated except when time splitting is used. The method was proved for strong 1D shock waves with inviscid equations by Tyler and Ellis (1970), for 3D embedded shocks by Kutler and Lomax (1971), and for quasi-1D flows with chemical nonequilibrium by Anderson (1970B). Li (1971) used it in conjunction with shock fitting in axisymmetric flow with chemistry. Thomas et al., (1971) used it with 3D shock fitting, "marching" in the time-like axial coordinate. Currently, MacCormack's method is very popular for aerodynamic calculations. However, Turkel (J.Comp.Phys.,Vol. 15, 1974, pp. 226-250) indicates that it may be unstable in some 2D flows without viscosity, and Taylor et al. (J.Comp.Phys., Vol. 9, 1972, pp.99-119) found it unstable in 1D inviscid calculations of rarefaction waves. These two papers and one by Anderson (J. Comp.Phys.,Vol. 15, 1974, pp.1-20) provide valuable comparative studies of various methods.

Exercises: Show that MacCormack's method applied to the 1D model equation (5-1) with $c \equiv \mathring{u}\Delta t/\Delta x = 1$ gives the exact solution, $u^{n+1}_i = u^n_{i-1}$. Show that MacCormack's method applied to equation (5-1) in non-conservation form is consistent with the PDE. Sketch the time-centering

Other "two-step Lax-Wendroff" methods and applications are as follows. Rubin et al. (1967) used Burstein's method (5-82) in one-dimensional flow of a radiating gas. Watkins

(1970) devised a new two-step method for treating the "stiff" chemical reaction equations (section III-F-5). Kentzner (1970B) experimented in shock-free flow with various weighted combinations and alternations of the Lax and midpoint leapfrog methods, as in Richtmyer's method (5-79). Strang (1963) presented a method similar to the original Lax-Wendroff method (5-72) through (5-74), and Gourlay and Morris (1968B) later presented a multi-step version of Strang's method using Marchuk time splitting (section III-A-13). Freudiger et al., (1967) devised a "flip-flop" scheme which depended on physical viscous terms (low Re) for conditional stability. Gourlay and Morris (1971) used two-step Lax-Wendroff differencing within the "hopscotch" method (see section III-A-18) for 1D shock calculations; see also Ames (1969). Bowley and Prince (1971) generalized a two-step Lax-Wendroff method to a trapezoidal mesh system.

Like the original Lax-Wendroff method, all these two-step versions may require additional explicit artificial viscosity in order to damp oscillations for a strong shock computation. Lapidus (1967) and Erdos and Zakkay (1969) added explicit Rusanov-type terms (see section V-D-3). The paper of Tyler and Ellis (1970) is important for comparisons of these methods and for Tyler's method of providing additional damping. For the one-dimensional model equation (5-1) Tyler noted the relations among several methods; for the value of ω in Rusanov's method (see section V-D-3) given by $\omega = 1/c$, the Rusanov method reduces to Lax's method; for $\omega = c$, it reduces to the Lax-Wendroff method, which is equivalent in this case to Leith's method, MacCormack's method, and the various two-step methods. For a shock-pressure ratio of about 10, the Richtmyer method (5-79) gave a shock thickness of about $3\Delta x$ and a maximum overshoot of 20%; the permuted MacCormack method (5-90) gave a shock thickness of about $6\Delta x$ measured to the nearly uniform flow or $3\Delta x$ measured just at the overshoot front, with a maximum overshoot of 8%. Other comparisons may be found in the original paper but, most importantly, Tyler then showed how excellent results may be obtained by adding an explicit artificial (bulk) viscosity term of the von Neumann-Richtmyer type, similar to Longley's method (section V-D-2), to the momentum and energy equations. Tyler's form is

$$q = - b_1 \Delta x \rho (|u| + a) \frac{\partial u}{\partial x} \tag{5-91A}$$

or

$$\alpha = b_1 \Delta x \rho (|u| + a) \tag{5-91B}$$

Forward time differencing is used on the artificial diffusion terms. Because of the implicit damping of the two-step methods, small values of b_1 were adequate. The shock thickness for the Richtmyer two-step method was 2 to $3\Delta x$, and for the permuted MacCormack method was 3 to $4\Delta x$; both had a maximum overshoot of only 0.18% with $b_1 = 0.15$ and $b_1 = 0.325$, respectively. (See Figure 5-1 on page 231.) Tyler and Ellis (1970) also tested rarefactions and shocks.

In view of the success of these tests and the obvious extension in dimensions, Tyler's method (5-91) is recommended for use with the class of two-step Lax-Wendroff methods.

V-E-7. The Method of Abarbanel and Zwas

Abarbanel and Zwas (1969) investigated a class of methods based on multi-step application of the original Lax-Wendroff method, (5-72) through (5-74). We denote the entire Lax-Wendroff method (5-72A) by the operator, L, writing

$$U^{n+1} = U^n + L\{U^n\} \tag{5-92}$$

Abarbanel and Zwas considered the general iterative form,

$$U^{(n+1)^{k+1}} = U^n + \Gamma \cdot L\left\{U^{(n+1)^k}\right\} + (1 - \Gamma)\cdot L\{U^n\} \tag{5-93}$$

They varied the maximum k and the implicit-explicit weighting factor, Γ. For max k → ∞ and Γ = 1, the method would be a fully implicit Lax-Wendroff form,

$$U^{n+1} = U^n + L\{\overline{U^{n+1}}\} \tag{5-94}$$

while for Γ = 0, it reduces to the original explicit form (5-92). The best results were obtained with the simplest form of Γ = 1 and one iteration. This two-step method is written as

$$\overline{U^{n+1}} = U^n + L\{U^n\} \tag{5-95A}$$

$$U^{n+1} = U^n + L\{\overline{U^{n+1}}\} \tag{5-95B}$$

Abarbanel and Zwas applied the method to one-dimensional shock propagation in Lagrangian coordinates. They found that iterations were more effective than the additional explicit artificial diffusion proposed by Lax and Wendroff (1960). The most demanding shock tests are cases with large shock pressure ratios and low specific heat ratios γ. For a pressure ratio of 4 and $\gamma = 1.2$, the method (5-95) gave a shock thickness $\delta_s = 6$ to $8\Delta x$ and, most significantly, zero-overshoot (monotonic) pressure profiles.

The method has not yet been applied in Eulerian coordinates but appears promising. It has been applied to the cylindrical shock problem (Abarbanel and Goldberg, 1971). As we have seen in section V-E-5, the Lax-Wendroff operator, L, becomes very complicated in multi-dimensional problems. An obvious extension of the method of Abarbanel and Zwas would be to replace L in equation (5-95) with one of the two-step methods of the previous section (V-E-6). The resulting four-step method in two dimensions would still require only about 1/4 the computer time, as in equation (5-95), and about 1/2 the time of the original Lax-Wendroff method. This obvious extension has not yet been attempted.

The method obviously has the same steady-state artificial viscosity as the Lax-Wendroff method.

V-E-8. Other Methods; Flux-Corrected Transport Algorithm of Boris

Various shock methods based on Lagrangian equations can be found in Richtmyer and Morton (1967).

A first-order two-step method which produces smooth shocks is the method of Godunov (1959); see also Godunov et al., (1961) and Richtmyer and Morton (1967). Like the Richtmyer two-step method and other methods of section V-E-6, it uses leapfrog differencing for the second step. The calculation of the provisional first step is very interesting. For the Lagrangian equations, provisional $(\overline{n+1})$ values of only u and p (not E) are needed. These are calculated by a solution of the Riemann shock-tube problem (e.g., see Owczarek, 1964) over the computational cell. The method is complicated, but it does treat any discontinuity in the flow (including contact surfaces) more realistically than do the artificial viscosity methods; it has been applied in two dimensions by Godunov et al., (1961) and Masson et al., (1969), and in three dimensions using simplified (small cross-flow) equations by Masson (1968).

McNamara (1966, 1967) used a modified Godunov method. The provisional first step is based on the linearized (weak) shock tube equations in a "floating" Eulerian 2D mesh which is periodically realigned to follow the movement of a contact surface. McNamara says that an inaccuracy in the form of a cusp in the shock near the stagnation streamline results from an inconsistency in the calculation of the mesh motion.

First- and second-order "hopscotch" methods (see section III-A-18) have been applied to 1D shock calculations by Gourlay and Morris (1971). Rusanov (1970) and Burstein and Mirin (1970, 1971) considered third-order shock methods. Lax (1969) descibes a method for Burger's equation devised by Glimm which is interesting from a pedagogical viewpoint.

In the 1D SHASTA code of Boris , the first stage has zeroth-order diffusion which gives

The *Flux-Corrected Transport* or *FCT* algorithm originally developed by Boris (NRL-MR-2357, Nov. 1971) has been refined and generalized* into a powerful method of treating shocks and other steep gradients. The first stage uses any of several methods, e.g. Lax-Wendroff, donor cell, or leapfrog, with explicit or implicit nonlinear diffusion present. In the second stage, the diffusive terms are partially cancelled (almost completely cancelled outside of shock regions with "phoenical" versions of FCT) by an *anti-diffusive step*. The crucial aspect of the algorithm is the prescription for *limiting the anti-diffusive magnitude so that the numerics add no new extrema to the solution*. Applications to complex problems like MHD shocks and two-fluid shear instability have met with impressive success. We feel that this technique, now refined and systematized, will find widespread future application.

As stated earlier, any of the incompressible methods of sections III-A and III-G are worth exploring for application to compressible flow. If the method has any temporal artificial viscosity, it may be applicable to compressible flow, provided that shocks are weak and/or sufficient physical viscosity (low Re) is present. Especially noted are the two-step method of Brailovskaya (section III-A-15) and Crocco's method (section III-A-12), which will be covered in the next section (V-F) on the viscous terms.

V-F. Viscous Terms in the Compressible Flow Equations

In this section, we consider methods of treating the physical viscous and heat conduction terms in the compressible Navier-Stokes equations. The methods may also apply to the explicit artificial diffusion terms considered in section V-D. Because of the confusingly large number of terms, the methods will actually be described by reference to simple model terms, in order to illustrate the concepts without a bewildering assemblage of subscripts and superscripts.

V-F-1. Spatial Difference Forms

All the viscous terms in equation (4-63) are comprised of the following two forms,

$$\frac{\partial}{\partial x}\left[f\,\frac{\partial g}{\partial x}\right] \quad , \quad \frac{\partial}{\partial x}\left[f\,\frac{\partial g}{\partial y}\right] \tag{5-97}$$

or the commuted $x \rightleftarrows y$ forms, wherein f and g are various combinations of normalized viscosity and heat conductivity coefficients, velocity components, and constants. For the simplest single-time-level differencing, these can be expressed by the following forms:

$$\frac{\partial}{\partial x}\left[f\,\frac{\partial g}{\partial y}\right]_{i,j} = \frac{\left[f\,\frac{\partial g}{\partial y}\right]_{i+1,j} - \left[f\,\frac{\partial g}{\partial y}\right]_{i-1,j}}{2\Delta x}$$

$$= \frac{f_{i+1,j}(g_{i-1,j+1} - g_{i+1,j-1}) - f_{i-1,j}(g_{i-1,j+1} - g_{i-1,j-1})}{4\Delta x\cdot\Delta y} + 0(\Delta x^2,\Delta y^2) \tag{5-98}$$

with a similar form applying to $\partial/\partial y\ [f\ \partial g/\partial x]$, and

$$\frac{\partial}{\partial x}\left[f\,\frac{\partial g}{\partial x}\right]_{i,j} = \frac{\left[f\,\frac{\partial g}{\partial x}\right]_{i+\frac{1}{2},j} - \left[f\,\frac{\partial g}{\partial x}\right]_{i-\frac{1}{2},j}}{\Delta x}$$

$$= \frac{f_{i+\frac{1}{2},j}\left[\frac{g_{i+1,j} - g_{i,j}}{\Delta x}\right] - f_{i-\frac{1}{2},j}\left[\frac{g_{i,j} - g_{i-1,j}}{\Delta x}\right]}{\Delta x}$$

$$= \frac{f_{i+\frac{1}{2},j}(g_{i+1,j} - g_{i,j}) - f_{i-\frac{1}{2},j}(g_{i,j} - g_{i-1,j})}{\Delta x^2} + 0(\Delta x^2) \tag{5-99}$$

---

*Book, D.L., Boris, J.P., and Hain. K. (1975), "Flux-Corrected Transport II: Generalizations of the Method", J. of Computational Physics, Vol. 18, pp. 248-283.

If f is constant in space, (5-98) reduces to the form of (3-12) and (5-99) reduces to (3-14). If f varies, the terms $f_{i\pm\frac{1}{2},j}$ in (5-99) may be evaluated in either of the two usual ways. For example, if $f \equiv \mu$ is temperature dependent, we could have either of the following.

$$f_{i\pm\frac{1}{2},j} = \frac{1}{2}\left\{f \text{ of } (T_{i,j}) + f \text{ of } (T_{i\pm1,j})\right\} \tag{5-100}$$

$$f_{i\pm\frac{1}{2},j} = f \text{ of}\left\{\frac{1}{2}\,(T_{i,j} + T_{i\pm1,j})\right\} \tag{5-101}$$

Both forms are second order and conserve the diffused quantity, g. If a linear temperature law is used, with $\omega = 1$ in equation (4-62) giving $f = T$, these forms are algebraically equal. But, if the Sutherland law (4-60) or if equation (4-62) with $\omega \neq 1$ is used, the computer time savings can be worthwhile if form (5-100) is used and f_{ij} is stored as an array (see discussion in section VII-A).

V-F-2. General Considerations

The complete linear stability analysis of advection and diffusion terms in the Navier-Stokes equations is extremely complicated and has been attempted in only one or two papers, and only for the simplest differencing schemes. Because interest currently focuses on the strong shock problems which are dominated by the inviscid terms, it is currently fashionable to do a stability analysis for only the inviscid equations and to make some hopeful statement that the addition of viscous terms will have little effect. This fashion is interesting because it brings us full circle from the fashion of a decade or two ago, when interest focused on the heat diffusion problem and it was believed that the addition of advection terms would not affect stability* (e.g., see Richtmyer, 1957). In truth, both inviscid and viscous terms are important, but the analysis is usually too difficult. Experience with the simple model equation (2-18) also indicates (see Chapter III) that it is not always possible to analyze the viscous and inviscid terms separately and then to just take the more restrictive condition as a limit. The addition of viscous terms may change an inviscid unstable method to a stable one (FTCS differencing of section III-A-4,5) or it may change an inviscid stable method to an unstable one (midpoint leapfrog method of section III-A-6). But the splitting of the stability analyses may give some clues and guidelines for experimentation. Further, the experience of Cheng (1968), Allen (1968), and Allen and Cheng (1970) showed that the analysis of the model equation (2-18) can provide a valuable indication of the stability behavior of the complete Navier-Stokes equations, at least for explicit methods.

There are also indications that the stability considerations of cross-derivative terms (equation 5-98) may be negligible. Kentzner (1967) has shown that cross-derivative terms do not affect stability, at least in the limit as $\Delta x \to 0$. Unlike the same statement made earlier for advection terms, this result seems to be borne out in computational experience with $\Delta x > 0$. At least, the experience has been that whatever stability restrictions arise from cross-derivatives have been overshadowed by other stability limits. Of course, this may not hold true in future methods,** but presently it allows a great simplification in our presentation of methods, because we need only describe the methods for model terms like $\partial/\partial x\,[f(\partial g/\partial x)]$ and $f(\partial^2 g/\partial x^2)$.

V-F-3. Methods for the Viscous Terms

The rule of thumb which we recognized (section III-A-14) for the model equation, that explicit time differencing schemes which are successful for the inviscid equations will not generally be successful when applied to the viscous terms alone, also seems to apply to the compressible flow equations. For example, two-step Lax-Wendroff differencing of the viscous terms is destabilizing (Rubin and Burstein, 1967; Freudiger et al., 1967) because

*In fact, this is true, as we have seen in Chapter III, if stability is defined only for the limiting situation of $\Delta x \to 0$, but is not true for $\Delta x > 0$.

**The ADI method of Mckee and Mitchell (1970) for the cross derivative terms may then be of interest.

of the effective time centering of the viscous terms.* Thommen (1966) suggested using simple FTCS differencing for just the diffusion term in both steps.** The result is conditional stability with a quite restrictive Δt at low cell Reynolds numbers R_c, as expected. The results of a one-dimensional analysis are presented graphically by Rubin and Burstein (1967). MacCormack (1970) does apply the two-dimensional version of his method (5-90) to both viscous and inviscid terms, using just the inviscid stability criteria in high-Re flows (MacCormack, 1970) and an additional viscous criteria of the form $\Delta t < bRe\Delta x^2$ for more viscous flows (MacCormack, 1969).

Pavlov (1969) used simple FTCS differencing (see section III-A-4,5) on the full Navier-Stokes equations achieving solutions for low-Re flows (Re = 50). Butler (1967) also used FTCS differencing for viscous terms in PIC and FLIC. Likewise, Scala and Gordon (1967) solved even lower Re flows, using upwind differencing for the advection terms and FTCS for the diffusion terms in the "hopscotch" method (see sections III-A-18, III-B-7) and using a transformed coordinate system. It should be noted that, although such studies are of value, the combination of high Mach number and low Reynolds number may put the physical flow outside the continuum flow regime, in which case the Navier-Stokes equations are no longer valid.

The two-step method (3-377) of Brailovskaya has been used to solve viscous shock-free flows (Brailovskaya, 1965). For the compressible flow equations (4-63), this method is succinctly described as follows. Subscripts I and V refer to the inviscid and viscous terms, respectively, of F and G.

$$\overline{U^{n+1}} = U^n - \Delta t[\delta F^n/\delta x + \delta G^n/\delta y] \tag{5-102A}$$

$$U^{n+1} = U^n - \Delta t \cdot \left[\frac{\delta \overline{F_I^{n+1}}}{\delta x} + \frac{\delta F_V^n}{\delta x} + \frac{\delta \overline{G_I^{n+1}}}{\delta y} + \frac{\delta G_V^n}{\delta y}\right] \tag{5-102B}$$

where $\delta/\delta x$ and $\delta/\delta y$ indicate space-centered differences, and $\overline{F_I^{n+1}}$ denotes F_I of $(\overline{U^{n+1}})$, etc. In two dimensions, the sufficient condition for linear stability as given by Brailovskaya (1965) is

$$\Delta t \leq \min\left(\frac{\Delta^2 Re}{8}, \frac{\Delta}{|u| + |v| + a\sqrt{2}}\right) \tag{5-103}$$

where Δ is the mesh spacing for $\Delta x = \Delta y = \Delta$, and (presumably) $\Delta = \sqrt{\Delta x \Delta y}$ otherwise. The second condition of equation (5-103) is the usual Courant-number restriction, $C \leq 1$, and the first condition is that

$$d \equiv \frac{\Delta t}{\Delta^2 Re} \leq \frac{1}{8} \tag{5-104}$$

which is *twice* as restrictive as the usual one-step diffusion limitation. These conditions were sufficient to assure stability for Brailovskaya's (1965) calculations, but a consideration of the energy equation (4-42D) and equations (4-44B) and (4-45) also suggests that the additional diffusion limitation of

$$d_2 \equiv \frac{\Delta t}{\Delta^2 N} \leq \frac{1}{8} \tag{5-105}$$

---

*The midpoint leapfrog method is unconditionally unstable for the simple diffusion equation; see section III-A-6.

**In Chudov's method (Brailovskaya et al., 1968), the viscous contribution is ignored in the first provisional step. Many authors have since used this method.

where

$$N = PrReM_o^2(\gamma - 1) \tag{5-106}$$

might be required. Since her calculations were made only for $M_o \leq 1$, $Pr \leq 1$ and (of course) $\gamma < 2$, Brailovskaya's condition (5-103) automatically satisfies equation (5-105) because $N < Re$. (For air, $N > Re$ for $M_o > 1.89$.)

Allen and Cheng (1970) changed the treatment of the diffusion terms in Brailovskaya's method, and so eliminated the diffusion stability limitation entirely, for their constant $\bar{\mu}$, $\bar{k}$ equations (i.e., $\mu = k = 1$). The method has already been described for the model equation in (3-395). For compressible flow, the pressure gradients in the momentum equations are differenced in the manner of the advection terms in equation (3-395). If the Cheng-Allen method is applied to variable property equations with terms like $\partial/\partial x[f(\partial g/\partial x)]$, the consistent evaluation of terms like $f_{i\pm\frac{1}{2},j}$ in equation (5-100) for the first provisional step would be

$$f_{i\pm\frac{1}{2},j} = \frac{1}{2}\left\{f \text{ of } (\overline{T_{ij}^{n+1}}) + f \text{ of } (T_{i\pm 1,j}^{n})\right\} \tag{5-107A}$$

For the second step,

$$f_{i\pm\frac{1}{2},j} = \frac{1}{2}\left\{f \text{ of } (T_{ij}^{n+1}) + f \text{ of } (\overline{T_{i\pm 1,j}^{n+1}})\right\} \tag{5-107B}$$

This would unfortunately result in an implicit set of equations. To retain explicitness, f must be evaluated in the first step as

$$f_{i\pm\frac{1}{2},j} = \frac{1}{2}\left\{f \text{ of } (T_{ij}^{n}) + f \text{ of } (T_{i\pm 1,j}^{n})\right\} \tag{5-108A}$$

and, in the second step, as either the above equation (5-108A) or as

$$f_{i\pm\frac{1}{2},j} = \frac{1}{2}\left\{f \text{ of } (\overline{T_{ij}^{n+1}}) + f \text{ of } (\overline{T_{i\pm 1,j}^{n+1}})\right\} \tag{5-108B}$$

For strong functional dependencies, $f = \mu(T)$ and $f = k(T)$, we suspect that this method will again give a diffusion time-step limitation,* and the original Brailovskaya method might be preferred because of its relative simplicity.

Allen (1968), Allen and Cheng (1970), and Cheng (1970) have successfully applied this method to the supersonic base flow problem, which contains weak shocks. A very significant virtue of the Brailovskaya and the Cheng-Allen two-step methods is that each of the two steps have identical forms, in contrast to all the two-step Lax-Wendroff methods (section V-E-6). In the steady-state analysis, this form does not have any artificial diffusion (although it does in the transient analysis). Thus, the Brailovskaya and Cheng-Allen methods seem preferable to the Lax-Wendroff methods for a steady flow problem with weak shocks.

Crocco's (1965) method for the advection terms has already been presented in section III-A-12. For the entire model equation (5-1), this three-level method may be described as follows.

$$\zeta_i^{n+1} = \zeta_i^n + \Delta t \cdot \left[-(1+\Gamma)\frac{\delta(\bar{u}\zeta)^n}{\delta x} - \Gamma\frac{\delta(\bar{u}\zeta)^{n-1}}{\delta x} + \alpha\frac{\zeta_{i+1}^n - 2\zeta_i^{n+1} + \zeta_{i-1}^n}{\Delta x^2}\right] \tag{5-109}$$

---

*The two-step method for the diffusion terms used by Roache and Mueller (1968) does allow the use of forms like (5-107). Unfortunately, further analysis shows that the method does not eliminate the diffusion limitation but merely reduces it.

Crocco applied this method to compressible flow with shock waves present, using the quasi-1D (area ratio) equations with constant properties. Pressure gradients are differenced like the advection terms in equation (5-109). The stability limitations are presented graphically (Crocco, 1965). Victoria and Steiger (1970) applied the method to 2D planar and axisymmetric flows with weak shocks, and included axisymmetry effects in the stability analysis. As in the Cheng-Allen method, variable properties in (5-107) render the equations implicit.

Briley and McDonald* and Baum and Ndefo** successfully applied ADI implicit methods (Section III-A-16) in supersonic viscous flow. The methods differ in details of linearization, which can severely compromise the implicit stability if inadequately handled.

A little-known method for treating the entire set of compressible flow equations was given by Nagel (1967). (Since the method has not been used for inviscid flows, it is included only in this section.) Nagel used the method with only the equations for the perturbation quantities (see Section VI-D) of the non-conservation variables ρ, u, v, and H, where H is the total enthalpy. The method would appear to be applicable in conservation variables, and we so describe it. We denote by terms like

$$\frac{\delta(\rho u)^{n+1}}{\delta t} = \frac{\delta(\rho u)}{\delta t}\left(\rho^{n+1},\ (\rho u)^{n+1},\ (\rho v)^{n+1},\ E_s^{n+1}\right) \tag{5-110}$$

the analog of $\partial(\rho u)/\partial t$ obtained from equation (4-63) using space-centered differences for all terms and using the time-level values indicated in equation (5-110). Similar equations apply for other variables. Then Nagel's method is described by the following 12 steps, each applied for all (i,j).

Stored from the previous 12-step cycle:

$$\rho^n,\ E_s^n,\ \frac{\delta\rho}{\delta t}^{n-\frac{1}{2}},\ \frac{\delta E_s}{\delta t}^{n-\frac{1}{2}},\ (\rho u)^{n+\frac{1}{2}},\ (\rho v)^{n+\frac{1}{2}},\ \frac{\delta(\rho u)}{\delta t}^{n},\ \frac{\delta(\rho v)}{\delta t}^{n}$$

Step:

1. $$\rho^{n+\frac{1}{2}} = \rho^n + \frac{\Delta t}{2}\frac{\delta\rho}{\delta t}^{n-\frac{1}{2}}$$

2. $$E_s^{n+\frac{1}{2}} = E_s^n + \frac{\Delta t}{2}\frac{\delta H}{\delta t}^{n-\frac{1}{2}}$$

3. $$\frac{\delta\rho}{\delta t}^{n+\frac{1}{2}} = \frac{\delta\rho}{\delta t}\left[(\rho u)^{n+\frac{1}{2}},\ (\rho v)^{n+\frac{1}{2}}\right] \tag{5-111}$$

4. $$\frac{\delta E_s}{\delta t}^{n+\frac{1}{2}} = \frac{\delta E_s}{\delta t}\left[\rho^{n+\frac{1}{2}},\ E_s^{n+\frac{1}{2}},\ (\rho u)^{n+\frac{1}{2}},\ (\rho v)^{n+\frac{1}{2}}\right]$$

5. $$\rho^{n+1} = \rho^n + \Delta t\,\frac{\delta\rho}{\delta t}^{n+\frac{1}{2}}$$

6. $$E_s^{n+1} = E_s^n + \Delta t\,\frac{\delta E_s}{\delta t}^{n+\frac{1}{2}}$$

7. $$(\rho u)^{n+1} = (\rho u)^{n+\frac{1}{2}} + \frac{\Delta t}{2}\frac{\delta(\rho u)}{\delta t}^{n}$$

8. $$(\rho v)^{n+1} = (\rho v)^{n+\frac{1}{2}} + \frac{\Delta t}{2}\frac{\delta(\rho v)}{\delta t}^{n}$$

* In Proc. Fourth International Conf. on Numerical Methods in Fluid Dynamics, R.D. Richtmyer, ed., Lecture Notes in Physics, Vol. 35, Springer-Verlag, New York, 1975, pp.105-110.
**In Proc. AIAA Computational Fluid Dynamics Conf., Palm Springs, CA, July 1973, pp.133-140.

9. $$\frac{\delta(\rho u)^{n+1}}{\delta t} = \frac{\delta(\rho u)}{\delta t}\left[\rho^{n+1}, E_s^{n+1}, (\rho u)^{n+1}, (\rho v)^{n+1}\right]$$

10. $$\frac{\delta(\rho v)^{n+1}}{\delta t} = \frac{\delta(\rho v)}{\delta t}\left[\rho^{n+1}, E_s^{n+1}, (\rho u)^{n+1}, (\rho v)^{n+1}\right]$$

11. $$(\rho u)^{n+3/2} = (\rho u)^{n+\frac{1}{2}} + \Delta t\,\frac{\delta(\rho u)^n}{\delta t}$$

12. $$(\rho v)^{n+3/2} = (\rho v)^{n+\frac{1}{2}} + \Delta t\,\frac{\delta(\rho v)^n}{\rho t}$$

A similar method was considered by Rozhdestvenskii and Lananko (1969). Nagel used the following stability criteria:

$$\Delta t < \frac{1.5\ \Delta x}{a} \tag{5-112A}$$

$$\Delta t < \frac{\Delta x}{\sqrt{2|u|}} \tag{5-112B}$$

$$\Delta t < \frac{\Delta x^2\ Re}{2} \tag{5-112C}$$

$$\Delta t < \left(\frac{\gamma}{\gamma - 1}\right)\frac{\Delta x}{|u|} \tag{5-112D}$$

Although Nagel verified these experimentally for his boundary layer flow, it would be reasonable to supplant the first two with the usual Courant-number restriction and to reduce the factor 1/2 in equation (5-112C) to 1/4, for more general flows.

Sakurai and Iwasaki (1970) used Saul'yev (ADE) differencing for the diffusion terms (see section III-A-17) in a 1D calculation, but the stability effect of the advection terms remains to be clarified.

We note that many workers in high M viscous flows now prefer to use a *non-conservative energy equation* (in T, instead of E_s) to avoid taking the difference of two large numbers (total energy - "dynamic" energy) when calculating temperatures and thence pressures.

V-G. Boundary Conditions for Compressible Flow

The remarks made in section III-C-1 on the dominant importance of computational boundary conditions apply equally to compressible flow. The discussion here will rely on some groundwork laid in section III-C on incompressible flow, but several aspects are unique to compressible flow. The most difficult boundary is, surprisingly enough, a simple wall.

V-G-1. Slip Walls

We first consider the inviscid equations with corresponding "slip" wall conditions.* The continuum boundary condition is that the flow is parallel to the wall at this point, i.e., $v_w = 0$. The other variables are *not* specified as boundary conditions. For example, it is inconsistent to specify T_w when using the inviscid equations.**

---

*These "slip" conditions correspond to the absence of viscous terms in the continuum equations. The term is also applied to the problem of the breakdown of the continuum approach near a wall in viscous flow. Whitehead and Davis (1969) consider the analytical formulation of surface conditions in this slip flow, including mass transfer (blowing) from the wall.

**Nor may $\partial T/\partial y$ be specified. As is well known, an adiabatic wall condition is consistent with inviscid flow, but the wall is adiabatic, with heat flux $Q_w = k(\partial T/\partial y)|_w = 0$, because the term $k = 0$ and not because $\partial T/\partial y = 0$.

It may also be useful to model a slip wall condition with the viscous equations. This corresponds to a retention of viscous terms in the area more than one cell away from the walls, implying that the boundary layer is less than Δy thick. It seems that additional conditions of zero gradient on U and T would be sufficient to determine a solution and would, in some vague sense, model a "slip" wall, but the validity of this approach is not clear at this time. (See discussion in section III-C-2.)

V-G-1-a. Slip Walls in the First Mesh System

In the first mesh system, node points lie on the surface, as depicted in Figure 5-2-a. For concreteness, we consider a wall aligned with a constant value of x, such as boundary B2 in Figure 3-22 (back on page 140).

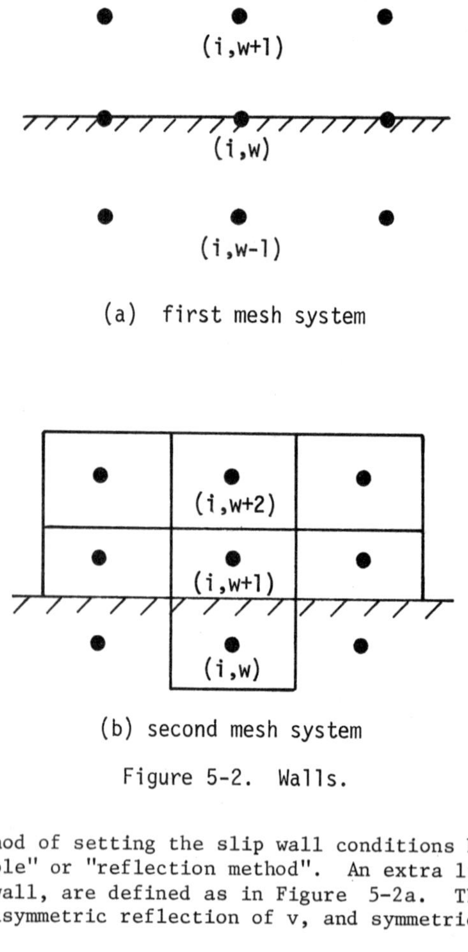

(a) first mesh system

(b) second mesh system

Figure 5-2. Walls.

The most popular method of setting the slip wall conditions has been by use of the so-called "reflection principle" or "reflection method". An extra line of fictitious points at $j = w - 1$, *inside* the wall, are defined as in Figure 5-2a. Then fictitious values are defined at $(w - 1)$ by antisymmetric reflection of v, and symmetric reflection of other variables.

$$v_{w-1} = - v_{w+1}$$

(5-113)

$$f_{w-1} = + f_{w+1} \quad , \quad f = \quad , u, E_s \text{ or } T$$

Additionally, the condition $v_w = 0$ is set at $j = 2$. New time-dependent values of the variables f_w are then computed, using the regular interior-point differencing at (i,w).

Equations (5-113) are *symmetry* conditions and are clearly applicable to the centerline, boundary B1 in Figure 3-22 (page 140). But generally, a wall is *not* a line of symmetry. It was forcibly pointed out by Moretti (1968A, 1968B), that the "reflection" boundary conditions (5-113) are erroneous. In Moretti's interpretation, these conditions introduce additional boundary conditions of $\partial f/\partial y = 0$ for $f = \rho$, u, E_s, which then actually over-specify the problem. However, the (w-1) points are to be interpreted as strictly fictitious values, defined only for the convenience of using regular interior-point differencing equations at (w). The question of whether or not the reflection treatment (5-113) is correct, and the nature of the errors, must be judged by reference to those equations.

First, the condition $v_w = 0$ is correctly set, of course. Next, consider the non-conservative form of the *inviscid* v-momentum equation from equation (4-3).

$$\frac{\partial v}{\partial t} = -u\frac{\partial v}{\partial x} - v\frac{\partial v}{\partial y} - \frac{1}{\rho}\frac{\partial P}{\partial y} \tag{5-114}$$

Since $v(x)|_w = 0$, then $\partial v/\partial x|_w = 0$ along a straight wall, and equation (5-114) reduces to

$$\left.\frac{\partial P}{\partial y}\right|_w = 0 \tag{5-115}$$

The reflection method (5-113) does produce $P_{w-1} = P_{w+1}$ and so gives $\delta P/\delta y|_w = 0$, which is correct for the inviscid equations.

Consider next the v-flux terms in the u-momentum equation. From the non-conservation form (4-2) for inviscid flow, we have

$$\frac{\partial u}{\partial t} = -u\frac{\partial u}{\partial x} - v\frac{\partial u}{\partial y} - \frac{1}{\rho}\frac{\partial P}{\partial x} \tag{5-116}$$

At the wall, $v_w = 0$, so the second term in equation (5-116), the v-flux of u-momentum, clearly should be zero at w. But consider the result of the reflection method (5-113) on the corresponding term in the conservation equation (4-55), which is $\partial(\rho uv)/\partial y$. At the wall,

$$\left.\frac{\delta(\rho uv)}{\delta y}\right|_w = \frac{(\rho uv)_{w+1} - (\rho uv)_{w-1}}{2\Delta y} = \frac{\rho_{w+1}u_{w+1}v_{w+1} - \rho_{w+1}u_{w+1}(-v_{w+1})}{2\Delta y} = \frac{(\rho uv)_{w+1}}{\Delta y} \neq 0 \text{ generally} \tag{5-117A}$$

As Δy is reduced, $v_{w+1} \to 0$, so equation (5-116) might appear to be a mathematically consistent approximation. But a more careful analysis shows

$$\left.\frac{\delta(\rho uv)}{\delta y}\right|_w = \frac{(\rho uv)_{w+1}}{\Delta y} = \frac{(\rho u)_{w+1}\left[v_w + \left.\frac{\partial v}{\partial y}\right|_w \Delta y + O(\Delta y^2)\right]}{\Delta y} = (\rho u)_{w+1}\left.\frac{\partial v}{\partial y}\right|_w + O(\Delta y) \tag{5-117B}$$

For a slip wall, it is *not* true that $\partial v/\partial y|_w = 0$ unless it is a symmetry boundary. From equation (5-117B), it thus appears that the error is persistent as $\Delta y \to 0$, so that the reflection method in the first mesh system does not produce a mathematically consistent FDE system. Moretti's (1968A,B) comparative calculations indicate that this error is severe for a coarse mesh, as expected.

An FDE solution with erroneous boundary conditions may still give an approximation to the PDE in some useful sense, but the method is *not* mathematically *consistent*; as $\vec{\Delta x} \to 0$, the FDE solution does not approach the PDE solution. It is curious that mathematicians, of all people, have been blythe about using these erroneous boundary conditions. Only comparatively recent papers have dealt with the global effects of such overspecified boundary conditions; see Kreiss and Lundquist (1968) and Osher (1969B). Something of the convenience of the reflection method may be salvaged by using it for the continuity and energy equations, and specifically zeroing the $\partial(\rho uv)/\partial y$ term in the u-momentum equation. This yields consistent boundary conditions for a straight wall.

An important and interesting problem in the consistency of the boundary conditions arises with the reflection method if a curved surface is represented by straight-line segments of width Δs. The reflection method still gives $\delta P/\delta n = 0$, but the correct value for steady slip flow should be

$$\left.\frac{\partial P}{\partial n}\right|_w = \rho(u^2 + v^2)/r \tag{5-118}$$

where r is the local radius of curvature. Thus, although the interior-point FDE and the surface geometry are consistent as $\Delta\vec{x} \to 0$, the boundary conditions (and therefore the global solution) are not consistent. The results of Kutler and Lomax (1971) show deterioration of agreement with experimental values as surface curvature increases, apparently due to the use of a zero pressure gradient instead of equation (5-118).

Kentzner (1970A) used equation (4-118), which is readily seen to be a "centripetal force" balance, to determine P_w by one-sided differencing. With the condition that the entropy, s, is constant along the inviscid wall streamline, this determined $\rho_w = \rho(P,x)$. (The method could perhaps be adapted to the case where a shock impinges on the wall, in which case s is not constant.)

Kentzner supplemented these with a relation based on characteristics, as Moretti (1968A) and Bastianon (1969) did earlier (see also Coakley and Porter, 1969). While these methods are correct, they are not adaptable to the viscous flow problem, are not conservative,* and are not simple to apply.

Another approach is to evaluate the v-flux terms in all the equations by one-sided differencing. In the continuity equation, for example, we could take

$$\left.\frac{\partial(\rho v)}{\partial y}\right|_w = \frac{(\rho v)_{w+1} - (\rho v)_w}{\Delta y} + O(\Delta y) = \frac{(\rho v)_{w+1}}{\Delta y} + O(\Delta y) \tag{5-119}$$

If $v_{w+1} > 0$, this amounts to upwind differencing for the y-flux at w. But if $v_{w+1} < 0$, it becomes *downwind* differencing, which is unstable and physically absurd (see section III-A-8). Although some converged solutions have been obtained with this approach, it generally is not to be recommended.

The boundary formulation of Lapidus (1967) appears to give a consistent FDE formulation, with node points on the wall. However, the method is not strictly conservative (although it is based on conservation concepts) because of the required re-definition and overlapping of cell boundaries which is "inconsistent" (in the common usage of the word) with the interpretation used at interior points. It is also somewhat complicated.

Bohachevsky and Kostoff (1971) and Emery and Ashurst (1971) treated surface blowing boundary conditions for inviscid flow, using the first mesh system.

Generally, we recommend the use of the second mesh system for slip walls.

V-G-1-b. Slip Walls in the Second Mesh System

Slip walls may be treated in a more straightforward manner in the second mesh system. The second mesh system is shown in Figure 5-2b. Cell *boundaries*, rather than node points (cell centers), are aligned with the wall. Thus, the first node point, at (w+1), is $\Delta y/2$ away from the wall. A fictitious point (w) may be defined inside the wall.

Gentry, Martin and Daly (1966) used slip-wall boundary conditions in this second mesh system. The only condition to be set is that the flux of all transported properties across

*Neither Moretti nor Kentzner use the conservation form, so this is not of any concern in their works.

the wall be zero. This condition can be set by formulating special equations with these flux terms = 0 for the cells at (w+1), or they may be set for some of the terms by the reflection method. Defining a fictitious cell at w inside the wall, we set

$$v_w = - v_{w+1} \tag{5-120A}$$

$$f_w = + f_{w+1} \text{ where } f = \rho,\ u,\ E_s,\ \text{or } T \tag{5-120B}$$

The result of this "reflection method" in the second mesh system is quite different from that of the first.* When second-order regular interior point differencing is used at (w+1), the flux terms at the wall (w+1/2) are identically zero for all f. This is easily seen by reference to the control volume derivation of the centered difference equations in section III-A-1. The interface flux values $(vf)_{w+\frac{1}{2}}$ are defined as

$$(vf)_{w+\frac{1}{2}} = \frac{1}{2}\left[(vf)_w + (vf)_{w+1}\right] = \frac{1}{2}\left[-(vf)_{w+1} + (vf)_{w+1}\right] = 0 \tag{5-121}$$

This interpretation is, of course, consistent with the evaluation of wall values of f at (w+1/2) by linear interpolation, but we emphasize here that we use linear interpolation not because we don't know of anything more exotic, but rather because only it is consistent with our second-order method for first derivatives at interior points.

Exercise: Show that the reflection method in the second mesh system does conserve mass and is compatible with a control-volume mass balance at the surface.

The reflection also gives the correct slip-wall value $\delta P/\delta y|_w = 0$ for a straight wall. As previously mentioned, it is not correct for a curved wall; see equation (5-118). There remains, moreover, a residual error in momentum flux, this time in the v-flux of v-momentum.

$$[v(\rho v)]_{wall} = \frac{1}{2}\left[(\rho v^2)_{w+1} + (\rho v^2)_w\right] = (\rho v^2)_{w+1} \neq 0 \tag{5-122}$$

But this error is not serious and the method would give a mathematically consistent approximation. That is,

$$\begin{aligned}\frac{\delta}{\delta y}(\rho v^2)_{w+1} &= \frac{\rho v^2|_{w+1\frac{1}{2}} - \rho v^2|_{w+\frac{1}{2}}}{\Delta y} = \frac{\rho v^2|_{w+1\frac{1}{2}} - (\rho v^2)_{w+1}}{\Delta y} \\ &= \frac{\rho v^2|_{w+1\frac{1}{2}} - \rho_{w+1}\left[v_{wall} + \Delta y/2 \left.\frac{\partial v}{\partial y}\right|_{wall} + 0(\Delta y^2)\right]^2}{\Delta y} \\ &= \frac{\rho v^2|_{w+1\frac{1}{2}} - \rho_{w+1}(\Delta y/2)^2 \left.\frac{\partial v}{\partial y}\right|^2_{wall} + 0(\Delta y^3)}{\Delta y} \\ &= \frac{\rho v^2|_{w+1\frac{1}{2}}}{\Delta y} - \left\{\rho_{w+1} \left.\frac{\partial v}{\partial y}\right|^2_{wall} \frac{\Delta y}{4} + 0(\Delta y^2)\right\}\end{aligned} \tag{5-123}$$

The bracketed term is the erroneous v-flux of v-momentum at the wall. It is merely a truncation error which vanishes as $\Delta y \to 0$, so the reflection method does give a mathematically consistent approximation for slip walls.

However, the error may easily be removed in a special treatment of the v-momentum equation at (w+1) cells adjacent to the boundary by simply setting the flux term = 0.

---

*Moretti (1968A,B) considered only the reflection method in the first mesh system.

Likewise, the pressure gradient evaluation may be corrected in accordance with equation (5-118) for a curved wall, or by using a one-sided difference for $\delta P/\delta y$, with the resultant first-order accuracy.

The example again demonstrates that the inside points are really fictitious. The setting of

$$v_w = - v_{w+1} \tag{5-124}$$

$$f_w = + f_{w+1} \tag{5-125}$$

assures zero v-flux of f across the wall at (w+1/2). But if f takes on not only the value of ρ but also ρv, an algebraic inconsistency appears. Conceptually, zero flux is obtained by setting the above conditions with $f = \rho$, ρu, ρv, and $(E_s + P)$, ignoring the algebraic relations.

It must be emphasized, however, that the reflection method must be modified for any curved wall, using equation (5-118) or one-sided differencing for $\delta P/\delta y|_w$.

So the reflection method is convenient and, with a little refinement, can be made to yield correct answers for slip walls in the second mesh system.

Molecular-level slip wall conditions, using continuum equations for the flow away from the wall, were treated heuristically by Butler (1967) in the second mesh system.

V-G-2. No-Slip Walls

No-slip walls for viscous flow present an untidy problem. The continuity equation is unchanged from inviscid flow, and density is best evaluated in the second mesh system. But the other variables are most accurately treated in the first mesh system. A staggered mesh system therefore suggests itself and is indeed successful. But a somewhat simpler solution is to evaluate ρ values near the wall as though the second mesh system were being used, but to relate these ρ values to values at the wall in the first mesh system. Although preferable methods may well be developed in the near future, we feel that this last method is currently the best available.

In order to justify the suggested method, it is necessary to consider the other possibilities in some detail.

V-G-2-a. No-Slip Walls in the First Mesh System

For viscous flow with the first mesh system, three of the requisite four boundary conditions are easy. The velocities are just

$$u_w = 0 \quad , \quad v_w = 0 \tag{5-126}$$

For a specified wall temperature, T_{wall}, we have

$$T_w = T_{wall} \tag{5-127}$$

For a specified wall Nusselt number (dimensionless heat flux) we have*

---

*Note that this Nusselt number does not contain any truncation error, but is specified as a parameter of the problem, just like M or Re; see discussion in section III-F-4.

If a non-zero heat transfer at the wall is specified, care must be taken that the correct Nusselt number is specified at the wall; allowance must be made for the change in normalizing $\bar{k}$ of the freestream to the wall value, if $\bar{k}$ is variable.

$$Nu_w = \frac{T_{w+1} - T_w}{\Delta y} \qquad (5\text{-}128)$$

or

$$T_w = T_{w+1} - Nu \cdot \Delta y \qquad (5\text{-}129)$$

The form (5-128) must be used if the second-order heat-conduction terms used at the interior point (w+1) are to yield a calculation which is computationally adiabatic (see section III-F-4).

The density condition is the difficulty. It may be possible to use one-sided differencing of the continuity equation, as in the inviscid case. If the term v_{w+1} is small enough *and if sufficient artificial damping is present*, a stable and converged solution may be obtained. Thus, Skoglund and Cole (1966) obtained a solution of the shock-wave/boundary-layer interaction problem using Rusanov's method* (section V-D-3) and one-sided differencing for $\partial\rho/\partial t|_w$, but the method would not work when the shock strength was sufficient to cause boundary layer separation. This was also the experience of Roache and Mueller (1970) and Allen and Cheng (1970) on the backstep problem. The reason is easily explained.

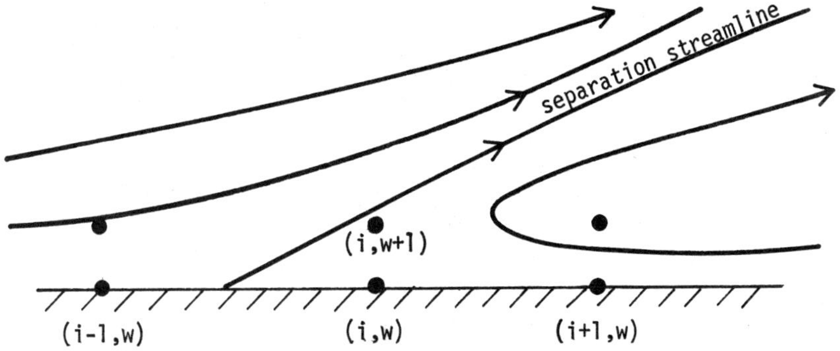

Figure 5-3. Separation region in the first mesh system.

Consider the separation streamline pattern shown in Figure 5-3. No matter how small Δy may be, it is clear that $v_{i,w+1} > 0$. Then for the continuity equation at (i,w) we have, by the no-slip conditions,

$$\left.\frac{\delta\rho}{\delta t}\right|_{i,w} = -\frac{\delta(\rho u)}{\delta x} - \frac{\delta(\rho v)}{\delta y} \qquad (5\text{-}130A)$$

$$= -\frac{(\rho v)_{w+1}}{\Delta y} \qquad (5\text{-}130B)$$

If a steady state is ever to be achieved, we must have $\delta\rho/\delta t = 0$ which, from equation (5-130B), requires $v_{w+1} = 0$. This is in fact compatible with the continuum conditions at a no-slip wall, arrived at from the continuity equation, of $\partial v/\partial y|_w = 0$ and $v_w = 0$. For a first-order determination of $\partial v/\partial y = 0$, this requires $v_{w+1} = 0$. But it is intuitively clear from Figure (5-3) that such a condition cannot be met computationally; as a result, the method

*They retained the explicit diffusion term in a one-sided difference form. On the other hand, Kessler (1968) zeroed this term at the wall slip-flow problem using the reflection method.

diverges.* The condition is further aggravated along the base in the backstep problem (boundary B5 in Figure 3-22).

The basic reason for the failure of the one-sided differencing method near separation is because there is no transport of mass *along* the wall; $\delta(\rho u)/\delta x|_w = 0$ always. From a control-volume point of view, the cell at (i,w) in Figure (5-3) can gain or lose mass only across the interface at (w+1/2). This situation may be changed by redefining the u-flux for cells along the wall. Using the interpretation indicated in Figure (5-4a), we have (for a centered-difference formulation) interface flux values of

$$(\rho u)_{i-\frac{1}{2},w} = \rho_{i-\frac{1}{2},w} v_{i-\frac{1}{2},w} = \frac{1}{2}(\rho_{i,w} + \rho_{i-1,w}) u_{i-\frac{1}{2},w} \tag{5-131}$$

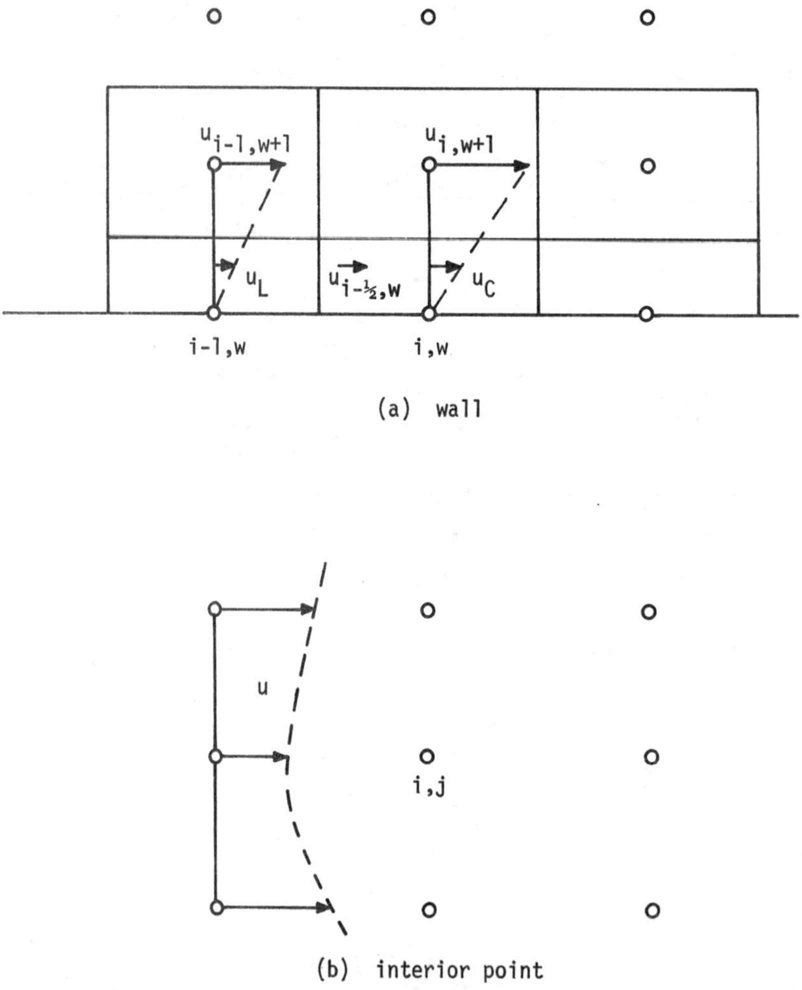

Figure 5-4. Possible cell interpretation of flux velocities at no-slip walls in first mesh system.

*It may be possible to achieve convergence in such a situation if an explicit artificial mass diffusion term is present, as in Rusanov's method (section V-D-3). The mass advection flux out of the wall node at separation might then be balanced by the artificial diffusion into the node. But the accuracy of the method, and even the consistency, is dubious.

where

$$u_{i-\frac{1}{2},w} = \frac{1}{2}(u_L + u_C) = \frac{1}{2}\left(\frac{1}{4}u_{i-1,w+1} + \frac{1}{4}u_{i,w+1}\right) \qquad (5\text{-}132)$$

A similar interpretation holds for the right-hand flux velocity, $u_{i+\frac{1}{2},w}$. This method does allow for a stable, conservative, and mathematically consistent calculation, even for separated flows (Roache and Mueller, 1968,1970; Skoglund and Gay, 1969). But it is "inconsistent" in the more common sense that this is not the interpretation of inter-cell flux velocities implied by the usual interior-point differencing. Such an interpretation is "consistent" with the evaluation of interior-point fluxes (Figure 5-4b) as in the following equation.

$$U_L = \frac{1}{2}\left\{\frac{1}{4}(u_{i-1,j+1} + 2u_{i-1,j} + u_{i-1,j-1}) + \frac{1}{4}(u_{i,j+1} + 2u_{i,j} + u_{i,j-1})\right\} \qquad (5\text{-}133)$$

But if this interpretation is used at interior points, the velocity field becomes smeared out and inaccurate.

Other approximations have been used to give a stable computation. One which has been commonly used is to approximate $\partial P/\partial y|_w = 0$ by setting

$$P_w = P_{w+1} \qquad (5\text{-}134)$$

Then ρ_w is determined from T_w and the equation of state (4-51). This evaluation may appear to be based on the boundary-layer approximation (see Schlichting, 1968) of $\partial P/\partial y \simeq 0$ through a boundary layer. In fact, it is a much milder approximation, since constant P is not applied throughout the boundary layer but only over the lowest Δy in the boundary layer. It has yielded stable computations for both non-separating boundary layers (Kurzrock and Mates, 1966) and for separating boundary layers (MacCormack, 1971), the separation being caused by a shock/boundary-layer interaction. MacCormack later repeated a calculation with a more accurate boundary condition and found virtually no difference in the solution (private communication).

Although this method is attractively simple and in some cases accurate, we do not recommend it generally. For one thing, it is non-conservative. But more fundamentally, although the solution of the FDE may in some sense be an approximation to the solution of the PDE, it is not a mathematically consistent approximation; that is, as $\Delta\vec{x} \to 0$, the FDE solution does not approach the PDE solution with correct boundary conditions. Most importantly, for cases of strong shock wave/boundary-layer interaction, low Re, separation, or small radius of surface curvature, it may not even be a reasonable approximation. Furthermore, it is unnecessary, since consistent methods are available, although they are somewhat more complicated.

It is obvious that such tricks as arbitrary extrapolations and curve fittings of ρ or P to the wall values, which have *no basis* in the physics even as an approximation, are also non-conservative and not mathematically consistent. Further, they are generally unstable for a separating flow problem.

The recommended solution is to use an evaluation of ρ near the walls in the *second* mesh system. The wall values of ρ in the *first* mesh system are then set to produce the same resultant near-wall pressure gradient. The complete method will be given in section V-G-2-c.

V-G-2-b. No-Slip Walls in the Second Mesh System

The no-slip conditions on velocity at the wall can be set in the second mesh system by the following method, but with a decrease in accuracy. It will be convenient to partially use the reflection method. If followed blindly, the reflection method will produce serious errors; however, we view it only as a technique by which we may conveniently set some of the true boundary conditions. Other of the conditions must be explicitly set, in a manner inconsistent with the reflection method. We thus use reflection as nothing more than a programming device. Practically speaking, it is a quite convenient programming device, but we agree with Moretti (1968A,B) that it does not deserve the dignity of the term "principle."

We first describe the reflection used to define the fictitious interior-point values, and then investigate the results of these definitions on the terms in the conservation equations. This investigation will indicate which terms would be incorrectly evaluated by the reflection and which, therefore, must be separately treated.

The reflection for the no-slip wall at (w+1/2), shown in Figure 5-2b, is

$$\rho_w = \rho_{w+1} \tag{5-135}$$

$$u_w = u_{w+1} \tag{5-136A}$$

$$v_w = -v_{w+1} \tag{5-136B}$$

Consistent with a linear interpolation, equation (5-136) sets $u_{wall} = 0$ and $v_{wall} = 0$. For a specified wall temperature, T_{wall}, we use the linear interpolation

$$T_{wall} = \frac{1}{2}(T_{w+1} + T_w) \tag{5-137}$$

to set

$$T_w = 2T_{wall} - T_{w+1} \tag{5-138}$$

For a specified wall Nusselt number, Nu, where

$$Nu = \frac{T_{w+1} - T_w}{\Delta y} \tag{5-139}$$

we set

$$T_w = T_{w+1} - \Delta y \cdot Nu \tag{5-140}$$

The combination of equations (5-135) and (5-136B) sets the v-mass flux across the wall at (w+1/2) = 0, as in the slip case. But the v-fluxes of both u- and v-momenta are now in error. Following the development of equation (5-123), we find

$$\frac{\delta}{\delta y}(\rho uv)_{w+1} = \frac{\rho uv|_{w+1\frac{1}{2}}}{\Delta y} - \left[\rho_{w+1}\left.\frac{\partial u}{\partial y}\right|^2_{wall}\frac{\Delta y}{4} + 0(\Delta y^2)\right] \tag{5-141}$$

which is again consistent, since the bracketed error term above vanishes as $\Delta y \to 0$. The error in the v-momentum may be even smaller; for steady-state, the continuity equation

$$\frac{\partial \rho}{\partial t} + \frac{\partial(\rho u)}{\partial x} + \frac{\partial(\rho v)}{\partial y} = 0 \tag{5-142A}$$

reduces to

$$\left.\frac{\partial(\rho v)}{\partial y}\right|_{wall} = 0 \tag{5-142B}$$

for a no-slip wall. Further expanding,

$$\rho \left.\frac{\partial v}{\partial y}\right|_{wall} + \cancelto{0}{v} \left.\frac{\partial \rho}{\partial y}\right|_{wall} = 0 \tag{5-143}$$

or

$$\frac{\partial v}{\partial y} = 0 \tag{5-144}$$

Thus, the first-order error in equation (5-123) is zero in the steady state; the steady-state error may be shown to have the form

$$\frac{\delta}{\delta y}(\rho v^2)_{w+1} = \frac{\rho v^2|_{w+1\frac{1}{2}}}{\Delta y} - \left\{\rho_{w+1}\frac{\Delta y^3}{64}\left.\frac{\partial^2 v}{\partial y^2}\right|^2_{wall} + O(\Delta y^5)\right\} \tag{5-145}$$

where the bracketed term is the erroneous v-flux of v-momentum at the wall.

While the momentum flux errors indicated in equation (5-141) and either (5-123) or (5-145) may be tolerable, they are unnecessary. It is recommended that the interior-point equations at (w+1) be changed to explicitly set these fluxes = 0. A similar statement applies to the energy equation, as indicated in the following exercise.

Exercise: Show that the reflection equations do correctly give zero v-flux at the wall of $(E_s + P)$ in the energy equation, but only for the special case of an adiabatic wall (Nu = 0). (For more general wall thermal conditions, this flux term should also be explicitly set = 0.)

Likewise, the pressure gradient, $\partial P/\partial y|_{w+1}$, will *not* be correctly given by regular differencing over the reflected values. If this second mesh system is to be used, the recommended method is to revert to one-sided differencing for $\partial P/\partial y$, unfortunately reducing this term to first-order accuracy.

$$\left.\frac{\delta P}{\delta y}\right|_{w+1} = \frac{P_{w+2} - P_{w+1}}{\Delta y} + O(\Delta y) \tag{5-146}$$

This method has proved stable, even for separated flow calculations (Allen, 1968; Allen and Cheng, 1970; Roache and Mueller, 1968, 1970). Allen (1968) and Skoglund and Gay (1968) suggested evaluating the pressure gradient using one-sided differencing of v-momentum equation at the wall, although this method does not appear to be straightforward.

Allen (1968) used an improved method for evaluating the advective flux terms near a no-slip wall in the second mesh system. For the backstep geometry of Figure 3-22 on page 140, he found that negative wall densities sometimes developed along the upper part of the base (boundary B5). The tendency was aggravated by low Re and by a fine mesh.* Allen reasoned that it was due to the inaccurate evaluation, by linear interpolation, of the mass flux term at the first cell off the wall, which would be $(w + 1\frac{1}{2})$ in Figure 5-2b. The quantity $(\rho v)|_{wall} = 0$; also, in the steady state, the continuity equation shows that $\partial(\rho v)/\partial y|_{wall} = 0$. This suggests a quadratic, rather than linear, variation in (ρv) near

*Roache and Mueller (1970) found no such occurrence, perhaps because of the coarse mesh used.

the wall. So Allen fit a quadratic form to (ρv), using not three node point values, but two node points (w+1), (w+2) and the known interface value at the wall, (w+1/2). This gave

$$(\rho v)_{w+1\frac{1}{2}} = \frac{1}{3}\left[(\rho v)_{w+2} + 3(\rho v)_{w+1} - (\rho v)_{w+\frac{1}{2}}\right] \tag{5-147}$$

where $(\rho v)_{w+\frac{1}{2}} = 0$ for the no-slip wall. The same form is used for the other flux quantities (ρuv), (ρv^2), and $v(E_s + P)$. Conservation is maintained because the same interface flux term (5-147) is used for calculations of the cell at (w+1) and at (w+2). Although some extra programming is required, the method corrected the development of negative densities when applied along the base.

The principal attraction of the reflection method in the second mesh system would be the convenient handling of the viscous terms, but the result is unfortunately a degradation of accuracy. All the desired values of ρ, u, v and T at the wall location are automatically set by reflection, consistent with their evaluation by linear interpolation. Although apparently reasonable, this approach in fact reduces the accuracy of second derivative terms (viscous stress and heat conduction) to first order, and may possibly cause boundedness errors. (See the discussion of the wall vorticity evaluation in section III-C-2.) Consequently, the use of the second mesh system, though convenient, is inaccurate and therefore not generally recommended.

However, it should be noted that Allen (1968) and Allen and Cheng (1970) modified the treatment of viscous terms at the first node off the wall in this second mesh system. They evaluated terms like $\partial u/\partial y|_{wall}$, needed in the viscous terms at (w+1), by the following second-order extrapolation from interior points.

$$\left.\frac{\partial U}{\partial y}\right|_{wall} = \frac{-8U_{wall} + 9U_{w+1} - U_{w+2}}{3\Delta y} + O(\Delta y^2) \tag{5-148}$$

For U = u or v, we have $U_{wall} = 0$, and equation (5-148) simplifies. Allen's (1968) tests on Burger's equations actually indicate that a first-order method is slightly preferable at high Re. In the present author's opinion, the accuracy of this method for multi-dimensional problems has not yet been completely proven.

V-G-2-c. Staggered Mesh Evaluation of Density

It is clear that, for no-slip walls, the variables u, v, and T are most conveniently and accurately evaluated in the first mesh system, in which node points are on the wall. Density, however, is most conveniently and accurately evaluated in the second mesh system, in which node values are $\Delta y/2$ off the walls. The use of a hybrid or staggered mesh system then suggests itself.

The geometry of the staggered mesh system is shown in Figure 5-5a. The variables u, v, and T are defined at the mesh nodes marked "•" and ρ is defined at the nodes marked "×." (This is not the same as the PIC mesh of section V-E-3, in which ρ and energy are defined in the same mesh, and u and v in another mesh.)

The continuity equation for an interior x-node $(ij)_x$ can be written readily from a control volume point of view. For example, the flux into the $(ij)_x$ node cell across the left boundary is written as

$$\overline{\rho}\overline{u}\Delta y = \frac{1}{2}\left(\rho^x_{i-1,j} + \rho^x_{ij}\right)\cdot\frac{1}{2}\left(u_{i,j}+u_{i,j+1}\right)\Delta y \tag{5-149}$$

where the x-superscript emphasizes that the variable is defined in the x-mesh. The definition of $\overline{\rho}$ in equation (5-149) is consistent with a space-centered, second-order method. For

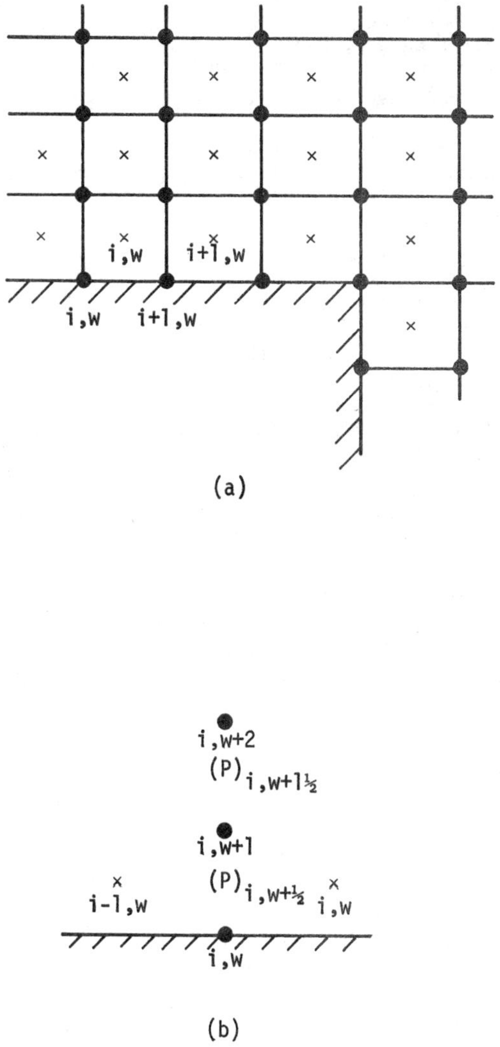

Figure 5-5. Staggered mesh system. u,v,T ate defined at • points, ρ is defined at × points.

the second upwind differencing approach, we would have

$$\overline{\rho u}\Delta y = \rho^x_{i-1,j}\,\frac{1}{2}\,(u_{ij} + u_{ij+1})\Delta y \tag{5-150}$$

The total continuity equation for upwind differencing, assuming $u > 0$ and $v > 0$, would be as follows. Superscript n is omitted for clarity.

$$\begin{aligned}\rho^{x,n+1}_{ij} = \rho^x_{ij} &- \left\{\frac{1}{2}\,(u_{i+1,j} + u_{i+1,j+1})\rho^x_{ij} - \frac{1}{2}\,(u_{ij} + u_{ij+1})\rho^x_{i-1,j}\right\}/\Delta x \\ &- \left\{\frac{1}{2}\,(v_{i+1,j+1} + v_{ij+1})\rho^x_{ij} - \frac{1}{2}\,(v_{i+1,j} + v_{ij})\rho^x_{ij-1}\right\}/\Delta y\end{aligned} \tag{5-151}$$

The same approach is easily followed to develop staggered-mesh continuity equations compatible with other finite-difference methods.

Exercise: Derive a staggered-mesh continuity equation compatible with Lax's method, section V-E-4.

When values of ρ in the ●-mesh are required, they are defined by averaging, as in

$$\rho_{ij} = \frac{1}{4}\left(\rho^x_{ij} + \rho^x_{i-1,j} + \rho^x_{i,j-1} + \rho^x_{i-1,j-1}\right) \tag{5-152}$$

Near the boundaries, the ρ^x evaluation is simple. At the $(i,w)_x$ node in Figure 5-5a, the left, right, and upper interfaces are the same as interior points, and the flux across the lower interface is zero because $1/2(u_{i,w} + u_{i+1,w}) = 0$. If an interior node at $(i,w-1)_x$ is defined and $\rho^x_{i,w-1}$ given an arbitrary finite value, regular interior-point differencing may be used* at $(i,w)_x$.

We have not yet settled the problem of evaluating ρ *at* the wall, in the ●-mesh. One could possibly use extrapolation from the interior x-mesh, but a simple and preferable approach is possible. The fact is that, because of the no-slip conditions, ρ *at* the wall is not required in any equation, except in the pressure-gradient term $\delta P/\delta y|_{i,w+1}$. The *easiest* approach is to use one-sided differencing for this term with resultant first-order accuracy. The *suggested* approach is to define this pressure gradient in terms of the calculated values, without ever needing a value of $(\rho_{i,w})$ in the ●-mesh. Instead of using

$$\left.\frac{\delta P}{\delta y}\right|_{i,w+1} = \frac{P_{i,w+2} - P_{i,w}}{2\Delta y} \tag{5-153}$$

we use a first difference form written across half-node spaces (see Figure 5-5b), which we define by a tilde overbar.

$$\left.\frac{\tilde{\delta} P}{\delta y}\right|_{i,w+1} = \frac{P_{i,w+1\frac{1}{2}} - P_{i,w+\frac{1}{2}}}{2\Delta y} \tag{5-154}$$

The upper value is set by

$$P_{i,w+1\frac{1}{2}} = \frac{1}{2}\left(P_{i,w+2} + P_{i,w+1}\right) \tag{5-155}$$

The lower value is set by interpolating ρ and T values from the two different mesh systems and using the normalized gas constant R_g.

$$P_{i,w+\frac{1}{2}} = R_g\, \frac{1}{2}\left(\rho^x_{i-1,w} + \rho^x_{i,w}\right) \cdot \frac{1}{2}\left(T_{i,w} + T_{i,w+1}\right) \tag{5-156}$$

Thus, the pressure gradient is readily calculated, and $\rho_{i,w}$ in the ●-mesh is not required.

This staggered-mesh method is quite effective, but does have some disadvantages (see section III-C-2). For example, inflow ρ values are defined on a line $\Delta x/2$ away from the inflow u, v, and T values.

We now present a method which is based on the staggered mesh idea concept but which is algebraically distinct from the method given above.

At the wall, the $\rho_{i,w}$ value in the ●-mesh may be wanted for plotting purposes. Any extrapolation would suffice, but a more meaningful approach is to set $\rho_{i,w}$ by the evaluation

---

*The use of a method similar to Lax's may be accommodated by setting $\rho_{i,w-1} = \rho_{iw}$.

of equation (5-154). However, once this approach is adopted, it is possible to dismiss with the x-mesh. Regular interior-point differencing is used for ρ in the ●-mesh. But near the boundaries, values of ρ in a *local* x-mesh are defined.

We define

$$\rho^{x,n}_{i-1,w} = \frac{1}{4}\left(\rho^n_{i,w} + \rho^n_{i,w+1} + \rho^n_{i-1,w} + \rho^n_{i-1,w+1}\right)$$

$$\rho^{x,n}_{i,w} = \frac{1}{4}\left(\rho^n_{i+1,w} + \rho^n_{i+1,w+1} + \rho^n_{i,w} + \rho^n_{i,w+1}\right)$$

$$\rho^{x,n}_{i+1,w} = \frac{1}{4}\left(\rho^n_{i+2,w} + \rho^n_{i+2,w+1} + \rho^n_{i+1,w} + \rho^n_{i+1,w+1}\right) \tag{5-157}$$

$$\rho^{x,n}_{i,w+1} = \frac{1}{4}\left(\rho^n_{i+1,w+1} + \rho^n_{i+1,w+2} + \rho^n_{i,w+1} + \rho^n_{i,w+2}\right)$$

Then an FDE continuity equation, such as (5-151), is used to calculate $\rho^{x,n+1}_{i,w}$ and other values adjacent to the wall. After new temperatures have been calculated from the energy equation, we evaluate $\tilde{\delta}P^{n+1}/\delta y|_{i,w+1}$ as in equation (5-154) and then set ρ^{n+1}_{iw} in the ●-mesh such that the pressure gradient over the half-node spaces equals that over the full node spaces. From (5-153) and (5-154),

$$\tilde{\delta}P/\delta y|_{i,w+1} = \delta P/\delta y|_{i,w+1} \tag{5-158}$$

This gives for the new (n+1) values,

$$P_{iw} = -P_{i,w+1} + R_g\left(\rho^x_{i-1,w} + \rho^x_{iw}\right)\left(T_{i,w+1} + T_{i,w}\right)/2 \tag{5-159}$$

and finally

$$\rho_{iw} = P_{iw}/R_g T_{ij} \tag{5-160}$$

To summarize: Old values of ρ^x are defined near the wall from ρ-values in the ●-mesh by equations like (5-157). Then new values of ρ^x are calculated by way of a staggered-mesh continuity equation such as (5-151) or by some method compatible with the interior-point method being used. Finally, the new pressure and density values in the ●-mesh are calculated by way of equations (5-159) and (5-160).

Although this method is certainly more cumbersome than some simple prescription like $P_w = P_{w+1}$, it is second-order accurate (Problem 5-13) and dependable, and is recommended.

V-G-3. Sharp Corners

As in the (ψ,ζ) incompressible flow, sharp corners require that some variables be double-valued.

At a corner in the first mesh system, shown in Figure 5-6a, the velocities $u_{ic,jc} = 0$ and $v_{ic,jc} = 0$ for a no-slip wall. A specified corner temperature may or may not be single-valued. If the wall temperature, T_w, along B2 is the same as the wall temperature along

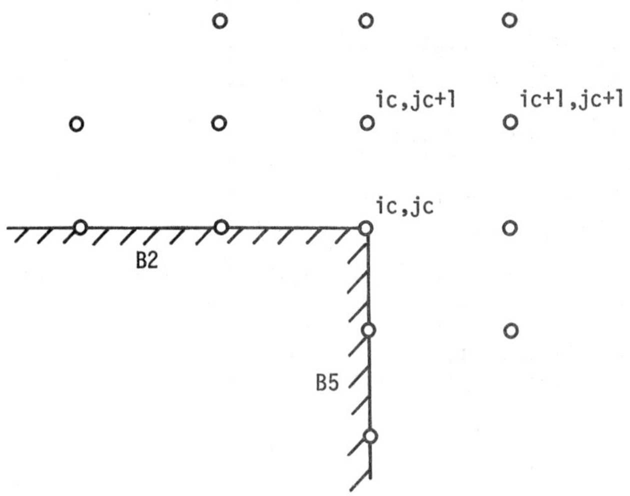

(a) first mesh system

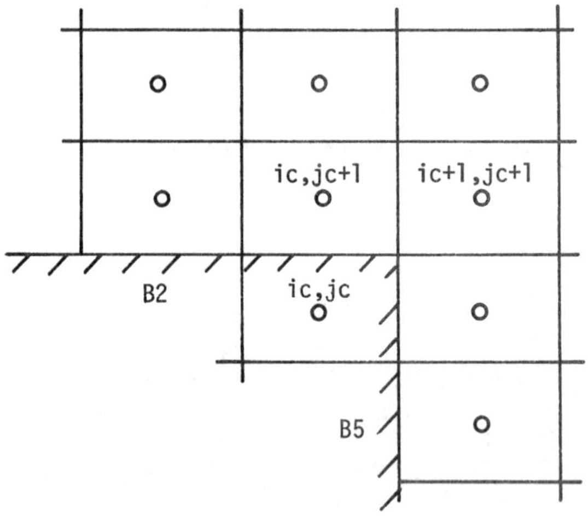

(b) second mesh system

Figure 5-6. Sharp corner cell treatment.

B5, then obviously $T_{ic,jc} = T_w$. But if B2 and B5 are held at different temperatures, then $T_{ic,jc}$ should accordingly be treated as multi-valued, with $T_{ic,jc} = T_A \equiv T(B2)$ when needed for the interior-point calculations at (ic,jc+1), and $T_{ic,jc} = T_B \equiv T(B5)$ when needed at (ic+1,jc). The use of 9-point equations for the heat-conduction term of the energy equation or the viscous cross-derivative term with temperature-dependent viscosity, will require a third value, $T_C = 1/2(T_A + T_B)$, for calculations at (ic+1,jc+1). The need for multivalued $T_{ic,jc}$ is also clear when adiabatic walls are considered. The wall B2 will not be computationally adiabatic unless $T_{ic,jc} = T_{ic,jc+1}$, and B5 will not be adiabatic unless $T_{ic,jc} = T_{ic+1,jc}$. Obviously, a single-valued $T_{ic,jc}$ is not sufficient. Likewise, double-valued $\rho_{ic,jc}$ are needed to give the proper pressure gradients, $\delta P/\delta y|_{ic,jc+1}$ and $\delta P/\delta x|_{ic+1,jc}$. A little programming effort is required.

For a slip wall, similar procedures may be followed. Explicit artificial viscosity terms may present a problem, since we want no-slip conditions for those terms. Kessler (1968) dropped the Rusanov terms at the wall (using the first mesh system), but this may not always give a stable calculation. Note that the true *bulk* viscosity terms, physical or artificial, do not require no-slip conditions, since they operate only on dilatational terms.

In the second mesh system, appropriate for inviscid flow, the sharp corner appears as shown in Figure 5-6b.

When new values at (ic+1,jc) and (ic,jc+1) are being calculated, the fluxes of (ρu), (ρv) and ($E_s + P$) across (ic+1/2,jc) and (ic,jc+1/2) are explicitly set to zero, as for other wall points. If reflection is used to obtain a zero flux of mass across these cell interfaces, it is necessary that a fictitious inner value at (ic,jc) be double-valued. (Since these are purely fictitious values set for programming convenience, the double values have no physical significance.) Similar to the double-valued treatment of corner vorticity in section III-C-12-a, we define

$$\rho_A = \rho(ic,jc+1) \tag{5-161A}$$

$$\rho_B = \rho(ic+1,jc) \tag{5-161B}$$

$$v_A = -v(ic,jc+1) \tag{5-161C}$$

$$u_B = -u(ic+1,jc) \tag{5-161D}$$

Then, if new values are being calculated at (ic,jc+1), we momentarily set

$$\rho_{ic,jc} = \rho_A \quad \text{and} \quad v_{ic,jc} = v_A \tag{5-162}$$

and use regular interior-point differencing at (ic,jc+1). Similarly, for new values at (ic+1,jc), we momentarily set $\rho_{ic,jc} = \rho_B$ and $u_{ic,jc} = u_B$.

This treatment is all that is required for the inviscid equations. This mesh system is not recommended for the viscous equations, as argued earlier. But if it is used, it requires the same double-valued treatment for the evaluation of diffusion terms like $\partial/\partial x[f(\partial g/\partial x)]$ at (ic,jc+1) and (ic+1,jc). When $f = \mu(T)$, then one of the double-valued T(ic,jc) is used, as above. Pressure gradients may still be evaluated by one-sided differencing.

When new values at (ic+1,jc+1) are being evaluated, a third value of f = u or v is needed at (ic,jc) for the cross-derivative terms. In this case, a value $f_{ic,jc}$ may be assigned in such a way as to give the physical value at (ic+1/2,jc+1/2) according to the following interpretation:

$$f_{ic+\frac{1}{2},jc+\frac{1}{2}} = \frac{1}{4}(f_{ic,jc} + f_{ic+1,jc} + f_{ic,jc+1} + f_{ic+1,jc+1}) \qquad (5\text{-}163)$$

From this equation, $f_{ic,jc}$ is solved as

$$f_{ic,jc} = 4f_{ic+\frac{1}{2},jc+\frac{1}{2}} - f_{ic+1,jc} - f_{ic,jc+1} - f_{ic+1,jc+1} \qquad (5\text{-}164)$$

For no-slip walls,* $u_{ic+\frac{1}{2},jc+\frac{1}{2}} = 0$, of course. We thus define the third values of

$$u_c = -u_{ic+1,jc} - u_{ic,jc+1} - u_{ic+1,jc+1} \qquad (5\text{-}165A)$$

$$v_c = -v_{ic+1,jc} - v_{ic,jc+1} - v_{ic+1,jc+1} \qquad (5\text{-}165B)$$

and use $u_{ic,jc} = u_c$, $v_{ic,jc} = v_c$ when new values are being calculated at (ic+1,jc+1). This evaluation is consistent with the interpretation of wall values according to linear interpolation; but, as argued earlier, it reduces the accuracy of the viscous terms to first order.

As discussed in section III-C-12 for incompressible flow, high accuracy resolution near the sharp corner would be met by the local use of polar coordinates centered on the sharp corner.

V-G-4. Symmetry Surfaces

Across symmetry surfaces, such as B1 in Figure 3-22 on page 140, symmetry conditions obviously prevail. If a line of node points j = s are on the line of symmetry, the reflection method may be used to give

$$\begin{aligned} v_{s-1} &= -v_{s+1} \\ v_s &= 0 \qquad (5\text{-}166) \\ f_{s-1} &= +f_{s+1} \quad \text{where} \quad f = \rho,\ u,\ E_s,\ \text{or } T \end{aligned}$$

If the nodes are $\Delta y/2$ off the symmetry line at s + 1/2, the symmetry condition gives

$$\begin{aligned} v_s &= -v_{s+1} \\ f_s &= +f_{s+1} \quad \text{where} \quad f = \rho,\ u,\ E_s,\ \text{or } T \end{aligned} \qquad (5\text{-}167)$$

Although both methods are unequivocally correct, it is interesting to note (Allen, 1968; Roache and Mueller, 1968) that the second method does give a residual truncation error in the v-flux of v-momentum across the symmetry line. This term is only $O(\Delta y^3)$ because $\partial v/\partial y = 0$ by symmetry, and equation (5-145) applies. It is not clearly preferable to specifically set the flux term equal to zero in this case.

*If a slip wall were being used with the viscous equations upstream of the corner, with $u_w = u_{w+1}$, it would seem plausible to use a single value of $u_{ic,jc}$ based on interpretation as part of the upstream wall, with $u_{ic,jc} = u_{ic,jc+1}$.

V-G-5. Upstream Boundary

The upstream boundary is shown as B4 in Figure 3-22 (page 140). All published 2D compressible flow solutions have completely specified the inflow properties, either as "freestream" conditions or, more appropriate to the backstep model of Figure 3-22, from a boundary-layer profile. Experience indicates (Skoglund et al., 1967; Allen, 1968; Allen and Cheng, 1970; Roache and Mueller, 1968; Skoglund and Gay, 1969) that the downstream flow can be quite sensitive to the inflow conditions and to the compatibility of the flow variables. Specifically, the v-component of velocity should be evaluated from the u-solution (although, at very high Re, the condition $v = 0$ might suffice) and the energy integrals should be compatible. See Schlichting (1968), Stewartson (1964), and Cohen and Reshotko (1956) for compressible boundary-layer solutions. See Skoglund and Gay (1969) for a method of determining compatible inflow variables from an experimental determination of inflow velocity, and Allen (1968) for the case when u(y) is given by a polynomial.

All these methods are based on similarity solutions, using either the Crocco or Busemann integrals. The limitation to $Pr = 1$ is common in similarity solutions, although the deviation of the solution from that for $Pr \simeq 0.74$ (air) is frequently considered insignificant.

An unusual evaluation of upstream properties was used by Anderson (1969), who used Lax-Wendroff-type time differencing on the quasi-1D equations (area ratio) including vibrational and chemical non-equilibrium. The value $i = 1$ was in the "reservoir" (see any gas dynamics text) with a large value of A/A (throat) ≈ 10. Then T_1 and ρ_1 were fixed at the reservoir values. But u was allowed to develop as part of the solution, by backward linear extrapolation.

$$u_1 = 2u_2 - u_3 \tag{5-168}$$

The choking condition at the throat determined the solution, and u_1 developed to a steady-state value compatible with the mass flux at the throat.

There is a fundamental incompatibility involved in the wall heat-flux condition for a specified inflow condition. The continuum adiabatic condition is $\partial T/\partial n = 0$, but when the continuum solution is used at discrete node points, it does not give the computationally adiabatic condition of $T_w = T_{w+1}$. Of course, it is approached as $\Delta y \to 0$, but there is a considerable difference for practical mesh sizes. One is left with the choice of setting the wall temperature equal to the continuum solution for an adiabatic wall, which gives a computationally non-adiabatic wall, or of using a computationally adiabatic wall temperature which is not equal to the continuum solution. The second choice seems preferable, more in keeping with the spirit of the conservation equations.

V-G-6. Downstream Boundary

There is a variety of acceptable methods for treating the downstream boundary, B6 in Figure 3-22 on page 140, for compressible flow. The plausible methods to try include all those incompressible flow methods for vorticity given in section III-C-7. These include linear extrapolation of the conservation variables,

$$U_I = 2U_{I-1} - U_{I-2} \qquad \text{where} \qquad U = \rho,\ \rho u,\ \rho v,\ (E_s + P) \tag{5-169}$$

or the non-conservative variables,

$$f_I = 2f_{I-1} - f_{I-2} \qquad \text{where} \qquad f = \rho,\ u,\ v,\ T \tag{5-170}$$

These two methods are not equivalent.

Other downstream methods will be given shortly; the important *general* point is that we cannot be entirely casual about the outflow treatment. As stated in section III-C-9 on the downstream paradox, a condition of supersonic inflow with quasi-1D equations makes the outflow boundary condition *more* important than with subsonic inflow condition. The outflow condition determines the shock location, as in the computational experiments of Crocco (1965) and in 2D calculations made by the present author.*

Even more fundamental, consider the often repeated statement that outflow boundary conditions are not important as long as the outflow is supersonic. This statement is incorrect. As we argued in section III-C-9, if the boundaries had no effect at all, it would be impossible to "turn off" an indraft supersonic wind tunnel. But the statement is not true *even* in the sense of smallness of errors. Allen (1968) found that use of the simple conditions

$$U_I = U_{I=1} \tag{5-171A}$$

or equivalently

$$f_I = f_{I-1} \tag{5-171B}$$

in the (viscous flow) backstep problem caused the solution to diverge monotonically, even with completely supersonic outflow. Conversely, Ruo (1967) found that this method did give converged results for a subsonic, compressible flow problem, and even said that it accelerated convergence compared with other methods. Also, Eaton and Zumwalt (1967) and Kessler (1968) found that the method gave good results even when a shock passed the boundary. The last three studies used Rusanov's method (section V-D-3). The reasons for these differing experiences are not known. Eaton and Zumwalt (1967) found that linear extrapolation as in equation (5-170) was unstable as a blast wave exited the mesh, even though the prior extablished flow was supersonic. It is also clear that linear extrapolation at the outflow boundary (B6) could introduce serious error, if not instability, for the case where a trailing shock wave intersected B6.

Barring the case of a shock wave exiting at B6, the linear extrapolation method (5-170) is generally acceptable, as in the computations of Saunders (1966), Lapidus (1967), Eaton and Zumwalt (1967), Skoglund and Cole (1967), Ruo (1967), Allen (1968), Allen and Cheng (1970), Roache and Mueller (1968), and Skoglund and Gay (1968).

Allen (1968) also experimented with quadratic and cubic extrapolations of the forms

$$f_I = 3f_{I-1} - 3f_{I-2} + f_{I-3} \tag{5-172}$$

$$f_I = 4f_{I-1} - 6f_{I-2} + 4f_{I-3} - f_{I-4} \tag{5-173}$$

He used equation (5-172) in final calculations (Allen and Cheng, 1970), but found that the differences between methods (5-172), (5-173) and (5-170) were insignificant. Allen (1968) also varied the position of the outflow boundary and found no significant difference between two solutions so obtained, one with partially subsonic outflow (near B1 in Figure 3-22 on

---

*Linear extrapolation (5-170) is generally acceptable for the backstep problem. But when used on a very coarse mesh with the downstream boundary four base heights behind the base, and with a no-slip splitter plate, the solution was destroyed. Although the mesh boundary was still outside the developing recompression region, the downstream error caused a raggedly defined strong shock wave to propagate forward from the downstream boundary, destroying the closed wake. Oscillations in the shock persisted, but its position generally settled down to the base corner; the wake became "open" with $u < 0$ all the way to the downstream boundary. This behavior is in agreement with the experimental phenomenon known as *diffuser stall*, occurring when the back pressure is raised. It also points out again the non-uniqueness of some computational problems.

page 140) and the other with entirely supersonic outflow, thus settling the question of whether supersonic outflow is necessary. (See also the outflow boundary tests of Ross and Cheng, 1971.) It therefore appears that supersonic outflow is *neither* necessary nor sufficient for either stability or accuracy.

Ruo (1967) also experimented with the average of equations (5-171B) and (5-170), and with a time shift method* given by

$$f_I^{n+1} = f_{I-1}^n \tag{5-174}$$

but found equation (5-171B) to be preferable, as stated earlier.

Erdos and Zakkay (1969) and Sinha et al., (1970) used a quadratic extrapolation, where the polynomial was fit to data over the last *five* grid points with the redundant data being used in a least-squares fit. This method appears to be no more accurate than simpler extrapolations.** It was unstable when shocks crossed the boundary.

In an inviscid flow problem using Rusanov's method (section III-D-3), Eaton and Zumwalt (1967) reverted to upwind differencing for first derivatives in the x-direction and evaluated the second derivatives of the explicit artificial diffusion terms at one point inside the boundary, as in

$$\frac{\partial}{\partial x}\left(\alpha_x \frac{\partial U}{\partial x}\right)\bigg|_I = \frac{\delta}{\delta x}\left(\alpha_x \frac{\delta U}{\delta x}\right)\bigg|_{I-1} \tag{5-175}$$

The y-derivatives at I are evaluated as at regular interior points. Like equation (5-171B), this method was even successful as the shock exited the outflow boundary.

Roache and Mueller (1968) used a similar method, patterned after their incompressible flow method (section III-C-7). The u-advection terms are evaluated by upwind differencing, presuming that the boundary is truly an outflow boundary. The x-diffusion terms, the cross-derivative terms, and the x-pressure gradient are evaluated at (I-1); by itself, this procedure produces a destabilizing tendency which is further counteracted by a time shift. Consistent with the incompressible flow method, the y-derivative terms may be evaluated by the regular interior-point method. As an example, the u-momentum equation (4-42B) would be differenced as follows:

$$(\rho u)_{I,j}^{n+1} = (\rho u)_{I,j}^n - \Delta t\left\{\frac{\delta P}{\delta x}\bigg|_{I-1,j}^{n-1} + \frac{(\rho u^2)_I^n - (\rho u^2)_{I-1}^n}{\Delta x} + \frac{\Delta(\rho uv)}{\Delta y}\bigg|_{I,j}\right\}$$

$$+ \frac{\Delta t}{Re}\left\{D_1\bigg|_{I-1,j}^{n-1} + \frac{\delta}{\delta y}\left(\mu\frac{\delta u}{\delta y}\right)\bigg|_I^n + \frac{\delta}{\delta y}\left(\mu\frac{\delta v}{\delta x}\right)\bigg|_{I-1}^{n-1}\right\} \tag{5-176}$$

where $\delta/\delta x$, $\delta/\delta y$ indicate centered differences, and $\frac{\Delta(\rho uv)}{\Delta y}$ indicates the y-flux contribution to $\partial(\rho u)/\partial t$ evaluated by the interior-point method being used. (This method has only been applied with upwind differencing at interior points.) In the work cited, all the viscous terms, including $\delta/\delta y[\mu(\delta u/\delta y)]$, were evaluated at (I-1) and (n-1), to avoid programming chores.

This method would appear to be accurate, but Allen's (1968) numerical tests indicate insensititivity between the various stable extrapolation methods. Accordingly, the much simpler method (5-170) is recommended for compressible high-Re flow, assuming that no strong shock crosses this outflow boundary B6.

---

*For the model equation (5-1), this method (5-174) is of course the exact solution, for a Courant number c = 1.

**As Cheng (1970) has pointed out, all extrapolations are based on computed, not exact, values, so their accuracy is limited.

If a large v-velocity existed at outflow, the accuracy might be improved by use of the simple wave approximation (next section) in the supersonic region, but this has not yet been accomplished.

V-G-7. Upper Boundary

As an incompressible flow, the upper boundary, B3 in Figure 3-22 on page 140 could be treated as a simple no-slip wall, or better, as a frictionless wind tunnel wall. However, we must expect gross effects such as wave reflections off the boundary (e.g., Wilkins, 1969). A much better method of representing the "free flight" case is available for supersonic flow. This method actually allows inflow across the lid, and is physically meaningful.

Near the lid, we make the assumption of a *simple wave* solution. We assume that characteristics of only one family are present and that these are approximately straight over Δy. Physically, we are approximating the lid inflow as inviscid, steady, homentropic flow resulting from the expansion of uniform flow over some curved surface, i.e., Prandtl-Meyer flow (e.g., Owczarek, 1964).

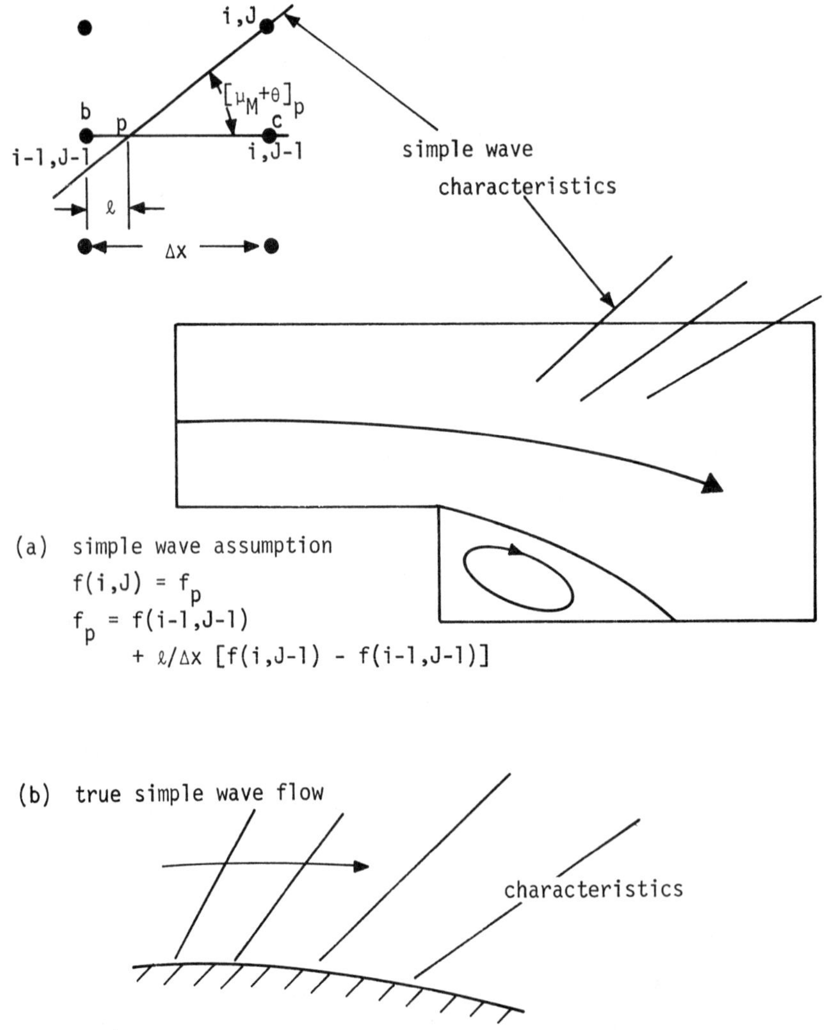

Figure 5-7. Upper boundary condition by the simple wave approximation.

The geometry of the method is shown in Figure 5-7. The point p and the lid mesh point (i,J) lie on the same "characteristic", or Mach line. This line is defined by the angle $(\mu_M + \theta)_P$ from the x-direction, where $\mu_M = \arcsin(1/M)$ is the Mach angle* and $\theta = \arctan(v/u)$

*Obviously, μ_M does not exist for $M < 1$.

is the flow direction angle. μ_M and θ are evaluated at p by linear interpolation between the values at (i-1,J-1) and (i,J-1), as are all the flow properties $f_p = \rho$, u, v, T at p. The simple wave solution states that flow properties are constant along the characteristic. Thus, once p is located, the upper boundary condition becomes

$$f_{i,J} = f_p = f_b + \frac{\ell}{\Delta x}(f_c - f_b) \qquad \text{where} \qquad f = \rho,\ u,\ v,\ \text{and}\ T \tag{5-177}$$

The location of p may be found as follows. The quantity $\omega = \tan(\mu_M + \theta)$ is evaluated at $b \equiv (i-1,J-1)$. If $\omega_b > \Delta y/\Delta x$, then p is as shown in Figure 5-7. The distance ℓ is then solved geometrically by evaluating $\omega' = \tan[90° - (\mu_M + \theta)]$ at b and c, solving

$$\ell = \frac{\dfrac{\Delta x}{\Delta y} - \omega_b'}{\dfrac{\omega_c' - \omega_b'}{\Delta x} + \dfrac{1}{\Delta y}} \tag{5-178A}$$

or if $\omega_b < \Delta y/\Delta x$, the point p lies a distance ℓ up along y from (i-1,J-1), and ℓ is solved from

$$\ell = \frac{\dfrac{\Delta y}{\Delta x} - \omega_b}{\dfrac{\omega_a - \omega_b}{\Delta y} + \dfrac{1}{\Delta x}} \tag{5-178B}$$

The ω_a is determined from a previously defined lid point, with the sweep proceeding from left to right. The *first* lid point, at (1,J), is defined with the inflow.

In the case of a shock wave or more gradual compression region crossing B3, two such Mach lines may intersect (i,J). In that case, one may prefer to use shock-patching methods, but the simpler method just described will still give a unique answer.

This method was used by Allen (1968), Allen and Cheng (1970), Roache and Mueller (1968), and Goodrich (1969). It produced stable and realistic results, with three different interior-point methods. However, it is essential that the upper boundary, B3, be outside of any boundary-layer region, or else some absurd answers will result.

For a blunt-body calculation, Lapidus (1967) used a method of linear extrapolation along the diagonals of the mesh, setting

$$f_{i,J} = 2f_{i-1,J-1} - f_{i-2,J-2} \tag{5-179}$$

which is similar to the simple wave method if the Mach lines are along the mesh diagonal. In the early stages of calculation, this method was destabilizing and was replaced by

$$f_{i,J} = f_{i-1,J-1} + r(f_{i-1,J-1} - f_{i-2,J-2}) \tag{5-180}$$

Lapidus allowed r to build up slowly, from r = 0.5 to r = 1 (which gives equation 5-179) over the first 500 time steps, to achieve stability. Eaton and Zumwalt (1967) found that quadratic extrapolation like (5-172) was unstable when a shock intersected B3, and they successfully used their outflow boundary condition (previous section, V-G-6) at B3. Likewise, Erdos and Zakkay (1969) used their downstream boundary condition at B3, provided that no shocks were present there.

V-H. *Convergence Criteria and Initial Conditions*

Many of the remarks of section III-D apply as well to compressible flow, especially the point that no satisfactory objective criterion exists for iteration convergence or for truncation convergence. The iteration convergence (to a steady state) for compressible flow is further complicated by the existence of more variables (e.g., pressure typically converges slower than density) and by the presence of another time constant, the time required for a pressure wave to traverse the mesh. Ross and Cheng (1970) indicate that a long-term transient decay may be expected in supersonic viscous flow, making it more difficult to judge convergence. In this regard, the checking of a final steady-state solution by repeating the calculation (at least for a representative test case) with widely different initial conditions would seem advisable.

Unfortunately, the initial condition problem seems to be more critical in supersonic flow than in incompressible flow. There are many examples in the literature where instabilities occurred for some initial conditions, but did not occur for better initial conditions. These are, of course, "nonlinear" instabilities by definition, although the source of the nonlinear instability appears in some cases to be due to maltreatment of boundary conditions, as suggested by Moretti (1968A,B). It is still true, however, that these instabilities are at least aggravated, and perhaps completely caused, by spurious shock wave propagations from poor initial conditions.

If only the steady-state solution is of interest, three methods are available for reducing the effect of initial conditions. (1) The problem may be started with low Δt. This method is often costly in computer time. (2) The problem may be started with an artificially low Reynolds number, and gradually increased to the desired value as the solution proceeds slowly. (3) The problem may be started with a different method which possesses more artificial damping. This is easily accomplished with the explicit artificial viscosity methods. Likewise, the Richtmyer two-step Lax-Wendroff method (section V-E-6), which uses a Lax method for the first step and a leapfrog method for the second, may be run for several hundred time steps using only the Lax method. If $C \ll 1$, this method is highly diffusing and will dependably give a smooth solution. Then the full two-step method may be introduced (Lapidus, 1967).

Since nonlinear instabilities may destroy solutions while one is experimenting, it is prudent to establish early solutions by short computer runs, storing the results of alternate runs on two magnetic tapes. If an instability ruins one set of data, one can pick up from the previous run with lower Δt, higher damping, etc.

Tyler and Zumwalt (1965) and Tyler and Ellis (1970) have shown that smoother shock profiles may be obtained in one-dimensional shock propagation problems if the initial discontinuity is spread over two cells. Instead of a jump from post-shock to pre-shock values a and b over one cell, as in

$$P_i = P_a \quad , \quad P_{i+1} = P_b \tag{5-181}$$

they defined an intermediate node value by the mean pressure

$$P_i = P_a \quad , \quad P_{i+1} = \frac{1}{2}(P_a + P_b) \quad , \quad P_{i+2} = P_b \tag{5-182}$$

(See the references cited for other details.) Watkins (1970) found that compatible initial conditions were especially important because of the transformed coordinates which he used. It has also been the author's experience, and that of L. D. Tyler (private communication), that the common two-step methods leave a persistent oscillation over three nodes at the initial shock position. This phenomenon has not yet been satisfactorily explained or cured.

Extrapolation methods to aid truncation convergence were applied to hyperbolic systems by Gourlay and Morris (1968C), Werner (1968), and Smith and McCall (1970).

V-I. Remarks on Subsonic and Supersonic Solutions

The thought commonly occurs that it is not worthwhile to use the incompressible flow equations, because one need only write a compressible flow program; the desired incompressible solution can then be obtained by using low M_o, say $M_o = 0.1$, in the compressible flow program. The compressible-flow program then appears more flexible.

Generally, this method is both highly inefficient and highly inaccurate.

The efficiency comes about from the obvious complication of viscous terms (see Chapter IV) and from Δt-limitations. The computational Δt of the compressible flow solution is limited by a condition on the Courant number like

$$C = \frac{(|u| + a)\Delta t}{\Delta x} < 1 \tag{5-183}$$

If $M_o \simeq 0.1$, we have $a \simeq 10|u|$. The Δt is then limited primarily by sound wave propagation in equation (5-183), reducing the allowable Δt by $\simeq 1/10$ compared to an incompressible formulation. Further, these unwanted sound waves in the mesh will increase aliasing error (section III-A-13) and, perhaps most important, will obscure and confuse the iteration convergence. It has been known since the paper of Charney et al., (1950) that it is preferable to effectively filter out these unwanted waves by using the incompressible flow equations.

The inaccuracy comes from a tendency toward indeterminacy as $M \to 0$. Even a superficial acquaintance with fluid dynamics shows that, in supersonic gas dynamics relations, one solves for pressure *ratios* as in a base pressure ratio $P_r = \bar{P}_b/\bar{P}_\infty$; whereas, in incompressible flow, one solves for pressure *differences** as in the base pressure coefficient $C_{PB} = (\bar{P}_B - \bar{P}_\infty)/\bar{q}_\infty$, where $\bar{q}_\infty$ is the freestream dynamic pressure. This $\bar{q}_\infty$ may be evaluated as

$$\bar{q}_\infty = \frac{1}{2}\bar{\rho}_\infty\bar{V}_\infty^2 = \frac{\gamma}{2}\bar{P}_\infty M_\infty^2 \tag{5-184}$$

Using this equation to relate the pressure difference to the pressure ratio, we obtain

$$C_{PB} = \frac{P_r - 1}{\frac{\gamma}{2}M_\infty^2} \tag{5-185}$$

Differentiating,

$$r \equiv \frac{d(C_{PB})}{d(P_r)} = \frac{2}{\gamma M_\infty^2} \tag{5-186}$$

For $\gamma = 1.4$ and $M_\infty = 0.1$, this gives a relative sensitivity ratio $r = 0.007$. That is, an error of less than 1% in the P_r solution using compressible flow equations can give a 100% error in C_{PB}.

For mixed flows, i.e., subsonic regions within supersonic flows, the pressure ratios are still the significant quantities, and the compressible flow equations are appropriate. But problems in which the large M range is traversed in time are very difficult. Examples are the acceleration of a body from rest to a supersonic speed, and late-time calculations of explosions.

Harlow and Amsden (1968) have developed the ICE method (Implicit-Continuous fluid-Eulerian) method which does allow accurate calculations from $\bar{M} = 0$ to $\bar{M} \gg 1$. The method

*Note that as $M \to 0$, the energy equation decouples from the momenta and continuity equations.

depends on an implicit formulation of the continuity equation, which introduces the required elliptic behavior into the equations (see Fromm, 1963, and Ruo, 1967). The method avoids the time-step limitation of the speed of sound. Further improvements and applications of ICE are given by Harlow and Amsden (1970) and Harlow et al., (1971).

V-J. Higher Order Systems

The remarks of section III-A-10 on the limitations of systems with higher order of truncation error apply even more strongly to supersonic flows. As mentioned in section V-E-2, spatial variables are not necessarily continuous in high-Re supersonic flows; in such cases, the Taylor series expansion used in the evaluation of truncation error is not valid.

Burstein and Mirin (1970, 1971) developed a splitting method for the inviscid equations which is third-order in space and time. The third-order method applied to a blunt-body detached shock problem gave improved accuracy for stagnation pressure, but worse accuracy for stagnation density, and was a factor of three slower than the second-order method. For another paper on a third-order method, see Rusanov (1971).

CHAPTER VI

OTHER MESH SYSTEMS, COORDINATE SYSTEMS, AND EQUATION SYSTEMS

Up to this point, we have presented the basic concepts and methods of computational fluid dynamics on the simplest possible problems, using some form of the Navier-Stokes equations, expressed in rectangular coordinates, and differenced in a regular mesh system with constant Δx and Δy. In this chapter, we shall treat very briefly some aspects of other coordinate and mesh systems, and flow equations other than the Navier-Stokes equations. We will not pursue these topics to any depth, due to the lack of space and time, and often of inclination and knowledge. All we will attempt to do here is to highlight some concepts and provide the reader with additional references. In order to make any sense out of the few comments offered, the reader should already be familiar with the topics.

Undoubtedly, the most important point of the chapter is that these considerations of coordinates, mesh, and flow equations *are* very important. It is not an entirely trivial extension to go from rectangular to spherical coordinates, for example. As in an analytical attack on a problem, the judicious choice of a coordinate system and of simplifications in the equations can make the difference between success and failure.

VI-A. *Special Mesh Systems*

The simplest variation of the rectangular mesh system is obtained by simply changing the mesh spacing in one direction at some point. This would typically be done for the purpose of obtaining higher resolution (and hopefully higher accuracy) in some region where the flow gradients were expected to change rapidly, e.g., in a boundary layer.

To illustrate, we consider the obvious method of changing from Δx_1 to Δx_2 between node points at some node i = m, as shown in Figure 6-1a.

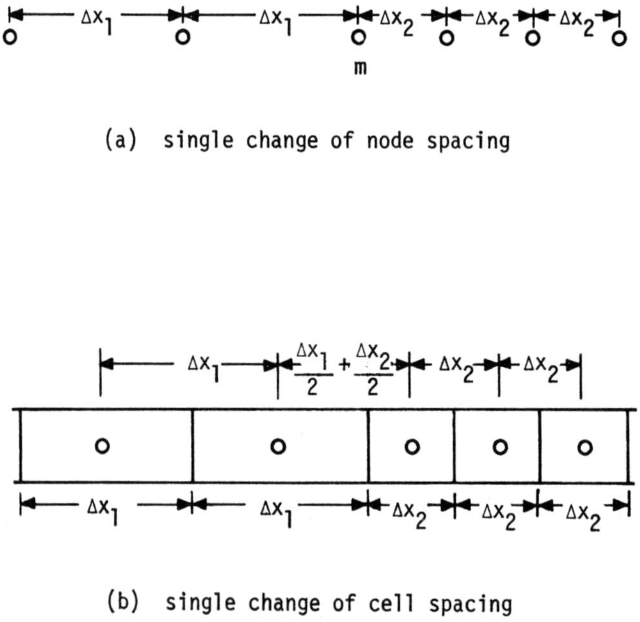

Figure 6-1. Change in mesh spacing.

Expanding a function f in a Taylor series forward and backward from i = m gives

$$f_{m+1} = f_m + \left.\frac{\partial f}{\partial x}\right|_m \Delta x_2 + \frac{1}{2}\left.\frac{\partial^2 f}{\partial x^2}\right|_m \Delta x_2^2 + \frac{1}{6}\left.\frac{\partial^3 f}{\partial x^3}\right|_m \Delta x_2^3 + O(\Delta x_2^4) \qquad (6\text{-}1)$$

$$f_{m-1} = f_m - \left.\frac{\partial f}{\partial x}\right|_m \Delta x_1 + \frac{1}{2}\left.\frac{\partial^2 f}{\partial x^2}\right|_m \Delta x_1^2 - \frac{1}{6}\left.\frac{\partial^3 f}{\partial x^3}\right|_m \Delta x_1^3 + O(\Delta x_1^4) \qquad (6\text{-}2)$$

The expression for $\left.\frac{\partial f}{\partial x}\right|_m$ is obtained by subtracting equation (6-2) from (6-1).

$$f_{m+1} - f_{m-1} = \left.\frac{\partial f}{\partial x}\right|_m (\Delta x_2 + \Delta x_1) + \frac{1}{2}\left.\frac{\partial^2 f}{\partial x^2}\right|_m (\Delta x_2^2 - \Delta x_1^2) + O(\Delta x^3) \qquad (6\text{-}3)$$

where by $O(\Delta x^3)$ we mean the largest of $O(\Delta x_1^3)$ or $O(\Delta x_2^3)$. Solving for $\left.\frac{\partial f}{\partial x}\right|_m$ gives

$$\left.\frac{\partial f}{\partial x}\right|_m = \frac{f_{m+1} - f_{m-1}}{\Delta x_2 + \Delta x_1} - \frac{1}{2}\left.\frac{\partial^2 f}{\partial x^2}\right|_m \frac{\Delta x_2^2 - \Delta x_1^2}{\Delta x_2 + \Delta x_1} + O(\Delta x^2) \tag{6-4}$$

This means that the form

$$\left.\frac{\delta f}{\delta x}\right|_m = \frac{f_{m+1} - f_{m-1}}{\Delta x_2 + \Delta x_1} \tag{6-5a}$$

is second-order accurate only if

$$O\left(\frac{\Delta x_2^2 - \Delta x_1^2}{\Delta x_2 + \Delta x_1}\right) \le O\left(\Delta x_1^2\right) \tag{6-5b}$$

Note that, for Δx_2 very *small,* the accuracy at m deteriorates to first order in Δx_1 (e.g., Blottner and Roache, 1971).

The expression for the second derivative is obtained by multiplying equation (6-2) by $s^2 = (\Delta x_2/\Delta x_1)^2$ and adding the result to (6-1).

$$f_{m+1} + (1 + s^2)f_m + s^2 f_{m-1} = \left.\frac{\partial f}{\partial x}\right|_m \Delta x_2(1 - s) + \left.\frac{\partial^2 f}{\partial x^2}\right|_m \Delta x_2^2$$

$$+ \frac{1}{6}\left.\frac{\partial^3 f}{\partial x^3}\right|_m \Delta x_2^2(\Delta x_2 - \Delta x_1) + O(\Delta x^4) \tag{6-6}$$

$$\left.\frac{\partial^2 f}{\partial x^2}\right|_m = \frac{f_{m+1} + (1 + x^2)f_m + s^2 f_{m-1}}{\Delta x_2^2} - \left.\frac{\partial f}{\partial x}\right|_m \left(\frac{1 - s}{\Delta x_2}\right) + O[(\Delta x_2 - \Delta x_1),\Delta x^2] \tag{6-7}$$

The resulting expression now requires $s = O(1 - \Delta x_1^2)$ just to be *first*-order accurate at $i = m$.

Salvadori and Baron (1961, page 67) derive the following expression by fitting a parabola through m,(m±1) and differentiating.

$$\frac{\delta^2 f}{\delta x^2} = \frac{2}{s(s + 1)} \frac{sf_{m+1} - (1 + s)f_m + f_{m-1}}{\Delta x_2^2} \tag{6-8}$$

The corresponding result for the first derivative is still equation (6-5).

The reason for the higher truncation error of these forms is easily interpreted from a control-volume point of view, as in the following exercise.

Exercise: By drawing cell boundaries between the nodes of Figure 6-1a, demonstrate that the point m is far from the cell center for $s << 1$. For $s \to 0$, the node m approaches the right-hand boundary of its cell.

The physical interpretation can be somewhat improved by considering not a single change in node spacing, Δx, but a single change in *cell size,* Δx, as shown in Figure 6-1b. Special equations are now required at both m and $(m + 1) = n$. These forms are similar to those already given.

Exercise: Show that the forms for $\delta f/\delta x$ and $\delta^2 f/\delta x^2$ given above are conservative.

It is clear from the above equations that, unless the mesh spacing is changed slowly, the formal truncation error is actually deteriorated, rather than improved.* Thus, Crowder and Dalton (1969) numerically experimented with five different changing (locally refined) meshes, and found that a constant mesh spacing gave the most accurate answers for their particular problem; see, however, Blottner and Roache (1971). Recall also the phenomenon of wave reflection off the change in mesh, mentioned in section V-D-4. However, it has been experimentally observed that the loss of accuracy of the total solution is not as bad as indicated by the formal truncation error, especially if an isolated mesh change is used. See, e.g., MacCormack (1971), Chavez and Richards (1970), and Magnus and Yoshihara (1970). Although much is unknown, it is true that the generally preferred method for increasing resolution locally is to use a *coordinate transformation,* which will be covered in the next section, VI-B. But first, we consider a few more aspects of mesh systems.

Geometric schemes for effecting simultaneous changes in Δx and Δy were given by Sinnott (1960), Runchal et al. (1969), Gillis and Liron (1969), Kacker and Whitelaw (1970), and Dawson and Marcus (1970). Bahvalov (1968) considered a fine mesh with second-order forms near boundaries, matched to a more coarse mesh with higher order forms at interior points. The relative merits of a single mesh-size change versus a continual variation were discussed by Roberts (1971). Interface mesh problems for the Lax-Wendroff method (section V-E-5) were considered by Ciment (1968). A discussion of ADI methods (sections III-A-16 and III-B-6) in a variable mesh was given by Spanier (1967). Beardsley (1971) considered a polar grid with equal angular spacing and parabolic radial spacing; modifications are required for the inviscid vorticity equation near $r = 0$.

It is often desirable to change the spatial mesh as a solution develops in time, in order to follow regions of high spatial change of gradients. The transference of the solution from one mesh to another is called "rezoning." Rezoning itself can affect the solution, for example by introducing a smoothing (artificial diffusion?) effect, or by introducing conservation errors. The development of automatic program rezoning dependent on the developing solution is an important and interesting problem; see Mason and Thorne (1970), Butler (1971), and Crowley (1971).

The forms given above, which are applicable to variable Δx and Δy, have also been used to describe irregular boundaries in a rectangular mesh. As shown in Figure 6-2, the point (i,j) adjacent to the boundary requires forms with changing Δx and Δy in order to accommodate boundary values b_1 and b_2. This procedure has long been recommended and sometimes gives apparently acceptable results. See, for example, Salvadori and Baron (1961), Thoman and Szewczyk, (1966), Singleton (1968), Tejeira (1966), and Dawson and Marcus (1970) for rectangular coordinates, and Lysen (1964) for the cylindrical stream function at an irregular boundary. Yet, we strongly recommend avoiding this procedure for the following reasons.

(1) The formal order of truncation error is deteriorated, as shown above. Note that for a general curved boundary such as an airfoil (Singleton, 1968) we may generally expect to find some interior point in the regular mesh which falls very close to the boundary, making $\Delta x_2/\Delta x_1$ and $\Delta y_2/\Delta y_1$ very small and therefore substantially increasing the formal truncation error.

---

*Methods for the advection terms which are only first-order accurate in a regular mesh (upwind differencing) are not further degraded by a change in mesh spacing, but the methods for the diffusion terms are. 4-point formulas are required to maintain second-order accuracy; see Southwell (1946). C.E. Pearson (MIT J. of Math. and Phys.,Vol.47,1968,pp.351-358) used 3-point equations in a self-adaptive changing mesh for quasi-1-D shock calculations.

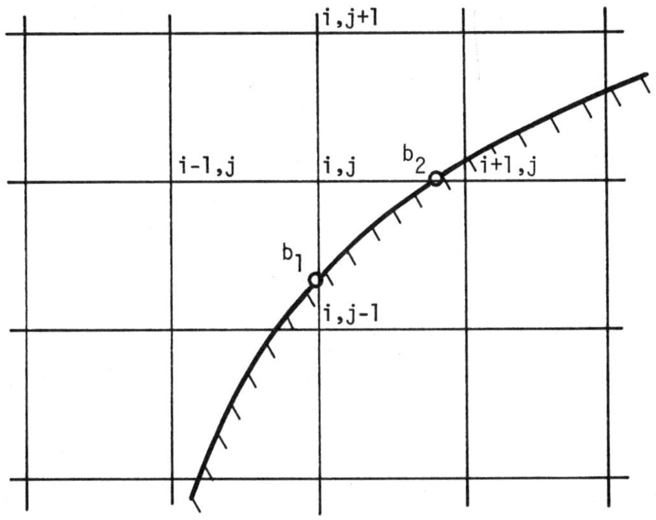

Figure 6-2. Irregular boundary points in a rectangular mesh.

(2) Iterative convergence will be slowed down (Tejeira, 1966), and near-optimum relaxation parameters (section III-B-4, 6, 7) may be difficult to determine.

(3) Stability considerations may become severe. Note that the usual Courant-number restriction gives $\Delta t \leq \Delta x/u$ (or a two-dimensional counterpart). When Δx is reduced near the boundary, the maximum Δt may be limited by this local condition. If no-slip boundaries are used, the restriction for the advection terms may not be excessive because u is also small locally; however, for viscous terms and for slip boundaries, these local conditions will likely be critical.

(4) Programming time and chances for error both increase with such treatments.

Truncation convergence problems of elliptic equations with irregular boundaries in rectangular meshes were considered by Thuraisamy (1969A,B). Iterative convergence of elliptic equations with gradient boundary conditions on a curved surface was considered by Matin (1968). The careful treatment of curved free surface boundaries in a rectangular mesh given by Chan et al. (1969) is recommended reading for anyone who uses this approach.

Another method of treating irregular boundaries is to locally align a curvilinear quadrilateral mesh with the boundary. An attractive method involves the use of automatic mesh generation by equipotential zoning (Winslow, 1963; Sackett and Healey, 1969) in which the mesh point locations are determined by the solution of an elliptic equation in a regular mesh. Godunov and Prokopov (1968) considered local curvilinear, non-orthogonal mesh methods for the "generalized" (variable coefficient) Laplace equation.

As with the case of obtaining locally high resolution, the preferred method for treating non-rectangular boundaries is again to select a non-rectangular coordinate system (or transformation) for the problem which aligns with the boundaries.

The triangular grid element has geometrical advantages for matching irregular boundaries. The elliptic Poisson equation is amenable to a triangular mesh and has been used in the "finite element" methods of structural analysis for this reason. Winslow (1966) has treated the quasi-linear Poisson equation in a non-uniform triangular mesh. Williamson (1969) considered both second- and fourth-order analogs of the frictionless vorticity transport equations in arbitrary triangular grids. He developed a class of methods which conserve mass, momentum, and energy. Bizzel et al. (1970) solved potential-flow free-surface problems in an irregular triangular mesh. Sadourny and Morel (1969) considered a hexagonal grid for a spherical surface (earth) and developed a method which conserves mass, total momentum, total kinetic energy, and squared vorticity for non-divergent flow. Problems with the use of quasi-homogeneous grids on a spherical surface (e.g., spherical triangles) were treated by Sadourny et al. (1968) and Williamson (1968, 1971). Bryan (1966) considered advection in an irregular, possibly polyhedral grid system and a method which conserves kinetic energy. To date, no satisfactory methods of differencing viscous terms in any of these grid systems have appeared. Bowley and Prince (1971) generalized a two-step Lax-Wendroff method to 9-point equations in a trapezoidal mesh; they included viscous terms, but no details were given.

Finally, special mention should be made of the hybrid grid system used by Thoman and Szewczyk (1966, 1969), shown in Figure 6-3. The desired resolution near the circular surface was obtained by an expanding radial scale in the polar coordinate mesh. The rectangular mesh spacings were adjusted so that, at the junction of the two meshes, cell centers are shared in both the polar-mesh and the rectangular-mesh systems. (The simple patching of rectangular and polar coordinate systems was used by Hurd and Peters (1970) for the simpler problem of a circular elbow bend between straight channels.)

VI-B. Coordinate Transformations

Coordinate transformations can be used for the purpose of aligning coordinates along physical boundaries, and/or increasing resolution in certain regions. The first purpose is met by describing the planar flow over a cylinder in polar coordinates (r,θ) rather than in rectangular coordinates, for example. This type of coordinate transformation will be covered in section VI-C.

The second purpose of a coordinate transformation, that of increasing resolution in certain areas, may sometimes be met by the same coordinate system used to align coordinates with boundaries, as in the use of planar elliptic coordinates for the flat-plate problem (Pao and Daugherty, 1969), for example. More often, the increased resolution can be achieved with a coordinate stretching transformation of some kind. As we have said (see also Taylor, 1969; Blottner and Roache, 1971) such coordinate transformations are generally more accurate than mesh changes. One of the most commonly used is an exponential stretch. Pao and Daugherty (1969) transformed the Cartesian (x,y) coordinates to (X,Y) by the relations

$$x = X \tag{6-9a}$$

$$y = b(e^{aY} - 1) \tag{6-9b}$$

where a and b are arbitrary constants, used to adjust the "stretch" to the flow regions. The (non-conservative) equations for vorticity and stream function in the untransformed (x,y) coordinates are

$$\frac{\partial\zeta}{\partial t} = -u\frac{\partial\zeta}{\partial x} - v\frac{\partial\zeta}{\partial y} + \frac{1}{Re}\nabla^2\zeta \tag{6-10}$$

$$\nabla^2\psi = \zeta \tag{6-11}$$

$$\nabla^2 = \frac{\partial^2}{\partial x^2} + \frac{\partial^2}{\partial y^2} \tag{6-12}$$

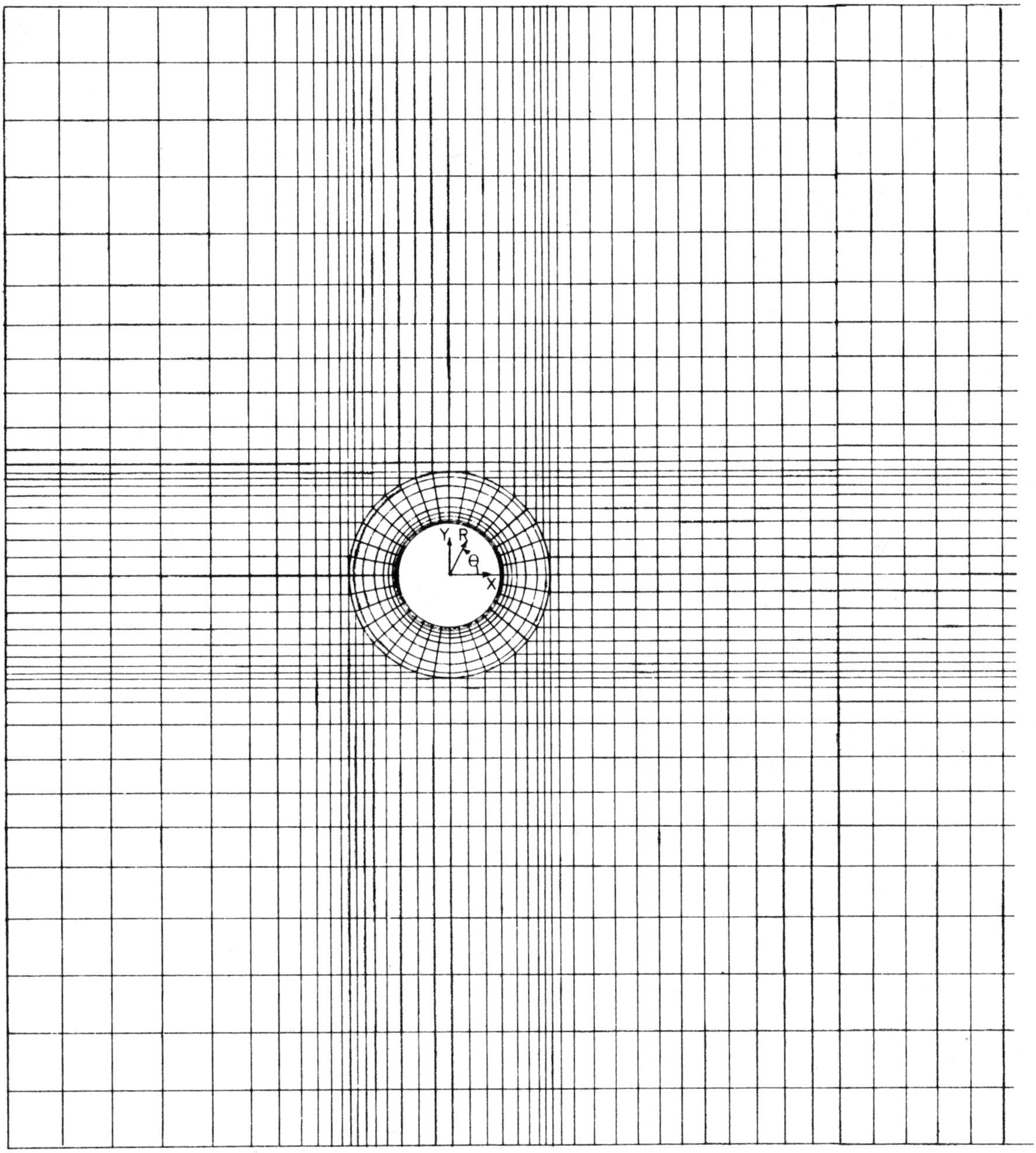

(a) hybrid mesh cell structure

Figure 6-3. Hybrid mesh system of Thoman and Szewczyk (1966,1969).

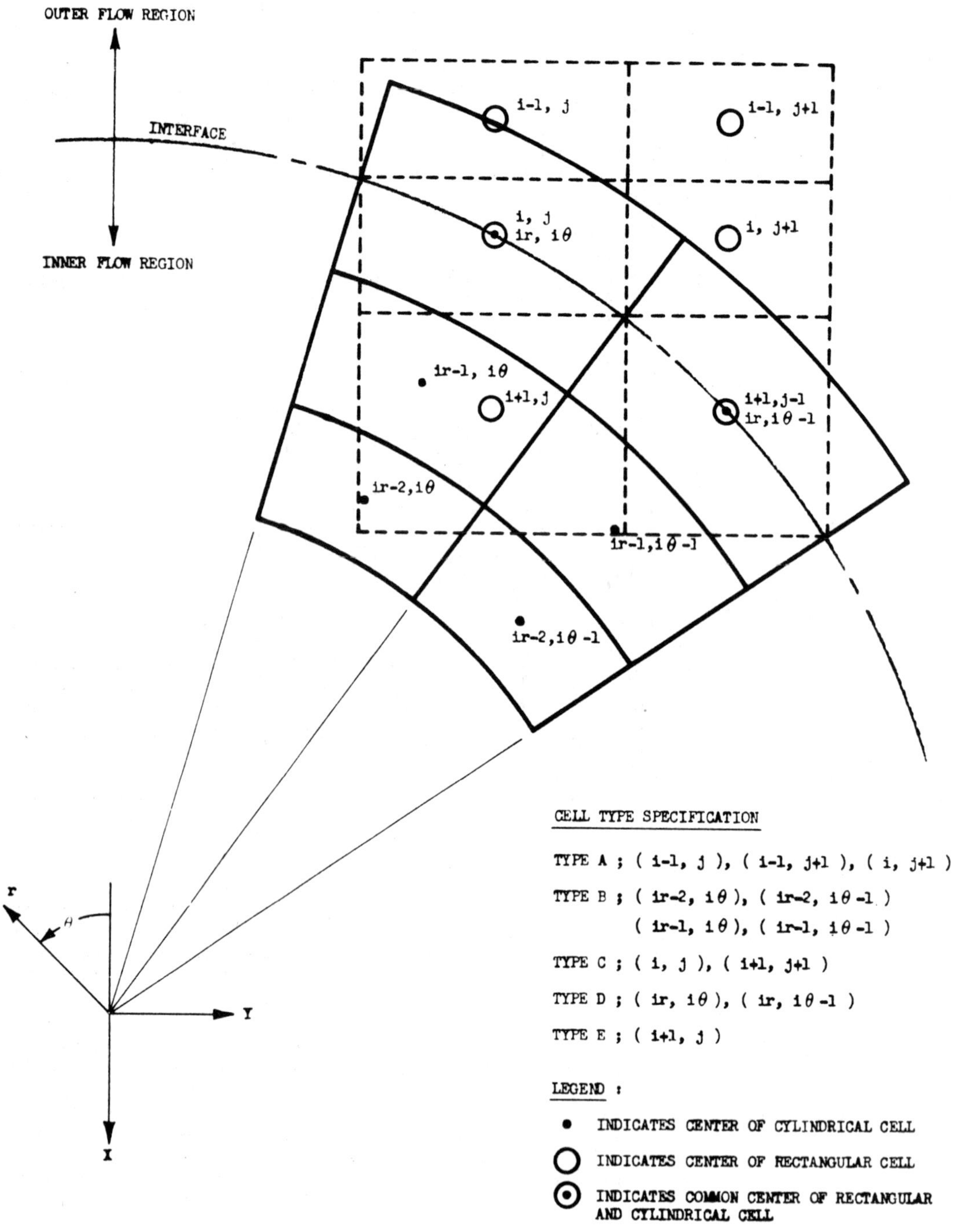

(b) dual interface cell definition

Figure 6-3. (continued)

where

$$u = \frac{\partial\Psi}{\partial y} \quad , \quad v = -\frac{\partial\Psi}{\partial x} \tag{6-13}$$

By equations (6-9), these transform to

$$\frac{\partial\zeta}{\partial t} = -\frac{e^{-aY}}{ab}\left(\frac{\partial\Psi}{\partial Y}\frac{\partial\zeta}{\partial X} - \frac{\partial\Psi}{\partial X}\frac{\partial\zeta}{\partial Y}\right) + \frac{1}{Re}\tilde{\nabla}^2\zeta \tag{6-14}$$

$$\tilde{\nabla}^2\Psi = \zeta \tag{6-15}$$

where

$$\tilde{\nabla}^2 = \frac{\partial^2}{\partial X^2} + \frac{e^{-2aY}}{a^2 b^2}\frac{\partial^2}{\partial Y^2} - \frac{e^{-aY}}{b}\frac{\partial}{\partial Y} \tag{6-16}$$

Apparently, the first use of such an exponential stretch was made by Jensen (1959) in studying incompressible sphere flow. Other uses were by Son and Hanratty (1969), Hamielec et al. (1967A,B) and Rimon and Cheng (1969) for incompressible spherical problems; Son and Hanratty (1969) and Hamielec and Raal (1969) for incompressible flow over cylinders; Skoglund et al. (1967) and Skoglund and Gay (1968) for the shock-wave/boundary-layer interaction problem on a flat plate; and by Pao and Daugherty (1969) for the flat plate problem in incompressible flow.

The purpose of such stretching transformations is the same as that of the expanding mesh systems discussed earlier in section VI-A, i.e., to increase resolution in a certain area. Note, however, that these two approaches are *fundamentally different*.* When the untransformed equations are differenced in the expanding mesh, the result is a deterioration of formal accuracy, as we have seen; but the transformed equations may be differenced in a regular mesh (such as constant ΔX, ΔY) with no deterioration in the formal order of truncation error, except that it will now be $O(\Delta Y^2)$ rather than $O(\Delta y^2)$. The transformation approach is therefore to be preferred. The potential of the transformation approach can be recognized when one considers that the exact solution for Poiseuille flow can be obtained with *one interior point* and a judicious parabolic transformation (Blottner and Roache, 1971).

Probably the most successful and significant finite transformation is that used by Moretti (Moretti and Abbett, 1966B; Moretti and Bleich, 1967, 1968) for 2D and 3D inviscid blunt-body shock-wave calculations, and by Moretti and Salas (1969, 1970) for viscous calculations.** See also Moretti (1969A,B). A general point between the body surface and the shock (Figure 6-4) is located by the polar coordinates (r,θ) centered in the body. The downstream extent is calculated to some θ_{max}, chosen so that the outflow is supersonic. The region bounded by the shock, centerline, body, and θ_{max} is then transformed to a rectangle in X and Y by way of

*Unfortunately, many authors say they are using an "expanding mesh" (which is true in the sense that the transformed location of calculated points would represent an expanding scale) when they actually are transforming the equations.

**In an earlier publication, Godunov et al. (1959) similarly maintained the shock as a discontinuity. Their method was intimately tied in to the elaborate interior-point method known as Godunov's method (see section V-E-8), and is essentially a floating *mesh* method, rather than a transformation.

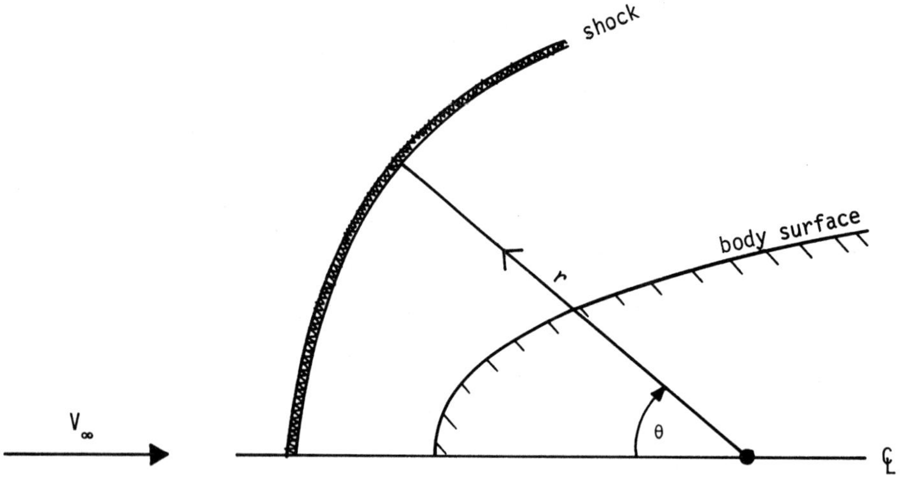

Figure 6-4. Moretti's transformation for blunt body shock calculations in two dimensions.

$$X = \frac{r - r_b(\theta)}{r_s(\theta) - r_b(\theta)} \quad , \quad X \in [0, 1] \tag{6-17a}$$

$$Y = \Pi - \theta \quad , \quad Y \in [\Pi - \theta_{max}, \Pi] \tag{6-17b}$$

where $r_b(\theta)$ and $r_s(\theta)$ are the radial positions of the body surface and the shock, respectively. The transformed equations are differenced with constant ΔX and ΔY using Moretti's method (section V-E-8), which is commonly referred to as a Lax-Wendroff type (section V-E-5) although terms like $\partial^2 f/\partial x \partial t$ are actually obtained by spatial differentiation of the original equations as in Leith's method (section III-A-13). The transformed equations are complicated by the non-orthogonality of X and Y. The shock coordinate $r_s(\theta)$ is part of the solution, of course, so $r_s(\theta)$ and the transformation change as the solution develops. The time-dependent $r_s(\theta)$ is calculated, starting from some initial guess, by calculating the spatial trajectory of the shock position from the Rankine-Hugoniot relations across the shock. (The X = 1 node point is behind the shock.) The whole procedure is quite elaborate, but the results make it all worthwhile. Because the shock is maintained as a discontinuity, the inaccuracies due to shock smearing (see section V-C, D, E) are not present. Likewise, the body surface lies along a coordinate which enhances accuracy. (The surface boundary values are calculated by recourse to the non-steady characteristics theory.) Excellent accuracy is obtained with a coarse mesh, allowing very reasonable computing times. Fairly accurate results can even be obtained with one internal point between body and shock, comparable to one-strip integral method solutions (e.g., see Homicz and George, 1970). D'Souza et al. (1971) used the method of toroidal coordinates to model an engine inlet.

The very considerable virtues of "Moretti's method" unfortunately have become confused in the literature. Moretti has not used conservation equations, and it has often been stated that shock calculations made with Moretti's shock-fitting transformation are preferable to those made with conservation equations. In fact, the method should be contrasted *not* with conservation equations, but with the shock-smearing approach (section V-C). The success of Moretti's method depends mainly on the shock-fitting transformation and the careful attention to surface boundary conditions, rather than on the lack of conservation form, nor even on the particular finite-difference form used at interior points. For example, Barnwell (1971) calculated 3D detached shocks using Moretti's transformation, but also using a version of Brailovskaya's method (section V-F-3) on the *conservation* equations. Also, Xerikos (1968) used the same shock-layer transformation in 3D using Lax's method (section V-E-4) for interior points, and Li (1971) calculated the axisymmetric blunt-body problem with chemistry using MacCormack's method (section V-E-6). Thomas et al. (1971) used a 3D shock-layer transformation, using MacCormack's method to "march" in a spatial coordinate direction.

Rather than perpetuate the animosity between the shock-fitting "school" and the "shock-smearing" or "shock-capturing" school, it would be more profitable to take advantage of both methods. Shock-fitting can be used to enhance the accuracy of the relatively simple bow shock, but conservation equation methods can be used at interior points to allow the "shock-capturing" of unexpected and/or complex embedded shocks, such as those which arose in the calculations of Kutler and Lomax (1971).

It must also be noted that Moretti's interior-point method, like all the Lax-Wendroff methods, does exhibit a Δt-dependent artificial viscosity effect when applied to a steady-state problem with outflow Courant numbers < 1 (Roache, 1971C; see Appendix B).

Gonidou (1967) used shock fitting with a transformation like Moretti's. The report by Xerikos (1968) includes calculations of a flare shock as well as the bow shock, and is recommended for details of the shock trajectory calculations and the centerline ($r = 0$) calculations for asymmetric flows. Pavlov (1969) also used the transformation (6-17) of the shock layer for low Re viscous flows. Migdal et al. (1969) used the same kind of transformation as equation (6-17a) to map the transverse coordinate of a nozzle into a rectangular region. Lapidus (1967) used a transformation from an arbitrary upstream boundary to the body, mapping this region into a rectangular region. He showed that such transformations preserve the conservative property. Aungier (1971) also used the Moretti shock-fitting approach.

Brailovskaya (1967) used a transformation like (6-17) to map a region near an expansion corner onto a rectangle, and followed with another "logarithmic condensation" ($\equiv$ exponential stretch) to achieve increased resolution near the wall; the calculations were made with Brailovskaya's method (section V-F-3).

A similar shock-layer floating *mesh* (rather than coordinate transformation) was used for inviscid 3D blunt-body calculations with Godunov's method (see section V-E-8) by Godunov et al. (1959), Dyakonov (1964), McNamara (1966, 1967), Masson et al. (1969), and Taylor and Masson (1970).

Another approach to resolving a discontinuity in an Eulerian mesh has been used by McNamara (1966, 1967). The discontinuity in this case was the contact surface formed from the collision of two oblique shocks. The Eulerian axisymmetric mesh was periodically realigned to follow the movement of the contact surface. An inaccuracy in the form of a cusp in the shock near the stagnation streamline resulted from an inconsistency in the calculation of this mesh motion. The development of methods to treat shock and contact surface discontinuities continues to occupy a major research effort.

Kalugin and Panchuk (1971) used a transformation to align coordinates along a moving wavy wall, in an exploratory simulation of a dolphin's skin.

When a wide range of parameters is involved, resolution may also be improved by making transformations of the dependent variables. Moretti and Abbett (1966B) solved compressible flow problems in terms of $\widetilde{P} = \ln P$ and $\widetilde{\rho} = \ln \rho$, and Scala and Gordon (1967) used similar transformations. (Other dependent variable transformations will be considered in section VI-C.)

The following points should be kept in mind when using stretching transformations.*

(1) Stability and convergence properties will be affected. For example, the one-dimensional Courant number restriction $u\Delta t/\Delta x \leq 1$ which is applicable to the untransformed equation (6-10) becomes, for the transformed equation (6-14),

*The general remarks also apply to mesh expansions.

$$\frac{e^{-aY}}{ab} \frac{\partial \Psi}{\partial Y} \Delta t/\Delta X \leq 1 \tag{6-18}$$

rather than $\frac{\partial \Psi}{\partial Y} \Delta t/\Delta X \leq 1$. Also, the system (6-15, 16) may be expected to converge differently than (6-11,12).

(2) Boundary conditions, notably the relation for wall vorticity (see section III-C-2), must be re-derived in the transformed coordinate system.

(3) Conservation properties may be lost, or at least altered in interpretation.

(4) Singularities may possibly be introduced (or removed) by the transformation.

(5) Increased resolution does not necessarily imply increased accuracy. This is an especially pertinent fact when boundary condition errors are present; see remarks in section III-C-1.

(6) Wave phenomena will be distorted, due to changes in damping, phase, and aliasing errors.

(7) The computation time per step is increased.

(8) Transformations like Moretti's for the shock region are *not* applicable to problems wherein shocks develop from the merging of continuous compression waves in viscous flow, as in the shock-wave/boundary-layer interaction problem, nor do they appear practical to problems with complex systems of reflected and intersecting shocks.

So far, we have discussed finite transformations. Another commonly used type of transformation is the infinite-to-finite mapping, apparently first used by Wang and Longwell (1964). They solved the problem of entrance flow in a duct using steady-flow equations; they located the (x,y) coordinate system with x = 0 at the plane of the duct entrance and transformed the x-coordinate by way of

$$X = 1 - \frac{1}{1 + cx} \tag{6-19}$$

The constant $c > 0$ for $x > 0$ maps the right-hand semi-infinite region $0 \leq x \leq +\infty$ onto $0 \leq X \leq 1$, whereas another value $c < 0$ for $x < 0$ maps $-\infty \leq x \leq 0$ onto another region, $0 \leq X \leq 1$. We define

$$a = \frac{dX}{dx} = \frac{c}{(1 + cx)^2} \tag{6-20}$$

$$b = \frac{d^2X}{dx^2} = \frac{-2c^2}{(1 + cx)^3} \tag{6-21}$$

$$\frac{\partial\zeta}{\partial t} = -a\left(\frac{\partial\Psi}{\partial y}\frac{\partial\zeta}{\partial X} - \frac{\partial\Psi}{\partial X}\frac{\partial\zeta}{\partial y}\right) + \frac{1}{RE}\tilde{\nabla}^2\zeta \tag{6-22}$$

$$\tilde{\nabla}^2\Psi = \zeta \tag{6-23}$$

$$\tilde{\nabla}^2 = b\frac{\partial}{\partial X} + a^2\frac{\partial^2}{\partial X^2} + \frac{\partial^2}{\partial y^2} \tag{6-24}$$

The rationale of such transformations is that the analytical boundary conditions at infinity may be used in the finite-difference equations (see remarks in sections III-C-10,11). Thus, for their entrance flow problem, Wang and Longwell (1964) could use the condition of uniform flow upstream and fully developed Poiseuille flow downstream. Moretti (1969A,B) also favors such transformations. Sills (1969) considered three classes of infinite-to-finite mappings, all of which have convenient explicit inverses. In the following, a and b are arbitrary positive constants, and the range is indicated.

$$X_1 = \frac{ax^b}{ax^b + 1}\ , \quad [0,\infty] \rightarrow [0,1] \tag{6-25}$$

$$X_2 = 1 - e^{-ax}\ , \quad [0,\infty] \rightarrow [0,1] \tag{6-26}$$

$$X_3 = \tanh(ax)\ , \quad [-\infty,+\infty] \rightarrow [-1,+1] \tag{6-27}$$

Sills (1969) also discussed the derivative forms and inverses, and noted that transformations based on arctan(x) and erf(x), while achieving the desired mapping, are unwieldly in application. Mehta and Lavan (1968) used

$$X = \frac{1 + \tanh[a(x+1/2)]}{1 + \tanh\left(\frac{a}{2}\right)} \tag{6-28}$$

which transforms $[-\infty,0]$ into $[0,1]$, with an inflection point at $x = -1/2$.

Lavan et al. (1969) mapped $[-\infty,+\infty]$ into $[0,1]$ using

$$X = \frac{1}{2}[1 + \tanh(ax)] \tag{6-29}$$

Migdal et al. (1969) transformed a nozzle transverse coordinate into a rectangular region by a transformation like (6-17A,B), and transformed the region from $x = 1$ (the nozzle exit) to $x = \infty$ onto $X \in [0,1)$ by way of

$$X = \frac{1 + e^{-2x_0}}{1 + e^{-2x/x_0}} \tag{6-30}$$

Taylor (1969) considered problems for $x \in [0,\infty]$ with a regular rectangular mesh for $x \in [0,1]$. For $x > 1$, the region $\langle 1,\infty]$ was mapped onto a finite region by any of the transformations

$$X_1 = -\frac{1}{x} \tag{6-31}$$

$$X_2 = -\frac{1}{(1+x)^2} \tag{6-32}$$

$$X_3 = -e^{-x} \tag{6-33}$$

all of which give $dX/dx > 1$. Taylor also considered the problem of matching the two coordinate systems.

Kentzner (1970A,B) calculated inviscid subsonic flow over a cylinder by numerically inserting the cylinder into uniform flow, using the compressible flow equations. A shock wave is formed and propagates outward. The shock is maintained as a discontinuity, and the shock-body region is transformed to a rectangular mesh by a non-orthogonal, curvilinear, time-dependent transformation. As time $\to \infty$, this method approaches an infinite-to-finite mapping, but Kentzner says that a simpler, non-time-dependent infinite-to-finite transformation failed because the disturbance (shock wave) reached infinity after a *finite number of time steps.* A fixed, finite mesh also failed in this problem, because the shock was reflected off the mesh boundary. The advantage of Kentzner's method is that the problem is mathematically well-posed, with no arm-waving required for the treatment of the far-field boundary.

In addition to the remarks already made about stretching transformations, the following points are pertinent to infinite-to-finite mappings.

(1) The use of analytical boundary conditions at "infinity" is not always preferable to computational outflow boundary conditions, as discussed in sections III-C-7 to 11 and V-G-6. (For example, consider the calculation of boundary-layer growth on a flat plate; if only the x-coordinate is transformed, the correct downstream condition at infinity for *any* finite y-dimension is $u = v = 0$.)

(2) The use of a transformed infinity condition seems intuitively to be a less accurate boundary condition than the best of section III-C-7 for cases where the mesh outflow is periodic, as in a vortex street wake. (For example, what happens when the de-transformed Δx becomes larger than the wake vortex spacing?)

(3) Aside from questions of relative accuracy of analytical versus computational *boundary conditions*, it is easy to show that stretching transformations can give *less* accurate results for *interior-point* calculations of periodic flows, as in the following exercise.

Exercise: Consider the inviscid model advection $\partial\zeta/\partial t = -u(\partial\zeta/\partial x)$ with constant u. Using upwind differencing, the FDE, like the PDE, requires *no* outflow boundary condition, and the exact solution can be obtained in the regular x mesh (see section III-A-8). Using the transformation as in equations (6-19) through (6-22), show that the exact answer cannot be obtained for u = constant.

In the above exercise, the exact answer can be obtained from the transformed problem if $u/(1 + x)^2$ is constant, but the case of u = constant is more meaningful to physical problems.

Several other transformations require mention. Allen and Southwell (1955) solved viscous incompressible flow over a cylinder by transforming to the orthogonal coordinates formed by the velocity potential and stream function of the (analytical) potential flow solution. Thom and Apelt (1961) and Apelt (1969) used conformal mappings (or what they called "symmetromorphic figures") to produce orthogonal curvilinear coordinates for incompressible flow problems, as did Brennen (1969) and Lee and Fung (1970). Freudiger et al. (1967) used conformal transformations for a compressible-flow problem.

The Crocco transformation (see Schlichting, 1968) for boundary-layer equations uses velocity as an independent variable. Crenshaw (1966) solved free shear-layer problems using the boundary-layer approximation (neglect of streamwise diffusion) and transforming the normal coordinate to a momentum coordinate, i.e., using momentum as the independent variable. Since momentum is a bounded property of the flow, the de-transformed finite-difference mesh grows automatically with the flow field. (See also Crenshaw and Hubbartt, 1969.) Van de Vooren and Dijkstra (1970) solved the incompressible Navier-Stokes equations over a flat plate by first writing the (Ψ,ζ) equations in parabolic coordinates and further transforming these to a finite rectangular region. The normal coordinate transformation depends on the first-order boundary-layer similarity solution (Blasius solution) while the streamwise coordinate is transformed with a logarithmic relation which removes the singularity at the leading edge. Armitage (1967) attempted a transonic swirl flow calculation in transformed "nozzle" coordinates. In his study of end-wall reflections in a shock tube, Watkins (1970) mapped the region between the shock and the wall with a linear transformation; a fixed number of computational cells (in the transformed coordinate) were maintained between the shock and the wall. Anderson et al. (1968) showed how to transform the inviscid compressible flow equations so that a conservation (divergence-free) form is maintained in any curvilinear coordinate system. Thompson et al.(J.Comp.Phys.,Vol. 15,No. 2, 1974,pp 226-250) have presented a powerful method for *numerically* generating body-fitted curvilinear coordinate systems.

In this section we present references for orthogonal coordinate systems, other than Cartesian, which may be "natural" for particular problems; e.g., planar flow over a parabolic shape is naturally calculated in parabolic coordinates. In some of the applications cited, further transformations (such as exponential stretching) were also used. In the following, the term "(Ψ,ζ) approach" refers to incompressible flow using the stream function and vorticity equations (Chapters II and III) or a modification thereof.

Planar problems using the (Ψ,ζ) approach were calculated in parabolic coordinates by van de Vooren and Djikstra (1970) and in elliptic coordinates by Pao and Daugherty (1969) and Leal (1969). Cylindrical (Ψ,ζ) equations were used by Kawaguti (1953), Thoman and Szewczyk (1966, 1969), and Richards (1970).

The (Ψ,ζ) equations in spherical coordinates were used by Jensen (1959), Brown (1967), and Rimon and Cheng (1969), and in cylindrical axisymmetric coordinates by Barakat and Clark (1966), Michael (1966), Torrance (1968), Shavit and Lavan (1971), Strawbridge and Hooper (1968), Friedmann et al. (1968), Friedman (1970), and Lee and Fung (1970). In the last three papers, the steady-state equations were used. Different forms of the equations are used in the different papers. The forms are, of course, equivalent in the continuum equations, but not necessarily in the finite-difference equations. Torrance (1968) uses a modified vorticity $\tilde{\zeta} = \zeta/r = \nabla x\vec{V}/r$ in cylindrical coordinates, so that the equations are not singular at $r = 0$. Note that axisymmetry makes $\zeta(r = 0) = 0$, but $\tilde{\zeta}(r = 0) \neq 0$.

Griffiths et al. (1969) studied elastico-viscous problems in cylindrical coordinates. Mancuso (1967) solved the Poisson equation with gradient boundary conditions on a spherical surface, using ADI methods (see section III-B-6). Eisen (1967B) treated truncation-error convergence problems for the spherically symmetric diffusion equation. Schulz (1964) considered conservation forms of the Lagrangian equations in cylindrical coordinates, with a tensor form of the artificial viscosity. Faccioli and Ang (1968) presented a discrete Eulerian model based on physical principles for the conservation laws with spherical symmetry. Eaton and Zumwalt (1967) used cylindrical coordinates in a blast-interaction computation.

Rimon (1968) and Rimon and Lugt (1969) used the (Ψ,ζ) equations in an oblate-spherical coordinate system, in which the body surface is aligned with ellipsoids of revolution, and the orthogonal coordinate lines are hyperbolas. Masliyah and Epstein (1970) used both oblate- and prolate-spheroidal coordinates. D'Souza et al. (1971) used toroidal coordinates to calculate compressible, inviscid flow at an engine inlet. There are other exotic coordinate systems which could be of use in fluid dynamics computations, and we present some pertinent references here. The book by Schelkunoff (1965) presents the gradient, divergence, Laplacian, and curl vector operators in the following coordinate systems: Cartesian, cylindrical, spherical, elliptic cylinder, prolate spheroidal, oblate spheroidal, biaxial, toroidal, bi-polar, parabolic cylinder, and paraboloidal. Larsen (1969) discussed the super-elliptic coordinate system, recently popularized in architecture by Piet Hein. This orthogonal system is based on the Lamé curves, defined by

$$\left(\frac{x}{a}\right)^n + \left(\frac{y}{b}\right)^n = 1 \tag{6-34}$$

For various values of the parameters a, b, and n, these curves comprise rectangles, ellipses, diamonds, asteroids, squares, circles, and their transition curves. For the ultimate in elaborate coordinate systems, Kopal (1969) presented the Roche coordinates, based on the equipotential surfaces of a rotating gravitational dipole, and suggested their use for the hydrodynamic calculations of gas streams in close binary (star) systems.

The text by Bird et al. (1960) is highly recommended as a reference for the fluid dynamics equations, including viscous terms, in rectangular, cylindrical, and spherical coordinate systems, plus a wealth of other information on fluid dynamics and other transport phenomena. Tsien (1958) gave the compressible, viscous Navier-Stokes equations in orthogonal curvilinear coordinates. (Neither Bird et al. nor Tsien used the conservation form, however.) Bohachevsky et al. (1965) gave the conservation form of the inviscid compressible flow equations in cylindrical and spherical coordinates. (We repeat the observation of Chapter IV, that the use of the conservation variables affects the form of the viscous terms in the compressible flow energy equation.)

The various kinds of radial coordinate systems (cylindrical, spherical, paraboloidal, etc.) require some comment. In regard to the conservative property, a semantic ambiguity exists. The usual form of the compressible flow continuity equation in cylindrical coordinates, assuming axisymmetry, is (Bird et al., 1960).

$$\frac{\partial\rho}{\partial t} + \frac{1}{r}\frac{\partial(\rho v)}{\partial r} + \frac{\partial(\rho z)}{\partial z} = 0 \tag{6-35}$$

Using any of several difference schemes based on upwind differences or on centered spatial differences (see Chapter 5), it may be shown (Problem 6-2) that this equation is "conservative" in the sense that mass is conserved as it flows from one cylindrical-section control volume to another (see section III-A-3). But it is not in a "divergence-free" form, which is also related to the conservation concept, and is important for theoretical reasons (Lax, 1954). The divergence-free form is obtained by writing the continuity equation in terms of the conservation variable $\eta = \rho r$, as in the following exercise.

Exercise: Show that equation (6-35) may be written in the divergence-free form

$$\frac{\partial\eta}{\partial t} + \frac{\partial(\eta v)}{\partial r} + \frac{\partial(\eta u)}{\partial z} = 0 \tag{6-36}$$

where $\eta = \rho r$.

Similar relations hold for the momentum and energy equations (see Bohachevsky et al., 1965). Note that, in this divergence-free form, the centerline (r = 0) value of the independent variable is just $\eta = 0$. However, the centerline values of the basic variables are still required for the radial pressure gradient at one node off the centerline, at j = 2, and also for the evaluation of some viscous terms. For the axisymmetric case, the *safest* approach is to write control-volume equations at the centerline for the advection terms.

Exercise: Show that the continuity equation written at the centerline (j = 1) in axisymmetric flow gives

$$\frac{\partial\rho}{\partial t} + \frac{\partial(\rho u)}{\partial x} + \frac{4(\rho v)_{1\frac{1}{2}}}{\Delta r} = 0 \tag{6-37}$$

(Also, we obviously have $v_1 = 0$.) This approach is conservative, but it does not provide an evaluation method for equations containing viscous terms. The question is even more confused for the non-axisymmetric cases in cylindrical or spherical coordinates. One clumsy but apparently acceptable approach is to revert to Cartesian coordinates near the centerline, patching in geometrically to the rest of the problem. At this time, a convenient method is not clear. It *is* clear, however, that the artifice of placing the node values at a half-space off the centerline ($r = 0$ at $j = 1+1/2$, for example) and then using only regular interior-point difference equations, is *incorrect*. The interior-point equations *cannot be written across the* $r = 0$ *line,* since the "direction" of decreasing r reverses at $r = 0$. A similar problem occurs in spherical coordinates at $\theta = 0$.

The computing time and algebraic complexity for the viscous terms is much greater for cylindrical and especially for spherical coordinates, compared with Cartesian coordinates. There are also open questions about the best form of explicit artificial viscosity terms in non-rectangular coordinates. Using Rusanov's method (section V-D-3), Eaton* has calculated axisymmetric swirling flows and has found errors to be reduced considerably by using centerline diffusion terms = 0.

VI-D. *Other Systems of Equations*

In this section we will consider, in order,

(a) systems of equations which are gross simplifications of the Navier-Stokes equations, such that the character of the equations is altered,

(b) simplifications of the Navier-Stokes equations which are not quite so radical,

(c) complications to the Navier-Stokes equations, and

(d) different means of expressing the physical laws of the Navier-Stokes and other equations of fluid dynamics.

The first simplication we consider is the *inviscid* assumption, which leads us to consider *potential flow,* the *method of characteristics,* and *transonic flow* calculations.

Potential flow results from the inviscid, incompressible flow assumption, and gives rise to a second-order linear boundary-value problem. This is historically the earliest class of solutions, most of which are closed form (see any basic fluid text). For complex shapes, it is often preferable to solve the potential flow equations numerically (e.g., Dwyer et al., 1971). But the classes of problems which virtually *require* a numerical approach are the axisymmetric and free-surface problems. Brennen (1969) solved axisymmetric cavity flows, Jeppson (1969) solved axisymmetric free-surface flows, Bizzel et al. (1970) treated axisymmetric tank draining problems, and Konstantinov (1970) and Whitney (1971) treated unsteady 2D free-surface flows. Another class of problems which may require numerical solution is the incompressible inviscid flow with vorticity (as in the classic problem of airfoil lift in a sheared free stream); see Chow et al. (1970), for example. Also, Howell and Spong (1969) and Gelder (1971) solved the velocity potential equation for subsonic compressible flow numerically; the former included sharp corner effects.

For steady supersonic inviscid flow, the equations become hyperbolic. The problem becomes an initial-value one, "marching" in space, and the well-known and powerful *method of characteristics* (MOC) is applicable. The basic theory and numerical procedures for two-dimensional homentropic flow are presented in several texts; see Shapiro (1953), Liepman and Roshko (1957), Abbott (1966), Owczarek (1964), and Chapman and Walker (1971). The original work by Tollmien (1949) is historically interesting. In the original MOC, the mesh-point locations are not predetermined, but are part of the solution. There are two of these methods, called the "method of waves" or "cell" method, and the "lattice point" method. The method of waves simplifies some arithmetic in hand calculations and is easier to understand physically, especially when a constant pressure boundary is present. But the method of waves requires a separate background (the velocity hodograph), presents

*Dr. R. R. Eaton, private communication.

computational pitfalls at a symmetry boundary line, and is not extendable to axisymmetric flow (because it is not really based on the mathematical theory of characteristics). The lattice point technique is to be preferred.

The lattice points are determined by the intersection of characteristics (Mach lines) of opposite families, i.e. lines laid off from the velocity vectors at $\pm\mu$, where $\mu = \arcsin(1/M)$. The text by Liepman and Roshko (1957) also treats the axisymmetric MOC, and the complications due to non-homentropy (i.e., variation of entropy normal to, but not along, streamlines). However, many other problems involved in obtaining solutions are not treated in current textbooks. Also, the more modern approach (of specifying a downstream increment and then locating the upstream Mach-line values by interpolation, in accord with the Courant-Friedrichs-Lewy (1928) stability requirement) is not covered in textbooks.

Shock development is indicated in a MOC calculation when characteristics (Mach lines) of the same family cross each other. In this case, an oblique shock wave must be "patched in," its angle determined by a match between the Rankine-Hugoniot relations and the post-shock characteristics relations; see Hartree (1958) and Richardson (1964) for planar MOC shock patching, including programming aspects, and Kennedy (1956), Weiss et al. (1966), Moreno (1967), and Abbett (1970) for axisymmetric MOC shock-patching details. The MOC is also fairly straightforward to apply to one-dimensional unsteady inviscid flow; see Shapiro (1953), Hoskin (1964), and Hoskin and Lambourn (1971) for details, including programming aspects in Hoskin (1964). The MOC does not use a "conservation" form, so conservation checks may be used as an index of truncation error. Powers and O'Neill (1963) noted that the conservation errors are worse at hypersonic speeds (small Mach angle and large gradients), and presented a method of determining the entropy at a mesh point from a mass flux calculation.

In steady 3D flow, a non-uniqueness is introduced which has given rise to 5 or 6 different 3D MOC methods. Chu (1964) gave a simple derivation of the 3D characteristic relations. Chuskin (1968) reviewed four 3D MOC methods, including nonequilibrium effects. Powers et al. (1967) gave a 3D MOC method and included boundary-layer calculations in their program. Magomedov (1966) discussed 3D steady and 2D unsteady MOC. Sauerwein and Diethelm (1967), Rakich (1967, 1969), Magomedov and Kholodov (1967), Grigoryev (1970), and Rakich and Cleary (1970) presented applications of their successful 3D MOC. The usual MOC approach is first-order accurate (see section V-E-2 for a relevent discussion); Ransom et al. (1970, 1971) discussed the stability and accuracy of a second-order 3D MOC. In a German-language reference, Roesner (1967) treated 3D unsteady MOC.

Other problems solved with a MOC approach are the following. Presley and Hanson (1969) and Fickett et al. (1970) calculated 1D unsteady flow of a reacting gas. Matthews (1969) calculated a steady quasi-1D flow. Huang and Chou (1968) calculated a point (spherical) explosion, including careful treatment of the initial singularity and the formation of a second shock. Adamson (1968) obtained some analytical solutions in the characteristics coordinate system. Numerical variations on MOC were considered by Paul and Ahmed (1970). Robertson and Willis (1971) calculated the expansion of a rarefied gas into a vacuum. Chushkin (1970B) calculated a supersonic flow combustion problem.

Some comparisons between the MOC and the regular mesh finite-difference approach were given by Fyfe et al. (1961) and Eaton (1970). The MOC solutions can also be "patched in" as part of an iterative procedure, together with numerical shear-layer solutions using boundary-layer approximations and flow model solutions; see Weiss and Weinbaum (1966), Weinbaum (1966), Ohrenberger and Baum (1970), Mueller et al. (1970), and Burggraf (1970).

The MOC must start from some initial line of data (not on a characteristic). The MOC cannot supply the upstream starting data, nor can it match up all physically possible downstream conditions. Flow over a blunt nose body will require a detached shock solution carried beyond the sonic line before MOC can be used; see Van Dyke (1958), Moretti and Bleich (1967, 1968), Moreno (1967), and Lewis et al. (1971). Even for sharp bodies, the planar oblique shock solution is required in planar flow, or the Taylor-Macoll conical shock solution in axisymmetric flow (e.g., Liepman and Roshko, 1957). For nozzles, the solution near the throat may be obtained by finite difference solutions of the transonic equations or from a semi-empirical theory (e.g., Ruptosh, 1952). In a nozzle calculation, the MOC will give a shock-free solution for the exit-plane pressure, P_e. But if the exhaust or back pressure P_b is such that $P_b/P_e > P_2/P_1$, where P_2/P_1 is the shock pressure ratio across a normal shock, then a shock in the nozzle in indicated. Its location cannot be determined by MOC. At even lower $P_b/P_e > 1$, the back pressure may induce significant upstream boundary-layer thickening or separation, which would also invalidate the MOC solution.

In very recent times, real progress has been made in inviscid *transonic* flow calculations, allowing development of embedded shocks. See AGARD (1968) for a review of earlier works, and Lipnitskii and Lifshits (1970), Murman and Krupp (1971), Murman and Cole (1971), Steger and Lomax (1971A,B), McDonald (1971), Gopalarkrishnan and Bozzola (1971), Cahn and Garcia (1971), Grossman and Moretti (1970), Kentzner (1970B), Belotserkovskii (1970), Magnus and Yoshihara (1970), and Krupp and Murman (1971) for recent successes. Test cases were presented by Lock (1970).

Without the assumption of inviscid flow, the Navier-Stokes equation may be simplified with the *boundary-layer* approximations, which involve disregarding diffusion terms in the streamwise direction. For the classical "first order" (Prandtl) boundary-layer analysis in incompressible flow, this approximation gives rise to a single equation, parabolic in space, and applicable in a thin region near a wall. The pressure throughout the boundary layer and the "edge" velocity are determined by the inviscid free-stream solution, leaving only one parabolic equation for incompressible flow, which may be "marched out" in the streamwise direction. See Schlichting (1968) or Rosenhead (1963) for the basic theory.

For some free-stream conditions, a "similarity" analysis reduces the PDE to a two-point, nonlinear ODE in the dimensionless stream function. An example is the well-known Blasius equation for flat-plate flows,

$$f''' + 2ff'' = 0 \tag{6-38a}$$

$$f(0) = f'(0) = 0 \ , \quad f'(\infty) = 1 \tag{6-38b}$$

Although the solutions of such equations are interesting, because of the nonlinearity and the boundary condition at infinity, the solutions are readily obtained by "shooting" methods. There are much more elegant techniques (e.g., Keller, 1968; Bailey et al., 1968; Cebeci and Keller, 1971) and, in some cases, the problem may be converted to a two-pass initial value problem (e.g., Mufti, 1969). But generally, the crude shooting method is satisfactory unless large parametric studies are sought, or unless singularities are approached.

The more challenging problem is the *non-similar* boundary layer, for which PDE's must be solved numerically. (Similarity solutions provide good checks for the PDE solutions.) There exists an extensive literature on this subject, which we will not cover in detail. Schlichting (1968) has a small section on this topic. Blottner (1970) reviews the references for laminar compressible and incompressible boundary layers. For laminar compressible boundary layers, see also Smith and Clutter (1965). Patankar and Spalding (1967B) covered heat and mass transfer in turbulent boundary layers in incompressible flows. The *turbulent* boundary-layer solution involves (1) a choice of a theory for the Reynolds stresses (either Prandtl mixing length, eddy viscosity, or the turbulent-energy equation approaches, most commonly) and (2) the use of a local Couette-flow solution near the wall, necessitated by the high shear rates of the turbulent boundary layer. The proceedings of the Stanford "Olympics" (Kline et al., 1968) provides an overview of the computing "state of the art" as of 1968.

A forthcoming book by Cebeci and Smith will treat turbulent compressible boundary layers (see also Laufer, 1969; Cebeci and Smith, 1970; Cebeci et al., 1970). Bradshaw and Ferriss (1971) presented the extension to compressible flow of their successful method based on the turbulent energy equation.

The method of Blottner and Flugge-Lotz (1963) (see also Blottner, 1968, 1969, 1970; Flugge-Lotz, 1969) uses implicit Crank-Nicolson differencing in the transverse (diffusion) direction and a coordinate transformation which is based on similarity solutions for the boundary layer equations. It has become a common basis of other modern methods. Cebeci (1969) calculated laminar and turbulent incompressible axisymmetric flow. Blottner (1969) and Dean and Eraslan (1971) included finite-rate chemistry and dissociation; Sibulkin and Dispaux (1968) and Pearce and Emery (1970) included radiation effects; and Kendall et al. (1966) included coupled charring wall ablator chemistry in their hypersonic boundary-layer calculations. Lewis (1970A,B; 1971) presented valuable comparisons of different methods. Keller and Cebeci (1971A) applied Richardson extrapolation to boundary-layer solutions.

The boundary-layer equations may also be used in free shear layer and wake calculations. Plotkin (1968) and Plotkin and Flugge-Lotz (1968) calculated the incompressible wake of a flat plate, Ghia et al. (1968) calculated the mixing of laminar coaxial jets, Crenshaw (1966) calculated incompressible wall-jets and free-jets, Patankar and Spalding (1967A) calculated free jets, Schechter (1967) used Pade integration for the wake, and El Assar (1969) solved the compressible wake using the von Mises transformation and a further infinite-to-finite mapping. Three-dimensional and non-steady boundary-layer calculations present special difficulties, not the least of which is finding a good free-stream inviscid solution. Three-dimensional boundary layers were treated by Der and Raetz (1962), Shevelev (1967), Dwyer (1968, 1971A, 1971B), Nash (1969), Krause (1969), Vaglio-Laurin and Miller (1971), Boericke (1971), Krause and Hirschel (1971), Powers et al. (1967), and Moore and DeJarnette (1971). The last two papers included 3D MOC calculations for the free-stream flow. Unsteady boundary layers were treated by Oleinik (1967), Dwyer (1968), Hall (1969), and Piquet (1970); see also Proceedings IUTAM Conference on Unsteady Boundary Layers, September 1971.

"Second-order" boundary-layer theories,* while still neglecting streamwise diffusion, may allow for pressure gradients across the boundary, the effect of curved bow-shock vorticity gradients, and/or the effect of the boundary-layer displacement thickness on the free stream (see Van Dyke, 1962A,B). Levine (1968) and Ohrenberger and Baum (1970) presented compressible second-order boundary-layer calculations. Werle and Wornom (1970) calculated the effects of second-order boundary-layer terms in incompressible flow.

There are other systems of equations which are not properly boundary-layer equations but which are related, in that the elimination of streamwise diffusion allows a marching solution in space. These include: equations for entrance flow in ducts (Cochrane, 1969; Bankston and McEligot, 1969; Loc, 1970; see also Billig and Gale, 1970 for a related problem); the higher-order approximations of Plotkin (1968) and Plotkin and Flugge-Lotz (1968); the highly successful shock-layer equations of R.T. Davis (1968, 1970A) with shock slip and (Davis, 1970B) chemical reactions; and "parabolic marching equations" (Patankar and Spalding, Intl. J. Heat and Mass Transfer,Vol.15,1972,pp.1787-1806, and Briley, J.Comp.Phys.,Vol.14,1974,pp.8-28.

Other papers of interest are the calculations of Burggraf and Stewartson (1971) of boundary layer induced by a potential vortex, Blottner's (1971) comparisons of several first- and second-order methods for boundary layers with chemical reaction, the quasi-linearization and Chebychev series combination of Jaffe and Thomas (1970), the "method of lines" calculations of Steiger and Sepri (1965), the stability definition and study for a class of methods by Murphey (1963), and the ingenious method of "local nonsimilarity" of Sparrow et al. (1970), which allows calculation at a desired downstream location without the solution of the entire upstream boundary-layer history.

Dwyer et al. (1971) have shown how a general potential flow numerical method may be combined with a non-similar incompressible boundary-layer solution to give a total high-Re solution (up to transition or separation) in very short time. Provided that human inertia can be overcome, this approach should have considerable impact on design practice and on teaching methods.

There are several other simplifications to the Navier-Stokes equations which do not change the character so drastically as those mentioned above. When the viscous terms completely dominate the advection terms, the incompressible (ψ,ζ) system can be written as a single, fourth-order, linear, *biharmonic equation* for steady flow, as in the following exercise.

Exercise: Using the equations of Chapter II, show that the (ψ,ζ) system for steady flow as $Re \to 0$ can be written

$$\nabla^4\psi \equiv \nabla^2(\nabla^2\psi) = 0 \tag{6-39}$$

A difficulty with this approach is in the application of boundary conditions. For fluid dynamics applications, see Thom (1953), Pearson (1964), Fairweather et al. (1967),

*"Second-order" here refers to the theory, i.e. the continuum equations used, rather than to the finite-difference method.

Bernal and Whiteman (1968, 1970), Distefano (1969), Bourcier and Francois (1969), and J. Smith (1970).

Another common simplification is the *Boussinesq* approximation, which is applicable to the calculation of thermally driven natural convection flows. Calder (1968) considered the derivation and interpretation of the Boussinesq equations in rectangular coordinates. Barnes (1967) used the linearized Boussinesq equations in cylindrical coordinates; Torrance and Rockett (1969), in rectangular and cylindrical coordinates; de Vahl Davis (1968) and Festa (1970), in rectangular coordinates; Brown (1967), in spherical coordinates; Williams (1969), the 3D equations in cylindrical coordinates; Cabelli and de Vahl Davis (1971) and Lipps and Somerville (1971), the 3D equations in rectangular coordinates. W. P. Crowley (1968B) solved both the mean-flow and perturbation-flow Boussinesq equations in a study of atmospheric weak cellular convection. The work of Daly and Pracht (1968) and Daly (1969A,B) established the range of applicability of the Boussinesq approximation.

The calculation of weak pressure waves can be greatly speeded if *linearized* compressible flow equations are used. Zumwalt (1967) calculated sonic booms and Lu (1967) calculated near-field acoustic radiation in this manner.

The *quasi-one-dimensional approximation,* in which the flow area is approximated as a function of one spatial dimension and flow properties are assumed constant over the area, can be profitably used in nozzle calculations; see Crocco (1965) and Anderson (1969A,B; 1970A,B).

Gunaratnam and Perkins (1970) developed high-order implicit methods for treating open-channel flow with the *St. Venant equations.*

The subject of *meteorological* and other *geophysical calculations* is extremely important and is a major subject in its own right. Numerical weather forecasting is certainly one of the most demanding calculations, and the methods used by meteorologists are quite elaborate. In some aspects, the various meteorological systems of equations may be more complicated than the usual equations used in aerodynamics, due to the presence of Coriolis terms, solute transport terms, evaporation and condensation, irregular terrain, radiation, etc. But certain simplifications are also used. For example, the vertical extent of the atmosphere may be calculated in surprisingly few "levels"; the 5-, 7- or 9-level models are relatively recent and sophisticated, whereas many features are calculable in even 1- or 2-level models (see, e.g., Thompson, 1961; Houghton and Isaacson, 1968). Certain geometric problems of the spherical sector may be removed by the "β-plane" approximation. The laminar viscosity terms are usually omitted, although atmospheric turbulence may introduce diffusion terms. (The "turbulence" may be two-dimensional turbulence, or what other fluid dynamicists might call "organized wave motion.") The vorticity approach is frequenty preferred to the primitive-equation approach; one of the most interesting approximations is that the time-dependent stream function, and therefore the advection velocity field, may be specified, with only the vorticity field to be calculated. The boundary conditions may be well specified for a local-area fine-mesh calculation from previous calculations on a coarse mesh. There is usually no interest in steady-state calculations in meteorological problems, although there is for other geophysical problems (e.g., solar convection cells). At least second-order formal accuracy in time is generally required. An interesting aspect is that the hydrostatic pressure, p, is sometimes taken as the independent variable replacing the vertical coordinate, h, which is solved as h(p).

Comparisons of various finite-difference methods for meteorological problems were given by Lilly (1965), Kasahara (1965), Grammeltvedt (1969), and Polger (1971). The β-plane approximation was used by Grammeltvedt (1969), Williamson (1969), and Polger (1971) and the use of spectral equations was described by Baer and King (1967), Baer and Simons (1968), and Grammeltvedt (1969).

Calculations on a global scale are naturally fascinating. Leith (1965) presented a complete description of the physical model and the difference methods which have been used (see section III-A-13) for global weather prediction, and Hardy (1968) performed numerical experiments on atmospheric tides using Leith's program. Bryan (1963) and W. P. Crowley (1970C) carried out studies of the wind-driven ocean, and Bryan (1969) presented a numerical model for the world ocean, including the effects of ocean bottom topography. Leith (1971) showed that global scale atmospheric dynamics obey some laws of *two-dimensional* turbulence, and gave estimates of atmospheric predictability as limited by finite-difference accuracy and by the accuracy and completeness of initial data.

The following references are a sampling of recent works in meteorology and oceanography. Leblanc (1967) experimented with the numerical prediction of the formation of stratus clouds. Arnason et al. (1968) numerically modeled roll clouds, including the rain stage. Arnason et al. (1967) showed that the extension of finite difference equations across a singularity in the inviscid baroclinic stability equation gave inaccurate results. Estoque and Bhumralkar (1969) calculated the two-dimensional flow over a localized heat source. Gary (1969) compared two difference methods and Sielecki and Wurtele (1970) compared three methods for the shallow-water equations (incompressible, inviscid flow with a free surface) on a hemisphere, and Gustafsson (1971) applied an ADI method. Houghton and Jones (1969) presented a numerical model for linearized vertically propagating gravity and acoustic waves in the atmosphere. Kasahara and Houghton (1969) showed that different (non-unique) weak discontinuous solutions of shallow-water equations for the flow over obstacles can be formed from the same initial conditions, depending on the form of the PDE's. Sobey (1970) compared four methods for calculating long sea waves. Mancuso (1967) gave a procedure for calculating the stream function and velocity potential from a given vorticity field on a spherical surface. McCreary (1967) performed finite-difference calculations of 3D atmospheric convection cells. Yanowitch (1969) numerically studied vertically propagating waves in a viscous isothermal atmosphere. Van de Hulst (1968) presented a method of solving the reflection and transmission of radiation from thick atmospheres. Williamson (1969) used the β-plane approximation to calculate frictionless flow using a nonhomogeneous triangular grid. Pao (1969A) proposed the use of finite-difference methods to study the origin of atmospheric turbulence (instability) in stably stratified media, either atmospheres or oceans. Polger (1971) studied the effect of initial conditions, boundary conditions, and averaging operators on long-term nonlinear instability. The textbooks by Thompson (1961) and Richardson (1965) cover numerical weather prediction, and the review paper of Bryan and Cox (1970) covers the calculation of large-scale ocean currents. Although not directly concerned with meteorological applications, Gillis and Liron (1969) and Briley and Walls (1971) calculated the spin-up of a closed cylinder and the establishment of the *Ekman layers*, which are important in many geophysical problems.

A few articles have been written with *soil dynamics* problems in mind. Dienes (1968) considered non-Newtonian equations and finite-difference forms applicable to soil dynamics. Faccioli and Ang (1968) presented methods for spherically-symmetric compressible flow, applicable to wave and shock propagation in earth media. One interesting aspect of soil dynamics is that wave discontinuities (earthquakes) may be propagated by a *shear* mechanism, rather than by a pressure mechanism. This would obviously change the "shock-smearing" approach considerably. Another geophysical problem is the *seepage flow;* see the paper by Budal and Vasiliev in Roslyakov and Chudov (1963), and Drake and Ellingson (1970).

The reference books of Ames (1969) and Richtmyer and Morton (1967) include some geophysical applications, the latter presenting a precis of the atmospheric-front calculations of Kasahara et al. (1965). The book by P. D. Thompson (1961) covers numerical weather analysis and prediction, and is recommended as an introduction for the non-meteorologist. Other papers on geophysical problems are to be found in the Physics of Fluids Supplement II (Frenkiel and Stewartson, 1969).

We now turn our attention to *complications* of the usual Navier-Stokes equations.

We first consider complications in the form of boundary conditions which require separate PDE solutions. *Free-surface problems* have most often been solved by the MAC method (see section III-G-4) and extensions thereof. Early free-surface treatments were coarse, but later refinements allow surface tension effects and wave breaking to occur. See Killeen (1966), Narang (1967), Hirt and Shannon (1968), Daly and Pracht (1968), Daly (1969A,B), Viecelli (1969), Easton (1969), Mitchell (1970), Nichols (1971), Vasiliev (1971), and Chan et al. (1969, 1971). Inviscid free-surface calculations were performed by Brady (1967). Lagrangian calculations of free-surface waves were performed by Brennen (1971). The boundary layer approximation of the neglect of streamwise diffusion has been used in free surface flows (e.g., liquid jets) by Pletcher and McManus (1965), Fagela-Alabastro and Hellums (1967), and Saggendorf (1971), the last including rheological effects. Two-layer flows are similar to free-surface flows. Two-stream mixing of different gases was studied by Ghia et al. (1968), and two-layer stratified flows by Gemmel and Epstein (1962) and Hwang (1968). Evaporation/absorption at the boundary of a vapor-gas mixture was treated by Ohman (1967). Chen and Collins (1971) calculated shock wave diffraction at a gas-liquid interface.

Walls which move in response to the fluid dynamics occur in such diverse situations as underground nuclear explosions, high-velocity guns, and the action of the esophagus in swallowing. Watson and Godfrey (1967) and Watson (1969) used a PIC-like (section V-E-3) mixed Lagrangian-Eulerian formulation, and Hirt (1971) used the ALE-code, or arbitrary Lagrangian or Eulerian code. In this code, the computational nodes may move with the fluid (Lagrangian), remain fixed (Eulerian), or move at other speeds. In particular, the formulation may be Eulerian in one direction and Lagrangian in another, which is well suited to the calculation of pulsatile flow in elastic-walled cylinders (arteries).

A useful complication to the Navier-Stokes equations for unsteady flow is to use *perturbation equations* (e.g., Nagel, 1967; W. P. Crowley, 1968B). For example, a study of incompressible boundary-layer shear instability could be carried out in the (Ψ,ζ) system, writing

$$\Psi = \Psi_o + \Psi' \qquad \text{(6-40a)}$$

$$\zeta = \zeta_o + \zeta' \qquad \text{(6-40b)}$$

The steady-state solution, Ψ_o and ζ_o, may be taken as the basic flow, say from a Blasius solution (Schlichting, 1968). Then, without approximations, the steady-state vorticity transport and stream function equations may be subtracted out, as in the following exercise.

Exercise: Show that the perturbation (Ψ', ζ') equations are

$$\frac{\partial \zeta'}{\partial t} = -\nabla\cdot\vec{V}_o\zeta' - \nabla\cdot\vec{V}'\zeta_o - \nabla\cdot\vec{V}'\zeta' + \frac{1}{Re}\nabla^2\zeta' \qquad \text{(6-41)}$$

$$\nabla^2\Psi' = \zeta' \qquad \text{(6-42)}$$

$$\vec{V}_o = \frac{\partial\Psi_o}{\partial y}\vec{i} - \frac{\partial\Psi_o}{\partial x}\vec{j} \qquad \text{(6-43)}$$

$$\vec{V}' = \frac{\partial\Psi'}{\partial y}\vec{i} - \frac{\partial\Psi'}{\partial x}\vec{j} \qquad \text{(6-44)}$$

Because there are no truncation errors in the evaluation of Ψ_o and ζ_o, the perturbation equations may be expected to yield more accurate results than the full equations. Also, terms like $\nabla\cdot\vec{V}'\zeta'$ may be selectively "turned off" in the computation to isolate nonlinear effects; likewise, v_o and v' may be set = 0 to test the effect of the parallel flow approximation of the classical Orr-Sommerfeld stability analysis (e.g., Schlichting, 1968). This flexibility and control is one of the most attractive aspects of computational studies of flow stability. (The unattractive and difficult aspects are artificial damping and phase errors.) Preliminary computational experiments (unpublished) by the present author indicate that new boundary treatments are required for the perturbation equations at outflow boundaries.

The most obvious kind of complication to the Navier-Stokes equations results from the addition of terms and equations, through the presence of *additional physical phenomena* like radiation, chemical reactions, magneto-hydrodynamic effects, coriolis forces, multiple fluids, ionization, relativistic effects, etc. These equations often alter the character of the solution and may dictate the choice of methods used; radiation may introduce an integral (rather than differential) character to the equations, chemical reactions may lead to "stiff" equations (see section III-F-5), mhd shocks may exist apart from the gas-dynamic shocks, etc.

Numerical aspects of *radiation* transport equations without flow were treated by Costello and Shrenk (1966), De Bar (1967), and Wendroff (1969). One-dimensional radiative gas dynamics problems were computed by Jischke and Baron (1969) and Rubin et al. (1967), and by Rubin (1970), Rubin and Khosla (1970), Carlson (1970), and Watkins (1970); the last four references also included chemical nonequilibrium and multi-fluid (two-temperature) effects. Sibulkin and Dispaux (1968) and Pearce and Emery (1970) calculated radiation effects using boundary-layer approximations. Nicastro (1968) presented a similarity study of radiative gas dynamics equations which would be important for check solutions. Finkleman (1968) used a method-of-characteristics approach to radiative gas dynamics calculations. Pomraning et al. (1969) discussed the basic equations for multi-dimensional radiative hydrodynamics. Bohachevsky and Kostoff (1971) considered axisymmetric flow with 3D radiation effects using Lax's method. Leith (1965) included the effect of solar radiative heating in his atmospheric calculations. Callis (1968, 1969, 1970) calculated radiation effects on the blunt body problem.

Petschek and Hanson (1968) treated *elastic flow*.

Chemical reactions in high temperature flows, sometimes including vibrational and/or chemical nonequilibrium, ionization, and multi-fluid effects, were treated in the following papers: Bohachevsky and Rubin (1966), Bohachevsky and Mates (1966), AGARD (1968), Presley and Hanson (1969), Kamzolov and Pirumov (1966), Anderson (1969A,B; 1970A,B), Blottner (1969, 1971), Lomax et al. (1969),Dellinger (1969, 1970), Lewis (1970A,B; 1971), Davis (1970B), Carlson (1969, 1970), Kyriss (1970), Watkins (1970), Spurk (1970), Evans et al. (1970), Shelton (1970), Stubbe (1970), Lewis et al (1971B), and Li (1971).

Plasma and/or *MHD effects* were modeled numerically by Vulis and Dzhaugashtin (1968), Apelt (1969), Dawson et al. (1969), Brandt and Gillis (1969), Birdsall and Fuss (1969), Shelton (1970), and Bowen and Park (1971). Andrews (1971) calculated 1D shock propagation with *phase changes*.

When the computational methods for complicated flow equations come to maturity, it seems that one of the great applications, beyond the calculation of specific flow fields, will be in the *testing* of candidate *constitutive equations*. Four prominent areas for which numerical experimentation with constitutive equations are required are fluid turbulence, soil dynamics, non-Newtonian fluids, and fluids containing small structures. The suggested constitutive equations for turbulence range from the simple eddy viscosity concept of Prandtl (see Schlichting, 1968) to higher order closure theories involving 10 or more coupled PDE's. The troubles with the eddy viscosity approach are that the coefficients lack universality and that energy is transferred only from the mean flow to the turbulent eddies, whereas experiments have disclosed several flow situations where the reverse occurs. The trouble with the more complicated theories is that they *are* complicated, and the closure assumptions become more obscure. Orszag (1970) has reviewed the shortcomings of closure schemes. Some of the most intriguing of the recent approaches have so intimately involved the finite-difference equations (e.g., using an eddy viscosity approach only at the scale of the mesh size) that they can be called "computational theories" of turbulence. (Orszag even refers to a "turbulence and *compulence*," i.e. computer-simulated turbulence.)

There are many computational methods for *turbulent boundary layers*. A review and comparison is provided by the proceedings of the "turbulence Olympics" held at Stanford (Kline et al., 1968). The book by Patankar and Spalding (1967B) describes in detail what is probably the simplest acceptable method for calculating non-similar incompressible turbulent boundary layers, and the forthcoming book by Cebeci and Smith will cover other methods. Recent papers include Patankar (1967), Harper and Kinder (1967), Laufer (1969), Pletcher (1969), Martellucci et al. (1969), Chan (1970), Cebeci and Smith (1970), Wolfshtein (1970), Kacker and Whitelaw (1970) and Keller and Cebeci (1970). Donaldson (1968) presented transition-like boundary-layer calculations based on Reynolds stress evolution.

Calculations of *turbulent flows* without the boundary-layer assumptions were performed using the elementary eddy-viscosity approach by Walker and Zumwalt (1966), Ruo (1967), Walker et al. (1968), and Bauer et al. (1968). Leith (1965) used different eddy viscosities in the horizontal and vertical directions for his global meteorological calculations. On the smaller scale of cloud modeling, Lilly (1966) concluded that only a 3D model was adequate; Leblanc (1967) considered three different formulations of turbulent exchange coefficients.

Maschek (1968) discussed the obviously prohibitive cell-size limitations of computing high-Re turbulence by direct, brute force methods. Gawain and Pritchett (1968) presented turbulence-like calculations. O'Brien (1970) and Gawain and O'Brien (1971) presented transition-like calculations of channel flow.

It appears that less direct methods are required to calculate turbulent flows. Work with "computational theories" of turbulence was presented by Amsden and Harlow (1968), Harlow and Romero (1969), Nakayama (1970), Hirt (1969, 1970), Lilly (1969, 1971), Orszag (1969), Leith (1969, 1971), Daly and Harlow (1970), Gawain and Pritchett (1970), Chorin (1971), and Deardorff (1971). Leith (1971) indicated that global-scale atmospheric turbulence obeys some laws of 2D turbulence; this indicates what we have previously noted, that atmospheric "turbulence" is often a more organized periodic flow which aerodynamicists would not call turbulence. Other computational and theoretical aspects of turbulence were treated in papers in the symposium proceedings edited to Pao (1969B).

In the area of *soil dynamics*, which encompasses creeping flows on geologic time scales for low speeds and "terradynamics" (Colp, 1968) at high speeds, it is surprising that there is no soil nor any soil-like laboratory material for which an adequate constitutive equation exists, even for 1D flows. Butcher (1971) presented a numerical technique for calculating 1D shocks in a rate-dependent porous material.

In the area of *non-Newtonian flows*, Giaquinta and Hung (1968) calculated recirculating flows for a Reiner-Rivlin fluid, and Saggendorf (1971) calculated the free-surface jet for a viscoelastic fluid. Kirwan (1968) considered constitutive relations for a *fluid containing small non-rigid structures*, e.g., blood cells in plasma.

Finally, we consider complications to flow equations that result from a *different choice of mathematical description:* Lagrangian methods, patching methods, and Monte Carlo methods.

In *Lagrangian methods*, the equations used result from focusing attention on a fixed fluid "particle" and following its course in the flow. They are contrasted with the Eulerian methods used throughout this book, in which attention is focused on a volume in space through which the fluid particles flow. We have already mentioned several schemes (such as PIC, section V-E-3) which combine aspects of Lagrangian and Eulerian viewpoints. For 1D flows, the Lagrangian approach is often considered simpler, but for multi-dimensional flows with large net distortions, the Lagrangian methods become inaccurate and/or highly complicated.*

In *patching methods*, the attempt is made to numerically patch together regions in which different simplifications to the Navier-Stokes equations are used. For example, the near wake region of a projectile can be calculated with an inviscid (method of characteristics) solution for the external flow, a boundary-layer solution for the separated shear layer, and perhaps an incompressible solution for the recirculation region. Besides the obvious programming complexities, these methods present conceptual difficulties in the matching conditions to be met (or selectively ignored) across the boundaries, the iterative location and description of the boundaries (e.g., is the dividing streamline to be fit to a quadratic equation? does it originate *at* the sharp base corner?), and the stability of the overall patching iteration. Despite these difficulties, some accurate solutions obtained with patching methods have been published.

In *Monte Carlo methods*, the statistical nature of the molecular viewpoint is simulated with a finite number of computational particles. With a proper formulation, a consistent approximation to the Boltzmann equations is obtained. Calculations have been performed with Knudsen numbers as low as 0.01, approaching the continuum flow regime.

Lagrangian and mixed Lagrangian-Eulerian approaches are found** in the works of Schulz (1964), Noh (1964), Frank and Lazarus (1964), Zuev (1966), Watson and Godfrey (1967), Watson (1969), B. K. Crowley (1967), Hicks and Pelzl (1968), Hicks (1969), Wilkins (1969), W. P.

*It has often been said that Lagrangian methods fail for problems with large "deformations." This statement is not precise; for example, a spherical explosion can produce large deformations, but is well suited to Lagrangian calculations. Rather, Lagrangian methods fail for problems with large "distortions" (i.e., shear) such as boundary layer calculations.

**The references of section V-E-3 are not repeated here.

Crowley (1970A), and Arkhangel'skii (1971). The book by Richtmyer and Morton (1967) is recommended for fundamentals. Blewett (1970) presented a 2D quasi-Lagrangian method in which cell energy, rather than mass, is followed. *Patching methods* were used by Weiss et al. (1966), Moretti and Abbett (1966A), Baum et al. (1964) and Baum and Ohrenberger (1970). Brailovskaya (1967) successfully patched a Navier-Stokes region near a corner to an inviscid region above and to a second-order boundary-layer region downstream. Briley (1970) discussed problems of patching viscous and inviscid solutions in incompressible flow. *Monte Carlo* and other "random-walk" methods were treated by Vogenitz et al., (1968, 1970), Vogenitz and Takata (1970, 1971), Bird (1969A,B,C; 1970A,B), Bugliarello and Jackson (1967), Talley and Whitaker (1969), Matthes (1970), and MacPherson (1971). The review by Yen (1969) described several Monte Carlo methods and other methods applicable to rarefied gas flows, and provides additional references. Another review of Monte Carlo methods was given by Halton (1970).

VI-E. Areas of Future Development

In this final section of this chapter, we offer some speculation on areas of future development in the computation of fluid dynamics problems, and we provide some references (far from comprehensive) in areas not previously covered.

In regard to basic methods, probably the most important development will be in the area of *"semi-analytic"* (or equivalently, "semi-discrete") methods. This term covers a variety of distinct methods (series truncation, integral methods, method of lines, mixed differential-difference method, etc.) which share the common procedure of reducing PDE's to sets of coupled ODE's, to which the efficient ODE methods (Runge-Kutta integration, invariant embedding, etc.) may be applied. This trend is away from the "simulation" approach of computational fluid dynamics. It is important to realize that the character of the ordinary differential equations generated may be radically different from the original PDE's; e.g., the laminar boundary layer PDE gives rise to "stiff" ODE's in the method of lines. References for these semi-analytic methods are as follows: Friedman (1956), Gershuni et al. (1966), Dennis et al. (1968), Dennis and Chang (1969, 1970), Kerr and Alexander (1968), Kerr (1968, 1969, 1970), Veronis (1968), Luckinbill and Childs (1968), Underwood (1969), Ndefo (1969), Bryan and Childs (1969), Meyer (1969), Takami and Keller (1969), Glese (1969A,B), Vemuri (1970), Nelson (1970), Jaffe and Thomas (1970), Pierce (1970), Klinger (1970), Easton and Catton (1970), Thompson (1971), Belotserkovskiy et al. (1970), Holt and Ndefo (1970), Chushkin (1970A), B. W. Thompson (1971), Dennis and Staniforth (1971), Holt and Masson (1971), and Melnik and Ives (1971).

Other techniques, of which more use is expected in the future, are *patching methods* (in spite of their limited success to date), *mixed Lagrangian-Eulerian* methods, and especially the extensive use of *self-adaptive transformation* and *shock fitting techniques*. Another possibility is the use of *"finite element"* methods* for inviscid subsonic flows; see Sackett and Healey (1969), the review by Zienkiewicz (1969), and the fluid dynamics applications by Argyris et al. (1970). The finite element method is also appropriate for pure diffusion problems, although the study of Emery and Carson (1971) indicated that finite-difference methods are preferable for time-dependent and for variable-property problems.**

The impact on computational fluid dynamics of future developments in *computer hardware and software* will obviously be significant. The class of machines with the size and speed of the CDC "Star" and the ILLIAC IV (see, e.g., Slotnick, 1971) will certainly expand the range of soluble problems to include many 3D problems. However, it is not anticipated in the foreseeable future that details of 3D problems with local regions of large second derivatives (e.g., complex internal flows with diamond shocks, oscillatory separations, turbulent bursts) will be resolvable with brute-force uniform-mesh methods. (The gross features may often be computed without resolving such details, but not always.) Interestingly, the parallel-processing feature of the new machines will tend to make explicit methods

*The term "finite element" is not very distinctive, and some authors have mistakenly applied it to PIC-like codes and to equations derived via the control-volume approach (section III-A-2). While these semantics are plausible, they are not historically meaningful. The historical fact is that "finite element" methods refer to those based on a variational principle.

**An excellent survey article on finite element methods by A. R. Mitchell was published too late to appear in the references of this book. Entitled "Variational Principles and the Finite Element Method," it appeared in the Journal of the Institute of Mathematics and its Applications, Vol. 9, 1972, pp. 378-389.

somewhat more attractive in comparison with implicit methods, thus possibly reversing the current trend towards implicit methods for transport equations. (The "hopscotch" methods of Gourlay (1970A) and Gourlay et al. (1971, 1971A,B) appear to be well suited for use on parallel-processing machines.) Another possibility of major hardware development is in *hybrid machines* (coupled analog-digital machines), which have always had great potential but so far have not seemed to live up to the promise of their youth. References to hybrid calculations of PDE's are Vichnevetsky (1968), Finn (1968), Nomura and Dieters (1968), Ung and Paul (1968), and Tsuboi and Ichikawa (1969). However, it is important to realize that the "pipeline" architecture of the CDC STAR, the "parallel" architecture of the ILLIAC IV, and the parallel-sequential wedding of hybrid computers, while giving machine efficiency, often lead to algorithmic inefficiency and programming complexity.

The area of computer *software* development seems to us even more promising than does hardware development. Just as the development of FORTRAN languages made shock-patching methods feasible, future developments are expected to allow more complex schemes. Initial work has already been done to establish a library of PDE routines using the IBM PL1 processor (Cardenas and Karplus, 1970) just as all computer centers now have library packages of Gauss elimination routines, ODE routines, common analytic functions, etc. Although the unique boundary conditions which define distinct fluid dynamics problems would seem to prohibit them from being relegated to such a routine level, it is possible that the selection (say) of a two-step Lax-Wendroff method for regular interior-point differencing may become as routine as a call to a SINE function. Boris (1970) is working on a symbolic style of Algol which allows an economy of notation rivaling the vector and tensor notation for continuum equations. Experience with such processors for ODE's (the MIMIC language) indicates that the compilation and execution are often quite inefficient of machine time and therefore may not be appropriate for "production" runs (see next chapter); however, subsequent optimization need not necessarily be done by a programmer but could be accomplished by software (Boris, 1970).

Also, some computation centers have established *interactive systems* for PDE's (Tillman, 1969), in which the user has hands-on operation and varies the problem parameters as the solutions are presented to him graphically on cathode ray tube displays. The continued development of *hybrid software* also appears necessary if that area is to mature.

Perhaps most significant is the developing ability of computer software to *perform basic algebra and calculus manipulations* in an interactive mode (see, e.g., Tobey, 1969). This capability may allow extensive experimentation with new computational methods, both for fully discrete methods and (more significantly) for semi-analytic methods. Currently, the processing time is prohibitively long for this approach. The impact of this software development on analytical methods such as matched asymptotic analysis may prove to be even greater than on computational methods. The generation of high-order perturbation solutions via computer algebra/calculus could become quite routine for "regular" perturbations, but fluid dynamics is "rich" in *singular* perturbation problems, so the analyst will not likely become obsolete. But the intriguing possibility exists that high-order analytic solutions generated with the aid of the computer may be used in the same way that finite-difference solutions are currently used. Commonly, the desirability of an analytic solution stems not only from the speed of computation but also from the insight to be gained from the study of the functional form of the solution. But if the "answer" is a 40-page equation, it becomes impossible to interpret. Instead, the computer analytic solution can be machine-translated directly into a FORTRAN program, which can then be used in the manner of a finite-difference program to simply calculate numerical answers for specific input parameters. The "analyst" may not even care to *read* the functional form of the answers !

In addition to speculating on methods and computers, we now speculate on the types of problems which will be solved in the near future. Most computational studies to date have been somewhat isolated, in keeping with the immaturity of the field. In the future, we expect to see *more comprehensive studies* of specific flow problems involving not only computational experiments, but also analysis and physical experiments. A few studies of this type are those of Burggraf (1966), Tejeira (1966), Leal (1969), Mueller and O'Leary (1970), Burggraf and Stewartson (1971), and Fanning and Mueller (1970). Continued intensive work on *"computational theories" of turbulence* is expected, as is the *testing of constitutive equations* for turbulence, soil dynamics, and plasmas. *Hydrodynamic stability* simulations are expected to increase in number and in elegance; previous studies are those of Desanto and Keller, (1961), Fromm and Harlow (1963), Dixon and Hellums (1967), Nagel (1967), Piacsek (1968, 1970), Veronis (1968), W. P. Crowley (1968B), Bellomo (1969), Donaldson (1969), Loer

(1969), DiPrima and Rogers (1969), and Crowder and Dalton (1971A). Hydromagnetic instabilities were studied by Morse and Nielson (1971).

The use of *special flow equations* will present new challenges to computational fluid dynamicists: e.g. the equation $u_t = - u_{xxx}$ which arises in the study of ocean currents near the equator (Keeping, 1968); *relativistic* fluid dynamics, important in astronomical calculations (Schwartz, 1967); hydrodynamic effects in *explosive welding* (Godunov et al., 1970). The *coupling* of fluid-dynamics computations to other physically important phenomena will become more common; see the pioneering work in coupling fluid dynamics with 3D ablation, shape change and internal heat conduction (Popper et al., 1970), and with the pitching motion of bodies (Thompson, 1968; Trulio and Walitt, 1969). Note that the latter type of problem is complicated by the inapplicability of the Galilean (wind tunnel) transformation; see Lighthill (1954).

Finally, we note that the subject of *information processing* and utilization, a subject discussed briefly in the next chapter, will continue to be an active area of research in its own right.

CHAPTER VII

RECOMMENDATIONS ON PROGRAMMING, TESTING, AND INFORMATION PROCESSING

In this chapter we present recommendations on computer programming and debugging practices, testing of computational methods, and processing of the information generated. Of course, these activities are not entirely distinct in practice, with feedback and iterations occuring among them. Many of these somewhat unconnected ideas are purely personal views, but my contacts with other laborers in the area indicate that none of these views are singular. Some people would dismiss these topics as mundane and trivial, but we consider an awareness of them to be essential to the successful orchestration of an entire research project.

VII-A. Computer Programming

It is necessary to distinguish between recommendations applicable to an *exploratory* program and to a *production* program. Many research projects require only exploratory programs. Indeed, it may be said that the research in computational *methods per se* is completed when a production program can be written, although such a program can then be used for research in the *fluid dynamics per se*, or development work. There is a clear analogy here with the design, development, check-out, and calibration of a wind tunnel, versus the systematic running of experiments.

1. For an exploratory program, do most of your own programming.

Of all the recommendations to be presented here, this will probably annoy the most people, but we feel strongly on this point. We quote Hamming (1962), "Chapter N+1. The Art of Computing for Scientists and Engineers." "The purpose of computing is insight, not numbers." And again, "It is not likely that great physical insights will arise in the mind of a professional coder who routinely codes problems." In the author's experience with multi-dimensional fluid dynamics problems, the most subtle aspects of the computation, particularly boundary-condition problems and initial-condition problems associated with multi-step methods, have emerged during programming. In fact, we find it difficult to distinguish the phases of programming and the actual method development, which often is left to the programmer.*

However, a professional programmer, if available, can provide assistance with input-output problems, tape handling, computer systems problems, and with neat subproblems such a Gaussian elimination routine, a tridiagonal matrix solver, etc. These latter jobs are sometimes referred to as "coding", as contrasted with the more creative "programming" jobs.

2. For a production program, call in a professional programmer, if available.

3. For an exploratory program, design the program modularly.

The advantage of modular programming is that each subprogram (SUBROUTINE's and FUNCTION's in FORTRAN) can be tested separately and can even be written by separate people.

4. For a production program, avoid modular programming.

It takes time to CALL a subprogram. For example, it takes 7 to 8 μsec to call a one-argument subprogram on a CDC 6600, compared with about 1 μsec for a * and 0.4 μsec for a ±. If the argument list exceeds 7, the call takes longer. The penalty is not excessive unless the call is made for every node-point calculation, however. Thus, a modularized subprogram to set initial conditions is efficient, but a subprogram to calculate a critical Δt by scanning the entire mesh, which in turn calls a subprogram to evaluate the speed of sound for compressible flow, can be time consuming if it is used at every time step. (But see No. 13 below.)

In this regard it is worthwhile to note the advantage of STATEMENT FUNCTIONS in Fortran. Their use is not widespread, apparently because their advantage is not widely known. A STATEMENT FUNCTION is *not* a subroutine and does not involve any "call" time. The processor merely substitutes the arithmetic of the statement into the "calling" statement. For example, consider the calculation of a Mach number M at the location (I,J) where the total velocity magnitude is VMAG and the temperature is T(I,J). The statement "VS(K,L)= ..." defines the statement function.

```
.
.
.
VS(K,L) = C1*SQRT(T(K,L))
REAL M
.
.
.
C1 = ...
```

*The relationship between a computational fluid dynamicist and a programmer is reminiscent of that older relationship between an aircraft designer and a draftsman, humorously described by K. D. Wood in his aerospace design books.

```
      .
      .
      .
      M = VMAG/VS(I,J)
      .
      .
```

This "call" to the (non-executable) statement function VS(K,L) results in the same machine-language instructions as the program

```
      .
      .
      .
      REAL M
      .
      .
      .
      C1 =
      .
      .
      .
      M = VMAG/(C1*SQRT(T(I,J)))
```

Yet the program is modular in the sense that the speed-of-sound computation is isolated to the reader of the program and can easily be changed.

For more involved calculations, statement functions can "call" other statement functions, and the physical cards can be extracted from the deck and tested modularly. The savings in programming effort, keypunching, and debugging can be considerable.*

5. *In early stages, be specific.*

For example, if your intent is to experiment with different outflow boundary treatments, design your first program with only the simplest method. Once that basic program is debugged, add an option selector for different treatments. This recommendation especially applies for multiple, interacting options. What usually happens is that some of the options are not tested.

6. *In early stages, be simple.*

It is not excessive even to suggest programming a different, simpler field-point method from the one to be used eventually. The other features of the program can then be debugged more easily. Complex methods make programs much harder to debug. (For example, in the author's experience in programming a two-step compressible flow method, a gross algebraic error in the density calculation in the first step first exhibited itself in a long-term divergence in the temperature.) It is also recommended to begin programming with the constant μ and k equations for compressible flow; to first program and debug only the gas-dynamics part of a radiation gas-dynamics problem; to withhold the complications of free-surface treatment until the interior-point program has been checked out, etc.

In connection with simplicity, we note the major difference between incompressible- and compressible-flow problems. Incompressible-flow equations are, in a sense, the more complicated, in that there are both a marching problem for ζ and an elliptic problem for ψ. But there are fewer variables, and the (ζ,ψ) equations are processed sequentially. For compressible flow, all four equations are processed in parallel with most methods and are therefore more difficult to debug.

7. *Don't start a compressible-flow project with an incompressible-flow problem.*

It appears plausible to solve an incompressible problem in the desired geometry and then graduate to the complications of compressible flow. In fact, the structures of the problems are so different that, although some of the experience with the incompressible problem will carry over, the whole process is not very efficient.

*Confusion and errors can arise if these chain "calls" become longer than two links, and the same characters are used for dummy arguments and "calling" arguments.

This remark applies to full Navier-Stokes solutions. For boundary-layer solutions, the structures of the incompressible and compressible problems are more similar.

8. Block out the program, and separate blocks with print statements.

As Hamming (1962) points out, it is a common mistake to rush into the details of arranging the computation. On the other hand, the present author does not hold formalized flow diagrams in very high esteem. In our experience, these flow diagrams are usually constructed in the report-writing stage from the program, rather than the program being written from the flow diagram. Except for the highest organizational levels of the program, the detailed flow diagrams are usually harder to read than the program itself, provided that COMMENT cards have been used well.

When something goes wrong in a calculation, it is important to know where in the program the malfunction became exhibited, what the current values of DO-loop indexes were, etc. In the debugging stage, extensive print-outs should be used to provide this information.

9. In estimating program development time, block out stages.

Estimating the time required to complete a computer project is not one of the present author's skills, although we have at least learned the well-known dictum, "it will take longer than you think." But it does help to block out different estimates for each stage of overall program block organization, initial phase programming (very specific program), debugging of the refined program, more exhaustive testing and verification of test cases, production run time, information processing time, etc.

10. Know your computer.

Machine idiosyncrasies are usually more important in "cleaning up" a production program. If a programming expert is available, he can allow for them best, but it helps if the program originator is knowledgeable.

Just as an illustrative example, we note the difference between a CDC 6600 and the old IBM 1620 in regard to mixed-mode arithmetic. The CDC 6600 converts mixed-mode arithmetic to floating point, but the IBM 1620 does not. On the IBM 1620, 2*X is faster than 2.*X, but just the opposite on the CDC 6600. On the IBM 1620, 2*X is equivalent to X+X; but on the CDC 6600, X+X is faster than 2.*X, since ± takes about 0.4 μsec but * takes about 1 μsec. (Note that X+X+X is slower than 3.*X.) Also, on the CDC 6600, * takes about 1 μsec but / takes about 2.9 μsec. Thus, it may be considerably more efficient to read in a Reynolds-number RE, compute (once) an inverse REI = 1./Re, and evaluate the contribution of diffusion terms as REI*(diffusion terms), rather than as (diffusion terms)/RE. The execution times for * and / are not different on all machines.

However, in the early stages of program development, readability should not usually be sacrificed for machine efficiency. Also, note that the more advanced processors (software) will often "optimize" such operations automatically. These examples do serve to point out how comparisons of the relative speed of different programs and different methods can be machine-, processor-, and programmer-dependent.

11. Avoid floating-point exponentiation.

The computing time penalties for floating-point exponentiation are so bad that they should be avoided, even in the early developmental stage of an exploratory program. On the CDC 6600, a floating-point multiplication X*X takes 1 μsec while the exponentiation X**2. takes 94 μsec. Even if X is zero, X**2. takes 12 μsec. (But note that the fixed-point exponentiation X**2 is equivalent in execution time to X*X.) The time penalty can often be diminished by making use of the SQRT function; X**0.5 takes 94 μsec, but SQRT(X) takes only 23 μsec on the CDC 6600.

The most common occurence of exponentiation in fluid dynamics is in the temperature-viscosity laws. The Sutherland gas law (e.g., Schlichting, 1968),

$$\frac{\mu}{\mu_o} = \left(\frac{T}{T_o}\right)^{3/2} \left(\frac{T_o + S}{T + S}\right)$$

where $\mu_o = \mu(T_o)$, and μ_o, T_o, S are constants of the gas, may appear in a FORTRAN statement of the normalized equations as

```
MU =A*(B*T)**1.5*C/(T + D)
```

where A, B, C, D are constants, or as the algebraic equivalent,

```
MU = A*SQRT(BB*T**3)*C/(T + D)
```

where BB is another constant. The first formulation on a CDC 6600 will require about 101.3 μsec, and the second about 31.3 μsec. However, the second formulation can give large intermediate values for (un-normalized) T >> 1, which can seriously diminish accuracy in a machine with a short word length.

Depending on the temperature range expected, the Sutherland law can often be approximated by

$$\mu/\mu_o = (T/T_o)^\omega$$

where ω ranges from 0.5 to 1. For $\omega = 0.5$, the statement

```
MU = A*SQRT(B*T)
```

will require about 25 μsec. For the simple case of $\omega = 1$, the statement

```
MU = AA*T
```

would only require 1 μsec. But in the total problem, the $\omega = 1$, or $\mu \propto T$, approximation saves even more time than the factor of 100 indicated above, since terms such as $\partial/\partial x\ (\mu \partial T/\partial x)$ are simplified to $\mu \partial^2 T/\partial x^2$.

If the accuracy of an elaborate viscosity relationship like the Sutherland law is required, a linear law in a STATEMENT FUNCTION may be used for the exploratory program. For the production program, the desired relation may be broken down in a subprogram into linear or square-root fits over subintervals.

For nonlinear viscosity laws, great amounts of computer time can be saved by storing a 2D array of viscosity values μ_{ij} instead of computing new μ for each T_{ij}. The first method requires IxJ calculations of the viscosity law, and terms like $\mu_{i+\frac{1}{2},j}$ are calculated as $\frac{1}{2}(\mu_{i+1,j} + \mu_{ij})$ using the form of equation (5-100). The alternate method, of calculating $\mu_{i+\frac{1}{2},j}$ by applying the viscosity law at each (i,j) location, either by the form of equation (5-100) or (5-101), requires 5 times as many calculations of the viscosity law for μ_{ij}, $\mu_{i\pm1,j}$ $\mu_{i,j\pm1}$. Further time may be saved if only a steady-state calculation is intended, by updating the array μ_{ij} only every 10 or so time steps.

The same observations apply to elaborate equations of state, which may be the dominant time consumer in studies of shock-wave propagation in solids.

12. *For a production program, group subscripted terms and use singly-subscripted operations.*

The FTCS differencing of the model advection-diffusion equation can be written as

$$\zeta_i^{n+1} = \zeta_i^n - c\left(\zeta_{i+1}^n - \zeta_{i-1}^n\right) + \alpha\left(\zeta_{i+1}^n - 2\zeta_i^n + \zeta_{i-1}^n\right)$$

A Fortran statement can be written as

```
ZP(I) = Z(I) - C*(Z(I+1) - Z(I-1)) + A*(Z(I+1) - 2*Z(I) + Z(I-1))
```

Counting the mixed-mode multiplication 2*Z(I) as an addition and the subscript indexing (I+1) and (I-1) as additions, the statement would require ten ± and two *, for about 6 μsec on a CDC 6600. If, instead, the terms are grouped as

```
ZP(I) = D1*Z(I) + D2*(Z(I+1)) + D3*(Z(I-1))
```

where D1 = (1.-2*A), D2 = (A-C), D3 = (A+C) are previously defined and therefore not calculated for all I, the statement gives four ± and three *, for about 4.6 μsec on a CDC 6600, for a ratio of 4.6/6 = 0.77.

Skoglund and Gay (1968, 1969) found that this grouping of terms, using the Lax-Wendroff method (Section V-E-5) and a linear viscosity law, reduced the computation time per iteration to 1/3 the original value. While this is of great value on a production program, the grouping is tedious and makes the program less readable; it is therefore not recommended for the early stages of an exploratory program.

The use of singly-subscripted operations can avoid repeated calculations which are internal to the FORTRAN language. Consider the simple replacement statement

```
W = A(I,J)
```

where A has been dimensioned A(IL,JL). The computer actually stores a single array which starts at the core memory location assigned to "A", and continues for a total of IL*JL locations. For a singly subscripted variable like P(IL), the location of P(I) is calculated as

```
loc P(I) = loc P + I-1
```

For the doubly subscripted variable, the machine must calculate the location

```
A(I,J) = loc A + I-1 + (J-1)*IL
```

For the common situation where many replacements and arithmetic operations will be made on many variables with the same double subscripts, considerable time can be saved by "preprocessing" the arithmetic to find the location, as in

```
L = I-1 + (J-1)*IL
W = A(L)
X = B(L)
Y = C(L)
Z = D(L)
```

Note that the arrays A, B, C, D may have been DIMENSIONED as doubly-subscripted variables, and yet can be treated as singly subscripted variables in FORTRAN operations.

The most significant opportunities for time saving by this technique arise in nested DO loops in 3D programs. The FORTRAN statements that result are not very readable, so the technique is not recommended for an exploratory program.

13. For a production program, avoid critical Δt calculations.

Provisions should be made for not calculating critical Δt every time iteration for a slowly varying problem, but only every 10th iteration or so. This consideration can significantly improve the running time of the program. For some problems, the critical Δt can be accurately approximated analytically, so that critical Δt calculations can be eliminated entirely.

VII-B. Debugging and Testing

A computer program is *debugged,* or corrected for programming errors; the program and the methods involved are *tested* for stability, convergence, accuracy, comparison, etc. Once the program is giving some kind of reasonable answers, with no compilation diagnostics, the two activities are often indistinguishable.

14. Debug and test extensively in a coarse mesh.

Obviously, accuracy cannot be tested in a coarse mesh. But stability and convergence of the FDE iterations usually can be tested in a ridiculously coarse mesh. A minimum requirement is that we have at least one regular interior node. Qualitatively reasonable solutions, sufficient for testing stability, iteration convergence, boundary treatments, option checks, plotting routines, etc., have been obtained for some problems in a 4 x 4 mesh with only 9 internal points. Avoid the decimal syndrome. It is not necessary to have the magic 10 points in a boundary layer; often 2 or even 1 will suffice for debugging.

If all users at a major computing center would adopt the policy of testing in as coarse a mesh as possible, the reduction in computing load would probably be comparable to the installation of a later-generation computer.

15. Recognize the importance of good debugging routines and of fast turn-around times.

There is a wide range of performance of FORTRAN processors with regard to diagnostics provided. Many computer centers have special debugging routines in addition to the basic processor. Even more important than these is the turn-around time, which is the correct characteristic time (rather than the computation time) to consider in program development time. The turn-around time is partially under the user's control, because most computer centers will give priority to short-run programs and, at a time-sharing facility, to programs requiring small storage. Both these items are aided primarily by testing in a coarse mesh. Sometimes, the user can test the program or components of it on older or slower machines which often have better turn-around times than faster machines at the same computer center.

For those aspects of turn-around time not directly under the user's control, he can at least pressure the computer center for faster response. More fundamentally, the decision of where to work or attend graduate school should be based in significant part on the turn-around time and overall computing situation at the candidate locations, if one if going to do serious work in computational fluid dynamics.

16. Isolate errors by turning off terms.

When computations fail to converge iteratively or give bad answers, and programming errors are suspected, the error can sometimes be isolated by "turning off" certain terms in the equations. For example, one can set A = 0 early in the program, and then re-punch a card with A*(y-diffusion terms) to test for boundary-layer behavior. The solution of the Poisson equation for stream function can be bypassed to linearize the vorticity transport equation in order to test for a nonlinear instability. Boundary values can be frozen. Any of the four variables of compressible flow can be turned off or the variables otherwise decoupled, although care must be taken that sneaky coupling does not appear through the equation of state and the conversion from conservation to non-conservation variables. A test run with $\Delta t = 0$ will often disclose errors in converting from conservation to non-conservation variables, although it is not appropriate to methods such as Lax's (section V-E-4).

Note also that an entire dimension can be turned off, to test either a method or a program. But as we have seen earlier, there are sometimes dimensional characteristics which are not retained in a reduced-dimension problem. Particularly, some authors have used one-dimensional problems to experimentally compare convergence rates of iterative schemes; but convergence is sensitive to dimensionality, and one-dimensional tests are not at all appropriate. The dimensional effect may be suppressed momentarily in a compressible-flow problem using explicit methods by starting with a uniform flow profile, in which case the effect of a wall should not propagate more than 1 node from the wall per time step for a single-step method (or m-nodes for an m-step method).

17. Try to exercise all option combinations in a production program.

This recommendation sounds obvious, but is actually impossible to follow in many programs because of the overwhelming number of possible combinations. Sometimes these options are not even fully tested in isolation, e.g., not all chemistry options are exercised, let alone all chemistry options in combinations with all boundary conditions, all viscosity laws, etc., In this regard, it is well to note that extreme complexity of a program (or a method) actually affects accuracy, or at least *faith* in the accuracy, since program errors can become practically impossible to find. It can even be said that such programs have an infinite debugging time. As use continues, more of the finite number of errors are being discovered and corrected; but usually options are also being added or changed, which can produce new errors which are not noticed until future uses occur with other option combinations. Even without adding options, the "quick fix" of one discovered error may mess up other option combinations which are not immediately disclosed. The present author has seen basic errors discovered regularly in elaborate multiple-option ballistic trajectory codes which had seen continual use for five years.

18. Test for stability and iterative convergence over a wide parameter range.

While still in the coarse mesh, test over the desired extremes of M, Re, etc. (This is not always possible, if stability depends on the mesh size, as in cell Re limitations.)

The suggestion of Miller (1967) may be useful in testing the stability of any computation with an evolutionary behavior. He suggests running two parallel computations, starting from initial conditions which differ by an amount that is small in some sense, but significantly larger than round-off errors. Instability is inferred by monitoring a functional (differences of squared independent variables) of the two parallel computations. But note that numerical and hydrodynamic instabilities are not distinguishable from this approach.

19. Avoid the nonlinear instability syndrome.

When a computation goes bad, nonlinear instability is the usual whipping boy. The term has been used to cover up programming errors, maltreatment of boundary conditions, and unanalyzed linear instabilities. In fact, truly nonlinear instabilities are relatively rare; the bounded oscillations in a strong shock region* and the difficulty with the "combined iteration" of (ζ,Ψ) (see section III-A-22) provide the only ready examples. More often, non-nonlinearities cover up errors and true instabilities by providing a mechanism of amplitude limitation.

True nonlinear instabilities may sometimes be cured (and errors may sometimes be obscured) by better initial conditions, which can be obtained with a less accurate but more stable method. Lapidus (1967) could practive this very easily, by just using the first step (Lax method) of a two-step Lax-Wendroff method (section V-E-6) to attain good initial conditions.

*Stetter's (1970) results on the inviscid Burger's equation indicate instability caused by a local inversion of the characteristic direction in the FDE.

20. Test stringently for the iterative convergence of the FDE.

This can usually be done in an extremely coarse mesh. As stated in sections III-D and V-H, there are no truly rigorous iterative convergence tests. But in a coarse mesh, it is often possible to attain identical convergence (i.e., zero changes) of an FDE computation to within the word length of the computer. (This is not possible for all methods, since a small amplitude oscillation sometimes persists.) If a second computation with different initial conditions then identically converges to within a round-off error tolerance of the first solution, convergence of the method can be claimed, not with actual certainty, but at least with a clear conscience.

21. Consider withholding some information as a check.

This is a suggestion of Hamming (1962). For example, known symmetry conditions are usually used to reduce the size of a problem. But for initial checks in a crude mesh, the development of symmetry from asymmetric initial conditions might be a meaningful check. Likewise, the single-valuedness of pressure at a corner (section III-E-3) can be withheld to check for truncation convergence. Other instances may occur to the reader.

22. When possible, test accuracy against exact solutions.

The accuracy of a method or of a program can sometimes be tested against exact solutions. These exact solutions need not necessarily be realistic problems to be of value; on the other hand, good comparisons with a particular exact solution do not necessarily imply that a method is generally accurate, or that a program is truly error-free.

Especially, the steady "viscometric" flows with parallel streamlines (Couette flow, Poiseuille flow, flow between rotating concentric cylinders, etc.) do not provide very meaningful tests of the accuracy of a method. The following are examples of useful exact solutions which have been used in the open literature to test computational methods.

P. J. Taylor (1968) used the impulsively started infinite flat-plate problem, called the Rayleigh problem or Stokes' first problem (Schlichting, 1968) to study stability of boundary formulations at outflow and inflow boundaries. The solution is

$$u(y,t) = U_w \operatorname{erfc} \frac{y}{2\sqrt{\nu t}}$$

where erfc is the complimentary error function. In addition to the impulsively started plate, an exact similarity solution also exists for the continuous velocity specification $U = at^b$. The solution is $U = at^b F(\eta)$, where $\eta = \frac{y}{2\sqrt{\nu t}}$ and $F(\eta)$ is the (numerical) solution of the following ODE.

$$F'' + 2\eta F' - 4bF = 0, \quad F(0) = 1, \quad F(\infty) = 0$$

Beardsley (1969) used the analogous problem in cylindrical coordinates of the impulsively stopped solid-body rotation in a cylindrical container of unity radius. The initial condition on azimuthal velocity v_θ is $v_\theta(r,o) = r$ for $r < 1$. For all time $t > 0$, $v_\theta(1,t) = 0$. the governing equations are

$$\frac{\partial \zeta}{\partial t} = \nu \nabla^2 \zeta , \quad v_\theta = \frac{\partial \Psi}{\partial r}$$

and

$$\zeta = \nabla^2 \Psi = \frac{1}{r} \frac{\partial}{\partial r} \left(r \frac{\partial \Psi}{\partial r}\right)$$

The exact solution in terms of Bessel functions is

$$v_\theta = -2\sum_{n=1}^{\infty} \frac{J_1(\lambda_n r)}{\lambda_n J_o(\lambda_n r_o)} e^{-\lambda_n^2 \nu t}$$

$$\Psi = 2\sum_{n=1}^{\infty} \frac{J_o(\lambda_n r) - J_o(\lambda_n r_o)}{\lambda_n^2 J_o(\lambda_n r_o)} e^{-\lambda_n^2 \nu t}$$

$$\zeta = -2\sum_{n=1}^{\infty} \frac{J_o(\lambda_n r)}{J_o(\lambda_n r_o)} e^{-\lambda_n^2 \nu t}$$

The flow near an infinite flat plate oscillating with $U_w = U_o \cos(\omega t)$, which is Stokes' second problem (Schlichting, 1968) could also be used. The solution is

$$u(y,t) = U_o e^{-\eta} \cos(\omega t - \eta)$$

where

$$\eta = u \frac{\omega}{2\nu}$$

A shortcoming of these one-dimensional solutions as test cases is the lack of advection terms in the governing equations. A non-physical problem which adds a non-homogeneous term was used to study accuracy by Greenspan (1967). The PDE is

$$\frac{\partial u}{\partial t} = \frac{\partial^2 u}{\partial x^2} + (\sin x)(\sin t + \cos t)$$

$$u(x,0) = 0 \;,\; 0 \leq x \leq \pi$$

$$u(0,t) = 0 = u(\pi,t) \text{ for } t > 0$$

The solution is

$$u(x,t) = (\sin x)(\sin t)$$

True nonlinear behavior in one dimension has been studied via the Burgers' equation (2-19).

$$\frac{\partial u}{\partial t} + u\frac{\partial u}{\partial x} = \alpha \frac{\partial^2 u}{\partial x^2}$$

which may be written in the conservation form

$$\frac{\partial u}{\partial t} + \frac{\partial}{\partial x}\left(\frac{u^2}{2}\right) = \alpha \frac{\partial^2 u}{\partial x^2}$$

Schroeder and Thomsen (1969) used the inviscid equation ($\alpha = 0$) with conditions

$$u(x,0) = x \;, \quad 0 \leq x \leq 1$$

$$u(0,t) = 0, \quad t > 0$$

The exact solution is $u = x/(1+t)$, which exhibits no shocks (see section IV-G) for $t > 0$. However, this solution gives $\partial^2 u/\partial x^2 = 0$ for all time, and does not appear to be an adequate test problem (for example, no artificial viscosity effect can appear). Cheng (1968) and Allen (1968) studied the accuracy of boundary outflow formulations and the conservation form using the viscous equation ($\alpha > 0$) with boundary conditions $u(0,t) = 0$, $u(-1/2,t) = 1$. The exact steady state solution over $-1/2 \leq x \leq 0$ is

$$u(x,\infty) = -\,2\alpha \tanh(Ax)$$

where

$$A = 2 \text{ arctanh}\left(\frac{1}{2\alpha}\right)$$

Taylor et al.(1972) used Lighthill's exact solution of the time-dependent problem of an initial step function in an infinite domain (no spatial boundary conditions). For initial conditions of

$$u(x,0) = a \text{ for } x \leq 0$$

$$u(x,0) = o \text{ for } x \geq 0$$

the exact solution is

$$u(x,t) = a/\left\{1 + \exp\left[\frac{a}{2\alpha}\left(x - \frac{1}{2}at\right)\right]b\right\}$$

where

$$b = \text{erfc}\,(-x/2\sqrt{\nu t})/\text{erfc}\,\left[(x - at)/2\sqrt{\nu t}\,\right]$$

and erfc is the complimentary error function.

Gourlay and Morris (1968A) used the following one-dimensional nonlinear PDE,

$$\frac{\partial u}{\partial t} + \frac{x^2 u^2}{t}\frac{\partial u}{\partial x} = \left(\frac{2x^3 u^2}{t} - 1\right)\cos(x^2 - t)$$

with conditions of

$$u(x,1) = \sin(x^2 - 1)$$

$$u(o,t) = -\sin t.$$

The problem was solved over $0 \leq x \leq 1$ and $1 \leq t$. The exact solution is

$$u(x,t) = \sin(x^2 - t)$$

A two-dimensional problem with truly nonlinear advection-like terms was used by Gourlay and Morris (1968B). The PDE is

$$\frac{\partial u}{\partial t} + \frac{\partial}{\partial x}\left(\frac{u^2}{4}\right) + \frac{\partial}{\partial y}\left(\frac{u^2}{4}\right) = f(u,x,y)$$

where

$$f(u,x,y) = 50uxy[y(1 - y)(2 - 3x) + x(1 - x)(2 - 3y)]$$

The PDE could also be used in a non-conservation form, with $\frac{\partial}{\partial x}\left(\frac{u^2}{4}\right) = \frac{u}{2}\frac{\partial u}{\partial x}$

Gourlay and Morris used initial conditions of

$$u(x,y,0) = 100x^2(1 - x)y^2(1 - y)$$

and boundary conditions of

$$u(0,y,t) = 0 \text{ for } t > 0$$

$$u(x,0,t) = 0 \text{ for } t > 0$$

They compared their numerical solutions to the exact solution for a *steady state,* which is

$$u = 100x^2y^2(1 - x)(1 - y)$$

The maximum value of $u(x,y,t)$ over $0 \leq x \leq 1$ and $0 \leq y \leq 1$ for all t is $\simeq 2.2$, on which stability considerations may be based.

Y. T. Chiu (1970) gave an exact solution for traveling waves in a stratified isothermal atmosphere, which would be a useful test problem for many atmospheric codes.

W. P. Crowley (1968A) tested methods for phase error with the following exact solution of the 1D, linear, variable-coefficient, inviscid model equation.

EXERCISE: Given

$$\frac{\partial \zeta}{\partial t} = -u\frac{\partial \zeta}{\partial x}$$

with initial conditions of

$$\zeta(x,0) = f[\ln(u)], \text{ f any function}$$

and $u = ax + b$, where a and b are arbitrary constants, show that the exact solution is

$$\zeta(x,t) = f(\ln u - at)$$

Greenspan (1967) used three one-dimensional problems with linear but variable coefficient advection-like terms. All three problems had initial conditions of $u(x,0) = x$ and had solutions of $u(x,t) = x\,e^{-t}$. These may be somewhat useful, but is should be noted that the exact solution gives $\partial^2 u/\partial x^2 = 0$ for all time. Thus, although the second derivative appears in the equation, the only contributions to this diffusion-like term are due to accumulated round-off and truncation errors, so the problems may not be representative. The same remarks apply to a two-dimensional nonlinear test problem, used by Gourlay and Morris (1968A), for which the solution $u = (1 - x - y)e^{-t}$ gives $\partial^2 u/\partial x^2 = \partial^2 u/\partial y^2 = 0$ for all time.

Realistic test problems with exact solutions for inviscid, one-dimensional, compressible, perfect gas flows are conveniently cataloged by Hicks (1968), who gives seven problems involving shocks, rarefactions, and wave interactions. These solutions were used by Hicks and Pelzl (1968) to compare the accuracy of methods in Lagrangian coordinates. Gordon and Scala (1969) used the planar piston, planar mass injection, and spherical explosion problems for testing. NiCastro (1968) has presented exact similarity solutions for the spherically symmetric radiative gas dynamic equations for both implosions and explosions. The restrictions on initial conditions and the form of the radiative transfer law are mild, and the solutions could be of great value as test cases for these computationally demanding problems. Similarly, Sternberg (1970) has presented similarity solutions for plane, cylindrical, and spherical shocks with chemical reaction.

B. K. Crowley (1967) tested her "almost Lagrangian" quasi-one-dimensional code on two idealized problems with mass addition and deletion. The mass source is assumed to provide mass at zero kinetic energy, with an energy sink used to balance the net energy flux to zero. For the mass deletion problem, the energy flux of the mass sink, which removes both internal and kinetic energy, is balanced with an energy source. Although these assumptions are of no physical interest, they allow exact solutions which are valuable for testing.

For boundary-layer methods and programs, a large number of exact similarity solutions exist (see Schlichting, 1968). Most of these require a power-law free-stream velocity distribution $u_e = ax^m$. (For incompressible flow, the class $u_e = ae^x$ may also be used for test purposes, although these flows are of no physical interest.) For compressible flow, additional restrictions of $Pr = 1$, $\mu \propto T$, and T_{wall} = constant are usually required. D. Weiss (1968) has presented a class of compressible-flow inviscid solutions which transform into the required power-law velocity relation under the Stewartson transformation. The similarity solutions for stagnation flows (see Schlichting, 1968) are well suited to testing finite-difference methods,* although they have not as yet been used for this purpose. Other test problems are the class of similar axisymmetric viscous compressible solutions of Shchennikov (1969), of which the viscous jet is a special case.

Of course, potential-flow analytic solutions provide a check for many aspects of a complete Navier-Stokes flow computer program; see, e.g., Kramer (1969).

23. When possible, test accuracy against proved approximate solutions.

The most common approximate solutions of the Navier-Stokes equations are the boundary-layer solutions (Schlichting, 1968). These may be analytical solutions, numerical similarity solutions of ordinary differential equations, or non-similar solutions of the PDE's. Note that the differences in the boundary-layer and Navier-Stokes problem are not only in the neglect of streamwise diffusion terms, but also in the application of far-field boundary conditions.

*A suggestion of Dr. F. G. Blottner.

The boundary-layer approximations provide good approximations at lower Re than generally assumed. Briley (1970) says his Navier-Stokes solutions with 15 points in the boundary layer just began to depart from the similar boundary-layer solutions at a momentum thickness Re of 15 to 30. Also, see Mueller and O'Leary (1970) for experimental comparisons at low Re.

24. When possible, test accuracy against experimental values.

Although this is obviously good practice, it is well to note that the accuracy of the experimental values is usually just as suspect as the accuracy of the computations. In particular, we note that an experimental three-dimensional flow along a plane of symmetry should *not* be mistaken for a truly two-dimensional flow. Although the z-velocity components may be zero, terms like $\partial^2/\partial z^2$ may be non-zero and may therefore contribute to a three-dimensional momentum flux. Also, note that pressure measurements within a compressible boundary layer are difficult.

25. When possible, test for accuracy using global solutions.

Cyrus and Fulton (1967), in their analytical study of FDE accuracy, demonstrate the danger of testing numerical data at a single location or at a few points to characterize the error in an entire problem. Also, Skoglund and Gay (1968, 1969), in their comparison of boundary formulations, found that local residuals in non-conservation forms were not sensitive to different techniques and were therefore not good indexes of accuracy.

It is worth noting that mere *qualitative* accuracy does not provide much of a recommendation for a particular interior-point method for a PDE. With good boundary conditions, bounded errors, and a smooth solution, how far wrong could a FDE solution be?

26. In comparing methods and programs, keep in mind machine-dependent aspects and programming aspects.

From a single comparison of two different programs of any appreciable complexity, little significance can be attached to small differences in computing speed, say less than 10 or 20%.

27. Initially, use simplified model equations to experiment with new methods.

VII-C. Information Processing

Once we start obtaining good solutions to a computational fluid dynamics problem, we are faced with a difficult question. What are we going to do with all those numbers?

The simulation of a typical problem, and especially of a three-dimensional transient problem, generates a tremendous amount of information, and the problem of interpreting and digesting the results is of great importance. As Williams (1969) has said, "An apparent simulation of physically observed characteristics does not in itself form an understanding of the flow." Currently available computers are adequate for the computation of many 2D flows, and even some 3D flows with high resolution appear achievable with the advent of computers like ILLIAC IV and the CDC "Star". But in the opinion of many workers in computational fluid dynamics, we are not even utilizing enough of the data which we can currently generate. Experiences does suggest the following recommendations.

VII-C-1. Numbers

28. Print out some answers at every iteration, and all the answers for some iterations.

This is especially important in the early development of an exploratory program. The vorticity or density at a particular location (preferably a sensitive area, such as an external corner region) may be printed out at each time step. (In order to reduce the printer line count, the values can be stored for 10-15 iterations and then printed out on a line.) In FORTRAN, an E-FORMAT is preferable.

Every 100 or so iterations, it is a good idea to print out all the answers. A tabulation of values in E-FORMAT provides complete information but may be difficult to digest. An effective device for a moderate-sized field is to convert to an I-FORMAT and print out the two or three significant figures with the sign for all the node points in the field. An example is given in Figure 7-1, which shows a plot of 100 × stream function for a

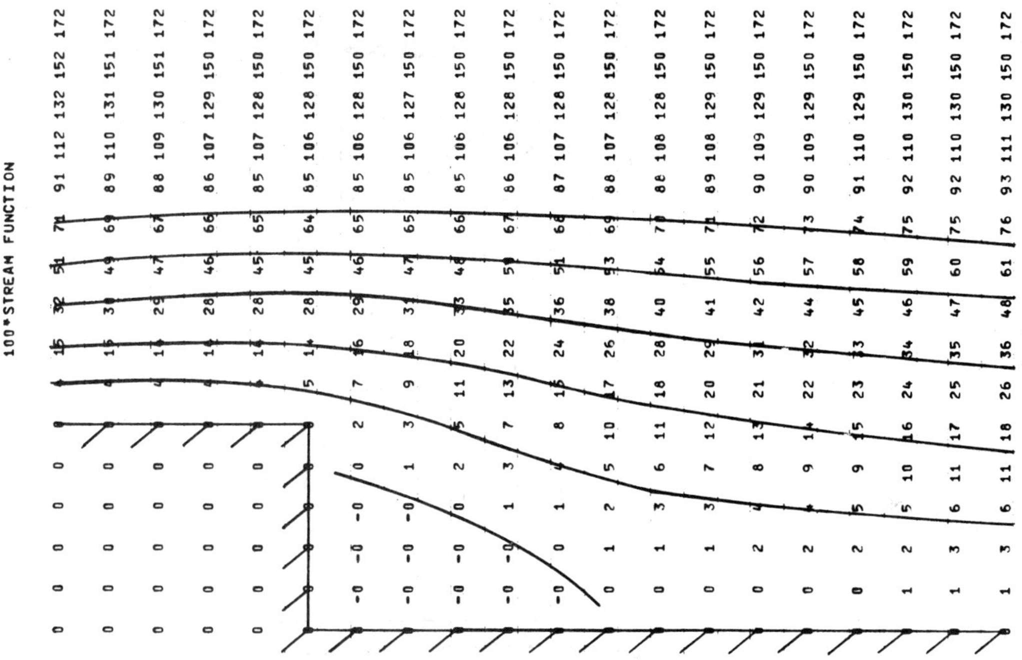

Figure 7-1. Digitized node value plot of backstep flow, Re = 10. Values are 100 x stream function, properly rounded, and printed by the computer on-line printer. Streamlines and wall symbols were sketched in by hand.

backstep flow. The flow direction is down the page. Using this node-point value map, one can sketch rough contour line plots by hand, to follow the flow development. The proper rounding is accomplished by plotting the integer variable LS(J), computed from the stream function S(I,J) as follows:

```
      DO 3 I = 1, IMAX
      DO 1 J = 1, JMAX
    1 LS(J) = 100.* S(I,J) + SIGN (0.5, S(I,J))
      PRINT 2 LS(J), J = 1, JMAX
    2 FORMAT (27(1XI4))
    3 CONTINUE
```

The SIGN (0.5, S(I,J)) is a library routine (or a user-written FUNCTION sub-program) which returns the magnitude of the first argument with the algebraic sign of the second argument.

It is worthwhile to print out the entire field, rather than just the area of interest. In a particular calculation* of separated flow over a V-notch, the view of vortex shedding in the region of the notch looked at least plausible, as shown in Figure 7-2A. But when the entire field was printed out (Figure 7-2B), plausibility lapsed. An error in downstream boundary conditions was then found and corrected.

*This striking illustration is provided through the courtesy of Prof. T. J. Mueller, University of Notre Dame; see Mueller and O'Leary (1970).

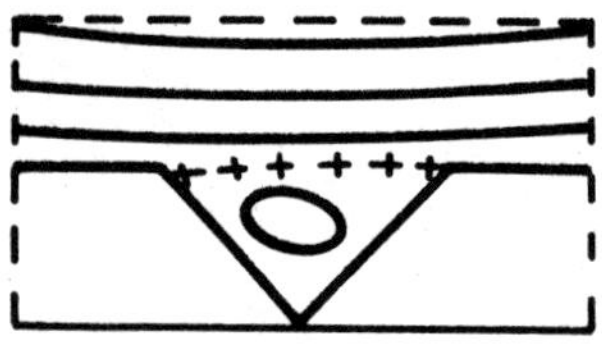

(a) partial field plot

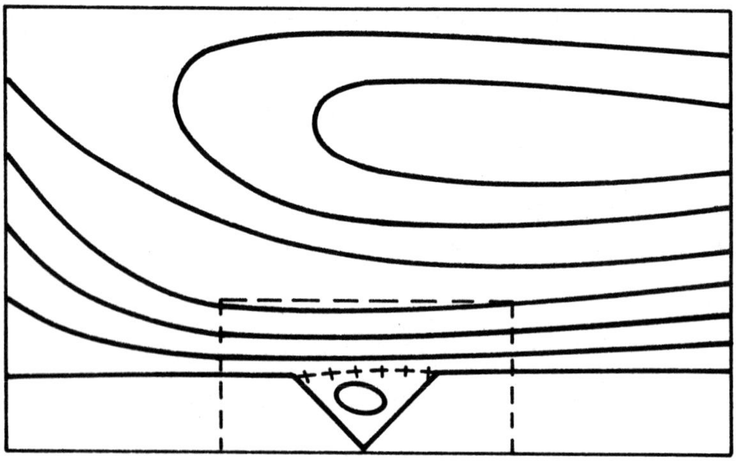

(b) complete field plot

Figure 7-2. Sketched streamline plots of an erroneous calculation of flow over a V-notch. The line ++++ is the dividing streamline.

It is also advisable to obtain complete field values at two consecutive time steps, say every 100 and 101 iterations; otherwise, the nature of a time-splitting instability (section III-A-6) may not be disclosed. When a new method is being developed or studied on a one-dimensional-model equation, it is advisable to print out all the information at every time step, especially the first few steps from the initial conditions. In multi-step methods, it may even be advisable to print out results of the intermediate steps. The added insight is usually worth the increased burden on output equipment. Of course, as confidence in a method and a program increases, the amount of output information used for monitoring can be reduced.

VII-C-2. Plots and Motion Pictures

29. Make extensive use of computer plotting for information processing.

Machine plots are the most commonly used information-processing device. Plots can be two-dimensional, three-dimensional, or contour (isogram) plots. They can be constructed on printers (typewriters) or on line plotters (pen-and-ink plotters or cathode ray tubes). The printer plots are best used for program debugging and "quick look" plots. They are limited in resolution to the horizontal and vertical spacing of the printer, but are usually much faster to obtain since they are produced in-line with the computation and require no additional tape handling. Line plotters give higher resolution plots, and offer more flexibility in plotting symbols and captions.

Two-dimensional plots can be used, for example, to display shock profiles in a one-dimensional shock-propagation code by plotting (say) density versus distance, at a particular time. Examples may be found in Figure 5-1 on page 231. Two-dimensional plots may also be used to portray iterative convergence by plotting some functional such as $\sum_{ij} \left|\zeta_{ij}^n\right|$ versus n. Examples may be found in Figure 3-32 on page 175. User-oriented software packages for both printer and line-plotter two-dimensional plots are becoming available at most computer centers. It is not very difficult to design such a printer program. To plot density R versus distance X = I*DX in a one-dimensional shock problem, let X run down the page and let R run from left to right. The integer number of spaces N which determines the height of the plot can be determined from

$$N = SC*R(I)$$

where SC is a scale factor determined to fit the density range into the number of horizontal spaces available. If density has been normalized so that the maximum expected R is 1, then SC = 100. can be used. (Most printers allow well over 100 horizontal spaces, but 100 is very convenient. Also, overshoot in the shock region may cause R > 0, requiring more than 100 spaces.) The alphanumeric type declaration can then be used in conjunction with Hollerith printing to type out (N - 1) blank characters, followed by an x, 0, + or other symbol to represent the function value. I is then indexed up, N is recalculated, and the next line is printed.

Three-dimensional plots and contour plots (isograms) are more qualitative in nature than two-dimensional plots. Examples of these complementary devices for presenting another dimension of information are given in Figure 7-3. Many excellent examples may be found in Harlow and Amsden (1971) and Thoman and Szewczyk (1966).

Contour plots are easy to construct on a printer. A vector V(J) of alphanumeric declaration is filled with blanks. For a particular value of I (which runs down the page), the normalized variable to be plotted, say R (I,J), is scanned in the J direction, and different alphanumeric characters are stored in V(J), depending on the range of R(I,J). An effective scheme is to store the sequence of characters (A,B,C,D,E,) or just asterisks for $R\varepsilon[0.,.1]$, $[.2,.3[$, $[.4,.5]$, $[.6,.7]$, $[.8,.9]$, and to store blanks for $R\varepsilon[.1, .2]$, $[.3,.4]$, $[.5,.6]$, $[.7,.8]$, $[.9,1.0]$. The A(J) are then printed, blanked out, and the scan repeated at (I+1). This type of plot was used by Thoman and Szewczyk (1966). An example is given in Figure 7-4.

If the mesh aspect ratio $\Delta x/\Delta y$ used does not agree with the printer ratio of line (vertical) spacing to character (horizontal) spacing, which is typically (1/6 inch)/(1/10 inch) = 5/3, preliminary interpolation in either the I or J direction will be required to produce an undistorted contour plot on the printer.

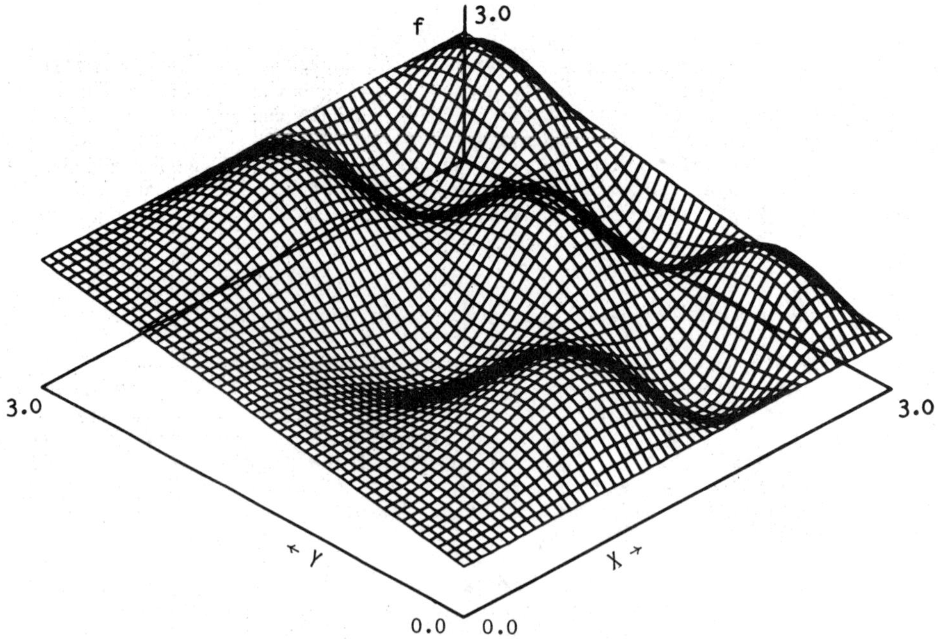

(a) three-dimensional plot of f(x,y)

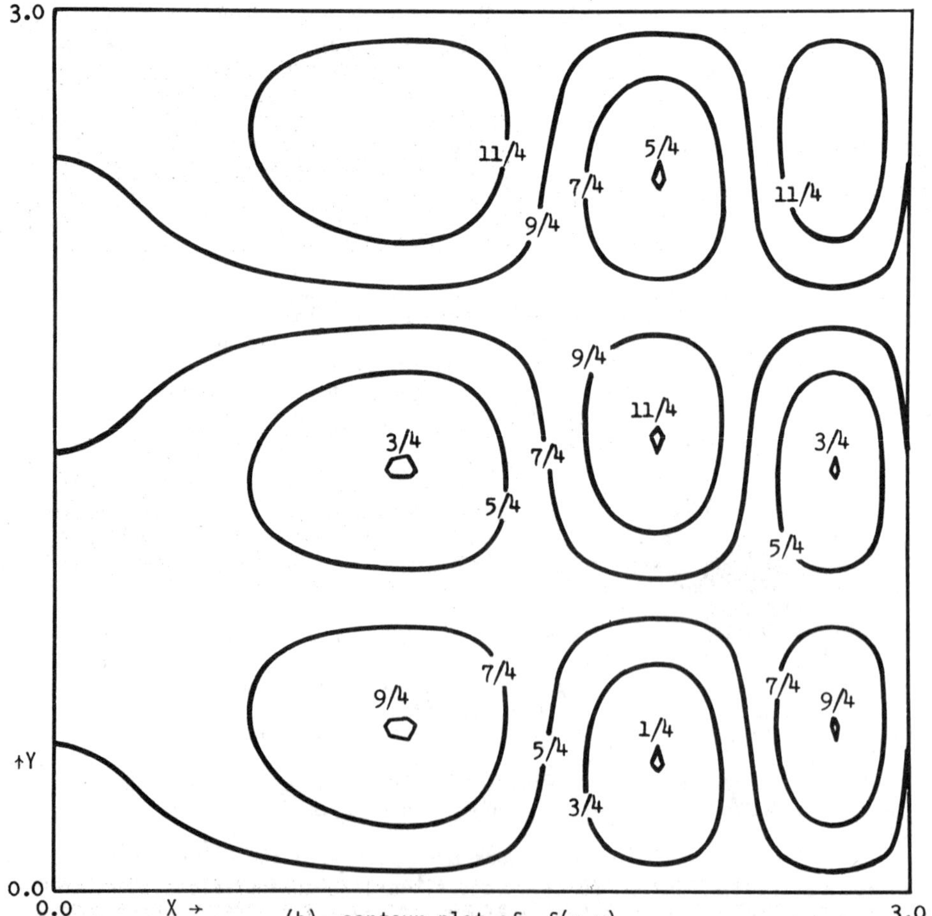

(b) contour plot of f(x,y)

Figure 7-3. Three-dimensional plot and contour plot of the function $f(x,y) = 1+y/2+\sin(\pi x^2/3)\sin(\pi y)$, obtained with a line plotter. After Sundberg (1970).

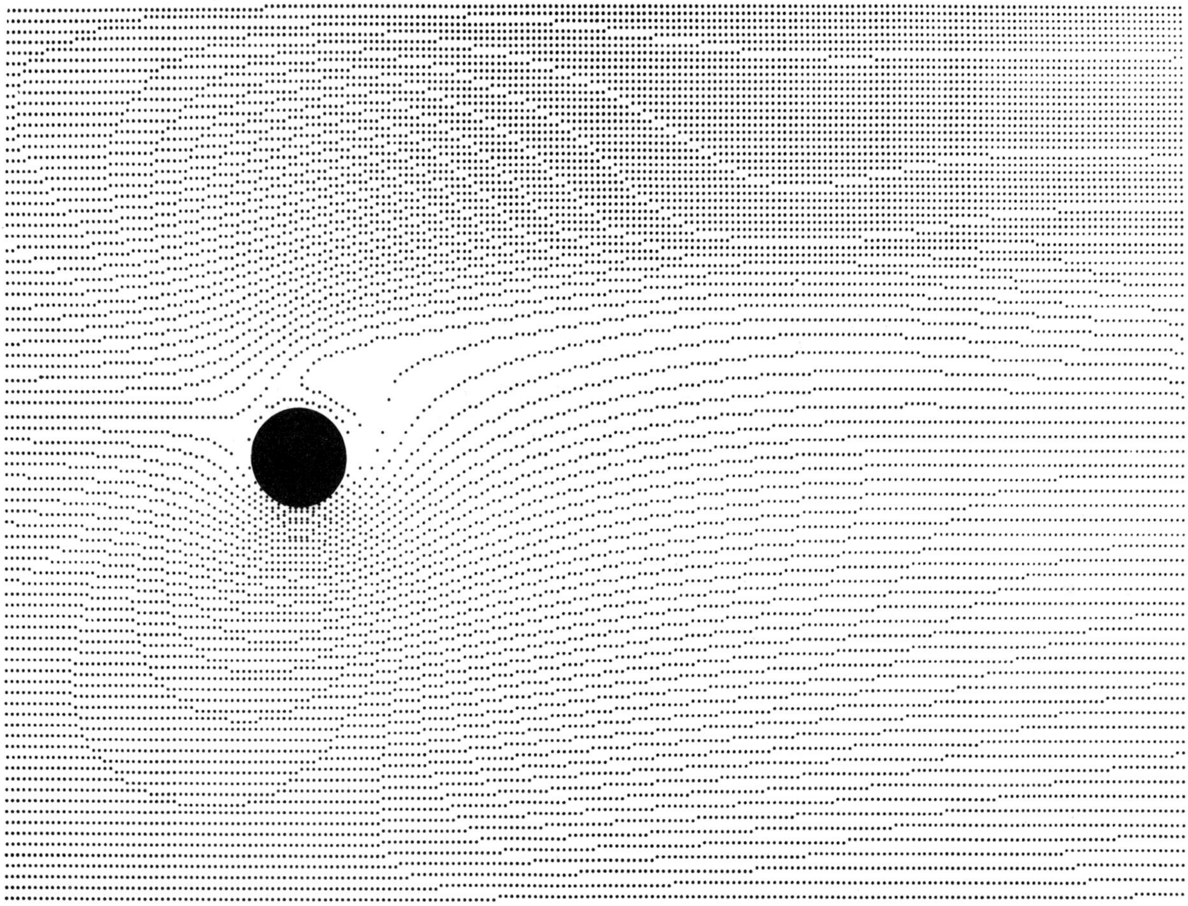

(a) streamlines, $\Delta\psi$ = 0.000196

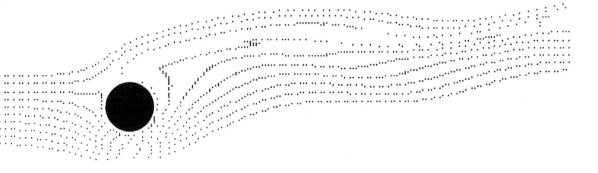

(b) streaklines

Figure 7 - 4. Stream function and streakline contour plots obtained from an on-line printer, after Thoman and Szewczyk (1966). Re = 200, lifting cylinder with $V_{tangential}/U_o = 2$, dimensionless time = 21.139.

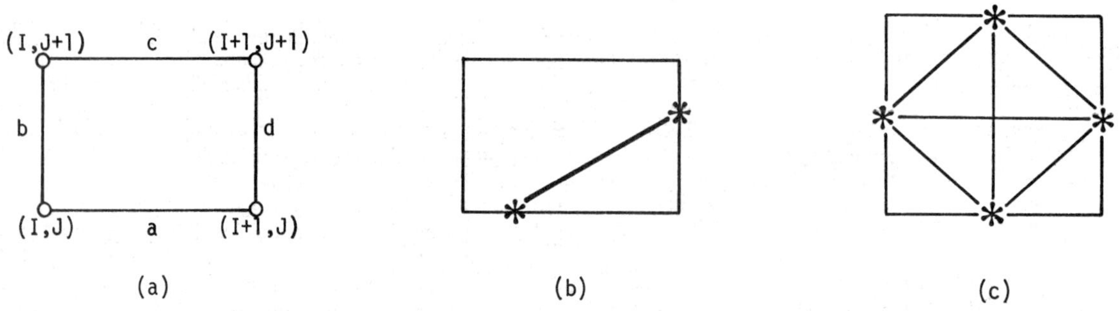

Figure 7-5. Contour plots on a line plotter.

The logic involved in producing a contour plot on a line plotter is interesting for its own sake. If the surface defined by the function value is monotonic in x and y, the logic is quite simple. But the presence of multiply-connected regions, of a directional extremum which causes a splitting of contour lines, and of saddle points, complicates the logic. The prescriptions given here are from Sundberg (1970).

The mesh is scanned for a desired contour value by focusing attention on a cell formed by four neighboring node positions. At (I,J), we focus our attention on the cell formed by (I,J), (I+1,J), (I,J+1), (I+1,J+1), depicted in Figure 7-5a. The sides of the cell are labeled a, b, c, d. Suppose we are plotting an S = 0 line. If the extreme values of S along the side a (i.e., S(I,J) and S(I+1,J)) differ in sign, then S= 0 somewhere between. The location along a is found by linear interpolation. Likewise for sides b, c, d. Then all these locations, denoted by *, are connected by straight lines on the line plotter as in Figure 7-5b, and the next cell is scanned.

This simple prescription nicely takes care of the following eventualities. If the contour line should pass through a corner point, say S(I+1,J) = 0, then two "points" are located, one on side a and one on side d. These two points have the same location, and are connected, in the logic of the program, by a zero-length line.

Using this definition of the number of contour value points, we now easily verify that the number of points per cell must be 0, 2, or 4. Following the circuit (a,b,c,d) around from S(I+1,J) with S = 0,there must be an even number of sign changes to return to S(I+1,J). Each sign change represents an S = 0 value along that line segment; hence, there must be an even number of S = 0 values.

If four locations are found, they are connected in all six possible ways, as in Figure 7-5c. This is a quite appropriate description in the total plot, because it is impossible to determine the surface shape inside the cell from just the data at the four corners; the region inside could be a ridge running along the diagonally opposed high values, a valley running along the diagonally opposed low values, or a saddle-point region. The correct character is indicated by the plots at adjacent cells, so the indeterminacy suggested by Figure 7-5c is appropriate.

An additional prescription is used when the contour value S = 0 occur (to within some tolerance) at two adjacent node values, say (I,J) and (I+1,J). Then S = 0 all along the side a is indicated, and no unique position can be located. In this case, no points are used along this line. In the case where a region of S = 0 extends over many cells, this treatment results in an S = 0 line being drawn along the edge of the region but not in the interior, which is again appropriate.

30. Don't plot with resolution much higher than the computation.

Instead of linear interpolation, the higher order techniques usually used in experimental data processing can be used without fear, since these arbitrary procedures do not feed back into the computation. However, these are frequently misleading because the computations are not carried out at high resolution. Linear interpolation plots represent the information in the least prejudiced manner, although smoother plots are more appealing. Quadratic interpolation is consistent with a second-order method.

A special problem arises in separated flow studies. The points of separation and reattachment cannot be located, except with semi-analytic methods (Underwood, 1969). If these data are to be presented, they should be unambiguously labeled as being inferred by extrapolation, etc. See Lavan et al., 1969; Roache and Mueller, 1970; Chavez and Richards, 1970. (The same remarks apply to separation and reattachment data from *physical* experiments.)

Computer centers are now developing libraries of user-oriented three-dimensional and contour-plot programs. An expert programmer, if available, can be requested to set up plotting routines.

Many computer centers now have facilities for producing motion pictures of the 3D plots or contour plots, with each frame being a contour plot at a different time. These motion pictures can be highly effective in conveying the dynamics of a time-dependent process. The Los Alamos group has pioneered this aspect of information processing, as well as the methods of computing the solutions. In written reports, frames from the motion pictures or sequences of the original plots can be used to present a sequence of events as in Figure 7-6, and the motion pictures themselves can be made available to interested people. See Hirt (1965), Harlow and Fromm (1965), Thoman and Szewczyk (1966), Donovan (1968, 1970), and Fromm (1970A) for examples.

Lax (1969) tells a story which illustrates the power of motion-picture information processing. The Kortweg-de Vries equation,

$$\frac{\partial u}{\partial t} + u\frac{\partial u}{\partial x} + \frac{\partial^3 u}{\partial x^3} = 0$$

describes long waves over water and some wave phenomena in plasma physics. The existence of certain solitary wave solutions were discovered by Kruskal and Zabusky, who first observed the emerging solitary waves by studying motion pictures of the computations. Once noted, careful computations isolated the phenomena and led to a pure mathematical solution. Thus, computations disclosed an unexpected property of a nonlinear PDE, a possibility which had fascinated von Neumann.

The choice of which variables to plot can be important. Usually, the variables of the computations are of interest, such as ρ, u, v, T for a compressible problem. For incompressible problems, all the variables ζ, ψ, u, v, P are of interest, no matter which combinations were used in the computation. (Pressures should be presented in coefficient form.) Other derived quantities which can be of interest are the source term of the pressure Poisson equation; the stagnation pressure coefficient, which is an indicator of viscosity effects (Burggraf, 1966; Macagno and Hung, 1967; Roache and Mueller, 1970); the dissipation function; the entropy; and the relative size of diffusion and advection terms. Allen (1968) plotted terms which indicate departure from boundary-layer behavior.

The $\psi = 0$ streamline, called the dividing or discriminating streamline, is of special significance in steady separated flow problems because it demarcates the recirculation flow regions. For emphasis, it can be plotted with a special symbol or distinctive line quantity. Dawson and Marcus (1970) accentuated the $\psi = 0$ line by simply plotting two lines, at $\psi = 0 \pm \varepsilon$, where ε is much less than the other intervals of the contour plot. (They used $\varepsilon = 0.001\ U_\infty L$.) Because recirculation velocities are small, a change in plotting increment $\Delta\psi$ is usually advisable for $\psi > 0$ and $\psi < 0$ (Roache and Mueller, 1968; Allen and Cheng, 1970). The direction of flow can also be indicated by plotting an arrow symbol, $\rightarrow$, available on most line plotters, for each line segment. It may also be possible to scale the length of the arrow symbol, so that it lies along a streamline, it indicates flow direction, and its length indicates velocity magnitude. Examples of ψ-plots are given in Figure 7-7a and b. The computed streamlines may be compared with patterns obtained experimentally by the surface aluminum-dust technique of Prandtl, as in Figure 7-8.

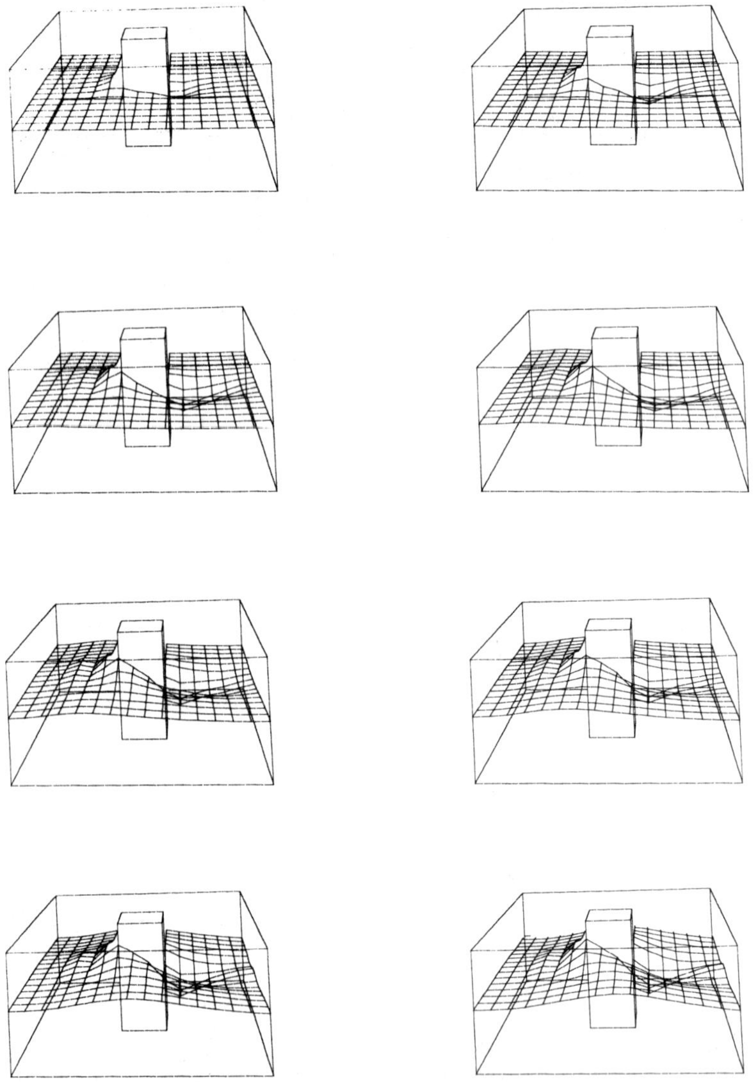

Figure 7-6. Frame sequences from a motion picture of contour plots, showing time development. The plots show, in perspective view, the surface of water flowing in the vicinity of a square pile. The calculations were carried out by W. Nichols of Los Alamos Scientific Labs using a 3D MAC code. (Courtesy of C.W. Hirt, LASL.)

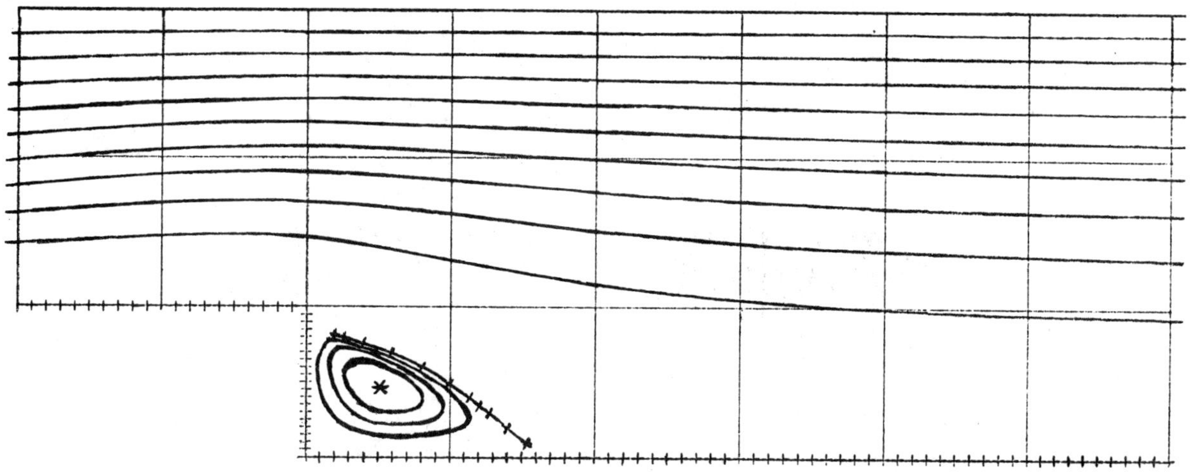

(a) stream function for incompressible flow over a backstep, Re = 10.

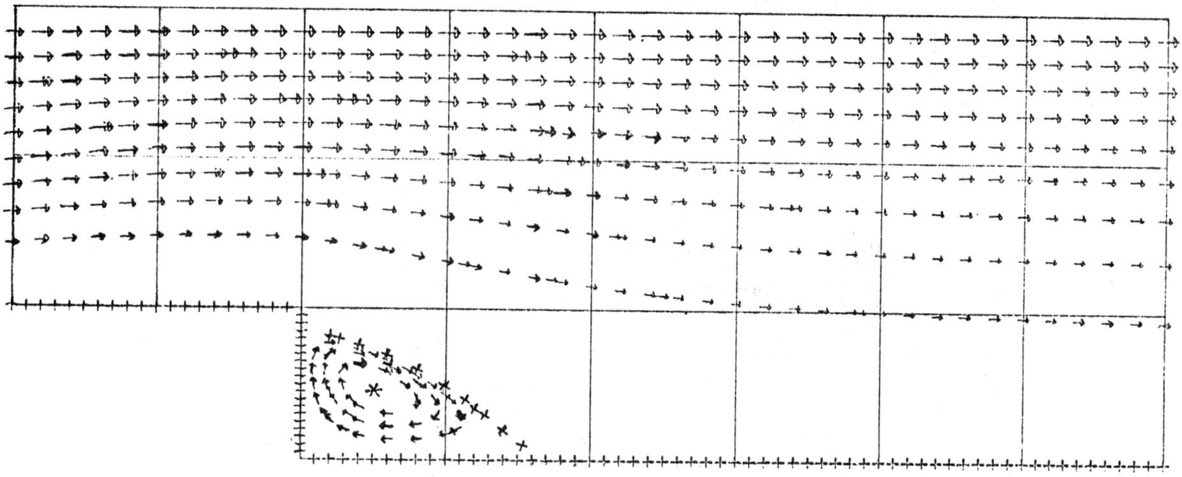

(b) stream function flow direction for incompressible flow over a backstep, Re = 10.

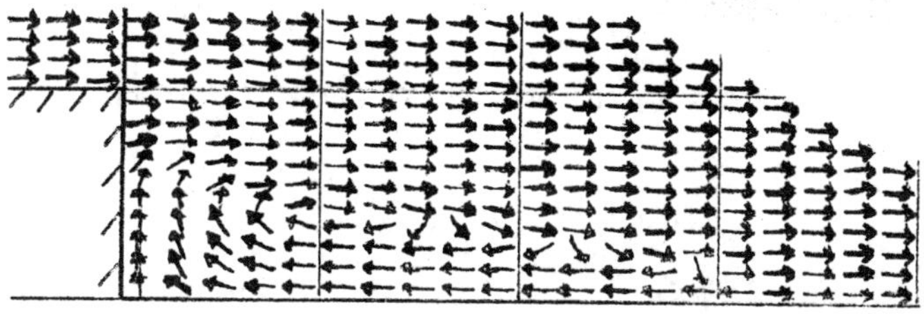

(c) flow direction for compressible flow in a base region, M_0 = 2.24, Re = 300, γ = 1.4.

Figure 7-7. Examples of stream function and flow direction plots. For parts (a) and (b), the ψ = 0 line is marked + , and $\Delta\psi$ = 0.17 above the + line and $\Delta\psi$ = 0.00127 below + line. (After Roache and Mueller, 1968.)

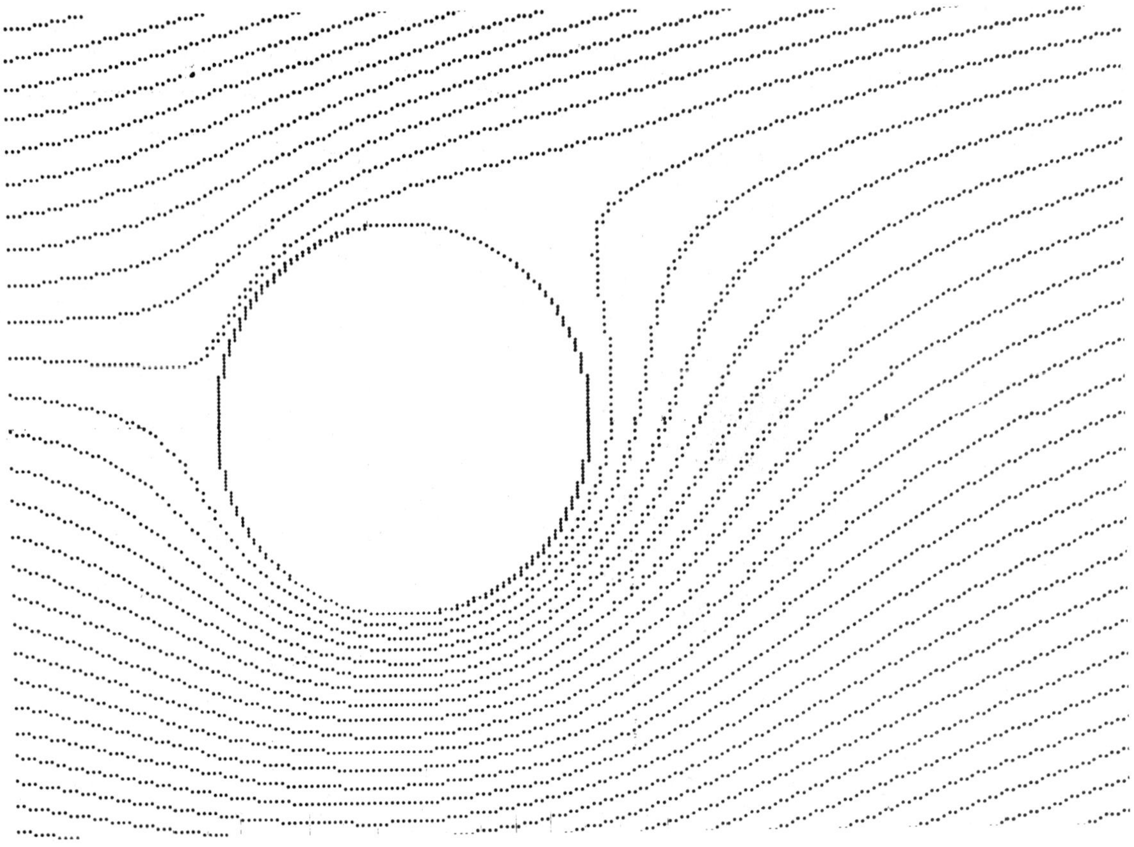

(a) observer moving with the body (wind tunnel reference frame)

Figure 7-8. Streamlines for unsteady flow around a circular cylinder. (After Thoman and Szewczyk, 1966)

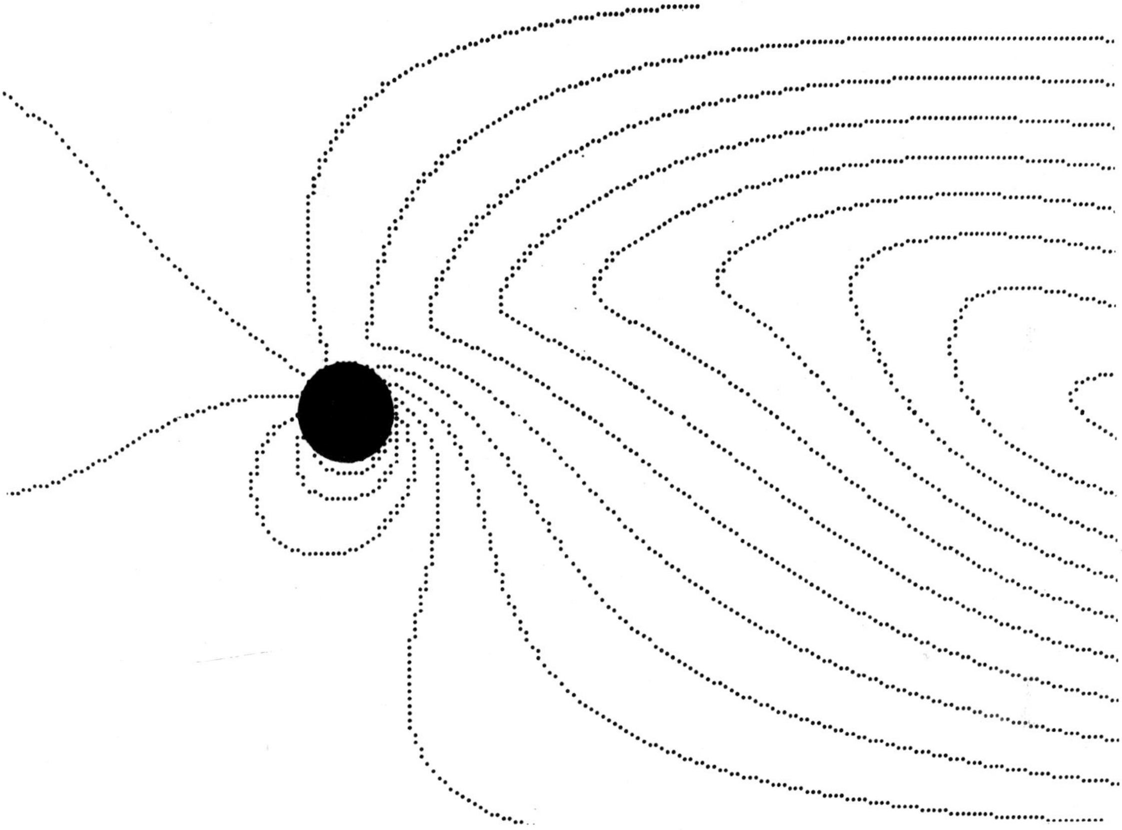

(b) observer moving with the free stream

Figure 7-8. (continued)

The stream function ψ must be calculated if the primitive variables have been used in the computation. The line integral $\psi_a - \psi_b = \int_a^b (u\,dy - v\,dx)$ should be evaluated by a quadrature scheme which is consistent with the velocity-field formulation. For second-order methods of the transport equations, this implies a Simpson's-rule integration rather than a higher order method. Because of the truncation error, the line integral will be path-dependent, and several path evaluations should perhaps be averaged. This ambiguity can be avoided by presenting, instead of a ψ-contour plot, a velocity vector map, in which the direction and magnitude of the velocity vector are represented by an arrow symbol plotted not along constant lines but simply at each node point (Roache and Mueller, 1970). An example is shown in Figure 7-7c.

In this Figure 7-7, the streamlines relative to the Eulerian-coordinate frame or wind-tunnel reference frame have been represented. It is also of interest to present the streamline pattern as viewed by an observer moving with the freestream. The velocities are transformed from (u,v) to $(u-U_\infty,v)$, and the streamline values from $\psi(x,y)$ to $[\psi'(x,y) - U_\infty y]$. Examples from Thoman and Szewczyk (1966) are shown in Figure 7-8; see also Michael (1966).

A special problem of compressible flow is the representation of shock and rarefaction wave positions. Contour plots of any variables can serve this purpose to some extent; in particular, iso-density plots are equivalent to interferometer photographs obtained from a physical experiment. Even in the physical experiment, the appearance of waves in a diffuse viscous region is variable, depending on the fringe shift of the interferometer. Computationally, the appearance analogously depends on the plotting increment, $\Delta\rho$, and the appearance of a shock is further diffused because resolution is limited by the mesh size (except in shock-fitting methods). Experimentally, the photographic appearance of shocks may be emphasized by using optical devices such as schlieren systems, which respond to density gradients ($\partial\rho/\partial m$ in some direction m, or $|\nabla\rho|$), or shadowgraphs, which roughly respond to $\nabla^2\rho$. The numerical equivalents of these, i.e., contour plots of $|\nabla\rho|$ or $\nabla^2\rho$, have failed to emphasize the shocks because the computation has distorted these derivative values. Wilkins (1969) produced the unambiguous plots of shock position shown in Figure 7-9 in a simple manner. These computations of a decelerating axisymmetric body were performed using a combination of a quadratic and a linear artificial viscosity (see section V-D). A search routine simply located the positions of local maxima of the artificial viscosity and plotted circles there. This shock-location indicator could also be used, even when no explicit artificial viscosity term is present, by locating maxima of $\nabla^2\rho$.

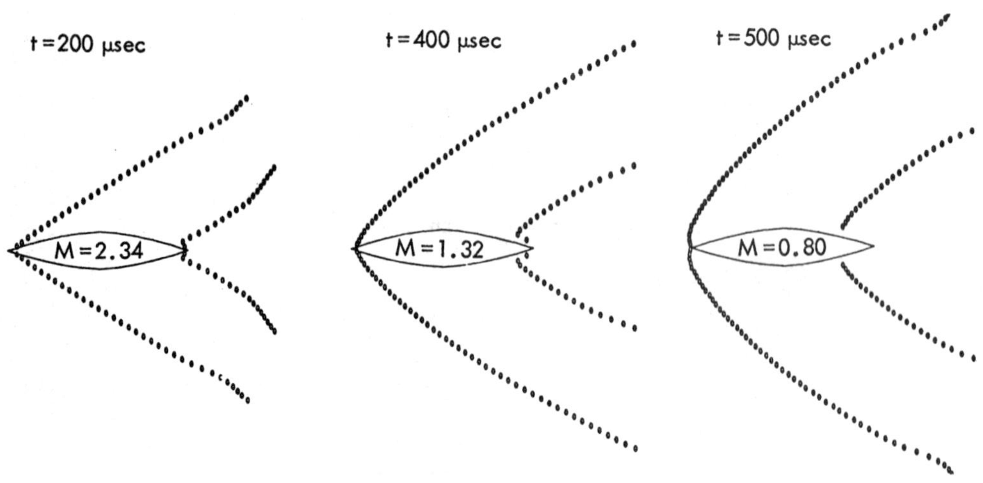

Figure 7-9. Shock positions of a decelerating axisymmetric body, after Wilkins (1969).

*Streak*line plots are readily available if MAC (section III-G-4) or PIC (section V-E-3) methods have been used, since marker particles are recorded with the computation. If other methods are used, marker particles can be introduced and these positions calculated as in section III-G-4. Streaklines are defined as the lines joining the same marker particles for all time. (In steady-state flow, streaklines and streamlines are identical.) The computed streaklines may be compared with physical streaklines produced by flow visualization techniques such as smoke visualization, dye visualization, the hydrogen bubble method, or the neutrally-buoyant glass bead techniques. An example from Harlow and Fromm (1965) is shown in Figure 7-10; see also Hirt (1965) and Thoman and Szewczyk (1966).

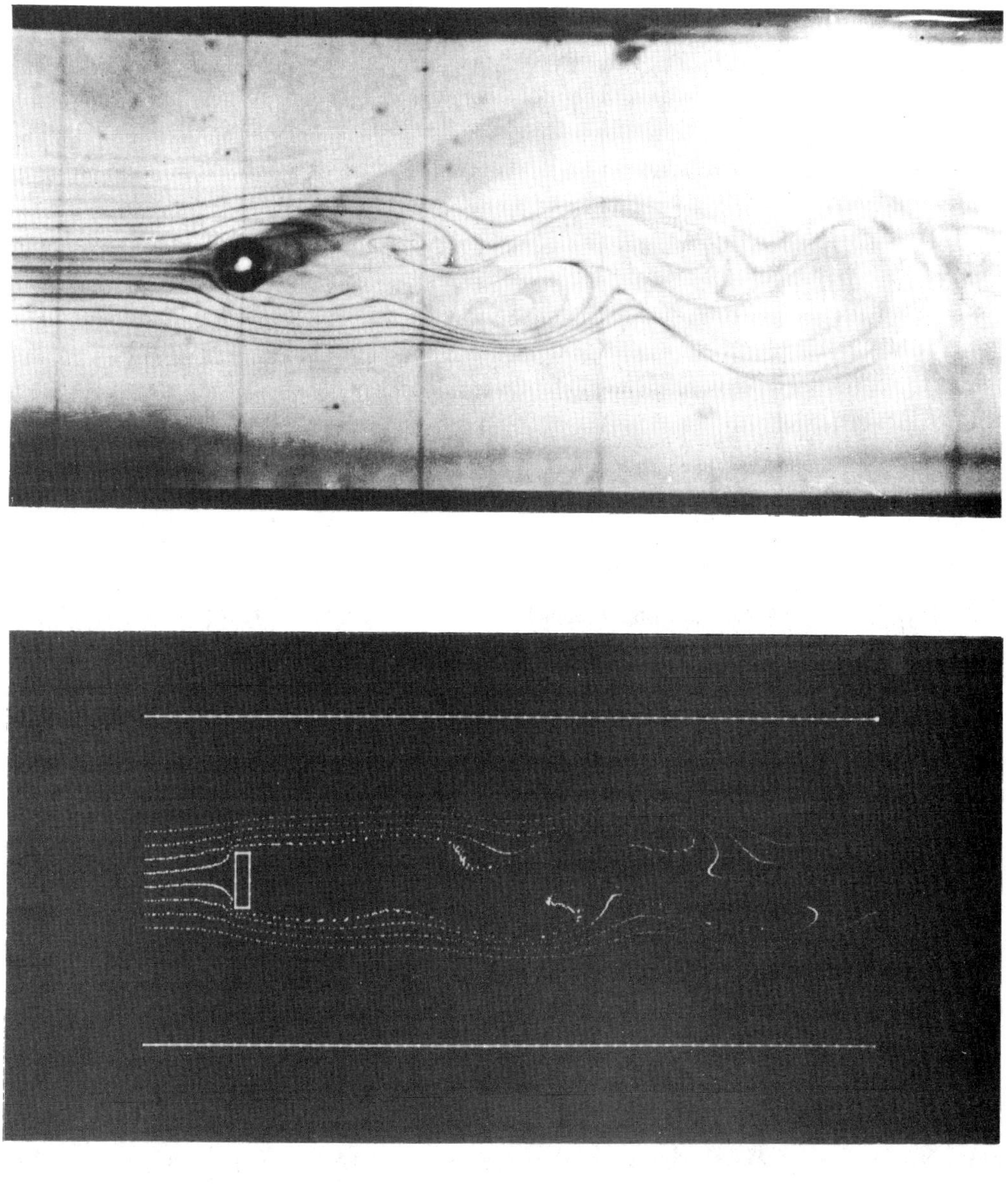

Figure 7-10. Comparison of computed streaklines with experimental values. Upper: dye streaklines in a water tunnel, photographed by A. Thom. Lower: MAC calculations. (After Harlow and Fromm, 1965.)

This work of Thoman and Szewczyk contains examples of many inventive and effective data presentations, including the representation of boundary-layer velocity profiles, as shown in Figure 7-11.

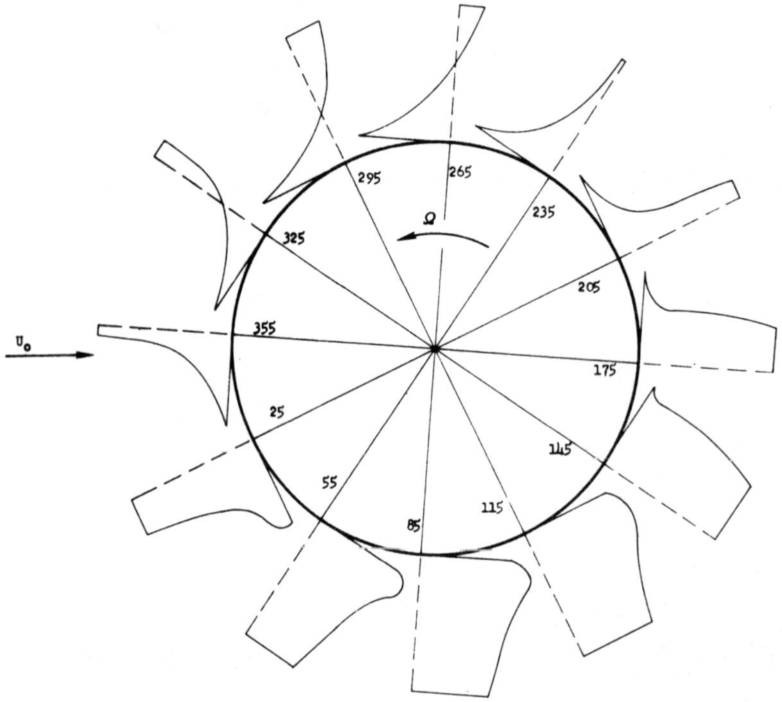

Figure 7-11. Presentation of boundary layer velocity profiles on a rotating cylinder, Re = 200, $V_{tangential}/U_o$ = 2. (After Thoman and Szewczyk, 1966.)

VII-C-3. Diagnostic Functionals

31. Define and compute diagnostic functionals appropriate to your problem.

Data interpretation and utilization are greatly helped by the presentation of diagnostic functionals of the solution. The simplest and most commonly used are the gross coefficients such as lift, moment, and drag coefficients, which may be further broken down as to contributions from friction, forebody pressure, base pressure, etc. Variations of skin-friction coefficient (shear stress), Nusselt number (heat transfer), and pressure coefficient along boundaries may also be presented. Again, it is recommended that the quadratures needed for coefficient evaluation be carried out by a method consistent with the fluid dynamics computation; for a second-order computation, this implies a Simpson's rule quadrature.

There are other diagnostic functionals which are becoming more widely used, particularly in geophysical studies. Some of these are integrals of certain quantities and others are just specific terms of the equations. The appropriate form varies from problem to problem. The intelligent use of diagnostic functionals with a flow computation can provide something of the insight normally gained from an analytical study, as pointed out by Williams (1969); he used global integrals of kinetic and potential energies and integrated forms of the governing PDE's.

W. P. Crowley (1968B), in his hydrodynamic stability study using the Bousinesq approximation, evaluated the perturbation kinetic energy and the total kinetic energy, and isolated the term $[u'v' (\partial u_o/\partial y)]$, which represents the transfer of energy between the perturbation flow (primes) and the mean flow (sub-zero). He then presented contour plots in space at selected stages in the flow development. He also isolated and presented contour plots of the source term for total kinetic energy (rising warm air is a kinetic-energy source), and the sink term representing irreversible dissipation; plotted the time derivative of global perturbation kinetic energy as a function of energy converted from the mean flow to the perturbation, the production of potential energy, and the conversion of perturbation kinetic energy into internal energy by dissipation; plotted the "available potential energy," i.e., that which can be converted into kinetic energy; and plotted the globally averaged perturbation kinetic energy, buoyancy term, shear source, and energy dissipation rate versus time. This work provides an excellent example of the thoughtful use of diagnostics; see also Smagorinsky et al. (1965).

32. Perform extensive information processing in a separate program.

The simulation of fluid turbulence is an extremely difficult problem which may or may not ever be accomplished in a satisfactory manner. Because the solutions are of a random nature, presentation of the primitive variables like u(x,y,z,t) are obviously useless. This is exactly the case with hot-wire anemometer data obtained in physical experiments.

The form of the functional diagnostic used must be guided by theory. The useful functionals should at least include local and global Reynolds stresses, the scale of the turbulence, the global dissipation term, and the various correlations such as time-averaged u'v' products and other higher correlations. (Note that the statistical theory of turbulence would make better use of ensemble averages than time averages, but the possibility of running a large number of computational experiments to obtain the ensemble average appears remote at this time.)

It is easy to see that the calculation of functional diagnostics for turbulence could be more time consuming than the flow computation itself. For this reason, it is recommended that the solutions be stored on magnetic tape and processed separately. Because of storage problems, it will not usually be possible to store the entire computational experiment (Gawain and Pritchett, 1968) but only time-sampled sections of it.

The same approach is being used in physical experimentation by Kovasnay and Frenkiel; the hot-wire anemometer data are recorded on magnetic tape, and subsequently sampled and analyzed. Gawain and Pritchett (1968) adopted the philosophy that the solution generation and extensive processing should not be confounded in the same program, but that *some* on-line processing should be done to provide a few statistical features necessary to indicate the state of the turbulence. (This is analogous to monitoring the physical experiment by way of an oscilloscope display of the hot-wire signal.) The further advantage of this approach, which is applicable to other transient studies besides turbulence, is that as a *theory* develops one can refine the analysis of the data without recomputing the solution.

VII-D. Closure

In this chapter, we have given some recommendations on computer programming, debugging and testing, and information processing. This is consistent with the stated objective of the book, which is to address those difficulties associated with actually obtaining and interpreting computational solutions to fluid dynamics problems. All these areas, and especially that of information processing, are research areas which will continue to be just as important and active as the research in computational methods.

APPENDIX A

TRIDIAGONAL ALGORITHM

We present here a method of solving a tridiagonal matrix problem. Richtmyer and Morton (1967, page 199) refer to this method as merely a special adaptation of the Gaussian elimination procedure, but it is commonly known as the Thomas algorithm. The description and notation given here somewhat follows Richtmyer and Morton, with more generality in the boundary conditions. Other references are Peacemen and Rachford (1955), von Rosenberg (1969), and Ames (1969). Error bounds are given by Fischer and Usmani (1969).

The interior-point equation is written as

$$-A_m W_{m+1} + B_m W_m - C_m W_{m-1} = D_m \tag{A-1}$$

For example, the one-dimensional Poisson equation

$$\frac{\partial^2 \psi}{\partial y^2} = \zeta(y) \tag{A-2}$$

gives

$$\frac{\psi_{m+1} - 2\psi_m + \psi_{m-1}}{\Delta y^2} = \zeta_m \tag{A-3}$$

or

$$A_m = +1 \quad , \quad B_m = +2 \quad , \quad C_m = +1 \quad , \quad D_m = -\zeta_m \Delta y^2 \tag{A-4}$$

To keep the size of the round-off errors down, Richtmyer and Morton (1967) show that it is sufficient that

$$A_m > 0 \quad , \quad B_m > 0 \quad , \quad C_m > 0 \tag{A-5}$$

and

$$B_m > A_m + C_m \tag{A-6}$$

We will consider m ranging from m = 1 to m = M, with various combinations of boundary conditions at either end.

We consider the set G of vectors W which are solutions to equation (A-1) with the specified left-hand (m = 1) boundary condition. This set G is a one-parameter family, the parameter being the value W_2. For example, if a Dirichlet condition is used with $W_1 = a_1$ specified, then, for each value of W_2, equation (A-1) may be solved for W_{m+1} and thus W_{m+1} marched out* by this recursion relation for all $m + 1 \geq 3$.

---

*In section III-B-8, we noted that this "march" procedure is not generally stable, since round-off error grows exponentially. The exception is the special case of the one-dimensional equation (A-3), for which the error growth is merely linear.

We postulate the existence of two vectors E and F such that, for any $W \in G$,

$$W_m = E_m W_{m+1} + F_m \tag{A-7}$$

The existence of such E and F will become evident. We now index equation (A-7) down in m to obtain

$$W_{m-1} = E_{m-1} W_m + F_{m-1} \tag{A-8}$$

Substituting for W_{m-1} from equation (A-8) into (A-1) and solving for W_m gives

$$W_m = \frac{A_m}{B_m - C_m E_{m-1}} W_{m+1} + \frac{D_m + C_m F_{m-1}}{B_m - C_m E_{m-1}} \tag{A-9}$$

Comparing equations (A-9) and (A-7) and noting that both equations must hold for all $W \in G$ (i.e., for all values W_2), we must have

$$E_m = \frac{A_m}{B_m - C_m E_{m-1}} \tag{A-10A}$$

$$F_m = \frac{D_m + C_m F_{m-1}}{B_m - C_m E_{m-1}} \tag{A-10B}$$

for $m \geq 2$. The left-hand boundary condition will determine E_1 and F_1, after which the recursion relations (A-10) can be used to calculate all E and F up tp $m = M-1$. Then W_M is set from the right-hand boundary condition, and equation (A-9) is used with the known A,B,C, D, and calculated E and F to solve recursively for W_m from W_{m+1}, marching down from $m = M-1$ to $m = 1$.

The left-hand boundary conditions are used to determine E_1 and F_1 as follows. Writing equation (A-7) at $m = 1$ gives

$$W_1 = E_1 W_2 + F_1 \tag{A-11}$$

For a Dirichlet condition $W_1 = a_1$, the relation (A-11) must hold for all possible values of W_2; thus,

$$E_1 = 0 \quad , \quad F_1 = a_1 \qquad \text{for} \qquad W_1 = a_1 \tag{A-12}$$

For a Neumann boundary condition, as in

$$\frac{\partial \psi}{\partial y} = s_1 \tag{A-13}$$

we have

$$\psi_2 - \psi_1 = s_1 \Delta y \tag{A-14}$$

or

$$W_1 = W_2 - s_1 \Delta y \tag{A-15}$$

Comparing equation (A-15) with (A-11) shows that

$$E_1 = 1 \quad , \quad F_1 = -s_1 \Delta y \quad \text{for} \quad \frac{\partial W}{\partial y} = s_1 \tag{A-16}$$

For a mixed (Robbin's) boundary condition, as in

$$\psi + p_1 \frac{\partial \psi}{\partial y} = q_1 \tag{A-17}$$

which gives

$$W_1 + p_1 \left(\frac{W_2 - W_1}{\Delta y} \right) = q_1 \tag{A-18}$$

we have

$$W_1 = - \left[\frac{p_1/\Delta y}{1 - p_1/\Delta y} \right] W_2 + \frac{q_1}{1 - p_1/\Delta y} \tag{A-19}$$

Comparing equation (A-19) with (A-11) shows that

$$E_1 = - \frac{p_1/\Delta y}{1 - p_1/\Delta y} \quad , \quad F_1 = \frac{q_1}{1 - p_1/\Delta y} \tag{A-20}$$

Obviously, $p_1/\Delta y \neq 1$ is required since, by equation (A-18), this would give $W_2 = q_1$ and W_1 indeterminate.

The right-hand boundary conditions are used to determine W_M as follows. For a Dirichlet condition $\psi = a_M$, obviously

$$W_M = a_M \tag{A-21}$$

For a Neumann condition $\partial\psi/\partial y = s_M$, we have

$$W_{M-1} = W_M - s_M \Delta y \tag{A-22}$$

Writing equation (A-8) at m = M,

$$W_{M-1} = E_{M-1} W_M + F_{M-1} \tag{A-23}$$

Equating (A-22) and (A-23) gives

$$W_M = \frac{F_{M-1} + s_M \Delta y}{1 - E_{M-1}} \tag{A-24}$$

For a mixed condition $\psi + p_M(\partial\psi/\partial y) = q_M$, we have

$$W_{M-1} = W_M \frac{1 + p_M/\Delta y}{p_M/\Delta y} - \frac{q_M}{p_M/\Delta y} \tag{A-25}$$

Equating (A-25) and (A-23) gives

$$W_M = \frac{F_{M-1} + q_M/(p_M/\Delta y)}{\dfrac{1 + p_M/\Delta y}{p_M/\Delta y} - E_{M-1}} = \frac{F_{M-1} + q_M\Delta y/p_M}{1 + \Delta y/p_M - E_{M-1}} \tag{A-26}$$

In the last two cases, equations (A-24) and (A-26), there exists the possibility of divide overflows as the denominator becomes small, dependent on initial conditions, the size of M, and the value of p_M. For example, if the Neumann condition is used at the left-hand boundary, then equation (A-16) shows $E_1 = 1$. For the one-dimensional Poisson equation (A-3), the parameters A, B, C from equation (A-4), substituted into (A-10A), give

$$E_w = \frac{1}{2 - E_1} = 1 \quad , \quad E_m = 1 \tag{A-27}$$

This causes the form (A-24) for Neumann conditions at the right-hand boundary to give $W_M = \infty$, unless it so happened that $s_M\Delta y = - F_{M-1}$. This is a proper behavior for Neumann conditions at both boundaries; we cannot arbitrarily specify gradient conditions at both boundaries, since

$$\left.\frac{\partial\psi}{\partial y}\right|_{y_M} = \left.\frac{\partial\psi}{\partial y}\right|_{y_1} + \int_{y_1}^{y_M} \frac{\partial^2\psi}{\partial y^2}\,dy = \left.\frac{\partial\psi}{\partial y}\right|_{y_1} + \int_{y_1}^{y_M} \zeta(y)dy \tag{A-28}$$

Thus the gradients at y_1 and y_M are related by the problem $\zeta(y)$ and cannot be independently specified as boundary conditions. If $\partial\psi/\partial y|_{y_M}$ is compatible with equation (A-28), the form (A-24) gives W = 0/0. This indeterminate form will destroy the computer calculations, but is compatible with the indeterminate form of the problem; that is, if W is a solution, so is (W + constant).

Summary

Set E_1 and F_1 from left-hand boundary conditions, using equation (A-12), (A-16), or (A-20). March out and store the vectors E_m and F_m, according to the recursion relations (A-10), up to m = M - 1. Set W_M from the right-hand boundary condition, using equation (A-21), (A-24), or (A-26). Finally, calculate the solution vector W_m by the recursion relation (A-7), marching down from m = M - 1 to m = 1.

A FORTRAN IV subroutine for this tridiagonal algorithm follows on the next page.

FORTRAN IV Subroutine

In the following subroutine, the correspondence between the algebraic names and the FORTRAN names is given below.

Algebra symbol	FORTRAN name
A, B, C, D, E, F, W	same
a_1	A1 if L1 = 1
a_M	AM if LM = 1
m	M
M	ML
$p_1/\Delta y$	A1 if L1 = 3
$p_M/\Delta y$	AM if LM = 3
q_1	Q1
q_M	QM
$s_1\Delta y$	A1 if L1 = 2
$s_M\Delta y$	AM if LM = 2

The dimension of the vectors is MD. All vectors must be dimensioned in the calling program as they are in the subroutine. The interior equations of the problem are stored in vectors A, B, C, D. L1 and LM are boundary-condition option indicators, and A1, Q1, AM, QM are the boundary values. L1 = 1,2,3 gives boundary conditions of first (Dirichlet), second (Neumann), or third (mixed) kind; likewise for LM. E and F are working vectors, storage for which must be provided in the calling program. The solution is in W.

Listing

```
      SUBROUTINE GTRI(A,B,C,D,MD,E,F,W,L1,A1,Q1,LM,AM,QM)
C     GENERAL TRI-DIAGONAL SOLVER
      DIMENSION A(MD),B(MD),C(MD),D(MD),E(MD),F(MD),W(MD)
      IF(L1.EQ.1) E(1) = 0.
      IF(L1.EQ.1) F(1) = A1
C     ABOVE OVERWRITTEN IF LM.NE.2
      IF(L1.EQ.2) E(1) = 1.
      IF(L1.EQ.2) F(1) = -A1
      IF(L1.EQ.3) E(1) = A1/(A1-1.)
      IF(L1.EQ.3) F(1) = Q1/(1.-A1)
      MM=MD-1
      DO 1 M=2,MM
      DEN=B(M)-C(M)*E(M-1)
      E(M)=A(M)/DEN
1     F(M)=(D(M)+C(M)*F(M-1))/DEN
      IF(LM.EQ.1) W(MD)=AM
      IF(LM.EQ.2) W(MD)=(F(MM)+AM)/(1.-E(MM))
      IF(LM.EQ.3) W(MD)=(F(MM)+QM/AM)/((1.+AM)/AM-E(MM))
      DO 2 MK=1,MM
      M=MD-MK
2     W(M)=E(M)*W(M+1)+F(M)
      RETURN
      END
```

APPENDIX B

ON ARTIFICIAL VISCOSITY

The following article is reprinted from the *Journal of Computational Physics*, Volume 10, pages 169-184, October 1972, by the kind permission of the publishers of that journal, Academic Press, Inc.

Reprinted from JOURNAL OF COMPUTATIONAL PHYSICS
All Rights Reserved by Academic Press, New York and London

Vol. 10, No. 2, October 1972
Printed in Belgium

On Artificial Viscosity*

PATRICK J. ROACHE

Numerical Fluid Dynamics Division, Aerothermodynamics Research Department, Sandia Laboratories, Albuquerque, New Mexico 87115

Received May 10, 1972

It is shown that the usual analysis for the implicit artificial viscosity of finite difference analogs of the linear advection equation is ambiguous, with different results obtained for transient and steady-state problems. The ambiguity is easily resolved for the inviscid equation, but for the advection-diffusion equation, the steady-state analysis is shown to be applicable to steady-state problems. It is demonstrated that the currently most popular methods, touted as having no artificial viscosity, actually do have such when applied to steady-state problems.

INTRODUCTION

"Artificial viscosity" is a particular kind of truncation error exhibited by some finite difference analogs of advection equations. The first use of the term was by von Neumann and Richtmyer [1], who explicitly added a viscosity-like term to the inviscid gas dynamic equations in order to allow the calculation of shock waves by what is now known as the "shock-smearing" or "through" method. Their explicit artificial viscosity term was deliberately made proportional to Δx^2, so as to assure mathematical consistency; that is, their *explicit* artificial viscosity term was indeed a second-order *truncation error*.

It has since been recognized that the same kind of artificially viscous behavior can be obtained, often inadvertently, just due to the truncation error of the FDE (finite difference equation). Noh and Protter [2] first presented an analysis of the *implicit* artificial viscosity of the upwind differencing method applied to the linear model advection equation

$$\zeta_t = -u\zeta_x . \tag{1}$$

For $u > 0$, the upwind differencing method for (1) gives the following FDE:

$$(\zeta_i^{n+1} - \zeta_i^{n})/\Delta t = -u[(\zeta_i^{n} - \zeta_{i-1}^{n})/\Delta x]. \tag{2}$$

* This work is supported by the U. S. Atomic Energy Commission.

The truncation error is $O(\Delta t, \Delta x)$. Rewriting (2) in terms of the Courant number $c = u\Delta t/\Delta x$ gives, for u = constant,

$$\zeta_i^{n+1} = \zeta_i^n - c(\zeta_i^n - \zeta_{i-1}^n). \tag{3}$$

For $c = 1$, the method gives $\zeta_i^{n+1} = \zeta_{i-1}^n$, which is the exact solution. The condition $c = 1$ is also the stability limit. For $c < 1$, the method introduces an artificial damping, in that the von Neumann stability analysis shows that the amplification matrix has eigenvalues $|\lambda| < 1$. Any method which has $|\lambda| < 1$ introduces such an artificial "damping," but a Taylor series expansion, as in the application of Hirt's stability analysis [3], shows that Eq. (3) is equivalent to

$$\zeta_t = -u\zeta_x + (u\Delta x/2)\,\zeta_{xx} - \tfrac{1}{2}\Delta t\zeta_{tt} + O(\Delta x^2, \Delta t^2). \tag{4}$$

The ζ_{tt} term in (4) is customarily evaluated from (1) for constant u as

$$\zeta_{tt} = -u\zeta_{xt} = -u(\zeta_t)_x = u^2\zeta_{xx}\,. \tag{5}$$

Using (5) in (4) gives

$$\zeta_t = -u\zeta_x + \alpha_e\zeta_{xx} + O(\Delta x^2, \Delta t^2), \tag{6}$$

where

$$\alpha_e = (u\Delta x/2) - (u^2\Delta t/2) = \tfrac{1}{2}u\Delta x(1 - c). \tag{7}$$

Since the method has introduced a nonphysical coefficient α_e of $\partial^2\zeta/\partial x^2$, we are justified in referring not only to the artificial damping, but more specifically, to artificial or numerical diffusion or numerical *viscosity* of the method. (Hirt [3] successfully uses $\alpha_e > 0$ as a necessary stability criterion.) For $c = 1$, (7) indicates $\alpha_e = 0$, a result consistent with the fact that the exact solution is obtained for $c = 1$.

Transient vs Steady-State Analyses

The above analysis has been used by many authors to describe the artificial viscosity of various methods, and the results are widely accepted as being applicable to multidimensional problems, with and without physical viscous terms. But the interpretation of α_e in multidimensional, viscous and/or steady-state problems is not as straightforward as it might appear. Suspicion arises when one considers the form of (7) which shows an α_e dependent on Δt through the Courant number c. Consider a problem in which a steady state has developed, with $\zeta_i^{n+1} = \zeta_i^n$. Once this condition is reached[1], both the FDE (2) and computational experience with the upwind differencing method in multidimensional problems indicate that a change in Δt does not change the steady-state solution. Yet (7) would indicate that a reduction in Δt increases α_e (through c). If the concept of artificial viscosity α_e means anything, it would appear that the FDE solution should depend on α_e; but we see that we can change α_e through Δt, and not change the steady-state solution.

Alternate to the above analysis of the transient equation, one can instead analyze for the α_e effect after assuming that a steady state exists. Setting $\zeta_i^{n+1} = \zeta_i^n$ in (2) and expanding in a Taylor series, we obtain a steady-state α_e, denoted by α_{es}, as

$$\alpha_{es} = \tfrac{1}{2} u \Delta x. \tag{8}$$

In this formulation, $\alpha_{es} \neq f(\Delta t)$ and the steady-state independence of Δt is not suspect.

The resolution of the ambiguity between the two different expressions (7) for α_e and (8) for α_{es} is readily accomplished by recognizing that, for the inviscid model equation, the only possible steady-state solution with $u =$ constant is the trivial solution $\zeta_i^n = \zeta_1^n =$ constant. In this case, $\partial^2\zeta/\partial x^2 = 0$, permitting an arbitrary form for α_e. The question is, which analysis (if either) is appropriate to problems with (a) diffusion terms present, (b) dimensions greater than one, (c) spatially varying or nonlinear advection velocities u?

The question may be easily and unambiguously answered for the addition of diffusion terms to (1), with a physical diffusion coefficient α

$$\zeta_t = -u\zeta_x + \alpha\zeta_{xx}. \tag{9}$$

Using upwind differencing on the advection term and forward-time centered-space differencing on the diffusion term gives

$$\zeta_i^{n+1} = \zeta_i^n - c(\zeta_i^n - \zeta_{i-1}^n) + d(\zeta_{i+1}^n - 2\zeta_i^n + \zeta_{i-1}^n), \tag{10}$$

where $d = \alpha\Delta t/\Delta x^2$. The steady-state analysis for Eq. (9) gives

$$0 = -u\zeta_x + (\alpha + \alpha_{es})\,\zeta_{xx} + O(\Delta x^2), \tag{11}$$

where α_{es} is again given by the steady-state form (8). The transient analysis is altered, because (5) must be replaced by

$$\zeta_{tt} = (-u\zeta_x + \alpha\zeta_{xx})_t = u^2\zeta_{xx} - 2u\alpha\zeta_{xxx} + \alpha^2\zeta_{xxxx} \tag{12}$$

and (6) must be replaced by

$$\zeta_t = -u\zeta_x + (\alpha + \alpha_e)\,\zeta_{xx} + O(\Delta x^2, \Delta t^2) + \text{HOD}, \tag{13}$$

[1] We do not wish to confuse the matter by considering iteration convergence criteria at this point.

where the higher-order derivative terms are

$$\text{HOD} = \Delta t[u\alpha\zeta_{xxx} - (\alpha^2/2)\,\zeta_{xxxx}] \tag{14}$$

and α_e is again the transient form given by (7). Hirt [3] ignores the HOD in (14) and in this way successfully predicts the transient stability behavior, but we are interested in the α_e appropriate for a steady-state solution, and we must retain the HOD.

For any steady-state solution, (9) gives

$$\zeta_{xxxx} = (u/\alpha)\,\zeta_{xxx} = (u/\alpha)^2\,\zeta_{xx} = (u/\alpha)^3\,\zeta_x\,. \tag{15}$$

We now apply these relations (15) for a steady state to the result of the transient analysis. Assuming a steady state in (13), using (14) and (15), and substituting (7) for α_e gives

$$\begin{aligned} 0 = &- u\zeta_x + \alpha\zeta_{xx} + (u\Delta x/2)\,\zeta_{xx} - (u^2\Delta t/2)\,\zeta_{xx} + \Delta t u\alpha(u/\alpha)\,\zeta_{xx} \\ &- \Delta t(\alpha^2/2)(u/\alpha)^2\,\zeta_{xx} + O(\Delta x^2, \Delta t^2), \end{aligned} \tag{16}$$

$$0 = -u\zeta_x + (\alpha + \alpha_{es})\,\zeta_{xx} + O(\Delta x^2, \Delta t^2), \tag{17}$$

where the steady state α_{es} is given by (8). It is thus clear that although the transient α_e of (7) may be appropriate for Hirt's stability analysis, the steady-state form α_{es} of (8) is appropriate when a steady-state condition has been reached, even though the transient equation is analyzed.

It may be argued that the last relation of (15) could be used to eliminate $\alpha_{es}\zeta_{xx}$ from (11), thus leading to the conclusion that no artificial viscosity coefficient α_{es} is present, but rather that an "artificial advection velocity" u_{es} is present, as in

$$0 = -(u - u_{es})\,\zeta_x + \alpha\zeta_{xx} + O(\Delta x^2), \tag{18}$$

where

$$u_{es} = \alpha_{es}(u/\alpha) = \tfrac{1}{2}u^2\Delta x/\alpha. \tag{19}$$

However, the "artificial velocity" term in (18) must still be interpreted as producing an artificial viscous effect, even though the $\alpha_{es}\zeta_{xx}$ term has been removed. The steady-state solution is not determined by α and u independently, but only by their ratio u/α, along with the boundary conditions. When the proper length normalizing of the spatial domain of definition is taken into account, this ratio u/α is a Reynolds number. An artificial viscous effect is then simply any effect which reduces the effective Reynolds number u/α. In (11), the artificial viscous effect is expressed as an artificial increase in α, which reduces u/α to $u/(\alpha + \alpha_{es})$. In (18), the artificial viscous effect is expressed as an artificial decrease in u, which reduces u/α to

$(u - u_{es})/\alpha$. Thus, both α_{es} in (11) and u_{es} in (18) act to reduce the effective Reynolds number and therefore have an artificial viscous effect.

There is, in fact, a quantitative ambiguity in these two steady-state anaylses, due to the use of (15) in the finite difference solution, whereas (15) is only applicable to the continuum solution. Equation (11) has a factor

$$\frac{u}{\alpha + \alpha_{es}} = \frac{u}{\alpha}\left(\frac{1}{1 + \frac{1}{2}u\Delta x/\alpha}\right), \tag{20}$$

whereas (18) has a factor

$$(u - u_{es})/\alpha = (u/\alpha)(1 - \tfrac{1}{2}u\Delta x/\alpha). \tag{21}$$

But since $1/(1 + \epsilon) = 1 - \epsilon + O(\epsilon^2)$, these two equations (20) and (21) for the artificial viscous effect are equal, to within a truncation error term of order Δx^2, provided that

$$\tfrac{1}{2}u\Delta x/\alpha \ll 1. \tag{22}$$

This is obviously true as $\Delta x \rightarrow 0$, in which case (15) becomes applicable to the FDE. [Equation (22) is the familiar requirement for formal accuracy of the upwind difference method, that the computational cell Reynolds number $u\Delta x/\alpha$ be $\ll 2$.]

Similarly, (15) might be used in (11) to express the first-order truncation error as a coefficient of ζ_{xxx} just as legitimately. But since we have no such term in the original continuum equation, this exercise does not lend itself to a fruitful interpretation of the physically analogous behavior of the FDE.

We also remark that if problems are considered with boundary conditions either of the form

$$\zeta(0) = a, \qquad \zeta_x(1) = b \tag{23}$$

or

$$\zeta(0) = a, \qquad \zeta(1) = b, \tag{24}$$

the resulting steady-state solution is

$$\zeta(x) = C_1 + C_2 e^{xu/\alpha} \tag{25}$$

with $C_2 \neq 0$. This solution gives nonzero values for all spatial derivatives. Unlike the situation for the inviscid equation, the distinction between the α_e of (7) and the α_{es} of (8) is then important.

For multidimensional problems with nonlinear coefficients, the resolution of the transient and steady-state analyses is not so neat. Both analyses give different values of α_e or α_{es} in different directions, each of the form (7) or (8). But the transient form α_e in (7) depends on (5), an equation which is not applicable to multidimensional and/or nonlinear problems. Further, the multidimensional transient analysis predicts that the steady-state solution for the upwind differencing method is a function of (Δt), which disagrees with computational experience. Thus, the steady-state analysis does appear to be appropriate for multidimensional nonlinear steady-state problems.

Analyses of Other Methods

In Table I, we present the results of both the transient and steady-state analyses for the artificial viscosity of various methods, based on the inviscid model equation (1). (Higher-order terms in the transient expansion have been given by Tyler [4].) The steady-state results for the inviscid equation are identical to those results obtained from the viscous equation (9), using for the viscous term any of the usual methods based on second-order space-centered differences; these include FTCS, fully implicit, ADI, Cheng–Allen, Crocco, Saul'yev, Adams–Bashforth methods [6], etc. For u = constant, the upwind difference method is equivalent to the "donor cell" [5] or "second upwind difference" method [6], which uses cell-averaged advection velocities at cell interfaces. It has nonzero artificial viscosity in both analyses, for $c < 1$. The forward-time, centered-space (FTCS) method is, of course, unstable in the absence of physical viscous terms, and accordingly has $\alpha_e < 0$ in the transient analysis [3]. The Lax method [7] is still frequently used, and also has nonzero artificial viscosity in both analyses, for $c < 1$.

Leith's method [8] (see also [2]) is very important. It is based on a second-order Taylor series expansion of (1) in time. For the model equation (1), Leith's method is algebraically identical to other methods based on the second-order time expansion, such as the Lax–Wendroff method [9], the Richtmyer [10] and other two-step Lax–Wendroff methods, Moretti's method [11], and MacCormack's method [12]. Leith's method also is involved in Fromm's method of zero average phase error [13], and even is related to Rusanov's method [14] for certain combinations of parameters. Significantly, $\alpha_e = 0$ is indicated only in the transient analysis.[2] From the steady-state analysis, $\alpha_{es} = \frac{1}{2}u^2\Delta t$ is indicated, implying that $\alpha_{es} = 0$ only as $\Delta t \to 0$. There is no danger of misinterpretation of higher-order terms here, because the method is *algebraically* equivalent to the FTCS method applied to the advection-diffusion equation (9) with the physical $\alpha = \frac{1}{2}u^2\Delta t$. Unlike the example of upwind differencing considered earlier, the FDE and computational experience now indicate that the steady-state solution will depend on Δt.

[2] Leith [8] was only concerned with the transient problem, of course. The present work is not to be construed as a criticism of Leith's work.

TABLE I

Implicit artificial viscosities from transient and steady-state analyses for various finite difference methods applied to $\zeta_t = -u\zeta_x$, with $c = u\Delta t/\Delta x$.

Description	Method	Transient	Steady	Formal truncation order
1. Upwind	$\zeta_i^{n+1} = \zeta_i^n - c(\zeta_i^n - \zeta_{i-1}^n)$	$\alpha_e = (u\Delta x/2)(1 - c)$	$\alpha_{es} = (u\Delta x/2)$	$O(\Delta t, \Delta x)$
2. FTCS	$\zeta_i^{n+1} = \zeta_i^n - (c/2)(\zeta_{i+1}^n - \zeta_{i-1}^n)$	$-(u^2\Delta t/2)$	0	$O(\Delta t, \Delta x^2)$
3. Lax	$\zeta_i^{n+1} = (1/2)(\zeta_{i+1}^n + \zeta_{i-1}^n)$ $-(c/2)(\zeta_{i+1}^n - \zeta_{i-1}^n)$	$(\Delta x^2/2\Delta t)(1 - c^2)$	$\Delta x^2/2\Delta t$	$O(\Delta t, \Delta x^2, \Delta x^2/\Delta t)$
4. Leith[a]	$\zeta_i^{n+1} = \zeta_i^n - (c/2)(\zeta_{i+1}^n - \zeta_{i-1}^n)$ $+ (c^2/2)(\zeta_{i+1}^n - 2\zeta_i^n + \zeta_{i-1}^n)$	0	$u^2\Delta t/2$	$O(\Delta t^2, \Delta x^2)$
5. Matsuno[b]	$\overline{\zeta_i^{n+1}} = \zeta_i^n - (c/2)(\zeta_{i+1}^n - \zeta_{i-1}^n)$ $\zeta_i^{n+1} = \zeta_i^n - (c/2)(\overline{\zeta_{i+1}^{n+1}} - \overline{\zeta_{i-1}^{n+1}})$	$u^2\Delta t$	0	$O(\Delta t, \Delta x^2)$

[a] Also Lax–Wendroff, 2-step Lax–Wendroff, Moretti, MacCormack.
[b] Also Brailovskaya, Cheng–Allen.

The two-step Matsuno method [15] of differencing the advection terms has also been used for compressible flow by Brailovskaya [16], using the same approach on the viscous terms, and by Allen and Cheng [17], using a special treatment of the physical diffusion terms which successfully removes the additional Δt restriction present in Brailovskaya's method due to the diffusion term. The Matsuno method requires special mention, because of a further ambiguity in the steady-state α_{es} analysis. This two-step method for (1) is written as

$$\overline{\zeta_i^{n+1}} = \zeta_i^{\,n} - (c/2)(\zeta_{i+1}^n - \zeta_{i-1}^n), \tag{26a}$$

$$\zeta_i^{n+1} = \zeta_i^{\,n} - (c/2)(\overline{\zeta_{i+1}^{n+1}} - \overline{\zeta_{i-1}^{n+1}}). \tag{26b}$$

The $\overline{(n+1)}$ values are provisional or intermediate values. The method may be interpreted as a first iterative approximation to the fully implicit method. For the purposes of stability analysis and artificial viscosity analysis, (26) may be rewritten as a single equation

$$\zeta_i^{n+1} = \zeta_i^{\,n} - (c/2)(\zeta_{i+1}^n - \zeta_{i-1}^n) + (c^2/4)(\zeta_{i+2}^n - 2\zeta_i^{\,n} + \zeta_{i-2}^n). \tag{27}$$

The equivalence of (27) to the two-step method (26) holds only for the model equation (1) at interior points; the presence of boundaries and nonlinearities destroys this equivalence. The last term of (27) is recognized as the usual 3-point expression for $\alpha\partial^2\zeta/\partial x^2$, but written over a mesh spacing of $2\Delta x$ rather than Δx. With this interpretation, the steady-state analysis would indicate $\alpha_{es} = 2u^2\Delta t$. However, the higher-order terms enter into the behavior of the equation in an unexpected and fortunate manner. Each of the two steps (26a) and (26b) has the same operator form, i.e.,

$$\overline{\zeta^{n+1}} = \zeta^n + L(\zeta^n), \tag{28a}$$

$$\zeta^{n+1} = \zeta^n + L(\overline{\zeta^{n+1}}). \tag{28b}$$

(This is in contrast to the two-step Lax–Wendroff methods, for example.) Allen and Cheng [17] noted the significant fact that, when a steady state is reached with this method, not only does $\zeta^{n+1} = \zeta^n$, but also $\overline{\zeta^{n+1}} = \zeta^n$. Using this information, the steady-state analysis for α_{es} can be applied to each step of (26) separately, rather than to (27). The result is $\alpha_{es} = 0$, as for the FTCS method. This conclusion has been verified in the present study by one-dimensional tests, which exhibit a steady-state solution which is not a function of Δt, in contrast to the analysis of (27) and in contrast to Leith's method.

A Two-Dimensional Experiment

To test the applicability of the results from the one-dimensional model equation (1) to the two-dimensional gas dynamic equations, a numerical experiment using Moretti's inviscid blunt body program [18] was run. A 6° half-angle sphere-cone was run at a free-stream Mach number $M = 10$, with an ideal gas and a ratio of specific heats $\gamma = 1.4$. The program utilizes shock patching in a curvilinear mesh system which adjusts as the solution develops. Since the shock is correctly maintained as a discontinuity in this program, the present results are not confused by the postshock oscillations of the "through" or shock-smearing calculation methods. An extremely coarse mesh was chosen to exaggerate the α_{es} effects; the mesh had only three mesh points (two intervals) between the body and the shock, and only five mesh points along the body. The object of the experiment was to show that the steady-state solution obtained with Moretti's method is a function of the Δt used, as indicated by the steady-state analysis for α_e. (This behavior is in contrast to that of the upwind difference method considered earlier, and indeed to most other finite difference methods.)

The most sensitive location was found to be the (2, 3) point, in the center of the mesh. The Δt was changed by the program input parameter STAB; for STAB = 1, the Δt used was about 0.94 of the linear stability limit for a square mesh. The first segment of solution A, shown in Fig. 1, was run out to 3000 time steps with STAB = 1, giving a dimensionless time $T = 15.82$. This represents a rather unequivocal steady state, with the normalized density ρ changing by only 2.5×10^{-6} in the last 200 time steps, or less than $2.74 \times 10^{-7}\%$ per time plane. Then the second segment of solution A was obtained by changing the critical time-step

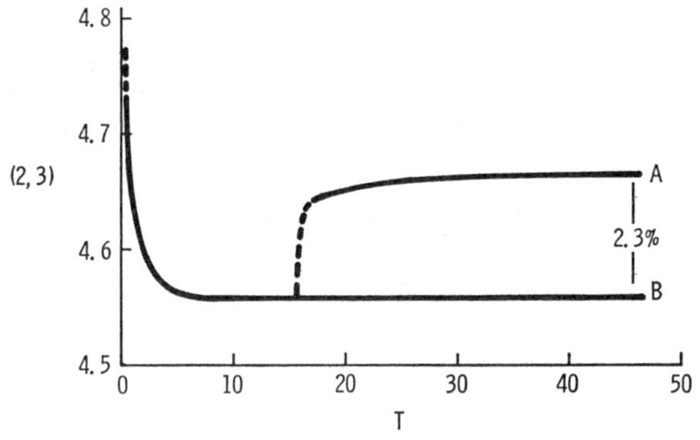

FIG. 1. Late time density solutions at point (2, 3) using Moretti's method. 6° sphere-cone, $M = 10$, $\gamma = 1.4$, 3×5 mesh. For $T < 15.82$, $\Delta t_A = \Delta t_B \simeq 0.94\, \Delta t$ crit. For $T \geqslant 15.82$, $\Delta t_A \simeq 1/5\, \Delta t_B$.

multiplier STAB to 1/5. Nothing else was altered. This computation was continued for an additional 28,000 time steps, at which $T = 46.45$. This gave a new steady-state solution, with ρ changing by only 4.32×10^{-4} in the last 1000 time planes. As a further check, a second solution B was run using the larger Δt (STAB = 1) all the way out to $T > 46.45$.

The difference between the two steady-state solutions at $T = 46.45$ is shown in Fig. 1, and is presented tabularly in Table II. At the most sensitive point (2, 3), the normalized densities differed by 2.3 %, the normalized pressures by 3 %, and the normalized shock stand-off distance by 0.6 %.

TABLE II

Steady-state solutions at point (2, 3) using Moretti's method. 6° sphere-cone, $M = 10$, $\gamma = 1.4$, 3×5 mesh, ρ, P, r_s = normalized values of density, pressure, and shock stand-off distance. $\Delta t_A \simeq 1/5\ \Delta t_B$.

	$\rho(2, 3)$	$P(2, 3)$	r_s
Solution A	4.664	76.54	1.142
Solution B	4.559	74.25	1.149
Percent difference	2.3 %	3.0 %	0.6 %

That the percentage difference between the two different solutions is small is to be expected, since the blunt body problem is known from physical and numerical experiments to be quite insensitive to Reynolds number. The FDE solution is then only a weak function of α_{es} and Δt, especially since inviscid boundary conditions are used on the surface so that no boundary layer develops, and since the shock is treated as a discontinuity. The numerical solutions obtained by this and other methods, using both implicit and explicit artificial viscosities, are certainly valid approximations. The significant point is that the two-dimensional steady-state solution obtained did depend on Δt, supporting the one-dimensional analysis for α_{es} which indicates that the method does exhibit an artificial viscosity effect in the steady state. A further indication of a viscosity effect was obtained from two solutions in a finer (5×7) mesh. The solution with STAB = 1 was steady to all four significant figures printed out for that test, whereas a "steady" solution obtained with STAB = 1/10 exhibited a persistent oscillation of ± 1 in the second significant figure of the density. This behavior is again consistent with the indication of the steady-state analysis for which $\alpha_{es} \propto \Delta t$.

Interpretation of the Lax–Wendroff Methods

The interpretation of the artificial viscosity for the Lax–Wendroff methods involves the resolution of paradoxical statements. On the one hand, we have the facts that (1) the transient analysis indicates $\alpha_e = 0$ and a formal truncation error for the inviscid equation of $O(\Delta x^2, \Delta t^2)$, and (2) the exact transient solution $\zeta_i^{n+1} = \zeta_{i-1}^n$ is obtained for $c = 1$ (and in $2D$ with time splitting, $\zeta_{ij}^{n+1} = \zeta_{i-1,j-1}^n$ for $c_x, c_y = 1$). On the other hand, we know that the steady-state analysis indicates $\alpha_{es} > 0$, with the Lax–Wendroff methods for the *inviscid* equation being *algebraically equivalent* to centered-space differencing of the steady *viscous* equation with $\alpha_{es} = u^2 \Delta t/2$.

The paradox is due to the effect of boundary conditions. In order to resolve this paradox, we consider the equation

$$-A\zeta_x + B\zeta_{xx} = 0, \tag{29a}$$

$$\zeta(0) = 0, \qquad \zeta(1) = 1. \tag{29b}$$

Using centered differences, we have

$$-(A/2\Delta x)(\zeta_{i+1} - \zeta_{i-1}) + (B/\Delta x^2)(\zeta_{i+1} - 2\zeta_i + \zeta_{i-1}). \tag{30}$$

The solution to (29) is

$$\zeta(x) = (1 - e^{xA/B})/(1 - e^{A/B}). \tag{31}$$

This solution to the continuum equation is plotted in Fig. 2a for various parameters A and B. The corresponding finite difference solutions for $\Delta x = 1/10$ are plotted in Fig. 2b. In order to interpret α_{es}, we must examine these solutions from the viewpoints of both the viscous and the inviscid equations.

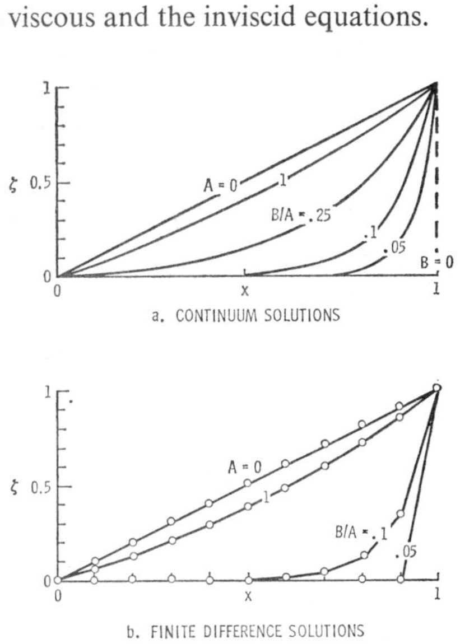

FIG. 2 Continuum and finite difference solutions to $-A\zeta_x + B\zeta_{xx} = 0$, $\zeta(0) = 0$, $\zeta(1) = 1$. Centered differences, $\Delta x = 1/10$.

We first consider the viscous equation. For $A/B = 0$, the first derivative (advection) drops out; we have the simple straight-line continuum solution $\zeta(x) = x$, and the exact finite difference solution $\zeta_i = (i - 1)\,\Delta x$. For $A/B > 0$, the ζ profile is blown downstream. As A/B becomes large, the continuum solution becomes $\zeta(x) \simeq 0$ up to the neighborhood of $x = 1$, where a rapid increase in ζ is required in order to meet the second boundary condition of $\zeta(1) = 1$. For $B = 0$, the continuum solution becomes $\zeta(0) = 0$ everywhere; the second boundary condition $\zeta(1) = 1$ cannot be met, and is extraneous since it would overspecify the problem. As this condition $B = 0$ is approached in the limit, we have the classical singular perturbation problem in the small parameter B/A, in which the order of the differential equation is reduced as $B/A \to 0$. For the finite difference equation, the behavior analogous to the singular perturbation problem occurs at $A/B = 20$ (more generally, $A\Delta x/B = 2$). At this condition, the FDE solution is $\zeta_i = 0$ for $i \leqslant 10$, and $\zeta_{11} = 1$. This FDE solution may be interpreted here as a qualitatively correct *viscous* behavior. But for $A\Delta x/B > 2$, oscillations and undershoot ($\zeta_{10} < 0$) develop as described in [6]. In terms of the *viscous* steady-state equation, this dividing condition $A\Delta x/B = 2$ corresponds to a cell Reynolds number $R_c \equiv u\Delta x/\alpha = 2$.

We next consider the inviscid equation using a Lax–Wendroff method. The condition $A\Delta x/B = 2$ now corresponds to $c = 1$. The exact transient solution is obtained as $\zeta_i^{n+1} = \zeta_{i-1}^n$; for $n > 10$, this gives the exact steady-state solution of $\zeta_i = \zeta_1 = 0$ for $i \leqslant 10$. For this condition of $c = 1$, the extraneous boundary value $\zeta_{11} = 1$ does not feed forward and influence the solution at interior points. The extraneous value $\zeta_{11} = 1$ is an error in this inviscid interpretation, but it is a purely *local* error for $c = 1$. Thus, there exists no contradiction with the formal truncation error of the method which implies an error of $O(\Delta x^2)$ for the steady-state problem. (This localness of the outflow error likewise removes the ambiguity for the upwind difference method and others which give the exact solution for $c = 1$.) The point is that the FDE solution for $A\Delta x/B$ can be validly interpreted as *either* a qualitatively correct viscous solution with $R_c = 2$, which solution includes $\zeta_{11} = 1$, or as an exactly correct inviscid solution at $c = 1$, which solution does not include the extraneous local boundary error $\zeta_{11} = 1$.

However, this interpretation is altered by two situations of practical importance; nonunity Courant numbers, and the addition of viscous terms to the equations.

The first situation which alters the usual analysis is the case of nonunity Courant number. For $c < 1$, the Lax–Wendroff methods no longer give the exact solution at interior points, but the formal truncation error is still $O(\Delta x^2, \Delta t^2)$. However, the FDE solution with a fixed outflow boundary value can only be interpreted as a viscous solution with $\alpha_{es} = u^2\Delta t/2$, and with only first order accuracy. The $O(\Delta x)$ error has been introduced by the requirement for the extraneous outflow boundary condition, which feeds forward for $c < 1$ and produces the artificially viscous behavior. Since the usual analysis for formal truncation error does not include boundary effects, it is inadequate in the present case of $c < 1$, and the true first order accuracy of the steady-state FDE solution is indicated by the steady-state analysis for artificial viscosity.

We have considered only a fixed outflow boundary condition, presumed to be in error. It is possible that this outflow error could actually be ordered (i.e., $\sim\Delta x^p$) using any of several methods [6]. The outflow error for $c < 1$ will almost certainly be $O(\Delta x)$, and the remarks made elsewhere in this paper are based on that assumption. But in the event that the ouflow error were only $O(\Delta x^2)$ then the Lax–Wendroff solution for $c < 1$ would be $O(\Delta x^2)$; however, the solution would still be artificially viscous in the sense that there exists a nonzero coefficient of ζ_{xx}.

The second situation which alters the usual analysis is the addition of viscous terms in the continuum equation. The viscous terms cannot be treated by the Lax–Wendroff time differencing, which would be unstable, but several authors have used FTCS differencing for the viscous terms, as in

$$\begin{aligned}\zeta_i^{n+1} = \zeta_i^n &- (c/2)(\zeta_{i+1}^n - \zeta_{i-1}^n) + (c^2/2)(\zeta_{i+1}^n - 2\zeta_i^n + \zeta_{i-1}^n)\\ &+ d(\zeta_{i+1}^n - 2\zeta_i^n + \zeta_{i-1}^n),\end{aligned} \tag{32}$$

where $d = \alpha\Delta t/\Delta x^2$, as before. (In the two-step Lax–Wendroff methods, the viscous term usually has been added in the second step only.) Here, the steady-state solution is clearly a viscous one, where the viscous term is the sum of the intended (physical) α and the artificial α_{es}. There is no contradiction of the usual truncation error analysis here, because the $O(\Delta x^2, \Delta t^2)$ result is not obtained from (32), which is readily seen to be only first order accurate. Note also that, with the physical viscous term present, large Courant numbers actually aggravate the α_{es} error. From Table I it is easily shown that the α_{es} of the Lax, Lax–Wendroff and upwind differencing methods are in the ratios $1/c{:}1{:}c$, respectively. For regions near the stability limit $c = 1$, the artificial viscosity of a Lax–Wendroff–FTCS method is virtually the same as the upwind difference method; within a boundary layer, $c < 1$, and the Lax–Wendroff–FTCS method will be more accurate, as in Ref. [12].

To summarize, the Lax–Wendroff methods do give the exact solution of the model equation, in both transient and steady-state cases, for $c = 1$ and no viscous terms present. The boundary error at outflow is purely local. The usual truncation error analysis is applicable, and indicates errors of only $O(\Delta x^2, \Delta t^2)$ with no artificial viscosity effect. The steady-state analysis showing $\alpha_{es} > 0$ is inappropriate. But for $c < 1$ in the inviscid equation, any outflow boundary error does have global effects which invalidate the usual truncation error analysis of the interior point equations and which introduce an artificially viscous behavior. Also, for the addition of viscous terms, the usual truncation error analysis is not applicable, and the method has an artificial viscosity effect which is aggravated by near-unity

Courant numbers. In both these cases of $c < 1$ and/or additional viscous terms, the steady-state analysis for artificial viscosity is appropriate, and shows $\alpha_{es} = u^2\Delta t/2$ and first-order formal accuracy.

Implications to Other Methods

Although we have not tested the following methods experimentally, the implications of the analyses on the model equation (1) are as follows (see Ref. [6] for references and details). The midpoint leapfrog, the Crocco, Adams–Bashforth, Heun, fully implicit, Crank–Nicholson, and the various ADI methods would have zero artificial viscosity in the steady state, except when upwind differencing is used for the advection terms as has been done in some ADI solutions. The multistep Strang, Abarbanel and Zwas, Fromm [13], and the Crowley methods would have a persistent nonzero α_{es} in the steady state. The only known methods for which the analyses indicate zero artificial viscosity in both the transient and steady-state analyses are the midpoint leapfrog method, the Arakawa method, the angled derivative method of Roberts and Weiss [19], and those ADI methods which have a truncation error of $O(\Delta x^2, \Delta t^2)$. Each of these has other disadvantages, of course.

It is interesting to note that the expression for α_{es} of the Leith method, $\alpha_{es} = \frac{1}{2}u^2\Delta t$, does not contain Δx directly. Thus, as $\Delta x \to 0$, the $\alpha_{es} \to 0$ only because of the Courant number restriction on stability, which requires $\Delta t \to 0$ as $\Delta x \to 0$. If a method were devised which used the second-order time expansion of the Leith (Lax–Wendroff, etc.) method but which was unconditionally stable, the α_{es} effect would persist even as $\Delta x \to 0$, for fixed Δt.

Final Remarks

We have four final remarks on the interpretation of artificial viscosity:

(1) The truncation error analysis indicates the *order* of the error, which is strictly applicable only as $\Delta x, \Delta t \to 0$. In a practical computation, we are generally interested not in the order of the truncation error, but in the *size* of the truncation error, for some Δx and Δt [6]. Thus, the addition of some miniscule viscous term (say $\text{Re} = 10^6$ for $\Delta x = 1/10$) *formally* deteriorates the truncation error of a Lax–Wendroff–FTCS calculation to $O(\Delta x)$, but the *size* of the error remains entirely negligible for $c = 1$. Note also that the *size* of the truncation error varies smoothly for $c \leqslant 1$, although the *order* jumps discontinuously (singularly) from the exact solution at $c = 1$ to $O(\Delta x)$ for $c < 1$.

(2) For multidimensional problems, the most important effect of viscosity, in the sense of producing a difference between viscous and inviscid solutions, is

usually not so much in the appearance of viscous terms at interior points, but in the enforcement of no-slip boundary conditions. Thus, Kentzner [20] has indicated that fairly accurate approximations to inviscid solutions can be obtained with Re as low as 300 in a reasonable mesh, provided that the inviscid (slip) boundary conditions are used. This means that inviscid solutions can be accurate even though artificial viscosity is present; however, the error may be somewhat more significant for viscous problems. (In assessing the α_{es} error of FDE solutions for drag coefficient C_D, for example, it is important to look not for some small error in C_D, but for a shift in Re to get the same C_D. This is obviously appropriate because of the usual weak sensitivity of flows to Re.)

(3) In multidimensional problems, the α_{es} terms depend on u and v, which are defined with respect to the Eulerian mesh. This means that different spatially varying α_{es} apply in the x and y directions, and tend to zero near stationary no-slip walls. Thus, "equivalent Re" interpretations are not possible even for viscous solutions, except in a qualitative sense, and viscous FDE solutions with nonzero α_{es} are often more accurate than might be expected from evaluating α_{es} based on freestream conditions. However, such solutions are not Galilean invariant [5]. Also, solutions for rotating bodies might exhibit anamolous behavior due to different α_{es} on the advancing and the retreating sides.

(4) Several methods are available [6] for freeing the outflow computational boundary condition in multidimensional flows. These will tend to reduce the upstream error associated with $c < 1$ in the inviscid equations.

Summary

It has been demonstrated that the usual method of analysis for the artificial viscosity of finite difference analogs for the advection terms is ambiguous, with different results being obtained for the transient and the steady-state analyses. The analysis indicates that many methods which have been touted as having no artificial viscosity, notably the Leith, Lax–Wendroff, two-step Lax–Wendroff, Moretti, and MacCormack methods, do have a Δt-dependent artificial viscosity effect when a steady-state solution is obtained for viscous flow and/or for $c < 1$. Viscous steady-state solutions obtained using these methods with Courant-numbers $c \simeq 1$ have only first order formal accuracy.

Acknowledgments

The author gratefully acknowledges the work of Dr. R. R. Eaton in obtaining the two-dimensional solutions cited, and the helpful discussions with Dr. Eaton, Dr. F. G. Blottner, and especially Dr. L. D. Tyler.

References

1. J. von Neumann and R. D. Richtmyer, A method for the numerical calculations of hydro dynamical shocks, *J. Appl. Phys.* **21** (1950), 232.
2. W. F. Noh and M. H. Protter, *J. Math. & Mech.* **12** (1963), 149.
3. C. W. Hirt, *J. Comput. Phys.* **2** (1968), 339.
4. L. D. Tyler, "Proc. Second International Conference on Numerical Fluid Dynamics," (M. Holt, Ed.), Springer-Verlag, Berlin, 1971.
5. R. A. Gentry, R. E. Martin, and B. J. Daly, *J. Comput. Phys.* **1** (1966), 87.
6. P. J. Roache, "Computational Fluid Dynamics," to be published.
7. P. D. Lax, *Commun. Pure Appl. Math.* **7** (1954), 159.
8. C. E. Leith, "Methods in Computational Physics" (B. Alder, Ed.), Vol. 4, p. 1, 1965.
9. P. Lax and B. Wendroff, *Commun. Pure Appl. Math.* **13** (1960), 217.
10. R. D. Richtmyer, "A Survey of Difference Methods for Non-Steady Fluid Dynamics," NCAR Tech. Note 63–2, 1963.
11. G. Moretti and M. Abbett, *AIAA J.* **4** (1966), 2136.
12. R. W. MacCormack, AIAA Paper No. 69-354 (1969). Also, R. W. MacCormack, "Proc. Second International Conference on Numerical Fluid Dynamics" (M. Holt, Ed.), Springer-Verlag, Berlin, 1971.
13. J. E. Fromm, *J. Comput. Phys.* **3** (1968), 176.
14. V. V. Rusanov, *Zh. Vych. Mat. Mat. Fiz.* **1** (1961), 267. Nat. Res. Council of Canada TT-1027, 1962.
15. D. K. Lilly, *U. S. Monthly Weather Review* **93** (1965), 11.
16. I. Yu. Brailovskaya, *Sov. Phys.—Dokl.* **10** (1965), 107.
17. J. S. Allen and S. I. Cheng, *Phys. Fluids* **13** (1970), 37.
18. G. Moretti and G. Bleich, "Three-Dimensional Inviscid Flow About Supersonic Blunt Cones at Angle of Attack," SC-CR-68-3728, Sandia Laboratories, Albuquerque, New Mexico, September, 1968.
19. K. V. Roberts and N. O. Weiss, *Math. Comp.* **20** (1966), 271.
20. C. P. Kentzner, *J. Comput. Phys.* **6** (1970), 168.

PROBLEMS

Problems marked with an asterisk are computer problems. Any of these may be solved within 5 to 10 minutes on modest computers such as CDC 3600, UNIVAC 1107, or IBM 360/50.

Chapter 2

2-1. Explore the possibility of deriving a vorticity transport equation for variable viscosity $\bar{\mu}$. Normalize with $Re = \bar{\rho}\bar{U}_o L/\bar{\mu}_o$, and $\mu = \bar{\mu}/\bar{\mu}_o$.

2-2. Show that the drag coefficient C_D of a thin flat plate of length L aligned parallel to the flow may be evaluated by quadrature of the skin-friction coefficient C_f as follows.

$$C_D = \frac{\text{drag force}}{\frac{1}{2}\rho V_\infty^2 \cdot (\text{plate area})}$$

$$C_D = 2C_f$$

$$C_f = \frac{1}{Re}\int_0^L \zeta_W dx$$

where ζ_W is the normalized vorticity at the plate surface.

Chapter 3

3-1. Derive the following finite-difference expressions, accurate to $O(\Delta x^2)$.

$$\zeta_{xxx}\big|_i = \frac{\zeta_{i+2} - 2\zeta_{i+1} + 2\zeta_{i-1} - \zeta_{i-2}}{2\Delta x^3} \quad \text{and} \quad \zeta_{xxxx}\big|_i = \frac{\zeta_{i+2} - 4\zeta_{i+1} + 6\zeta_i - 4\zeta_{i-1} + \zeta_{i-2}}{\Delta x^4}$$

3-2. Consider the compressible flow continuity equation for mass density ρ.

$$\frac{\partial \rho}{\partial t} = -\nabla\cdot(\rho\vec{V}) = -\rho\nabla\cdot\vec{V} - \vec{V}\cdot\nabla\rho$$

In 1D,

$$\frac{\partial \rho}{\partial t} = -\frac{\partial(\rho u)}{\partial x} = -\rho\frac{\partial u}{\partial x} - u\frac{\partial \rho}{\partial x}$$

(a) Show that the use of centered-space differences on the first form gives a conservative system.

(b) Show that the second (expanded) form is *also* conservative, and is consistent with the "zip" differencing (e.g., see Hirt, 1968) interpretation of the interface flux terms as

$$(\rho u)_{i+\frac{1}{2}} = \frac{1}{2}\left[\rho_i u_{i+1} + \rho_{i+1}u_i\right]$$

(c) In order to dampen some nonlinear oscillations using the second (expanded) form, the following modification is proposed. The advection velocity u_i in the term $u_i\left(\frac{\rho_{i+1} - \rho_{i-1}}{2\Delta x}\right)$ will be replaced by its spatial average, giving $\frac{1}{2}(u_{i+1} + u_{i-1})$ $\left(\frac{\rho_{i+1} - \rho_{i-1}}{2\Delta x}\right)$. Show that this method is still second-order accurate but is not conservative.

3-3. Consider the second-derivative diffusion-like term, $\partial/\partial x[\alpha(\partial\zeta/\partial x)]$. With $\zeta = ax^2$ and $\alpha = bx^2$, the continuum solution is $\partial/\partial x[\alpha(\partial\zeta/\partial x)] = 6abx^2$. Show that the use of the conservation differencing form gives

$$\frac{\delta}{\delta x}\left(\alpha\,\frac{\delta\zeta}{\delta x}\right)_i = \frac{\frac{1}{2}(\alpha_{i+1}+\alpha_i)(\zeta_{i+1}-\zeta_i) - \frac{1}{2}(\alpha_i+\alpha_{i-1})(\zeta_i-\zeta_{i-1})}{\Delta x^2} = 7abx^2$$

whereas the use of the expanded, non-conservation form

$$\alpha\,\frac{\delta^2\zeta}{\delta x^2} + \left(\frac{\delta\alpha}{\delta x}\right)\left(\frac{\delta\zeta}{\delta x}\right)$$

gives the exact answer. (This example is due to Dr. F. G. Blottner.)

3-4. Consider the second-derivative diffusion-like term, $\partial/\partial x[\alpha(\partial\zeta/\partial x)]$. For the conservation form (see problem 3-3), show that the lowest order truncation error term is

$$-\frac{1}{6}\Delta x^2\left(\zeta_x\alpha_{xxx} + \frac{3}{2}\alpha_{xx}\zeta_{xx} + \alpha_x\zeta_{xxx}\right)$$

For the expanded, non-conservation form, the lowest order error is

$$-\frac{1}{6}\Delta x^2\left(\alpha_{xxx}\zeta_x + \alpha_x\zeta_{xxx} + \frac{1}{2}\alpha\zeta_{xxxx}\right)$$

For the forms

$$\zeta = a_1 + b_1x + c_1x^2$$

$$\alpha = a_2 + b_2x + c_2x^2$$

this error analysis indicates that the *non-conservation* form is *more* accurate, giving the $O(\Delta x^2)$ term = 0, whereas the conservation form gives the $O(\Delta x^2)$ term $= -c_1c_2$. (This example is due to Dr. F. G. Blottner.)

*3-5. Program the FTCS method for the 1D model equation. Experiment with various outflow computational boundary conditions, both for fixed and cyclic (sine wave) inflow values of ζ. The outflow treatments should at least include the zero-gradient, linear-extrapolation, and upwind-difference methods.

3-6. Consider the midpoint leapfrog method applied to the 1D inviscid model advection equation far from boundaries. Using the approach of the discrete perturbation stability analysis, let $\zeta_i^n = \varepsilon$, all other $\zeta^n = 0$. Show that the farthest downstream effect is given by

$$\zeta_{i+k}^{n+k} = (c^k)\varepsilon$$

where c is the Courant number. For points farther than this "front", $\zeta_{i+1}^{n+k-1} = 0 = \zeta_{i+k+1}^{n+k}$. In the upstream direction, show that a false solution propagates as

$$\zeta_{i-k}^{n+k} = (-c)^k$$

Plot the solutions for c = 1, 1/2, 1/10.

*3-7. Investigate the behavior of the leapfrog method on the one-dimensional problem.

(a) Study the effect of Courant number c and Δt on the problem $\zeta(x,0) = 0$, $\zeta(0,t) = \sin(t)$, with the computational boundary condition $\zeta_{i\ell}^{n+1} = \zeta_{i\ell-1}^{n+1}$

(b) With $\zeta_i^1 = \zeta_i^2 = 0$ for i odd, $\zeta_i^1 = +1$ and $\zeta_i^2 = -1$ for i even, and $\zeta_1^n = 0$, study the behavior with two different downstream treatments: (1) $\zeta_{i\ell}^{n+1} = \zeta_{i\ell-1}^{n+1}$, and (2) $\zeta_{i\ell}^{n+1} = \zeta_{i\ell-1}^{n}$. A 10-cell mesh ($i\ell = 11$) is sufficient. The reader may wish to scan Chapter VII first.

*3-8. Using the 1D-model advection-diffusion equation, design a computational test program to determine the effects of a spatially varying advection speed, u, on stability. (This open-ended problem can cover many possible combinations of basic interior-point schemes, initial conditions, boundary conditions, etc. It could be assigned for a weekend problem or for a PhD dissertation. It is especially effective as a class project, in which different students study different methods.)

3-9. For the Blasius boundary layer on a flat plate (see Schlichting, 1968), the following relation may be derived:

$$v_e/U_e = 4.302/Re_\delta$$

wherein v_e and U_e are the velocity components at the edge of the boundary layer, and Re_δ is the Reynolds number based on boundary-layer thickness, δ.

$$Re_\delta = U_e\delta/\nu$$

Show that the use of upwind differencing may be expected to give roughly a 20% decrease in effective Reynolds number through the artificial viscosity effect, for 10 points in the boundary layer.

3-10. Consider the "continuous time, discrete space" CTDS approach to stability analysis, in which time is not discretized. This approach would be appropriate to a hybrid computer (digital-analog) in which the problem time ran in scaled analog computer time. CTDS analyses have also been used to study classes of difference methods which are dimensional composites of ODE methods; e.g., FTCS differencing is in this class, but Leith's method is not.

Consider the general FDE form of the inviscid model equation,

$$\left.\frac{d\zeta}{dt}\right|_i = \frac{c}{2\Delta x}\left[(1+\beta)\zeta_{i-1} - 2\beta\zeta_i + (\beta - 1)\zeta_{i+1}\right]$$

For $\beta = 0$, this gives centered-space differencing, whereas $\beta = 1$ gives upwind differencing. Show that the above FDE gives the exact solution of the following PDE.

$$\frac{\partial\zeta}{\partial t} = -c\frac{\partial\zeta}{\partial x} + \frac{c\beta}{\Delta x}\sum_{\substack{n=2\\ n\text{ even}}}^{\infty}\frac{\Delta x^n}{n!}\frac{\partial^n\zeta}{\partial x^n} - \frac{c}{\Delta x}\sum_{\substack{n=3\\ n\text{ odd}}}^{\infty}\frac{\Delta x^n}{n!}\frac{\partial^n\zeta}{\partial x^n}$$

(Lomax, et al., 1970)

Show the importance of considering the time-differencing method in the analysis by contrasting the above analysis with that in the text for upwind differencing.

3-11. Devise a method for the diffusion equation based on Adams-Bashforth time-differencing. Prove at least conditional stability. For the combined advection-diffusion equation, prove that at least conditional stability exists for u/α small.

3-12. Apply Hirt's stability analysis to Crocco's method for inviscid flow, and show that $\alpha_e = u^2 \Delta t(\Gamma - 1/2)$.

3-13. (A) Using the results of Problem (3-1) for the spatial derivative ζ_{xxx}, design a method like Leith's method, but using a third-order Taylor-series expansion in time.
(B) Investigate the method for stability. Use the Neumann stability analysis, and assume $c << 1$ if necessary.

3-14. Consider an implicit upwind-difference method for the inviscid model equation,

$$\zeta_i^{n+1} = \zeta_i^n - c\left(\zeta_i^{n+1} - \zeta_{i-1}^{n+1}\right)$$

Analyze for stability. Does the method possess the transportive property?

3-15. The FTCS method applied to the 1D diffusion results in the necessary and sufficient stability condition on $d \equiv \alpha\Delta t/\Delta x^2$ of $d \leq d_{max} = 1/2$. The Matsuno-Brailovskaya two-step method results in a sufficient condition of $d_{max} \leq 1/4$. If the iteration is continued indefinitely, will the time step continue to be further restricted as $d_{max} \to 0$, or will the iterations make the method approach the fully implicit method, for which $d_{max} = \infty$?

3-16. (A) Diagram the two steps of the Kurihara method in the x-t plane. Show how time and space centering is achieved.
(B) Rewrite the method as a one-step procedure.
(C) Demonstrate that the method will not exhibit the time-splitting instability of the leapfrog method, but that the $\Lambda = 2\Delta x$ component *is* stationary.
(D) Demonstrate that the method does *not* give the exact solution for $c = 1$.
(E) Perform a von Neumann stability analysis and show that $|\lambda| = 1$.
(F) Analyze for artificial viscosity effects in the transient and steady-state cases.
(G) Prove that the truncation error $E = 0(\Delta x^2, \Delta t^2)$.

3-17. The following two-step, second-order method has been suggested by Blottner (personal communication).

$$\zeta^{n+\frac{1}{2}} = \zeta^n + \frac{\Delta t}{2} L(\zeta^{n+\frac{1}{2}})$$

$$\zeta^{n+1} = \zeta^n + \Delta t L(\zeta^{n+\frac{1}{2}})$$

Note that the first step is implicit, while the second is explicit. Analyze it for stability using the von Neumann method, for the cases of

(A) $L(\zeta) = -\dfrac{(u\zeta_{i+1} - u\zeta_{i-1})}{2\Delta x}$

(B) $L(\zeta) = \alpha\dfrac{(\zeta_{i+1} - 2\zeta_i + \zeta_{i-1})}{\Delta x^2}$

3-18. Analyze the stability of the one-sweep Saul'yev method for the diffusion equation, using the von Neumann analysis.

3-19. Show that $|c| \leq 1$ is required to avoid spatial amplification of the angled-derivative ADE method for the inviscid advection equation.

3-20. Show that the artificial viscosity coefficient for steady-state solutions in 1D for Fromm's method is 1/2 that for Leith's method.

3-21. Show that the SOR method, sometimes called the extrapolated Liebman method, is indeed a linear *extrapolation* of Liebman's method with $\omega = 1$.

*3-22. Experimentally determine the optimum relaxation parameter ω_o for a rectangular region with a cutout (see Figure 3-22 on page 140). Note ω_o will depend on the size of the region used and on $\Delta x, \Delta y$.

3-23. Consider the following suggested method for evaluating wall vorticity at point w. An artificial point at (w-1), located Δy *inside* the wall, is defined, and the value $\psi_{w-1} = \psi_{w+1}$ is set. Then centered differences are used to evaluate the Laplacian $\nabla^2\psi_w = \zeta_w$. Show that this method is arithmetically equivalent to using the first-order method for ζ_w.

3-24. Given values of f at mesh points f_i, $f_{i\pm1}$, $f_{i\pm2}$, etc., define new node values at $i\pm1/2$, etc., by averaging

$$f_{i\pm\frac{1}{2}} = \frac{1}{2}(f_i + f_{i\pm1})$$

Denote by $\delta f/\delta x$, $\delta^2 f/\delta x^2$ the usual second-order differences over node values $(i\pm1)$, such as

$$\left.\frac{\delta f}{\delta x}\right|_i = \frac{f_{i+1} - f_{i-1}}{2\Delta x}$$

Denote by $\tilde{\delta} f/\tilde{\delta} x$, $\tilde{\delta}^2 f/\tilde{\delta} x^2$ the second-order differences over $(i\pm1/2)$, such as

$$\left.\frac{\tilde{\delta} f}{\tilde{\delta} x}\right|_i = \frac{f_{i+\frac{1}{2}} - f_{i-\frac{1}{2}}}{\Delta x}$$

$$\left.\frac{\tilde{\delta} f}{\tilde{\delta} x}\right|_{i+\frac{1}{2}} = \frac{f_i - f_{i-1}}{\Delta x}$$

Show that

(A)

$$\left.\frac{\tilde{\delta} f}{\tilde{\delta} x}\right|_i = \left.\frac{\delta f}{\delta x}\right|_i$$

(B)

$$\left.\frac{\tilde{\delta}^2 f}{\tilde{\delta} x^2}\right|_i = 2\left.\frac{\delta^2 f}{\delta x^2}\right|_i$$

$$\left.\frac{\tilde{\delta}^2 f}{\tilde{\delta} x^2}\right|_{i\pm\frac{1}{2}} = 0$$

3-25. When mesh values of f are linearly interpolated to a finer (1/2) mesh, the values of $\delta f/\delta x$ at original node points are not changed, but the values of $\delta^2 f/\delta x^2$ *are* changed (see previous problem). Devise a method of defining the fine-mesh values of f such that $\tilde{\delta}^2 f/\tilde{\delta} x^2|_i = \delta^2 f/\delta x^2|_i$. Note that $\tilde{f}_i \neq f_i$ is required and that $\tilde{\delta} f/\tilde{\delta} x|_i \neq \delta f/\delta x|_i$, generally.

3-26. Consider the 1D problem $\delta^2 f/\delta x^2 = q$, with Dirichlet boundary conditions $f(0) = 0$, $f(1) = 1$. Consider using the usual second-order form, but with the lower boundary located at a half-node space $i = 1\text{-}1/2$. The first boundary condition, $f_{1\frac{1}{2}} = 0$, is then set by requiring $f_1 = -f_2$, where f_1 is "below" the lower boundary. Using only 3 or 4 nodes so that hand calculation may be used, show that (A) this mesh system produces boundedness errors, in that $\psi_2 < 0$, which is impossible in the continuum equations, and (B) the mesh system results in only first-order accuracy.

Using a desk calculator or a simple computer program, it may be verified that the boundedness error persists as $\Delta x \to 0$.

3-27. Consider the problem of *which* fluid dynamics boundary conditions at a no-slip wall should be used with the (ψ,ζ) equations. Consider flow in driven cavity, in which case all boundaries are no-slip walls. The wall conditions of $u = 0$ (or $u = 1$ on the "lid") and $v = 0$ then give ψ boundary conditions of

$$\psi = 0 \quad \text{(or other constant)} \tag{a}$$

and

$$\frac{\partial\psi}{\partial n} = 0 \quad \text{(or 1 on the lid)} \tag{b}$$

To render the problem quite simple, consider the Re = 0 (Stokes flow) case, in which the steady-state equations reduce to the linear biharmonic equation,

$$\nabla^4\psi = 0 \tag{c}$$

We solve the biharmonic by splitting it into two Poisson equations:

$$\nabla^2\psi = \zeta \tag{d}$$

$$\nabla^2\zeta = 0 \tag{e}$$

Show that the use of (a) in (d) and (b) in (e) is possible; but that (a) in (e) and (b) in (d) is not. Refer to section III-C-2.

*3-28. (A) Solve the 2D problem of driven-cavity flow on the computer. Normalize velocities by the lid velocity. Use a coarse mesh with $\Delta x = \Delta y = 1/4$ and Re = 10. Use FTCS differencing on the vorticity transport equation, SOR on the Poisson equation, and the first-order wall vorticity evaluation. Justify your convergence decision.

(B) Repeat problem (3-28A) with a "wind stress" boundary condition on the driving "lid" with $\zeta_{lid} = c$.

3-29. Consider the steady-state 1D model equation. For $u > 0$, $\alpha > 0$, show that boundary conditions of

$$u(0) = a \quad , \quad u^N = b$$

where the notation u^N means the N-th order derivative, are sufficient to specify the solution. Consider the special cases of $u = 0$ and of $\alpha = 0$ in regard to the sufficiency of the boundary conditions.

3-30. For the 1D linear steady-state model equation, investigate the appearance of "wiggles" when a gradient condition, $\partial\zeta/\partial x = S \neq 0$, is used at outflow.

3-31. Derive equation (3-581A), the Poisson equation for pressure.

3-32. Derive the relation between pressure gradient tangential to a wall and vorticity gradient normal to a wall (see Pearson, 1965A)

$$\frac{\partial P}{\partial S} = \frac{1}{Re}\frac{\partial \zeta}{\partial n}$$

Note that the single-valuedness of P then implies the following condition on a line integral around the surface of any body.

$$\oint \frac{\partial \zeta}{\partial n}\, ds = 0$$

The failure of a numerical solution to meet the corresponding line quadrature around a body may be viewed as a kind of conservation error in the wall vorticity evaluation, and may be used as an index of truncation-error convergence.

3-33. For the stiff equation $dT/dt = aT$, set up the explicit, "fully implicit", and Crank-Nicolson methods. Analyze for stability and overshoot conditions.

3-34. Derive equation (3-609), section III-G-6, for a variable-property vorticity transport equation.

Chapter 4

4-1. Check the normalizing in equations (4-42) and (4-43).

4-2. Derive equation (4-52).

Chapter 5

5-1. Show that $u + a \geq a_0$ for adiabatic flow and that, for such flows, the critical stability restriction on Courant number may be expected to occur at the maximum velocity condition.

*5-2. Program Lax's method for a 1D shock-propagation problem. Use fixed inflow conditions and zero gradient at outflow. For $\gamma = 1.4$, the following conditions apply to a shock-strength pressure ratio of 4.978.

Pre-shock: $\rho = 1$, $E = 2.5$, $u = 0$

Post-shock: $\rho = 2.812$, $E = 16.0489$, $u = 1.601$

The post-shock values may be set at $i = 1$, and pre-shock for $i = 2$ to $IL = 50$. Experiment with different Courant numbers; particularly, note the diffusion for $\Delta t = 0$.

*5-3. Repeat problem (5-2) for upwind differencing. Examine the behavior of post-shock oscillations as Δt and the shock pressure ratio are varied.

*5-4. Repeat problem (5-2) for any of the explicit artificial viscosity methods given in section V-D. (For class projects, different methods may be assigned to different students.)

*5-5. Repeat problem (5-2) for any of the implicit artificial viscosity methods given in section V-E.

*5-6. Using any of the codes developed for problems (5-2), experiment with different initial conditions in which the initial shock is spread over several nodes.

5-7. Show that the artificial viscosities of the Lax, upwind, and Lax-Wendroff (Moretti, MacCormack, etc.) methods in the steady-state analysis are in the ratios $1/c:1:c$, respectively. Note that, when Courant numbers near the stability limit $c = 1$ are used, all these methods have about the same artificial-viscosity effect in steady state and all are first order (Roache, 1971B).

5-8. Devise a 1D shock-propagation method using an explicit artificial viscosity and based on Crocco's method for viscous flows.

*5-9. Test the method of problem (5-8) computationally, and compare it with other methods.

*5-10. Apply one of the explicit methods for the incompressible vorticity transport equation (such as DuFort-Frankel, ADE, or Heun) to 1D shock propagation, using an explicit artificial viscosity.

5-11. Determine the accuracy of using the reflection method for setting boundary conditions at the end wall of a closed shock tube in 1D flow. Consider both inviscid and viscous flow equations, in both the first and second mesh systems. (See Tyler and Ellis, 1970; Watkins, 1970.)

5-12. Determine the effect of the use of the reflection method in the second mesh system on the viscous cross-derivative terms.

5-13. Demonstrate formally the second-order accuracy of the $\delta P/\delta x$ solution adjacent to a no-slip wall, using the staggered-mesh solution for density as recommended in section V-G-2-c.

*5-14. Solve the driven cavity flow problem (3-28) for compressible subsonic flow. Set $M = 1$ at the lid, with wall temperatures fixed at T_o. Use constant-property equations with Brailovskaya's method or the Cheng-Allen method. Be sure to keep the cell Reynolds number < 2. (This problem requires a fairly large programming effort.)

Chapter 6

6-1. Perform the exponential stretch transformation of equation (6-9) on the conservative form of the vorticity transport equation (2-12). Is the transformed equation "conservative" in some sense?

6-2. Show that FTCS differencing (which would be unstable) of the cylindrical axisymmetric continuity equation,

$$\frac{\partial \rho}{\partial t} + \frac{1}{r}\frac{\partial(\rho v)}{\partial r} + \frac{\partial(\rho u)}{\partial z} = 0$$

is "conservative" in the control volume sense of section III-A-3. Note that it is not conservative in the sense of being "divergence-free".

REFERENCES AND BIBLIOGRAPHY

The following books have material pertinent to computational fluid dynamics: Forsythe and Wasow (1960), Richtmyer (1957), Saul'yev (1964), Patankar and Spalding (1967B), Richtmyer and Morton (1967), Ames (1965,1969), Zienkiewicz (1967), Mitchell (1969), Gosman et al. (1969), and Yanenko (1971). See also the compilation of Russian papers edited by Roslyakov and Chudov (1963), the published proceedings of two meetings on numerical fluid dynamics edited by Frenkiel and Stewartson (1969) and Holt (1971), the AIAA reprint compilation by Chu (1968), and especially the continuing Methods of Computational Physics Series (currently nine volumes) published by Academic Press and edited by Alder, et al.

Current research papers may be found in the following journals: Journal of Computational Physics, AIAA Journal, Journal of Fluid Mechanics, The Physics of Fluids, Numerical Methods in Engineering, AIChE Journal, ASME Journal of Heat Transfer, Mathematics of Computation, and Computer and Fluids.

Aalto, S.K. (1967), "An Iterative Procedure for the Solution of Nonlinear Equations in a Banach Space," MRC-TSR-774, Mathematics Research Center, Madison, Wisconsin.

Abarbanel, S., and Zwas, G. (1969), "An Iterative Finite-Difference Method for Hyperbolic Systems," Mathematics of Computations, pp. 549-565.

Abarbanel, S., Bennett, S., Brandt, A., and Gillis, J. (1970), "Velocity Profiles at Low Reynolds Numbers," J. of Applied Mechanics, Transactions of the ASME, Vol. 37, Ser. E., No. 1, March, 1970, pp. 2-4.

Abarbanel, S., and Goldberg, M. (1971), "Numerical Solution of Quasi-Conservative Hyperbolic Systems: The Cylindrical Shock Problem," Tel-Aviv University, Israel, January, 1971.

Abbett, Michael (1970), "The Mach Disc in Underexpanded Exhaust Plumes," AIAA Paper No. 70-231, AIAA 8th Aerospace Sciences Meeting, New York, New York, January 19-21, 1970.

Abbott, Michael B. (1966), An Introduction to the Method of Characteristics, American Elsevier Publishing Company, Inc., New York.

Ackerman, R.C. (1971) "Boundary-Layer Separation at a Free Streamline-Finite Difference Calculations." See Holt (1971).

Adamson, T.C. (1968), "Solutions for Supersonic Rotational Flow Around a Corner Using a New Co-ordinate System," J. Fluid Mechanics, Vol. 34, Part 4, pp. 735-758.

Agard (1968), "Transonic Aerodynamics."

Ahamed, S.V. (1970), "Application of the Acceleration of Convergence Technique to Numerical Solution of Linear and Non-Linear Vector Field Problems with Numerous Sources," J. of Engineering Science, Vol. 8, No. 5, pp. 403-413.

Alder, B., Fernbach, S., and Rotenberg, M. (1970), "Methods in Computational Physics," Plasma Physics, Vol. 9.

Aleksidze, M.A., and Pertaia, K.V. (1969), "Universal Program for the Finite-Difference Solution of the Dirichlet Problem of the Poisson Equation with the Aid of Enhanced-Accuracy Formulas," Tiflis, Izdatel'stvo Metsniereba.

Allen, D.N. de G., and Southwell, R.V. (1955), "Relaxation Methods Applied to Determine the Motion, in Two Dimensions, of a Viscous Fluid Past a Fixed Cylinder," Quarterly J. of Mechanics and Applied Mathematics, Vol. 8, pp. 129-145.

Allen, John S. (1968), "Numerical Solutions of the Compressible Navier-Stokes Equations for the Laminar Near Wake in Supersonic Flow," Princeton University, Ph.D. Dissertation, Princeton, New Jersey.

Allen, J.S., and Cheng, S.I. (1970), "Numerical Solutions of the Compressible Navier-Stokes Equations for the Laminar Near Wake," Physics of Fluids, Vol. 13, No. 1, January, 1970, pp. 37-52.

Ames, W.F. (1965), Nonlinear Partial Differential Equations in Engineering, Academic Press, New York.

Ames, W.F. (1969) Numerical Methods for Partial Differential Equations, Barnes and Noble, Inc., New York, New York.

Amsden, A.A., and Harlow, F.H. (1964), "Slip Instability," Physics of Fluids, Vol. 7, pp. 327-334.

Amsden, A.A., and Harlow, F.H. (1965), "Numerical Calculation of Supersonic Wake Flow," AIAA J., Vol. 3, No. 11, November, 1965, pp. 2081-2086.

Amsden, A.A., and Harlow, F.H. (1968), "Transport of Turbulence in Numerical Fluid Dynamics," J. of Computational Physics, Vol. 3, pp. 94-110.

Amsden, A.A., and Harlow, F.H. (1970A), "The SMAC Method: A Numerical Technique for Calculating Incompressible Fluid Flows," LA-4370, Los Alamos Scientific Laboratory, Low Alamos, New Mexico, May, 1970.

Amsden, A.A., and Harlow, F.H., (1970B), "A Simplified MAC Technique for Incompressible Fluid Flow Calculations," J. of Computational Physics, Vol. 6, pp. 322-325.

Anderson, J.D., Jr. (1969A), "A Time-Dependent Quasi-One-Dimensional Analysis of Populations Inversions in an Expanding Gas," NOLTR 69-200, United States Naval Ordnance Laboratory, White Oak, Maryland, December, 1969.

Anderson, J.D., Jr. (1969B), "A Time-Dependent Analysis for Vibrational and Chemical Nonequilibrium Nozzle Flows," AIAA Paper No. 69-668, AIAA Fluid and Plasma Dynamics Conference, San Francisco, California, June 16-18, 1969.

Anderson, J.D., Jr. (1970A), "A Time-Dependent Analysis for Vibrational and Chemical Nonequilibrium Nozzle Flows," AIAA J., Vol. 8, No. 3, March, 1970, pp. 545-550.

Anderson, J.D., Jr. (1970B), "Time-Dependent Solutions of Nonequilibrium Nozzle Flows - A Sequel," AIAA J., Vol. 8, No. 12, pp. 2280-2282.

Anderson, J.L., Preiser, S., and Rubin, E.L. (1968), "Conservation Form of the Equations of Hydrodynamics in Curvilinear Coordinate Systems," J. of Computational Physics, Vol. 2, pp. 279-287.

Anderson, O.L. (1967), "Numerical Solutions of the Compressible Boundary Layer Equations for Rotating Axisymmetric Flows," Rensselaer Polytechnic Inst., Ph.D. Dissertation, Troy, New York.

Andrews, D.J. (1971), "Calculation of Mixed Phases in Continuum Mechanics," J. of Computational Physics, Vol. 7, pp. 310-326.

Angel, E. (1968A), "Discrete Invariant Imbedding and Elliptic Boundary-Value Problems over Irregular Regions," J. of Mathematical Analysis and Applications, Vol. 23, pp. 471-484.

Angel, E. (1968B), "Dynamic Programming and Linear Partial Differential Equations," J. of Mathematical Analysis and Applications, Vol. 23, No. 3, September, 1968, pp. 628-638.

Angel, E., and Kalaba, R. (1970), "One Sweep Numerical Method for Vector Matrix Difference Equations with Two Point Boundary Conditions," Tech. Rept. No. 70-16, Electrical Engineering Dept., University of Southern California, Los Angeles, California.

Anon. (1953), "Equations, Tables and Charts for Compressible Flow," NACA TR-1135.

Anucina, N.N. (1964), "Difference Schemes for Solving the Cauchy Problem for Symmetric Hyperbolic Systems," Soviet Mathematics, Vol. 5, pp. 60-64.

Apelkrans, M.Y.T. (1968), "On Difference Schemes for Hyperbolic Equations With Discontinuous Initial Values," Mathematics of Computation, Vol. 22, No. 103, pp. 525-540.

Apelt, C.J. (1969), "Hartmann Flow in Annular Channels. Part 2. Numerical Solutions for Low to Moderate Hartmann Numbers," J. Fluid Mechanics, Vol. 37, Part 2, pp. 209-229.

Arakawa, A. (1966), "Computational Design of Long-Term Numerical Integration of the Equations of Fluid Motion: I. Two-Dimensional Incompressible Flow," J. Computational Physics, Vol. 1, No. 1, August, 1966, pp. 119-143.

Argyris, J.H., Mareczek, G., and Scharpf, D.W. (1970), "Two- and Three-Dimensional Flow Analysis Using Finite Elements," Nuclear Engineering and Design, Vol. 11, No. 2, pp. 230-236.

Arkhangel'skii, N.A. (1971), "An Algorithm for Numerical Solution of the Cylindrical Blast Problem with Allowance for Counterpressure by the Network Method," Zhurnal Vychislitel'noi Matematiki i Matematicheskoi Fiziki, Vol. 11, January-February, 1971, pp. 222-236.

Armitage, J.V. (1967), "The Lax-Wendroff Method Applied to Axial Symmetric Swirl Flow," Blanch Anniversary Volume, Aerospace Research Laboratories, United States Air Force, February, 1967.

Armstrong, T.P., and Nielson, C.W. (1970), "Initial Comparison of Transfer and Particle-in-Cell Methods of Collisionless Plasma Simulation," Physics of Fluids, Vol. 13, No. 7, July, 1970, pp. 1880-1881.

Arnason, G., Brown, P.S., and Newburg, E.A. (1967), "A Case Study of the Validity of Finite Difference Approximations in Solving Dynamic Stability Problems," J. of Atmospheric Sciences, Vol. 24, pp. 10-17.

Arnason, G., Greenfield, R.S., and Newburg, E.A. (1968), "A Numerical Experiment in Dry and Moist Convection Including the Rain Stage," J. of Atmospheric Sciences, Vol. 25, No. 3, pp. 404-415.

Aungier, R.H. (1971), "A Computational Method for Unified Solutions to the Viscous Flow Field About Blunt Bodies," Air Force Weapons Lab., Kirtland Air Force Base, New Mexico. In NASA-SP-252, "The Entry Plasma Sheath and Its Effects on Space Vehicle Electromagnetic Systems," pp. 241-260.

Avleeva, V. Kh. (1970), "Experimental Study of Heat Transfer in a Sphere and a Plate in a Supersonic Rarefied Gas Flow," Akademiia Nauk SSSR, Izvestiia, Mekhanika Zhidkosti i Gaza, March-April, 1970, pp. 191-196.

Aziz, K. (1966), "A Numerical Study of Cellular Convection," Ph.D. Thesis, Rice University, Houston, Texas.

Aziz, K., and Hellums, J.D. (1967), "Numerical Solution of the Three-Dimensional Equations of Motion for Laminar Natural Convection," Physics of Fluids, Vol. 10, No. 2, February, 1967, pp. 314-324.

Babenko, K.I., and Voskresenskii, G.P. (1961), "A Numerical Method of Calculating the Three-Dimensional Flow of a Supersonic Gas Current Around Solids," Zh. Vyeh. Mat., Vol. 1, No. 6, pp. 1051-1060.

Baer, F., and King, R.L. (1967), "A General Computational Form for a Class of Nonlinear Systems Incorporating Both Spectral and Finite-Difference Approximations," J. of Computational Physics, Vol. 2, pp. 32-60.

Baer, F., and Simons, T.J. (1968), "Computational Stability and Time Truncation of Coupled Nonlinear Equations with Exact Solutions," Atmospheric Science Paper No. 131, Department of Atmospheric Science, Colorado State University, Fort Collins, Colorado.

Babuska, I. (1970), "The Finite Element Method for Elliptic Equations with Discontinuous Coefficients," Computing, Vol. 5, No. 3, pp. 207-213.

Bahvalov, N.S. (1968), "Numerical Solution of the Dirichlet Problem for Laplace's Equation," Vestn. Mosk. Univ., Ser. II: Khim. (Moscow), Vol. 14, No. 5, 1959, pp. 171-195.

Bailey, H.E. (1969), "Numerical Integration of the Equations Governing the One-Dimensional Flow of a Chemically Reactive Gas," Physics of Fluids, Vol. 12, No. 11, pp. 2292-2300.

Bailey, P.B., Shampine, L.F., and Waltman, P.E. (1968), Nonlinear Two Point Boundary Value Problems, Academic Press, New York, New York.

Bankston, C.A., and McEligot, D.M. (1969), "A Numerical Method for Solving the Boundary Layer Equations for Gas Flow in Circular Tubes with Transfer and Property Variations," LA-4149, Los Alamos Scientific Laboratory, Los Alamos, New Mexico.

Barakat, H.Z., and Clark, J.A. (1966), "Analytical and Experimental Study of the Transient Laminar Natural Convection Flows in Partially Filled Liquid Containers," Proceedings of the Third International Heat Transfer Conference, Vol. II, Chicago, Illinois, p. 152.

Barnes, S.L. (1967), "Effects of Large-Scale Subsidence on Cellular Convection in the Atmosphere: A Numerical Experiment," Ph.D. Dissertation, The University of Oklahoma, Norman, Oklahoma, 1967.

Barnwell, R.D. (1967), "Numerical Results for the Diffraction of a Normal Shock Wave by a Sphere and for the Subsequent Transient Flow," NASA TR R-268, September, 1967.

Barnwell, R.W. (1971), "Three-Dimensional Flow Around Blunt Bodies with Sharp Shoulders," AIAA Paper No. 71-56, AIAA 9th Aerospace Sciences Meeting, New York, New York, January 25-27, 1971.

Bashforth, F., and Adams, J.C. (1883), An Attempt to Test the Theories of Capillary Action by Comparing the Theoretical and Measured Forms of Drops of Fluid, Cambridge Press.

Bastianon, Ricardo A. (1969), "Unsteady Solution of the Flowfield over Concave Bodies," AIAA J., Vol. 7, No. 3, pp. 531-533.

Bauer, P.J., Zumwalt, G.W., and Fila, L.J. (1968), "A Numerical Method and an Extention of the Korst Jet Mixing Theory for Miltispecie Turbulent Jet Mixing," AIAA Paper No. 68-112, AIAA 6th Aerospace Sciences Meeting, New York, New York, January 22-24, 1968.

Baum, E., King, H.H., and Denison, M.R. (1964), "Recent Studies of the Laminar Base-Flow Region," AIAA. J., Vol. 2, pp. 1527-1534.

Bazzhin, A.P. (1971), "Some Results of Calculations of Flows Around Conical Bodies at Large Incidence Angeles." See Holt (1971).

Beardsley, R.C. (1969), "Alternative Finite Different Approximations for the Viscous Generation of Vorticity at a Stationary Fluid Boundary," Report GRD/69-2, Massachusetts Institute of Technology, October, 1969.

Beardsley, R.C. (1971), "Integration of the Planetary Vorticity Equation on a Parabolic Circular Grid," J. of Computational Physics, Vol. 7, pp. 273-288.

Bellman, R., Cherry, I., and Wing, G.M. (1958), "A Note on the Numerical Integration of a Class of Non-Linear Hyperbolic Equations," Quarterly J. Applied Mathematics, Vol. 16, pp. 181-183.

Bellomo, E. (1969), "Transverse Disturbances in Plane Poiseuille Flow. Two Examples as a Check on a Numerical Program for Finite-Amplitude Disturbances in Parallel Flows," Meccanica, J. of the Italian Association of Theoretical and Applied Mechanics, Vol. 4, No. 2, June, 1969, pp. 109-121.

Belotserkovskii, O.M., Popov, F.D., Tolstykh, V.N. Fomin, and Kholodov, A.S. (1970), "Numerical Solution of Some Problems in Gas-Dynamics," Zhurnal Vychislitel'noi Matematiki i Matematicheskoi Fiziki, Vol. 10, March-April, 1970, pp. 401-416.

Belotserkovskiy, O.M., Popov, F.D., Tolslykh, A.I., Fomin, V.N., and Kholodov, A.S. (1970), "Numerical Solution of Certain Gas Dynamic Problems," NASA-TT-F-13229, October, 1970.

Belotserkovskii, O.M. (1970), "Numerical Methods of Some Transsonic Aerodynamics Problems," J. Of Computational Physics, Vol. 5, pp. 587-611.

Belotserkovskii, O.M. (1971), "On the Calculation of Gas Flows with Secondary Floating Shocks." See Holt (1971).

Bengtsson, L. (1964), "Some Numerical Experiments on the Effect of the Variation of Static Stability in Two-Layer Quasi-Geostrophic Models," Tellus, Vol. 16, No. 3, pp. 327-348.

Benison, G.I., and Rubin, E.L. (1969), "A Difference Method for the Solution of the Unsteady Quasi-One-Dimensional Viscous Flow in a Divergent Duct," PIBAL Report No. 69-9, Polytechnic Institute of Brooklyn, New York.

Bernal, M.J.M., and Whiteman, J.R. (1968), "Numerical Treatment of Biharmonic Boundary Value Problems with Re-Entrant Boundaries," MRC TSR 914, Mathematics Research Center, Madison, Wisconsin. Also, The Computer Journal (1970), pp. 87-91.

Billig, L.O., and Galle, K.R. (1970), "A Numerical Method for Calculating Fully Developed Laminar Velocity Profiles from Temperature Profiles," J. of Heat Transfer, Vol. 92, Ser. C, No. 2, pp. 245-251.

Biot, M.A. (1956), "Applied Mathematics: An Art and A Science," J. Aeronautical Sciences, May, 1956, pp. 406-410.

Bird, G.A. (1969A), "The Structure of Rarefied Gas Flows Past Simple Aerodynamic Shapes," J. Fluid Mechanics, Vol. 36, Part 3, pp. 571-576.

Bird, G.A. (1969B), "Numerical Studies in Gas Dynamics," Sydney University, Australia, Rept. on Research Grant AF-AFOSR-915-67.

Bird, G.A. (1969C), "The Formation and Reflection of Shock Waves," Proceedings of the Symposium on Problems in Rarefied Gas Dynamics, Vol. 1, pp. 301-311.

Bird, G.A. (1970A), "Breakdown of Translational and Rotational Equilibrium in Gaseous Expansions," AIAA J., Vol. 8, No. 11, pp. 1998-2003.

Bird, G.A. (1970B), "Direct Simulation and the Boltzmann Equation," Physics of Fluids, Vol. 13, No. 11, November, 1970, pp. 2676-2681.

Bird, R.B., Stewart, W.E., and Lightfoot, E.N. (1960), Transport Phenomena, John Wiley and Sons, Inc., New York.

Birdsall, C.K., and Fuss, D. (1969), "Clouds-in-Clouds, Clouds-in-Cells Physics for Many-Body Plasma Simulation," Journal of Computational Physics, Vol. 3, No. 4, April, 1969, pp. 494-511.

Birkoff, G., Varga, R.S., and Young, D. (1962), "Alternating Direction Implicit Methods," Advances in Computers, Vol. 3., F.L. Alt, ed., Academic Press.

Bizzell, G.D., Concus, P., Crane, G.E., and Satterlee, H.M. (1970), "Low Gravity Draining from Hemispherically Bottomed Cylindrical Tanks," NASA CR-72718, June 1, 1970.

Blewett, P.J. (1970), "Method of Computing Fluid Motion in Two-Dimensional Cartesian or Cylindrical Coordinates by Following Lagrangian Energy Cells," LA-4464, Los Alamos Scientific Laboratory, Los Alamos, New Mexico.

Blottner, F.G., and Flugge-Lotz, I. (1963), "Finite-Difference Computation of the Boundary Layer with Displacement Thickness Interaction," J. de Mecanique, Vol. 22, No. 4, December, 1963, pp. 397-423.

Blottner, F.G. (1968), "Prediction of the Electron Number Density Distribution in the Laminar Air Boundary Layer on Sharp and Blunt Bodies," AIAA Paper No. 68-733, AIAA Fluid and Plasma Dynamics Conference, Los Angeles, California, June 24-26, 1968.

Blottner, F.G. (1969), "Viscous Shock Layer at the Stagnation Point with Nonequilibrium Air Chemistry," AIAA J., Vol. 7, No. 12, December, 1969, pp. 2281-2288.

Blottner, F.G. (1970), "Finite Difference Methods of Solution of the Boundary Layer Equations," AIAA J., Vol. 8, No. 2, February, 1970, pp. 193-199.

Blottner, F.G. (1971), "Finite-Difference Methods for Solving the Boundary Layer Equations with Second-Order Accuracy." See Holt (1971).

Blottner, F.G., and Roache, P.J. (1971), "Nonuniform Mesh Systems," J. Computational Physics, Vol. 8, pp. 498-499.

Boericke, R.R. (1971), "Laminar Boundary Layer on a Cone at Incidence in Supersonic Flow," AIAA J., Vol. 9, No. 3, March, 1971, pp. 462-468.

Boers, J.E. (1969), "Digital Computer Solution of Laplace's Equation Including Dielectric Surfaces," SC-RR-69-446, December, 1969, Sandia Laboratories, Albuquerque, New Mexico.

Bohachevsky, I.O., Rubin, E.L., and Mates, R.E. (1965), "A Direct Method for Computation of Nonequilibrium Flows with Detached Shock Waves," AIAA Paper No. 65-24, AIAA 2nd Aerospace Sciences Meeting, New York, New York, January 25-27, 1965.

Bohachevsky, I.O., and Mates, R.E. (1966), "A Direct Method for Calculation of the Flow About an Axisymmetric Blunt Body at Angle of Attack," AIAA J., Vol. 4, No. 5, May, 1966, p. 766.

Bohachevsky, I.O., and Rubin, E.L. (1966), "A Direct Method for Computation of Nonequilibrium Flows with Detached Shock Waves," AIAA J., Vol. 4, No. 4, April, 1966, p. 600.

Bohachevsky, I.O., and Kostoff, R.N. (1971), "Hypersonic Flow Over Cones with Attached and Detached Shock Waves," AIAA Paper No. 71-55, AIAA 9th Aerospace Sciences Meeting, New York, New York, January 25-27, 1971.

Boris, J.P. (1970), "The Solution of Partial Differential Equations Using a Symbolic Style of Algol," NRL Memorandum Report 2168, Naval Research Lab, Washington, D.C., November, 1970.

Bossel, H.H. (1971), "Study of Vortex Flows at High Swirl by an Integral Method Using Exponentials." See Holt (1971).

Boswell, W.C., Jr., and Werle, M.J. (1971), "Numerical Solutions of the Navier-Stokes Equations for Incompressible Uniform Flow Past a Parabola: A Study of Finite Grid Size Effects," VPI-E-71-5, Virginia Polytechnic Institute and State University, February, 1971.

Boujot, J., Soule, J.L., and Temam, R. (1971), "Traitement Numerique d'um Probleme de Magnetohydrodynamique." See Holt (1971).

Bourcier, M., and Francois, C. (1969), "Numerical Integration of the Navier-Stokes Equations in a Square Domain, La Recherche Aerospatiale, No. 131, July-August, 1969, pp. 23-33.

Bourot, J. (1969), "The Numerical Calculation of Oseen Hydrodynamic Fields Around A Sphere," Serie A-Sciences Mathematiques, Vol. 269, No. 21, November 24, 1969, pp. 1017-1020.

Bowen, S.W., and Park, C. (1971), "Computer Study of Nonequilibrium Excitation in Recombining Nitrogen Plasma Nozzle Flows," AIAA J., Vol. 9, No. 3, March, 1971, pp. 403-499.

Bowley, W.W., and Prince, J.F. (1971), "Finite Element Analysis of General Fluid Flow Problems," AIAA Paper No. 71-602, AIAA 4th Fluid and Plasma Dynamics Conference, Palo Alto, California, June 21-23, 1971.

Boynton, F.P., and Thomson, A. (1969), "Numerical Computation of Steady, Supersonic, Two-Dimensional Gas Flow in Natural Coordinates," J. of Computational Physics, Vol. 3, pp. 379-398.

Bradshaw, P., and Ferriss, D.H. (1971), "Calculation of Boundary-Layer Development Using the Turbulent Energy Equation: Compressible Flow on Adiabatic Walls," J. Fluid Mechanics, Vol. 46, Part 1, pp. 83-110.

Brady, W.G. (1967), "Numerical Analysis of Two Free-Boundary Problems in Fluid Dynamics," AIAA Paper No. 67-217, AIAA 5th Aerospace Sciences Meeting, New York, New York, January 23-26, 1967.

Brailovskaya, I. (1965), "A Difference Scheme for Numerical Solution of the Two-Dimensional, Nonstationary Navier-Stokes Equations for a Compressible Gas," Soviet Physics-Doklady, Vol. 10, No. 2, pp. 107-110.

Brailovskaya, I.Y. (1967), "Calculation of Viscous Compressible Gas Flow Past a Corner," Izv. AN SSSR. Mekhanika Zhidkosti i Gaza, Vol. 2, No. 3, pp. 82-92.

Brailovskaya, I.Y. (1968), "Difference Method for the Numerical Solution of Two-Dimensional Unsteady Navier-Stokes Equations for a Compressible Gas," Sb. Rabot. Vychisl. Tsentra Mosk Un-ta, Vol. 7, pp. 3-15.

Brailovskaia, I.Y., Kuskova, T.V., and Chudov, L.A. (1968), "Difference Methods of Solving the Navier-Stokes Equations (Survey)," Vychisl. Metody i Programmirovanie, No. 11, Moscow Univ., 1968, pp. 3-18.

Brailovskaya, I.Y., Kuskova, T.V., and Chudov, L.A. (1970), "Difference Methods of Solving the Navier-Stokes Equations," International Chemical Engineering, Vol. 10, No. 2, pp. 228-236.

Bramlette, T.T. (1971), "Plane Poiseuille Flow of a Rarefied Gas Based on the Finite Element Method," Physics of Fluids, Vol. 14, No. 2, February, 1971, pp. 288-293.

Brandt, A., and Gillis, J. (1969), "Asymptotic Approach to Hartmann-Poiseiulle Flows," J. Computational Physics, Vol. 3, No. 4, pp. 523-538.

Brazier-Smith, P.R., and Latham, J. (1969), "Numerical Computations of the Dynamics of the Disintegration of a Drop Situated in an Electric Field," Proceedings, Royal Society of London, Series A, Vol. 312, September 2, 1969, pp. 277-289.

Brennen, C. (1968), "A Numerical Solution of Axisymmetric Cavity Flows, Parts I and II," Ship Report No. 114, National Physical Laboratory.

Brennen, C. (1969), "A Numerical Solution of Axisymmetric Cavity Flows," J. Fluid Mechanics, Vol. 37, Part 4, pp. 671-688.

Brennen, C. (1971), "Some Numerical Solutions of Unsteady Free Surface Wave Problems Using the Lagrangian Description of the Flow." See Holt (1971).

Brian, P.L.T. (1961), "A Finite-Difference Method of High-Order Accuracy for the Solution of Three-Dimensional Transient Heat Conduction Problems," American Institute of Chemical Engineering J., Vol. 7, No. 3, pp. 367-370.

Briggs, B.R. (1960), "The Numerical Calculation of Flow Past Conical Bodies Supporting Elliptic Conical Shock Waves at Finite Angles of Incidence," NASA TN D-340, National Aeronautics and Space Administration, Ames Research Center, Moffett Field, California.

Briley, W.R. (1968), "Time-Dependent Rotating Flow in a Cylindrical Container," Ph.D. Thesis, Texas University, Austin, Texas.

Briley, W.R. (1970), "A Numerical Study of Laminar Separation Bubbles Using the Navier-Stokes Equations," Report J110614-1, United Aircraft Research Laboratories, East Hartford, Connecticut.

Briley, W.R., and Walls, H.A. (1971), "A Numerical Study of Time-Dependent Rotating Flow in a Cylindrical Container at Low and Moderate Reynolds Numbers." See Holt (1971).

Brown, R.J. (1967), "Natural Convection Heat Transfer Between Concentric Spheres," Ph.D. Thesis, University of Texas at Austin, August, 1967.

Bryan, J.L., and Childs, B. (1969), "Numerical Solutions of Diffusion Type Equations," RE 4-69, University of Houston, Houston, Texas.

Bryan, K. (1963), "A Numerical Investigation of a Nonlinear Model of a Wind-Driven Ocean," J. Atmospheric Sciences, Vol. 20, November, 1963, pp. 594-606.

Bryan, K. (1966), "A Scheme for Numerical Integration of the Equations of Motion on an Irregular Grid Free of Nonlinear Instability," U.S. Monthly Weather Review, Vol. 94, No. 7, pp. 39-40.

Bryan, K. (1969), "A Numerical Method for the Study of the Circulation of the World Ocean," J. of Computational Physics, Vol. 4, pp. 347-376.

Bryan, K., and Cox, Michael D. (1970), "The Calculation of Large-Scale Ocean Currents," AIAA Paper No. 70-4, AIAA 8th Aerospace Sciences Meeting, New York, New York, January 19-21, 1970.

Buckingham, A.C., Birnbaum, N., and Hofmann, R. (1970), "VISELK 3, Computational Studies of Transient Viscous Flows, Part 2: Development of Planar Shock Wave; A Shock Tube Starting Simulation," Report No. PITR-70-4, Part 2, Physics International Company, 2700 Merced Street, San Leandro, California, September, 1970.

Buckingham, R.A. (1962), Numerical Methods, Sir Isaac Pitman and Sons, Ltd., London.

Bugliarello, G., and Jackson, E.D., III (1967), "Random Walk Simulation of Convective Diffusion from Instantaneous Point Sources in a Laminar Field," Developments in Mechanics, Vol. 3, Part 2, Dynamics and Fluid Mechanics (Proceedings of the Ninth Midwestern Conference held at the University of Wisconsin, Madison, August 16-18, 1965), pp. 461-472.

Buleev, N.I., and Timukhin, G.I. (1969), "Numerical Solution to the Hydrodynamic Equations of a Plane Flow of Incompressible Viscous Fluid," Akademiia Nauk SSSR, Sibirskoe Otdelenie, Izvestiia, Seriia Tekhnicheskikh Nauk, February, 1969, pp. 14-24.

Buneman, O. (1967) "Time-Reversible Difference Procedures," J. Computational Physics, Vol. 1, pp. 517-535.

Buneman, O. (1969), "A Compact Non-Iterative Poisson Solver," SUIPR Report No. 294, Stanford University, Stanford, California, May, 1969.

Burgers, J.M. (1948), "A Mathematical Model Illustrating the Theory of Turbulence," Advances in Applied Mechanics, Vol. 1, R. von Mises and T. von Karman, ed., Academic Press, New York, pp. 171-199.

Burgess, W.P. (1971), "Extrapolation Techniques Applied to Parabolic Partial Differential Equations," Ph.D. Dissertation, Princeton University, March, 1971.

Burggraf, O.R. (1966), "Analytical and Numerical Studies of the Structure of Steady Separated Flows," J. Fluid Mechanics, Vol. 24, Part 1, pp. 113-152.

Burggraf, O.R. (1970), "Computational Study of Supersonic Flow over Backward-Facing Steps at High Reynolds Number," ARL-70-0275, Aerospace Research Lab., Wright-Patterson Air Force Base, Ohio, November, 1970.

Burrgraf, O.R., and Stewartson, K. (1971), "The Structure of the Laminar Boundary Layer Under a Potential Vortex." See Holt (1971).

Burstein, S.Z. (1964), "Finite-Difference Calculations for Hydrodynamic Flows Containing Discontinuities," J. Computational Physics, Vol. 1, No. 2, November, 1966, pp. 198-222.

Burstein, S.Z. (1965), "Finite Difference Calculations for Hydrodynamic Flows Containing Discontinuities," NYO-1480-33, Courant Institute of Mathematical Sciences, New York University, New York, New York.

Burstein, S.Z. (1967), "Finite-Difference Calculations for Hydrodynamic Flows Containing Discontinuities," J. Computational Physics, Vol. 2, pp. 198-222.

Burstein, S.Z., and Mirin, A.A. (1970), "Third Order Difference Methods for Hyperbolic Equations," J. of Computational Physics, Vol. 5, pp. 547-571.

Burstein, S.Z., and Mirin, A.A. (1971), "Difference Methods for Hyperbolic Equations Using Space and Time Split Difference Operators of Third Order Accuracy." See Holt (1971).

Butcher, B.M. (1971), "Numerical Techniques for One-Dimensional Rate-Dependent Porous Material Compaction Calculations," SC-RR-710112, Sandia Laboratories, Albuquerque, New Mexico, April, 1971.

Butler, T.D. (1967), "Numerical Solutions of Hypersonic Sharp-Leading-Edge Flows," The Physics of Fluids, Vol. 10, No. 6, pp. 1205-1215.

Butler, T.D. (1971), "Linc Method Extensions." See Holt (1971).

Buzbee, B.L., Golub, G.H., and Nielson, C.W. (1969), "The Method of Odd/Even Reduction and Factorization with Application to Poisson's Equation," LA-4141, Los Alamos Scientific Laboratory, Los Alamos, New Mexico.

Buzbee, B.L., Dorr, F.W., George, J.A., and Golub, G.H. (1970A), "The Direct Solution of the Discrete Poisson Equation on Irregular Regions," STAN-CS-71-195, Computer Science Department, Stanford University, Stanford, California.

Buzbee, B.L., Golub, G.H., and Nielson, C.W. (1970B), "The Method of Odd/Even Reduction and Factorization with Application to Poisson's Equation, Part II," LA-4288, Los Alamos Scientific Laboratory, Los Alamos, New Mexico, May, 1970.

Cabelli, A., and Davis, G.D. (1971), "A Numerical Study of the Benard Cell," J. Fluid Mechanics, Vol. 45, Part 4, pp. 805-829.

Cahn, M.S., and Garcia, J.R. (1971), "Transonic Airfoil Design," J. Aircraft, Vol. 8, No. 2, pp. 84-88.

Calder, K.L. (1968), "In Clarification of the Equations of Shallow-layer Thermal Convection for a Compressible Fluid Based on the Boussinesq Approximation," Quarterly J. of the Royal Meteorological Society, Vol. 94, pp. 88-92.

Callens, E. Eugene, Jr. (1970), "A Time-Dependent Approach to Fluid Mechanical Phenomenology," AIAA Paper No. 7--46, AIAA 8th Aerospace Sciences Meeting, New York, New York, January 19-21, 1970.

Callis, L.B. (1968), "Time Asymptotic Solutions of Blunt-Body Stagnation-Region Flows With Nongray Emission and Absorption of Radiation," AIAA Paper No. 68-663.

Callis, L.B. (1969), "Solutions of Blunt-Body Stagnation-Region Flows With Nongray Emission and Absorption of Radiation by a Time-Asymptotic Technique," NASA TR R-299, Langley Research Center, Langley Station, Hampton, Virginia.

Callis, L.B. (1970), "Time Asymptotic Solutions of Flow Fields with Coupled Nongray Radiation About Long Blunt Bodies," AIAA Fifth Thermophysics Conference, Los Angeles, California, June 29-July 1, 1970.

Cameron, I.G. (1966), "An Analysis of the Errors Caused by Using Artificial Viscosity Terms to Represent Steady-State Shock-Waves," J. Computational Physics, Vol. 7, No. 7, pp. 1-20.

Campbell, C.M., and Keast, P. (1968), "The Stability of Difference Approximations to a Self-Adjoint Parabolic Equation, Under Derivative Boundary Conditions," Mathematics of Computation, Vol. 22, No. 102, pp. 336-347.

Campbell, D.R., and Mueller, T.J. (1968), "A Numerical and Experimental Investigation of Incompressible Laminar Ramp-Induced Separated Flow," UNDAS TN-1068-M1, Department of Aero-Space Engineering, University of Notre Dame, Notre Dame, Indiana.

Cardenas, Alfonso, F., and Karplus, Walter J. (1970), "PDEL -- A Language for Partial Differential Equations," Communications of the ACM, Vol. 13, No. 3, pp. 184-191.

Carlson, L.A. (1969), "Radiative Transfer, Chemical Nonequilibrium and Two Temperature Effects Behind a Reflected Shock Wave in Nitrogen," Ph.D. Dissertation, Ohio State University, Columbus, Ohio.

Carlson, L.A. (1970), "Radiative-Gasdynamic Coupling, Two-Temperature, and Chemical Non-equilibrium Effects Behind Reflected Shock Waves," AIAA Paper No. 70-774, AIAA 3rd Fluid and Plasma Dynamics Conference, Los Angeles, California, June 29-July 1, 1970.

Carnahan, B., Luther, H.A., and Wilkes, J.O. (1969), Applied Numerical Methods, John Wiley and Sons, Inc., New York, New York.

Carré, B.A. (1961), "The Determination of the Optimum Accelerating Factor for Successive Over-Relaxation," Computer J., Vol. 4, No. 1, pp. 73-78.

Caspar, J.R. (1968), "Applications of Alternating Direction Methods to Mildly Nonlinear Problems," (NASA-CR-73865), Maryland University, College Park, Maryland.

Cebeci, T. (1969), "Laminar and Turbulent Incompressible Boundary Layers on Slender Bodies of Revolution in Axial Flow," Transactions of the ASME, pp. 1-6.

Cebeci, T., and Smith, A.M.O. (1970), "A Finite-Difference Method for Calculating Compressible Laminar and Turbulent Boundary Layers," Transactions of the ASME, pp. 1-13.

Cebeci, T., Smith, A.M.O., and Mosinskis, G. (1970), "Calculation of Compressible Adiabatic Turbulent Boundary Layers," AIAA J., Vol. 8, No. 11, pp. 1974-1982.

Cebeci, T., and Keller, H.B. (1971), "Shooting and Parallel Shooting Methods for Solving the Falkner-Skan Boundary-Layer Equation," J. of Computational Physics, Vol. 7, pp. 289-300.

Chalenko, P.I. (1970), "Engineering Formulas for Solving Boundary Value Problems for the Helmholtz and Poisson Equations by the Method of Summary Representations," NASA TT-F-13059, August, 1970.

Chan, R.K.-C., Street, R.L., and Strelkoff, T. (1969), "Computer Studies of Finite-Amplitude Water Waves," Tech. Report No. 104, Department of Civil Engineering, Stanford University, Stanford California, June, 1969.

Chan, R.K.-C., and Street, R.L. (1970), "A Computer Study of Finite-Amplitude Water Waves," J. of Computational Physics, Vol. 6, pp. 68-94.

Chan, R.K.-C., Street, R.L., and Fromm, J.E. (1971), "The Digital Simulation of Water Waves - An Evaluation of SUMMAC." See Holt (1971).

Chapman, A.J., and Walker, W.F. (1971), Introductory Gas Dynamics, Holt, Rinehart, and Winston, New York.

Chapman, D.R., Wimbrow, W.R., and Kester, R.H. (1952), "Experimental Investigation of Base Pressure on Blunt-Trailing-Edge Wings at Supersonic Velocities," NACA TR 1109, 1952.

Charney, J.G., Fjortoft, R., and von Neumann, J. (1950), "Numerical Integration of the Barotropic Vorticity Equation," Tellus, Vol. 2, No. 4, pp. 237-254.

Charney, J.G. (1962), "Integration of the Primitive and Balance Equations," Proc. International Symposium on Numerical Weather Prediction, Tokyo, Japan, pp. 131-152.

Chavez, S.P., and Richards, C.G. (1970), "A Numerical Study of the Coanda Effect," 70-FIcs-12, The American Society of Mechanical Engineers, United Engineering Center, New York, New York.

Chen, H.-T., and Collins, R. (1971), "Shock Wave Propagation Past an Ocean Surface," J. of Computational Physics, Vol. 7, pp. 89-101.

Cheng, S.I. (1968), "Accuracy of Difference Formulation of Navier-Stokes Equations," A.M.S. Department, Princeton University, Princeton, New Jersey.

Cheng, S.I. (1970), "Numerical Integration of Navier-Stokes Equations," AIAA J., Vol. 8, No. 12, December, 1970, pp. 2115-2122.

Chiu, Y.T. (1970), "Traveling Waves of Arbitrary Amplitude in Compressible Hydrodynamics under Gravity: An Exact Solution," The Physics of Fluids, Vol. 13, No. 12, p. 2950.

Chorin, A.J. (1967), "A Numerical Method for Solving Incompressible Viscous Flow Problems," J. of Computational Physics, Vol. 2, pp. 12-26.

Chorin, A.J. (1968), "Numerical Solution of Incompressible Flow Problems," in Studies in Numerical Analysis 2, Society for Industrial and Applied Mathematics, Philadelphia, Pennsylvania, pp. 64-70.

Chorin, A.J. (1969), "On the Convergence of Discrete Approximations to the Navier-Stokes Equations," Mathematics of Computation, Vol. 23, No. 106, April, 1969, pp. 341-353.

Chorin, A.J. (1971), "Computational Aspects of the Turbulence Problem." See Holt (1971).

Chou, P.C., and Mortimer, R.W. (1966), "Numerical Integrations of Flow Equations Along Natural Coordinates," AIAA J., Vol. 4, No. 1, pp. 26-30.

Chow, C.L. (1969), "Explicit Heat Conduction Equations at Thermally Insulated Surface," J. of Heat Transfer, Transactions of the ASME Series C, Vol. 91, No. 3, pp. 446-447.

Chow, F., Krause, E., Liu, C.H., and Mao, J. (1970), "Numerical Investigations of an Airfoil in a Nonuniform Stream," AIAA J. of Aircraft, p. 531.

Chu, C.K. (1968), ed., Computational Fluid Dynamics, American Institute of Aeronautics and Astronautics, 1290 6th Avenue, New York, New York.

Chu, Chong-Wei (1964), "A Simple Derivation of Three-Dimensional Characteristic Relations," AIAA J., Vol. 2, No. 7, pp. 1336-1337.

Chu, W.-H. (1970), "Development of A General Finite Difference Approximation for a General Domain, Part 3: A Direct and Reverse Successive Overrelaxation Method," Southwest Research Institute, San Antonio, Texas.

Chushkin, P.I. (1968), "Method of Characteristics for Spatial Supersonic Streams," Moscow, Vychisletelnyi Tsentr AN SSR.

Chushkin, P.I. (1968), "Method of Characteristics for Three-Dimensional Supersonic Flows," Moskva, VTs AN SSSR.

Chushkin, P.I. (1970A), "Supersonic Flows About Conical Bodies," J. of Computational Physics, Vol. 5, pp. 572-586.

Chushkin, P.I. (1970B), "Numerical Analysis of Combustion in Supersonic Flows," Paper ICAS 70-52, International Council of the Aeronautical Sciences, Congress, 7th, Rome, Italy, September 14-18, 1970.

Ciment, Melvyn (1968), "Stable Difference Schemes with Uneven Mesh Spacings," New York Univ., New York, Courant Institute of Mathematical Sciences, Rept. NYO 1480-100.

Coakley, J.D., and Porter, R.W. (1969), "Characteristics at Boundaries in Numerical Gas Dynamics," PDL Note 2-69, Plasmadynamics Laboratory, Dept. of Mech. and Aerosp. Eng., Illinois Institute of Technology, Chicago, Illinois, 60616, November, 1969.

Cochrane, G.F., Jr. (1969), "A Numerical Solution for Heat Transfer to Non-Newtonian Fluids with Temperature-Dependent Viscosity for Arbitrary Conditions of Heat Flux and Surface Temperature, Ph.D. Thesis, Oregon State University, June, 1969.

Cohen, C.B., and Reshotko, E. (1956), "Similar Solutions for the Compressible Laminar Boundary Layer with Heat Transfer and Pressure Gradient," NACA TR 1293, 1956.

Collins, R., and Chen, H.-T. (1970), "Propagation of a Shock Wave of Arbitrary Strength in Two Half Planes Containing a Free Surface," J. of Computational Physics, Vol. 5, pp. 415-442.

Collins, R., and Chen, H.-T. (1971), "Motion of a Shock Wave Through a Nonuniform Fluid." See Holt (1971).

Colony, R., and Reynolds, R.R. (1970), "An Application of Hockney's Method for Solving Poisson's Equation," American Federation of Information Processing Societies, Spring Joint Computer Conference, Atlantic City, New Jersey, May 5-7, 1970.

Colp, J.L. (1968), "Terradynamics: A Study of Projectile Penetration of Natural Earth Materials," SC-DR-68-215, Sandia Laboratory, Albuquerque, New Mexico, June, 1968.

Cooley, J.W., and Tukey, J.W. (1965),"An Algorithm for the Machine Calculation of Complex Fourier Series," Mathematics of Computation, Vol. 19, p. 297.

Costello, F.A., and Shrenk, G.L. (1966), "Numerical Solution to the Heat Transfer Equations with Combined Conduction and Radiation," J. of Computational Physics, Vol. 1, pp. 541-543.

Courant, R., Friedrichs, K.O., and Lewy, H. (1928), "Uber die Partiellen Differenzengleichur-gen der Mathematischen Physik," Mathematische Annalen, Vol. 100, pp. 32-74. (See Courant et. al, 1967.)

Courant, R., Isaacson, E., and Rees, M. (1952), "On The Solution of Nonlinear Hyperbolic Differential Equations by Finite Differences," Communications on Pure and Applied Mathematics, Vol. V., pp. 243-255.

Courant, R., and Friedrichs, K.O. (1948), Supersonic Flow and Shock Waves, Interscience Publishers, Inc., New York.

Courant, R., Friedrichs, K., and Lewy, H. (1967), "On The Partial Difference Equations of Mathematical Physics," IBM Journal, March, 1967, pp. 215-234.

Crandall, S.H. (1956), Engineering Analysis - A Survey of Numerical Procedures, McGraw-Hill Book Company, New York, New York.

Crane, R.I. (1968), "Numerical Solutions of Hypersonic Near Wake Flow by the Particle-In-Cell Method," N69-35664, Oxford University, August, 1968.

Crank, J., and Nicolson, P. (1947), "A Practical Method for Numerical Evaluation of Solutions of Partial Differential Equations of the Heat-Conduction Type," Proceedings of the Cambridge Philosophical Society, Vol. 43, No. 50, pp. 50-67.

Crenshaw, J.P. (1966), "Two-Dimensional and Radial Laminar Free Jets and Wall Jets," Ph.D. Thesis, Georgia Institute of Technology, July, 1966.

Crenshaw, J.P., and Hubbartt, J.E. (1969), "Comments on 'Transformations for Infinite Regions and their Applications to Flow Problems'," AIAA J., Vol. 7, No. 11, p. 2189.

Crider, J.E., and Foss, A.S. (1966), "Computational Studies of Transients in Packed Tubular Chemical Reactors," American Institute of Chemical Engineers J., Vol. 12, No. 3, pp. 514-525.

Crocco, Luigi (1965), "A Suggestion for the Numerical Solution of the Steady Navier-Stokes Equations," AIAA J., Vol. 3, No. 10, October, 1965, pp. 1824-1832.

Crowder, H.J., and Dalton, C. (1969), "On the Stability of Poiseuille Pipe Flow," Tech. Rept. No. 1, University of Houston, Department of Mechanical Engineering, Houston, Texas, 77004, August, 1969.

Crowder, H.J., and Dalton, C. (1971A), "On the Stability of Poiseuille Flow in a Pipe," J. of Computational Physics, Vol. 7, pp. 12-31.

Crowder, H.J., and Dalton, C. (1971B), "Errors in the Use of Nonuniform Mesh Systems," J. of Computational Physics, Vol. 7, pp. 32-45.

Crowley, B.K. (1967), "PUFL, An 'Almost-Lagrangian' Gasdynamic Calculation for Pipe Flows with Mass Entrainment," J. of Computational Physics, Vol. 2, pp. 61-86.

Crowley, W.P. (1967), "Second-Order Numerical Advection," J. of Computational Physics, Vol. 1, pp. 471-484.

Crowley, W.P. (1968A), "Numerical Advection Experiments," Monthly Weather Review, Vol. 96, No. 1, January, 1968, pp. 1-11.

Crowley, W.P. (1968B), "Numerical Calculations of Viscous Incompressible Stratified Shear Flows," UCRL-50538, University of California, Lawrence Radiation Laboratory, Livermore, California, November 1, 1968.

Crowley, W.P. (1970A), "Some Numerical Experiments with Rayleigh-Taylor Instability for a Compressible Inviscid Fluid," UCRL-50845, Lawrence Radiation Laboratory, Livermore, California, April, 1970.

Crowley, W.P. (1970B), "An Empirical Theory for Large Amplitude Rayleigh-Taylor Instability," UCRL-72650, Lawrence Radiation Laboratory, Livermore, California, August 7, 1970.

Crowley, W.P. (1970C), "A Numerical Model for Viscous, Nondivergent, Barotropic, Wind-Driven, Ocean Circulations," J. of Computational Physics, Vol. 6, pp. 183-199.

Crowley, W.P. (1971), "FLAG: A Free-Lagrange Method for Numerically Simulating Hydrodynamic Flows in Two Dimensions." See Holt (1971).

Cryer, C.W. (1969), "Topological Problems Arising when Solving Boundary Value Problems for Elliptic Partial Differential Equations by the Method of Finite Differences," Tech Rept. 69, Math. Dept., Wisconsin University, Madison, Wisconsin, August, 1969.

Curtiss, C.F., and Hirschfelder, J.O. (1952), "Integration of Stiff Equations," Proceedings of the National Aeronautical Society, Vol. 38, pp. 235-243.

Cyrus, N.J., and Fulton, R.E. (1967), "Accuracy Study of Finite Difference Methods," NASA TN D-4372, National Aeronautics and Space Administration, Langley Research Center, Langley Station, Hampton, Virginia.

Dahl, G. (1969), "Fortran Program for the Calculation of the Variables of State of Real Gas Mixtures with Applications," DLR Mittelung 69-27, Deutsche Forschungs-und Versuchsanstalt fuer Liftund Raumfahrt, Brunswick, West Germany, December, 1969.

Daly, B.J. (1967), "Numerical Study of Two Fluid Rayleigh-Taylor Instability," Physics of Fluids, Vol. 10, February, 1967, pp. 297-307.

Daly, B.J., and Pracht, W.E. (1968), "Numerical Study of Density-Current Surges," Physics of Fluids, Vol. 11, No. 1, pp. 15-30.

Daly, B.J. (1969A), "Numerical Study of the Effect of Surface Tension on Interface Instability," Physics of Fluids, Vol. 12, No. 7, pp. 1340-1354.

Daly, B.J. (1969B), "A Technique for Including Surface Tension Effects in Hydrodynamic Calculations," J. of Computational Physics, Vol. 4, pp. 97-117.

Daly, B.J., and Harlow, F.H. (1970), "Transport Equations in Turbulence," The Physics of Fluids, Vol. 13, No. 11, pp. 2634-2649.

Daly, B.J., and Harlow, F.H. (1971), "Inclusion of Turbulence Effects in Numerical Fluid Dynamics." See Holt (1971).

Denard, M.B. (1969), "Numerical Studies of Effects of Surface Friction on Large-Scale Atmospheric Motions," Monthly Weather Review, Vol. 97, No. 12, p. 835.

Davis, G. De Vahl (1968), "Laminar Natural Convection in an Enclosed Rectangular Cavity," International J. Heat and Mass Transfer, Vol. 11, pp. 1675-1693.

Davis, P., and Rabinowitz, P. (1960), Advances in Computers, Vol. II, Academic Press.

Davis, R.T. (1968), "The Hypersonic Fully Viscous Shock-Layer Problem," SC-RR-68-840, Sandia Laboratories, Albuquerque, New Mexico, December, 1968.

Davis, R.T. (1970A), "Numerical Solution of the Hypersonic Viscous Shock-Layer Equations," AIAA J., Vol. 8, No. 5, May, 1970.

Davis, R.T. (1970B), "Hypersonic Flow of a Chemically Reacting Binary Mixture Past a Blunt Body," AIAA Paper No. 70-805, AIAA 3rd Fluid and Plasma Dynamics Conference, Los Angeles, California, June 29-July 1, 1970.

Davis, R.T. (1972), "Numerical Solution of the Navier-Stokes Equations for Laminar Incompressible Flow Past a Parabola," J. Fluid Mechanics, Vol. 51, part 3, pp. 417-433.

Dawson, C., and Marcus, M. (1970), "DMC - A Computer Code to Simulate Viscous Flow About Arbitrarily Shaped Bodies," Proceedings of the 1970 Heat Transfer and Fluid Mechanics Institute, pp. 323-338.

Dawson, J.M., Kruer, W.L., Boris, J.P., Orens, J.H., and Oberman, C. (1969), "Numerical Simulation of Plasmas at Princeton," Tech. Rept. 1 April - 13 December, 1969, Princeton University, Princeton, New Jersey, December, 1969.

Dean, C.F., and Eraslan, A.H. (1971), "Chemically Reacting Viscous Supersonic Flow in a Two Dimensional Divergent Plane Channel," AIAA Paper No. 71-44, AIAA 9th Aerospace Sciences Meeting, New York, New York, January 25-27, 1971.

Deardorff, J.W. (1970), "A Numerical Study of Three-Dimensional Turbulent Channel Flow at Large Reynolds Numbers," J. of Fluid Mechanics, Vol. 41, Part 2, April, 1970, pp. 453-480.

Deardorff, J.W. (1971), "On the Magnitude of the Subgrid Scale Eddy Coefficient," J. of Computational Physics, Vol. 7, pp. 120-133.

DeBar, R.B. (1967), "Difference Equations for the Legendre Polynomial Representation of the Transport Equation," J. of Computational Physics, Vol. 2, pp. 197-205.

Dellinger, T.C. (1969), "Computation of Nonequilibrium Merged Stagnation Shock Layers by Successive Accelerated Replacement," AIAA Paper No. 69-655, AIAA Fluid and Plasma Dynamics Conference, San Francisco, California, June 16-18, 1969.

Dellinger, T.C. (1970), "Nonequilibrium Air Ionization in Hypersonic Fully Viscous Shock Layers," AIAA Paper No. 70-806, AIAA 3rd Fluid and Plasma Dynamics Conference, Los Angeles, California, June 29-July 1, 1970.

Dellinger, T.C. (1971), "Computation of Nonequilibrium Merged Stagnation Shock Layers by Successive Accelerated Replacement," AIAA J., Vol. 9, No. 2, pp. 262-269.

Dem'ianovich, Iu. K. (1968), "Convergence Rate Estimates for Some Projection Methods of Solving Elliptic Equations," USSR Computational Mathematics and Mathematical Physics, Vol. 8, No. 1, pp. 102-128.

Dennis, S.C.R., Hudson, J.D., and Smith, N. (1968), "Steady Laminar Forced Convection from a Circular Cylinder at Low Reynolds Numbers," Physics of Fluids, Vol. 11, No. 5, pp. 933-940.

Dennis, S.C.R., and Chang, Gau-Zu (1969), "Numerical Integration of the Navier-Stokes Equations in Two Dimensions," MRC Tech. Summary Rept. #859, Mathematics Research Center, The University of Wisconsin, Madison, Wisconsin.

Dennis, S.C.R., and Chang, Gau-Zu (1970), "Numerical Solutions for Steady Flow Past a Circular Cylinder at Reynolds Numbers up to 100," J. Fluid Mechanics, Vol. 42, Part 3, pp. 471-489.

Dennis, S.C.R., and Staniforth, A.N. (1971), "A Numerical Method for Calculating the Initial Flow Past a Cylinder in a Viscous Fluid." See Holt (1971).

Der, J., Jr., and Raetz, G.S. (1962), "Solution of General Three-Dimensional Laminar Boundary-Layer Problems by an Exact Numerical Method," IAS Paper No. 62-70, IAS 30th Annual Meeting, New York, New York, January 22-24, 1962.

Desanto, D.F., and Keller, H.B. (1961), "Numerical Studies of Transition from Laminar to Turbulent Flow over a Flat Plate," New York University, April, 1961. See also SIAM J., Vol. 10, 1962, p. 569.

Dias, G.F. (1970), An Investigation of a New Class of Linear Finite Difference Operators to be Used in Solution of Partial Differential Equations," M.S. Thesis, Naval Postgraduate School, Monterey, California, June 1970.

Dickman, D.O., Morse, R.L., and Nielson, C.W. (1969), "Numerical Simulation of Axisymmetric, Collisionless, Finite-β Plasma," Physics of Fluids, Vol. 12, No. 8, pp. 1708-1716.

Didenko, V.I., and Liashenko, I.M. (1964), "On the Numerical Solution of Boundary-Value Problems for Elliptic Differential Equations with Constant Coefficients," Ukrainskii Matematicheskii Zhurnal, Vol. 16, No. 5, pp. 681-690.

Dienes, J.K. (1968), "An Eulerian Method for Calculating Strength Dependent Deformation, Part I," GAMB-8497, Gulf General Atomic, Inc., San Diego, California 92112, February 2, 1968.

DiPrima, R.C., and Rogers, E.H. (1969), "Computing Problems in Nonlinear Hydrodynamic Stability." See Frenkiel and Stewartson (1969).

Distefano, G.P. (1968), "Stability of Numerical Integration Techniques," American Institute of Chemical Engineers J., Vol. 14, No. 6, November, 1968, pp. 946-955.

Distefano, Nestor (1969), "Dynamic Programming and the Solution of the Biharmonic Equation," California University, Berkeley, California, March, 1969.

Dixon, T.N., and Hellums, J.D. (1967), "A Study on Stability and Incipient Turbulence in Poiseuille and Plane-Poiseuille Flow by Numerical Finite-Difference Simulation," American Institute of Chemical Engineers J., Vol. 13, No. 5, pp. 866-872.

Donaldson, C. DuP. (1968), "A Computer Study of an Analytical Model of Boundary Layer Transition," AIAA Paper No. 68-38, AIAA 6th Aerospace Sciences Meeting, New York, New York, January 22-24, 1968.

Donovan, L.F. (1968), "A Numerical Solution of Unsteady Flow in a Two-Dimensional Square Cavity," NASA TM X-52459, Lewis Research Center, Cleveland, Ohio, 1968.

Donovan, L.F. (1970), "A Numerical Solution of Unsteady Flow in a Two-Dimensional Square Cavity," AIAA Journal, Vol. 8, No. 3, March 1970, pp. 524-529.

Dorr, F.W. (1969A), "Remarks on the Iterative Solution of the Neumann Problem on a Rectangle by Successive Line Over-Relaxation," Mathematics of Computation, Vol. 23, No. 105, pp. 177-179.

Dorr, F.W. (1969B), "The Direct Solution of the Discrete Poisson Equation on a Rectangle," LA-4132, Los Alamos Scientific Laboratory of the University of California, Los Alamos, New Mexico.

Dorr, F.W. (1970), "The Direct Solution of the Discrete Poisson Equation on a Rectangle," SIAM Review, Vol. 12, No. 2, March, 1970, pp. 248-263.

Douglas, J., Jr. (1955), "On the Numerical Integration of $\partial^2u/\partial x^2 + \partial^2u/\partial y^2 = \partial u/\partial t$ by Implicit Methods," J. Society of Industrial Applied Mathematics, Vol. 3, No. 1, March, 1955, pp. 42-65.

Douglas, J., and Rachford, H.H. (1956), "On the Numerical Solution of Heat Conduction Problems in Two and Three Space Variables," Transactions of the American Mathematical Society, Vol. 82, pp. 421-439.

Douglas, J., Jr. (1957), "A Note on the Alternating Direction Implicit Method For the Numerical Solution of Heat Flow Problems," Proceedings of the American Mathematical Society, Vol. 8, pp. 409-412.

Douglas, J. (1962), "Alternating Direction Methods for Three Space Variables," Numerische Mathematik, Vol. 4, pp. 41-63.

Douglas, J., and Gunn, J.E. (1964), "A General Formulation of Alternating Direction Implicit Methods, Part I, Parabolic and Hyperbolic Problems," Numerische Mathematik, Vol. 6, pp. 428-453.

Douglas, J., Jr., and DuPont, T. (1968), "The Numerical Solution of Water Flooding Problems in Petroleum Engineering by Variational Methods," in Studies in Numerical Analysis 2, Society for Industrial and Applied Mathematics, Philadelphia, Pennsylvania, pp. 53-63.

Drake, R.L., and Ellingson, M.B. (1970), "The Application of a Local Similarity Concept in Solving the Radial Flow Problem," J. of Computational Physics, Vol. 6, pp. 200-218.

D'Souza, N., Molder, S., and Moretti, G. (1971), "A Time-Dependent Method for Blunt Leading Edge Hypersonic Internal Flow," AIAA Paper No. 71-85, AIAA 9th Aerospace Sciences Meeting, New York, New York, January 25-27, 1971.

DuFort, E.C., and Frankel, S.P. (1953), "Stability Conditions in the Numerical Treatment of Parabolic Differential Equations," Math. Tables and Other Aids to Computation, Vol. 7, pp. 135-152.

DuPont, T., Kendall, R.P., and Rachford, H.H., Jr. (1968), "An Approximate Factorization Procedure for Solving Self-Adjoint Elliptic Difference Equations," SIAM J. Numerical Analysis, Vol. 5, No. 3, pp. 559-573.

Dwyer, H.A. (1968), "Calculation of Three Dimensional and Time Dependent Boundary Layer Flows," AIAA Paper No. 68-740, AIAA Fluid and Plasma Dynamics Conference, Los Angeles, California, June 24-26, 1968.

Dwyer, H.A. (1971A), "Hypersonic Boundary Layer Studies on a Spinning Sharp Cone at Angle of Attack," AIAA Paper 71-57, AIAA 9th Aerospace Sciences Meeting, New York, New York, January 25-27, 1971.

Dwyer, H.A. (1971B), "Boundary Layer on a Hypersonic Sharp Cone at Small Angle of Attack," AIAA J., Vol. 9, No. 2, p. 277.

Dwyer, H.A., Doss, E.D., and Goldman, A. (1971), "Rapid Calculation of Inviscid and Viscous Flow over Arbitrary Shaped Bodies," J. Aircraft, Vol. 8, No. 2, pp. 125-127.

Dyakonov, Yu. N. (1964), "Three Dimensional Flow Around Blunt Bodies by a Supersonic Stream of Perfect Gas," Izvestiya Akademii Nauk SSSR, Mekhanika i Mashinostroyeniye, No. 4, pp. 150-153.

Dzhakupov, K.B., and Kuznetsov, B.G. (1969), "On the Numerical Computation of Stationary Viscous Incompressible Fluid Problems," Lockheed Missiles and Space Co. Translation, N69-26293. Izv. AN SSSR, Mekh. Zhidk. i Gaza, No. 1, 1969, pp. 95-99.

Dzhakupov, K.B. (1969), "Two Numerical Methods of Solving the Steady-State Navier-Stokes Equations for a Viscous Compressible Fluid," Akademiia Nauk SSSR, Feb. 1969, pp. 14-24.

Easton, C.R. (1969), "Applications of the Marker-and-Cell Numerical Program," N69-13117, QAC-63140, McDonnell-Douglas Astronautics Co., Santa Monica, California.

Easton, C.R., and Catton, I. (1970), "Initial Value Techniques in Free-Surface Hydrodynamics," MDAC Paper WD 1459, McDonnell-Douglas Astronautics Company, West, Santa Monica, California, November, 1970.

Eaton, R.R., and Zumwalt, G.W. (1967), "A Numerical Solution for the Flow Field of a Supersonic Cone-Cylinder Entering and Leaving a Blast Sphere Diametrically," SC-CR-67-2532, Sandia Laboratories, Albuquerque, New Mexico.

Eaton, R.R. (1970), "Three-Dimensional Numerical and Experimental Flowfield Comparisons for Sphere-Cones," J. Spacecraft, Vol. 7, No. 2, pp. 203-204.

Eddy, E.P. (1949), "Stability in the Numerical Solution of Initial Value Problems in Partial Differential Equations," NOLM 10232, Naval Ordnance Laboratory, White Oak, Silver Spring, Maryland.

Edwards, A.L. (1969), "TRUMP Computer Program: Calculation of Transient Laminar Fluid Flow in Porous Media," UCRL-50664, Lawrence Radiation Laboratory, University of California, Livermore, California.

Ehrlich, R., and Hurwitz, H., Jr. (1954), "Multigroup Methods for Neutron Diffusion Problems," Nucleonics, Vol. 12, No. 2, February, 1954, pp. 23-30.

Eisen, D. (1967A), "The Equivalence of Stability and Convergence for Finite Difference Schemes with Singular Coefficients," Numerische Mathematik, Vol. 20, pp. 20-29.

Eisen, D. (1967B), "On the Numerical Solution of $u_t = u_{rr} + 2/r(u_r)$," Numerische Mathematik, Vol. 10, pp. 397-409.

El Assar, R.J. (1969), "Compressible Laminar Wake of a Thin Flat Plate," Paper 69-WA/FE-6, American Society of Mechanical Engineers, Winter Annual Meeting, Los Angeles, California, November 16-20, 1969.

Emery, A.F. (1968), "An Evaluation of Several Differencing Methods for Inviscid Fluid Flow Problems," J. of Computational Physics, Vol. 2, pp. 306-331.

Emery, A.F., and Ashurst, W.T. (1971), "The Numerical Computation of Supersonic Flows About Finite Bodies with Application to Massive Surface Blowing," SCL-DR-69-159, Sandia Laboratories, Livermore, California, January, 1971.

Emery, A.F., and Carson, W.W. (1971), "An Evaluation of the Use of the Finite-Element Method in the Computation of Temperature," Transactions of the ASME, J. of Heat Transfer, May, 1971, pp. 136-145.

Erdos, J., and Zakkay, V. (1969), "Numerical Solution of Several Steady Wake Flows of the Mixed Supersonic/Subsonic Type by a Time-Dependent Method and Comparison with Experimental Data," AIAA Paper No. 69-649, AIAA Fluid and Plasma Dynamics Conference, San Francisco, California, June 16-18, 1969.

Estoque, M.A., and Bhumralkar, C.M. (1969), "Flow Over a Localized Heat Source," U.S. Monthly Weather Review, Vol. 97, No. 12, pp. 850-859.

Evans, J.S., Schnexnader, C.J., and Huber, P.W. (1970), "Computation of Ionization in Re-Entry Flowfields," AIAA J., Vol. 8, No. 6, June, 1970, pp. 1082-1089.

Evans, M.E., and Harlow, F.H. (1957), "The Particle-in-Cell Method for Hydrodynamic Calculations," Los Alamos Scientific Lab., Rept. No. LA-2139, Los Alamos, New Mexico.

Evans, M.E., and Harlow, F.H. (1958), "Calculation of Supersonic Flow Past an Axially Symmetric Cylinder," J. Aeronautical Sciences, Vol. 25, pp. 269-270.

Evans, M.W., and Harlow, F.H. (1959), "Calculation of Unsteady Supersonic Flow past a Circular Cylinder," American Rocket Society J., pp. 46-48.

Evans, M.W., Harlow, F.H., and Meixner, B.D. (1962), "Interaction of Shock or Rarefraction with a Bubble," Physics of Fluids, Vol. 5, No. 6, pp. 651-656.

Faccioli, E., and Ang, A.H.-S. (1968), "A Discrete Eulerian Model of Spherically Symmetric Compressible Media," J. of Computational Physics, Vol. 3, pp. 226-258.

Fagela-Albastro, E.B., and Hellums, J.D. (1967), "Laminar Gas Jet Impinging on an Infinite Liquid Surface: Numerical Finite Difference Solution Involving Boundary and Free Streamline Determinations," I & EC Fundamentals, Vol. 6, No. 4, pp. 580-587.

Fairweather, G., Gourlay, A.R., and Mitchell, A.R. (1967), "Some High Accuracy Difference Schemes with a Splitting Operator for Equations of Parabolic and Elliptic Type," Numerische Mathematik, Vol. 10, pp. 56-66.

Fairweather, G. (1969), "A Note on a Generalization of a Method of Douglas," Mathematics of Computation, Vol. 23, No. 106, pp. 407-409.

Fanning, A.E., and Mueller, T.J. (1970), "A Numerical and Experimental Investigation of the Oscillating Flow in the Wake of a Blunt Body," UNDAS TN-1970-M2, Dept. of Aerospace and Mechanical Engineering, University of Notre Dame, Notre Dame, Indiana. See also AIAA Paper No. 71-603, 1971.

Festa, J.F. (1970), "A Numerical Model of a Convective Cell Driven by Non-Uniform Horizontal Heating," Massachusetts Institute of Technology, Cambridge, Massachusetts.

Fickett, W., Jacobson, J.D., and Wood, W.W. (1970), "The Method of Characteristics for One-Dimensional Flow with Chemical Reaction," LA-4269, Los Alamos Scientific Laboratory, July, 1970.

Filler, L., and Ludloff, H.F. (1961), "Stability Analysis and Integration of the Viscous Equations of Motion," Mathematics of Computation, Vol. 15, pp. 261-274.

Finkleman, D. (1968), "A Characteristics Approach to Radiation Gasdynamics," AIAA Paper No. 68-163, AIAA 6th Aerospace Sciences Meeting, New York, New York, January 22-24, 1968.

Finn, D.L. (1968), "Bibliography on Techniques for Solving Partial Differential Equations by Hybrid Computation and Other Methods," NASA-CR-102218, Georgia Institute of Technology, April 22, 1968.

Fischer, C.F., and Usmani, R.A. (1969), "Properties of Some Tridiagonal Matrices and their Application to Boundary Value Problems," SIAM J. on Numerical Analysis, Vol. 6, No. 1, pp. 127-142.

Fischer, G. (1965A), "A Survey of Finite Difference Approximations to the Primitive Equations," U.S. Monthly Weather Review, Vol. 93, No. 1, pp. 1-10.

Fischer, G. (1965B), "On a Finite Difference Scheme for Solving the Non-Linear Primitive Equations for a Barotropic Fluid with Application to the Boundary Current Problem," Tellus, Vol. 17, No. 4, pp. 405-413.

Flugge-Lotz, I.I. (1969), "Computation of the Laminar Compressible Boundary Layer," Final Rept. on Contract AF49(638)-1385, Dept. of Applied Mechanics, Stanford University, Stanford, California.

Forsythe, G.E. (1956), "Difference Methods on a Digital Computer for Laplacian Boundary Value and Eigenvalue Problems," Communications of Pure and Applied Mathematics, Vol. 9, pp. 425-434.

Forsythe, G.E., and Wasow, W. (1960), Finite Difference Methods for Partial Differential Equations, Wiley, New York.

Forsytne, G.E. (1970), "Pitfalls in Computation, or Why a Math Book Isn't Enough," Tech. Rept. No. CS-147, Computer Science Dept., Stanford University, Stanford, California.

Fortin, M., Peyret, R., and Teman, R. (1971), "Calcul des Ecoulements d'un Fluide Visqueux Incompressible." See Holt (1971).

Fox, L. (1948), "A Short Account of Relaxation Methods," Quarterly J. Mechanics and Appl. Math., Vol. 1, pp. 253-280.

Foy, L.R. (1964), "Steady State Solutions of Hyperbolic Systems of Conservation Laws with Viscosity Terms," Communications on Pure and Applied Mathematics, Vol. 17, pp. 177-188.

Frank, R.M., and Lazarus, R.B. (1964), "Mixed Eulerian-Lagrangian Method," Methods in Computational Physics, Vol. 3, pp. 47-68.

Frankel, S.P. (1950), "Convergence Rates of Iterative Treatments of Partial Differential Equations," Math Tables and Other Aids to Computation, Vol. 4, pp. 65-75.

Frankel, S.P. (1956), "Some Qualitative Comments on Stability Considerations in Partial Difference Equations," Proc. Sixth Symposia in Applied Mathematics, AMS, Vol. 6 - Numerical Analysis, pp. 73-75.

Frenkiel, F.N., and Stewartson, K. (ed.) (1969), High-Speed Computing in Fluid Dynamics, Physics of Fluids Supplement II, American Institute of Physics, New York, New York.

Freudiger, F.A., Gallaher, W.H., and Thommen, H.U. (1967), "Numerical Calculation of the Low Reynolds Number Flow Over a Blunt Wedge with Rearward Facing Step," ARL 67-0151, Aerospace Research Laboratories, United States Air Force, Wright-Patterson AFB, Ohio.

Fried, Isaac (1969), "Some Aspects of the Natural Coordinate System in the Finite-Element Method," AIAA J., Vol. 7, No. 7, pp. 1366-1368.

Friedman, Menaheim (1970), "Flow in a Circular Pipe with Recessed Walls," J. of Applied Mechanics, Vol. 37, No. 1, pp. 5-8.

Friedman, N.E. (1956), "The Truncation Error in a Semi-Discrete Analog of the Heat Equation," J. Math and Physics, Vol. 35, No. 3, pp. 299-308.

Friedmann, M., Gillis, J., and Liron, N. (1968), "Laminar Flow in a Pipe at Low and Moderate Reynolds Numbers," Applied Scientific Research, Vol. 19, pp. 426-438.

Froehlich, R. (1966), "On Stable Methods of Matrix Factorization for Block-Tri-Diagonal Matrices, Parts I and II," GA 7164, General Atomic, Division General Dynamics, San Diego, California, October 31, 1966.

Fromm, J.E. (1961), "Lagrangian Difference Approximations for Fluid Dynamics," Los Alamos Scientific Lab. Rept. No. 2535, Los Alamos, New Mexico.

Fromm, J.E., and Harlow, F.H. (1963), "Numerical Solution of the Problem of Vortex Street Development," Physics of Fluids, Vol. 6, No. 7, pp. 975-982.

Fromm, J.E. (1963), "A Method for Computing Non-Steady Incompressible Viscous Fluid Flows," Los Alamos Scientific Lab., Rept. No. LA-2910, Los Alamos, New Mexico.

Fromm, Jacob (1964), "The Time Dependent Flow of an Incompressible Viscous Fluid," Methods of Computational Physics, Vol. 3, pp. 345-382.

Fromm, J.E. (1965), "Numerical Solutions of the Nonlinear Equations for a Heated Fluid Layer," Physics of Fluids, Vol. 8, No. 10, pp. 1757-1769.

Fromm, J.E. (1967), "Finite Difference Methods of Solution of Non-Linear Flow Processes with Application to the Benard Problem," Los Alamos Scientific Lab. Rept. LA-3522, Los Alamos, New Mexico.

Fromm, J.E. (1968), "A Method for Reducing Dispersion in Convective Difference Schemes," J. of Computational Physics, Vol. 3, pp. 176-189.

Fromm, J.E. (1969), "Practical Investigation of Convective Difference Approximations of Reduced Dispersion." See Frenkiel and Stewartson (1969).

Fromm, J.E. (1970A), "A Numerical Study of Buoyancy Driven Flows in Room Enclosures," IBM Research Laboratory, San Jose, California.

Fromm, J.E. (1970B), "A Numerical Method for Computing the Non-Linear, Time-Dependent, Buoyant Circulation of Air in Rooms," RJ 732, IBM Research Report, San Jose, California.

Fromm, J.E. (1970C), "Lectures on Large Scale Finite Difference Computation of Incompressible Fluid Flows," An Introduction to Computer Simulation in Applied Science, Palo Alto Scientific Center, January, 1970, pp. 34-94.

Fromm, J.E. (1971), "A Numerical Study of Buoyance Driven Flows in Room Enclosures," See Holt (1971).

Fujita, T., and Grandoso, H. (1968), "Split of a Thunderstorm into Anticyclonic and Cyclonic Storms and their Motion as Determined from Numerical Model Experiments," J. of the Atmospheric Sciences, Vol. 25, pp. 416-439.

Fulford, G.D., and Pei, D.C.T. (1969), "A Unified Approach to the Study of Transfer Processes," Industrial and Engineering Chemistry, Vol. 61, No. 5, pp. 47-69.

Fyfe, I.M., Eng, R.C., and Young, D.M. (1961), "On the Numerical Solution of the Hydrodynamics Equations," SIAM Review, Vol. 3, No. 4, October, 1961, pp. 298-308.

Gaier, D., and Todd, J. (1967), "On the Rate of Convergence of Optimal ADI Processes," Numerische Mathematik, Vol. 9, pp. 452-459.

Garabedian, P.R., and Lieberstein, H.M. (1958), "On the Numerical Calculation of Detached Bow Shock Waves in Hypersonic Flow," J. of Aeronautical Sciences, Vol. 25, p. 109.

Gary, J. (1964), "On Certain Finite Difference Schemes for Hyperbolic Systems," Math. of Computation, pp. 1-18.

Gary, J.M. (1969), "A Comparison of Two Difference Schemes Used for Numerical Weather Prediction," J. of Computational Physics, Vol. 4, pp. 279-305.

Gawain, T.H., and Pritchett, J.W. (1968), "Turbulence in an Unbounded, Uniform-Shear Flow: A Computer Analysis," NRDL-TR-68-86, May, 1968, U.S. Naval Radiological Defense Laboratory, San Francisco, California.

Gawain, T.H., and Pritchett, J.W. (1970), "A Unified Heuristic Model of Fluid Turbulence," J. of Computational Physics, Vol. 5, pp. 383-405.

Gawain, T.H., and O'Brien, G.D., Jr. (1971), "Numerical Simulation of Transition and Turbulence in Plane Poiseuille Flow." See Holt (1971).

Geiringer, H., (1959) "On the Solution of Systems of Linear Equations by Certain Iterative Methods," Reissner Anniversary Volume, Ann Arbor, Michigan, pp. 365-393.

Gelder, D., (1971) "Solution of the Compressible Flow Equations," International J. for Numerical Methods in Engineering, Vol. 3, January-March, 1971, pp. 35-43.

Gemmel, A.R., and Epstein, N. (1962), "Numerical Analysis of Stratified Laminar Flow of Two Immiscible Newtonian Liquids in a Circular Pipe," Canadian J. of Chemical Engineering, pp. 215-224.

Gentry, R.A., Martin, R.E., and Daly, B.J. (1966), "An Eulerian Differencing Method for Unsteady Compressible Flow Problems," J. of Computational Physics, Vol. 1, pp. 87-118.

George, J.A. (1970), "The Use of Direct Methods for the Solution of the Discrete Poisson Equation on Non-Rectangular Regions," STAN-CS-70-159, Computer Science Department, Stanford University, Stanford, California.

Gershuni, G.Z., Zhukhovitskii, E.M., and Tarunin, E.L. (1966), "Numerical Study of Convection of a Liquid Heated From Below," Izv. AN SSSR. Mekhanika Zhidkosti i Gaza, Vol. 1, No. 6, pp. 93-99.

Ghia, K.N., Torda, T.P., and Lavan, Z. (1968), "Laminar Mixing of Heterogeneous Axisymmetric Coaxial Confined Jets," NASA CR 72480, NASA, Office of Scientific and Technical Information, Washington, D.C., November, 1968.

Giaquinta, A.R., and Hung, T.-K. (1968), "Slow Non-Newtonian Flow in a Zone of Separation," J. of the Engineering Mechanics Division, Proc. of the American Society of Civil Engineers, December, 1968, pp. 1521-1537.

Gilinskiy, S.M., Telenin, G.F., and Tinyakov, G.P. (1970), "A Method of Calculating Supersonic Flow Around Blunted Bodies with Detached Shock Waves," NASA TT F-13, 026, National Aeronautics and Space Administration, Washington, D.C., 20546.

Gillis, J., and Liron, N. (1969), "Numerical Integration of Equations of Motion of a Viscous Fluid," Final Rept. Contract No. F61052-68-C-0053, Weizmann Institute of Science, Department of Applied Mathematics, Rehovot, Israel.

Glese, J.H. (1969A), "A Unified Account of Two-Level Methods for the One-Dimensional Heat Equation," Memo. Rept. No. 1956, Ballistic Research Laboratories, Aberdeen Proving Ground, Maryland.

Glese, J.H. (1969B), "A Bibliography for the Numerical Solution of Partial Differential Equations," Memo. Rept. No. 1991, Ballistic Research Laboratories, Aberdeen Proving Ground, Maryland, July, 1969.

Glowinski, R. (1971), "Methodes Numeriques pour l'Ecoulement Stationnaire d'un Fluide Rigide Visco-Plastique Incompressible." See Holt (1971).

Goad, W.B. (1960), "WAT: A Numerical Method for Two-Dimensional Unsteady Fluid Flow," Los Alamos Scientific Lab., Rept. No. LAMS-2365, Los Alamos, New Mexico.

Godunov, S.K. (1959), "Finite Difference Method for Numerical Computation of Discontinuous Solutions of the Equations of Fluid Dynamics," Matematicheskii Sbornik, Vol. 47, No. 3, pp. 271-306.

Godunov, S.K., Zabrodin, A.V., Prokopov, G.P. (1959), "A Computational Scheme for Two-Dimensional Nonstationary Problems of Gas Dynamics and Calculation of the Flow from a Shock Wave Approaching a Stationary State," USSR J. Computational Mathematics and Mathematical Physics, pp. 1187-1219.

Godunov, S.K., Zabrodyn, A.W., and Prokopov, G.P. (1961), "A Difference Scheme for Two-Dimensional Unsteady Problems of Gas Dynamics and Computation of Flow with a Detached Shock Wave," Zh. Vychyslitelnoi Matematiki i Matematicheskoi Fiziki, Vol. I, No. 6, pp. 1020-1050.

Godunov, S.K., and Semendyayev, K.A. (1962), "Difference Methods for the Numerical Solution of Problems in Gas Dynamics," Zh. Vychyslitelnoi Matematiki i Matematicheskoi Fiziki, Vol. 2, No. 1, pp. 3-14.

Godunov, S.K., and Prokopov, G.P. (1968), "The Solution of Differential Equations by the Use of Curvilinear Difference Networks," Zh. Vychislitelnoi Matematiki i Matematicheskoi Fiziki, Vol. 8, pp. 28-46.

Godunov, S.K., Deribas, A.A., Zabrodin, A.V., and Kozin, N.S. (1970), "Hydrodynamic Effects in Colliding Solids," J. of Computational Physics, Vol. 5, pp. 517-539.

Gol'din, V. Ya., Ionkin, H.I., and Kalltkin, N.N. (1969), "Entropy Scheme for Gas Dynamics Calculation," NASA TT-F-13, 018, Washington, D.C., July, 1970.

Gonidou, R. (1967), "Supersonic Flows Around Cones at Incidence," NASA-TT-F-11473, National Aeronautics and Space Administration, Washington, D.C. 20546.

Gonor, A.L., Lapygin, V.I., and Ostapenko, N.A. (1971), "The Conical Wing in Hypersonic Flow." See Holt (1971).

Goodrich, W.D. (1969), "A Numerical Solution for Problems in Laminar Gas Dynamics," Ph.D. Dissertation, The University of Texas at Austin, Austin, Texas.

Gopalarkrishnan, S., and Bozzola, R. (1971), "A Numerical Technique for the Calculation of Transonic Flows in Turbomachinery Cascades," Paper 71-GT-42, American Society of Mechanical Engineers, Gas Turbine Conference and Products Show, Houston, Texas, March 28-April 1, 1971.

Gopalsamy, K., and Aggarwala, B.D. (1970), "Monte Carlo Methods for Some Fourth Order Partial Differential Equations," Zeitschrift fur angewandte Mathematik und Mechanik, Vol. 50, December, 1970, pp. 759-767.

Gordon, P. (1968), "A Note on a Maximum Principle for the DuFort-Frankel Difference Equation," Mathematics of Computation, Vol. 22, No. 102, pp. 437-440.

Gordon, P., and Scala, S.M. (1969), "The Calculation of the Formation of Discontinuities in Planar and Spherically-Symmetric Nonisentropic Inviscid Flows," General Electric Company, Space Sciences Laboratory, Valley Forge, Pennsylvania, unpublished ms.

Gosman, A.D., Pun, W.M., Runchal, A.K., Spalding, D.B., and Wolfshtein, M. (1969), Heat and Mass Transfer in Recirculating Flows, Academic Press.

Gosman, A.D., and Spalding, D.B. (1971), "Computation of Laminar Recirculating Flow Between Shrouded Rotating Discs." See Holt (1971).

Gourlay, A.R., and Mitchell, A.R. (1966A), "Alternating Direction Methods for Hyperbolic Systems," Numerische Mathematik, Vol. 8, pp. 137-149.

Gourlay, A.R., and Mitchell, A.R. (1966B), "A Stable Implicit Difference Method for Hyperbolic Systems in Two Space Variables," Numerische Mathematik, Vol. 8, pp. 367-375.

Gourlay, A.R., and Morris, J.L. (1968A), "Finite-Difference Methods for Nonlinear Hyperbolic Systems," Mathematics of Computation, Vol. 22, No. 101, pp. 28-39.

Gourlay, A.R., and Morris, J.L. (1968B), "A Multistep Formulation of the Optimized Laz-Wendroff Method for Nonlinear Hyperbolic Systems in Two Space Variables," Mathematics of Computation, Vol. 22, No. 104, pp. 715-720.

Gourlay, A.R., and Morris, J.L. (1968C), "Deferred Approach to the Limit in Non-Linear Hyperbolic Systems," Computing J., Vol. 2, pp. 95-101.

Gourlay, A.R., and Mitchell, A.R. (1969A), "The Equivalence of Certain Alternating Direction and Locally-One Dimensional Difference Methods," SIAM J. on Numerical Analysis, Vol. 6, No. 1, March, 1969, pp. 37-46.

Gourlay, A.R., and Mitchell, A.R. (1969B), "A Classification of Split Difference Methods for Hyperbolic Equations in Several Space Dimensions," SIAM J. on Numerical Analysis, Vol. 6, No. 1, March, 1971, pp. 62-71.

Gourlay, A.R., and Morris, J. L. (1970), "On the Comparison of Multistep Formulations of the Optimized Lax-Wendroff Method for Nonlinear Hyperbolic Systems in Two Space Variables," J. of Computational Physics, Vol. 5, pp. 229-243.

Gourlay, A.R. (1970A), "Hopscotch: A Fast Second-Order Partial Differential Equation Solver," J. Inst. Maths. Applics, Vol. 6, pp. 375-390.

Gourlay, A.R. (1970B), "Time Dependent Problems," Report 0002, IBM Scientific Centre, Peterlee, County Durham, United Kingdom.

Gourlay, A.R., and Morris, J.L. (1971), "Hopscotch Methods for Non-linear Hyperbolic Systems," IBM (United Kingdom) Ltd., Peterlee, County Durham, England.

Gourlay, A.R., and McGuire, G.R. (1971A) "General Hopscotch Algorithms for the Numerical Solution of Partial Differential Equations," IBM (United Kingdom) Ltd., Peterlee, County Durham, England.

Gourlay, A.R., and McGuire, G.R. (1971B), "General Hopscotch Algorithm for the Numerical Solution of Partial Differential Equations," J. Inst. Maths. Applics., Vol. 7, pp. 216-227.

Gourlay, A.R., and McKee, S. (1971), "On a Numerical Comparison of Hopscotch with A.D.I. and L.O.D. Methods for Parabolic and Elliptic Equations in Two Space Dimensions with a Mixed Derivative," IBM (United Kingdom) Ltd., Peterlee, County Durham, England. To be published.

Gradowczyk, M.H., and Folguera, H.C. (1965), "Analysis of Scour in Open Channels by Means of Mathematical Models," La Houille Blanche, No. 8, pp. 761-768.

Grammeltvedt, A. (1969), "A Survey of Finite-Difference Schemes for the Primitive Equations for a Barotropic Fluid," Monthly Weather Review, Vol. 97, No. 5, pp. 384-404.

Greenspan, D. (1967), "Approximate Solution of Initial-Boundary Parabolic Problems by Boundary Value Techniques," MRC-TSR-792, Mathematics Research Center, The University of Wisconsin, Madison, Wisconisn, August, 1967.

Greenspan, D. (1969A), "Numerical Studies of Steady, Viscous, Incompressible Flow in a Channel with a Step," J. of Engineering Mathematics, Vol. 3, No. 1, pp. 21-28.

Greenspan, D. (1969B), "Numerical Studies of Prototype Cavity Flow Problems," The Computer J., Vol. 12, No. 1, pp. 88-93.

Griffiths, D.F., Jones, D.T., and Walters, K. (1969), "A Flow Reversal due to Edge Effects," J. Fluid Mechanics, Vol. 36, Part 1, pp. 161-175.

Grigoryev, Y.I. (1970), "One Direct Diagram of a Method of Characteristics for the Calculation of a Three Dimensional Gas Flow," NASA-TT-F-13019.

Grossman, B., and Moretti, G. (1970), "Time-Dependent Computation of Transonic Flows," Paper 70-1322, American Institute of Aeronautics and Astronautics, Annual Meeting and Technical Display, 7th, Houston, Texas, October 19-22, 1970.

Gunaratnam, D.J., and Perkins, F.E. (1970), "Numerical Solution of Unsteady Flows in Open Channels," Hydrodynamics Report No. 127, School of Engineering, Massachusetts Institute of Technology, Cambridge, Massachusetts.

Gururaja, J., and Deckker, B.E.L. (1970), "Numerical Solutions of Flows Behind Shock Waves in Non-Uniform Regions," Thermodynamics and Fluid Mechanics Convention, University of Glasgow, March 23-25, 1970, The Institution of Mechanical Engineers, 1 Birdcage Walk, Westminster, London, SWI, pp. 103-111.

Gustafsson, B. (1969), "On Difference Approximations to Hyperbolic Differential Equations over Long Time Intervals," SIAM J. Numerical Analysis, Vol. 6, No. 3, p. 508.

Gustafsson, B. (1971), "An Alternating Direction Implicit Method for Solving the Shallow Water Equations," J. of Computational Physics, Vol. 7, pp. 239-254.

Gwynn, J.M., Jr. (1967), "A Study of Six- and Nine-Point Finite-Difference Analogues of the One-Dimensional Heat Equation," Ph.D. Thesis, The University of North Carolina at Chapel Hill, North Carolina.

Hadjidimos, Apostolos (1969), "On Some High Accuracy Difference Schemes for Solving Elliptic Equations," Numerical Mathematics, Vol. 13, pp. 396-403.

Hahn, S.G. (1958), "Stability Criteria for Difference Schemes," Communications on Pure and Applied Mathematics, Vol. XI, pp. 243-255.

Hall, M.G. (1969), "A Numerical Method for Calculating Unsteady Two-Dimensional Laminar Boundary Layers," Ingenieur-Archiv, Vol. 38, No. 2, pp. 97-106.

Halton, J.H. (1970), "A Retrospective and Prospective Survey of the Monte Carlo Method," SIAM Review, Vol. 12, No. 1, January, 1970, pp. 1-63.

Hamielec, A.E., Hoffman, T.W., and Ross, L.L. (1967A), "Numerical Solution of the Navier-Stokes Equation for Flow Past Spheres: Part I, Viscous Flow Around Spheres With and Without Radial Mass Efflux," American Institute of Chemical Engineering J., Vol. 13, No. 2, pp. 212-219.

Hamielec, A.E., Johnson, A.I., and Houghton, W.T. (1967B), "Numerical Solution of the Navier-Stokes Equation for Flow Past Spheres: Part II, Viscous Flow Around Circulating Spheres of Low Viscosity," American Institute of Chemical Engineering J., Vol. 13, No. 2, pp. 220-224.

Hamielec, A.E., and Raal, J.D. (1969), "Numerical Studies of Viscous Flow Around Circular Cylinders," The Physics of Fluids, Vol. 12, No. 1, pp. 11-17.

Hamming, R.W. (1962), Numerical Methods for Scientists and Engineers, McGraw-Hill Book Company, Inc., New York, New York.

Hardy, J.W. (1968), "Tides in a Numerical Model of the Atmosphere," UCRL-50368, Lawrence Radiation Laboratory, University of California, Livermore, California, April 18, 1968.

Harlow, F.H., Dickman, D.O., Harris, D.E., and Martin, R.E. (1959), "Two-Dimensional Hydrodynamic Calculations," Rept. No. LA-2301, Los Alamos Scientific Laboratory, Los Alamos, New Mexico.

Harlow, F.H. (1963), "The Particle-In-Cell Method for Numerical Solution of Problems in Fluid Dynamics," Proceedings of Symposium In Applied Mathematics, Vol. 15, pp. 269-288.

Harlow, F.H. (1964), "The Particle-In-Cell Computing Method for Fluid Dynamics," Methods in Computational Physics, Vol. 3, p. 319.

Harlow, F.H., and Fromm, J.E. (1964), "Dynamics and Heat Transfer in the von Karman Wake of a Rectangular Cylinder," The Physics of Fluids, Vol. 7, No. 8, pp. 1147-1156.

Harlow, F.H., and Fromm, J.E. (1965), "Computer Experiments in Fluid Dynamics," Scientific American, Vol. 212, No. 3, pp. 104-110.

Harlow, F.H., and Welch, J.E. (1965), "Numerical Calculation of Time-Dependent Viscous Incompressible Flow of Fluid with Free Surface," Physics of Fluids, Vol. 8, No. 12, pp. 2182-2189.

Harlow, F.H., and Welch, J.E. (1966), "Numerical Study of Large Amplitude Free Surface Motions," Physics of Fluids, Vol. 9, pp. 842-851.

Harlow, F.H., and Amsden, A.A. (1968), "Numerical Calculation of Almost Incompressible Flow," J. of Computational Physics, Vol. 3, pp. 80-93.

Harlow, F.H., and Hirt, C.W. (1969), "Generalized Transport Theory of Anisotropic Turbulence," Rept. No. LA-4086, Los Alamos Scientific Laboratory, Los Alamos, New Mexico.

Harlow, F.H., and Romero, N.C. (1969), "Turbulence Distortion in a Nonuniform Tunnel," Rept. No. LA-4247, Los Alamos Scientific Laboratory, Los Alamos, New Mexico.

Harlow, F.H. (1969), "Numerical Methods for Fluid Dynamics, an Annotated Bibliography," Rept. No. LA-4281, Los Alamos Scientific Laboratory, Los Alamos, New Mexico.

Harlow, F.H., and Amsden, A.A. (1970A), "Fluid Dynamics: An Introductory Text," Rept. No. LA-4100, Los Alamos Scientific Laboratory, Los Alamos, New Mexico.

Harlow, F.H., and Amsden, A.A. (1970B), "A Numerical Fluid Dynamics Method for All Flow Speeds," Los Alamos Scientific Laboratory, New Mexico, LADC-12190. Also, J. Computational Physics, Vol. 8, 1971, pp. 197-213.

Harlow, F.H., and Amsden, A.A. (1971), "Fluid Dynamics," Rept. No. LA-4700, Los Alamos Scientific Laboratory, Los Alamos, New Mexico, June 1971.

Harlow, F.H., Amsden, A.A., and Hirt, C.W. (1971), "Numerical Calculation of Fluid Flows at Arbitrary Mach Number." See Holt (1971).

Harper, L.R., Kinder, J. (1967), "Second Report on Turbulent Mixing Studies," Rept. No. A.P. 5451, Bristol Sideley Engines Ltd., United Kingdom.

Hartree, D.R. (1958), Numerical Analysis, Oxford University Press, London, 2nd Edition, Chapter X, pp. 257-263.

Hayes, J.K. (1970), "Four Computer Programs Using Green's Third Formula to Numerically Solve Laplace's Equation in Inhomogeneous Media," Rept. No. LA-4423, Los Alamos Scientific Laboratory, Los Alamos, New Mexico.

Heie, H., and Leigh, D.C. (1965), "Numerical Stability of Hyperbolic Equations in Three Independent Variables," AIAA J., Vol. 3, No. 6, pp. 1099-1103.

Heywood, J.G. (1970), "On Stationary Solutions of the Navier-Stokes Equations as Limits of Nonstationary Solutions," Archive for Rational Mechanics and Analysis, Vol. 37, No. 1, pp. 48-60.

Hicks, D. (1968), "Hydrocode Test Problems," AFWL-TR-67-127, Air Force Weapons Laboratory, Kirtland Air Force Base, New Mexico.

Hicks, D., and Pelzl, R. (1968), "Comparison Between a von Neumann-Richtmyer Hydrocode (AFWL's Puff) and a Lax-Wendroff Hydrocode," AFWL-TR-68-112, Air Force Weapons Laboratory, Kirtland Air Force Base, New Mexico.

Hicks, D.L. (1969), "One-Dimensional Lagrangian Hydrodynamics and the IDLH Hydrocode," SC-RR-69-728, Sandia Laboratories, Albuquerque, New Mexico.

Hicks, D.L. (1971), "Extending Courant, Friedrichs, and Lewy's (1928) Compactness Result to Nonlinear Equations of State and Non-Differentiable Initial Data," SC-RR-710302, Sandia Laboratories, Albuquerque, New Mexico, July, 1971.

Higbie, L.C., and Plooster, M.N. (1968), "Variable Pseudoviscosity in One-Dimensional Hyperbolic Difference Schemes," J. of Computational Physics, Vol. 3, pp. 154-156.

Hildebrand, F.B. (1952), "On the Convergence of Numerical Solutions of the Heat-Flow Equation," J. Math and Physics, Vol. 31, No. 1, pp. 35-41.

Hildebrand, F.B. (1968), Finite-Difference Equations and Simulations, Prentice-Hall, Englewood Cliffs, New Jersey.

Hill, L.R., and Larsen, G.E. (1970), "Users' Manual for the TOIL Code at Sandia Laboratories," SC-DR-70-61, Sandia Laboratories, Albuquerque, New Mexico.

Hirasaki, G.J. (1967), "A General Formulation of the Boundary Conditions on the Vector Potential in Three-Dimensional Hydrodynamics," Ph.D. Thesis, Rice University, Houston, Texas.

Hirota, I., Tokioka, T., and Nishiguchi, M. (1970), "A Direct Solution of Poisson's Equation by Generalized Sweep-Out Method," J. of the Meteorological Society of Japan, Ser. II, Vol. 48, No. 2, April, 1970, pp. 161-167.

Hirt, C.W. (1965), "Multidimensional Fluid Dynamics Calculations with High Speed Computers," AIAA Paper No. 65-3, AIAA 2nd Aerospace Sciences Meeting, New York, New York, January 25-27, 1965.

Hirt, C.W., and Harlow, F.H. (1967), "A General Corrective Procedure for the Numerical Solution of Initial-Value Problems," J. of Computational Physics, Vol. 2, pp. 114-119.

Hirt, C.W. (1968), "Heuristic Stability Theory for Finite-Difference Equations," J. of Computational Physics, Vol. 2, pp. 339-355.

Hirt, C.W., and Shannon, J.P. (1968), "Free-Surface Stress Conditions for Incompressible-Flow Calculations," J. of Computational Physics, Vol. 2, pp. 403-411.

Hirt, C.W. (1969), "Computer Studies of Time-Dependent Turbulent Flows." See Frenkiel and Stewartson (1969).

Hirt, C.W. (1970), "Generalized Turbulence Transport Equations," LA-DC-10526, Los Alamos Scientific Laboratory, Los Alamos, New Mexico.

Hirt, C.W. (1971), "An Arbitrary Lagrangian-Eulerian Computing Technique." See Holt (1971).

Hockney, R.W. (1965), "A Fast Direct Solution of Poisson's Equation Using Fourier Analysis," J. Association of Computing Machinery, Vol. 12, p. 95.

Hockney, R.W. (1971), "The Potential Calculation and Some Applications," Methods in Computational Physics, Vol. 9.

Hodgkins, W.R. (1966), "On the Relation Between Dynamic Relaxation and Semi-Iterative Matrix Methods," Numerische Mathematik, Vol. 9, pp. 446-451.

Hoeper, P.S., and Wenstrup, R.S., Jr. (1970), "Numerical Solution to Laplace's Equation in Spherical Coordinates with Axial Symmetry," J. of Applied Physics, Vol. 41, No. 5, pp. 1879-1882.

Hofmann, R., and Reaugh, J. (1968), "Shock Hydrodynamic Calculations in Wires," Report No. PIFR-104, Physics International Company, 2700 Merced Street, San Leandro, California, May, 1968.

Holt, M., and Ndefo, D.E. (1970), "A Numerical Method for Calculating Steady Unsymmetrical Supersonic Flow Past Cones," J. of Computational Physics, Vol. 5, pp. 463-486.

Holt, M., ed. (1971), Proceedings of the Second International Conference on Numerical Methods in Fluid Dynamics, Lecture Notes in Physics, Vol. 8, Springer-Verlag, New York.

Holt, M., and Masson, B.S. (1971), "The Calculation of High Subsonic Flow Past Bodies by the Method of Integral Relations." See Holt (1971).

Homicz, G.F., and George, A.R. (1970), "A Comparison between the Method of Integral Relations and the Method of Lines as Applied to the Blunt Body Problem," J. Spacecraft, Vol. 7, No. 12, December, 1970, pp. 1483-1484.

Hopf, E. (1950), "The Partial Differential Equation $u_t + uu_x = \mu u_{xx}$," Communications in Pure and Applied Mathematics, Vol. 3, pp. 201-230.

Hoskin, N.E. (1964), "Solution by Characteristics of the Equations of One-Dimensional Unsteady Flow," Methods in Computational Physics, Vol. 3, pp. 265-294.

Hoskin, N.E., and Lambourn, B.D. (1971), "The Computation of General Problems in One Dimensional Unsteady Flow by the Method of Characteristics." See Holt (1971).

Houghton, D., Kasahara, A., and Washington, W. (1966), "Long-Term Integration of the Barotropic Equations by the Lax-Wendroff Method," U.S. Monthly Weather Review, Vol. 94, No. 3, pp. 141-150.

Houghton, D.D., and Isaacson, E. (1968), "Mountain Winds," in Studies in Numerical Analysis 2, Society for Industrial and Applied Mathematics, Philadelphia, Pennsylvania, pp. 21-52.

Houghton, D.D., and Jones, W.L. (1966), "A Numerical Model for Linearized Gravity and Acoustic Waves," J. of Computational Physics, Vol. 3, pp. 339-357.

Howell, R.H., and Spong, E.D. (1969), "Numerical Solution of Subsonic Compressible Flow at Two-Dimensional Inlets," AIAA J., Vol. 7, No. 7, pp. 1392-1394.

Huang, S.L., and Chou, P.C. (1968), "Calculations of Expanding Shock Waves and Late-Stage Equivalence, DIT Rept. 125-12, Drexel Institute of Technology, Philadelphia, Pennsylvania, April, 1968.

Hubbard, B. (ed.) (1971), Numerical Solution of Partial Differential Equations - II, Academic Press, New York, New York, Proceedings of the Second Symposium on the Numerical Solution of Partial Differential Equations, University of Maryland, College Park, Maryland, May 11-15, 1970.

Hung, T.-K., and Macagno, E.O. (1966), "Laminar Eddies in a Two-Dimensional Conduit Expansion," La Houille Blanche, Vol. 21, No. 4, pp. 391-400.

Hurd, A.C., and Peters, A.R. (1970), "Analysis of Flow Separation in a Confined Two-Dimensional Channel," J. of Basic Engineering, ASME Paper No. 70-FE-29.

Hwang, J.D. (1968), "On the Numerical Solutions of the General Navier-Stokes Equations for Two-Layered Stratified Flows," Ph.D. Thesis, Oregon State University, Corvallis, Oregon.

Ingham, D.B. (1967), "Note on the Numerical Solution for Unsteady Viscous Flow Past a Circular Cylinder," J. Fluid Mechanics, Vol. 31, Part 4, pp. 815-818.

Ishizaki, H. (1957), "On the Numerical Solutions of Harmonic, Biharmonic and Similar Equations by the Difference Method not Through Successive Approximations," Disaster Prevention Research Institute, Kyoto University Bulletins, Bulletin No. 18, August, 1957.

Ivanov, K.P., Ladyzhenskaia, O.A., and Rivkind, V. Ia. (1970), "A Network Method of Solving the Navier-Stokes Equations in Cylindrical Coordinates," Leningradskii Universitet, Vestnik, Matematika, Mekhanika, Astronomiia, Vol. 25, July, 1970, pp. 37-41.

Jaffe, N.A., and Thomas, J. (1970), "Application of Quasi-Linearization and Chebyshev Series to the Numerical Analysis of the Laminar Boundary-Layer Equations," AIAA J., Vol. 8, No. 3, March, 1970, p. 483.

Jain, P. (1967), "Numerical Study of the Navier-Stokes Equations in Three Dimensions," MRC-TSR-751, Mathematics Research Center, Madison, Wisconsin.

Jamet, P. (1968), "Numerical Methods and Existence Theorems for Parabolic Differential Equations whose Coefficients are Singular on the Boundary," Mathematics of Computation, Vol. 22, No. 104, pp. 721-744.

Jamet, P., Lascaux, P., and Raviart, P.-A. (1970), "Une Methode de Resolution Numerique des Equations de Navier-Stokes," Numerical Mathematics, Vol. 16, pp. 93-114.

Jensen, V.G. (1959), "Viscous Flow Round a Sphere at Low Reynolds Numbers ($\leq$ 40)," Proceedings of the Royal Society of London, Series A, Vol. 249, pp. 346-366.

Jenssen, D., and Straede, J. (1968), "The Accuracy of Finite Difference Analogues of Simple Differential Operators," Proceedings of the WMO/IUGG Symposium on Numerical Weather Prediction in Tokyo, November 26-December 4, 1968.

Jeppson, R.W. (1969), "Finite Difference Solutions to Free Jet and Confined Cavity Flows Past Disks with Preliminary Analyses of the Results," PRWG-76-1, Utah State University, Logan, Utah, November 1969.

Jische, M.C., and Baron, J.R. (1969), "Application of the Method of Parametric Differentiation to Radiative Gasdynamics," AIAA J., Vol. 7, No. 7, p. 1326.

John, F. (1952), "On the Integration of Parabolic Equations by Difference Methods," Communications on Pure and Appl. Math., Vol. 5, p. 155.

John, F. (1953), "A Note on 'Improper' Problems in Partial Differential Equations," Communications on Pure and Appl. Math., Vol. 8, p. 591.

Johnson, W.E. (1967), "TOIL - A Two-Material Version of the Oil Code," GAMD-8073, General Dynamics, P.O. Box 608, San Diego, California, July 7, 1967.

Jones, D.J. (1968), "Numerical Solutions of the Flow Field for Conical Bodies in a Supersonic Stream," Aeronautical Report, LR-507, National Research Council of Canada, Ottawa 7, Canada.

Kacker, S.C., and Whitelaw, J.H. (1970), "Prediction of Wall-Jet and Wall-Wake Flows," J. Mechanical Engineering Science, Vol. 12, No. 6, pp. 404-420.

Kalugin, V.N., and Panchuk, V.I. (1971), "Viscous Incompressible Fluid Flow Along a Traveling Wave (Numerical Experiments)," Hydrodynamic Problems of Bionics, USSR, JPRS 52605, March 12, 1971.

Kamzolov, V.N., and Pirumov, U.G. (1966), "Calculation of Nonequilibrium Flows in Nozzles," Mekhanika Zhidkosti i Gaza, Vol. 1, No. 6, pp. 25-33.

Karplus, W.J. (1958), "An Electric Circuit Theory Approach to Finite Difference Stability," Transactions of the AIEE, Vol. 77, Part I, May, 1958, pp. 210-213.

Kasahara, A. (1965), "On Certain Finite Difference Methods for Fluid Dynamics," U.S. Monthly Weather Review, Vol. 93, No. 1, pp. 27-31.

Kasahara, A., Isaacson, E., and Stoker, J.J. (1965), "Numerical Studies of Frontal Motion in the Atmosphere - I", Tellus, Vol. 17, No. 3, pp. 261-277.

Kasahara, A., and Houghton, D.D. (1969), "An Example of Nonunique, Discontinuous Solutions in Fluid Dynamics," J. of Computational Physics, Vol. 4, pp. 377-388.

Katsanis, T. (1967), "A Computer Program for Calculating Velocities and Streamlines for Two-Dimensional Incompressible Flow in Axial Blade Rows," NASA TN D-3762, January, 1967.

Kawaguti, M. (1953), "Numerical Solution of the Navier-Stokes Equations for the Flow Around a Circular Cylinder at Reynolds Number 40," J. Physical Society of Japan, Vol. 8, pp. 747-757.

Kawaguti, M. (1961), "Numerical Solution of the Navier-Stokes Equations for the Flow in a Two-Dimensional Cavity," J. Physical Society of Japan, Vol. 16, pp. 2307-2315.

Kawaguti, M. (1965), "Numerical Solutions of the Navier-Stokes Equations for the Flow in a Channel with a Step," MRC TSR 574, Mathematics Research Center, Madison, Wisconsin.

Keast, P., and Mitchell, A.R. (1967), "Finite Difference Solution of the Third Boundary Problem in Elliptic and Parabolic Equations," Numerische Mathematik, Vol. 10, pp. 67-75.

Keeping, A.J. (1968), "The Numerical Solution of a Third Order Partial Differential Equation," Ph.D. Thesis, Stevens Inst. of Tech., Hoboken, Jew Jersey.

Keller, H.B. (1958), "On Some Iterative Methods for Solving Elliptic Difference Equations," Quarterly of Applied Mathematics, Vol. 16, No. 3, pp. 209-226.

Keller, H.B. (1961), "On the Solution of Semi-Linear Hyperbolic Systems by Unconditionally Stable Difference Methods," Communications of Pure and Applied Mathematics, Vol. 14, pp. 447-456.

Keller, H.B. (1968), Numerical Methods for Two-Point Boundary-Value Problems, Ginn-Blaisdell Publishing Co., Waltham, Massachusetts.

Keller, H.B. (1969), "Accurate Difference Methods for Linear Ordinary Differential Systems Subject to Linear Constraints," SIAM J. Numerical Analysis, Vol. 6, No. 1, pp. 8-30.

Keller, H.B., and Cebeci, T. (1971A), "Accurate Numerical Methods for Boundary Layer Flows - II. Two-Dimensional Turbulent Flows," AIAA Paper No. 71-164, AIAA 9th Aerospace Sciences Meeting, New York, New York, January 25-27, 1971.

Keller, H.B., and Cebeci, T. (1971B), "Accurate Numerical Methods for Boundary Layer Flows - I. Two Dimensional Laminar Flows." See Holt (1971).

Kellog, R.B. (1969), "A Nonlinear Alternating Direction Method," Mathematics of Computation, Vol. 23, No. 105, pp. 23-27.

Kendall, R.M., Bartlett, E.P., Rindal, R.A., and Moyer, C.B. (1966), "An Analysis of the Coupled Chemically Reacting Boundary Layer and Charring Ablator, Part I, Summery Report," NASA CR-1060, National Aeronautics and Space Administration.

Kennedy, E.C. (1956), "Calculation of the Flow Fields Around a Series of Bi-Conic Bodies of Revolution Using the Method of Characteristics as Applied to Supersonic Rotational Flow," OAL-CM-873, Ordnance Aerophysics Laboratory Daingerfield, Texas, June 7, 1956.

Kentzer, C.P. (1967), "Stability of Finite Difference Methods for Time Dependent Navier-Stokes Equations," Fluid Dynamics Transactions, Vol. 4, pp. 45-51.

Kentzer, C.P. (1970A) "Computations of Time Dependent Flows on an Infinite Domain," AIAA Paper No. 70-45, AIAA 8th Aerospace Sciences Meeting, New York, New York, January 19-21, 1970.

Kentzer, C.P. (1970B) "Transonic Flows Past a Circular Cylinder," J. of Computational Physics, Vol. 6, pp. 168-182.

Kentzer, C.P. (1971), "Discretization of Boundary Conditions on Moving Discontinuities." See Holt (1971).

Kerr, A.D., and Alexander, H. (1968), "An Application of the Extended Kantorovich Method to the Stress Analysis of a Clamped Rectangular Plate," Acta Mechanica, Vol. 6, pp. 180-196.

Kerr, A.D. (1968), "An Extension of the Kantorovich Method," Quarterly of Applied Mathematics, pp. 219-229.

Kerr, A.D. (1969), "An Extended Kantorovich Method for the Solution of Eigenvalue Problems," International J. Solids Structures, Vol. 5, pp. 559-572.

Kerr, A.D. (1970), "A New Iterative Scheme for the Solution of Partial Differential Equations," NYU-AA-70-10, New York University, University Heights, New York, New York.

Kessler, T.J. (1968), "Numerical Experiments of Plane Shock Diffraction from Two-Dimensional Rectangular Obstacles," MM 68-5425-24, Bell Telephone Labs, Whippany, New Jersey.

Killeen, D.B. (1966), "The Numerical Solution of Equations Describing an Unsteady Draining Process," Ph.D. Thesis, Tulane University, New Orleans, Louisiana.

Kirwan, A.D., Jr. (1968), "Constitutive Equations for a Fluid Containing Nonrigid Structures," The Physics of Fluids, Vol. 11, No. 7, pp. 1440-1446.

Klein, D.D. (1967), "An Investigation of the Accuracy of Numerical Solutions of the Two-Dimensional Diffusion Equation," M.S. Thesis, Air Force Institute of Technology Air University, Wright-Patterson Air Force Base, Ohio.

Kline, S.J., Morkovin, M.V., Sovran, G., and Cockrell, D.S. (1968), Computation of Turbulent Boundary Layers, AFOSR-IFP-Stanford Conferences, Vol. 1 and II.

Klinger, A. (1970), "Generating Functions, Difference-Differential and Partial-Differential Equations," IEEE Transactions on Education, pp. 46-48.

Kofoed-Hansen, O. (1968), "Error Indicators for the Numerical Solution of Non-Linear Wave Equations," Riso Rept. No. 177, Danish Atomic Energy Commission Research Establishment Riso, Roskilde, Denmark.

Konstantinov, G.A. (1970), "Numerical Method for Solving Non-Stationary Axisymmetric Problems of Hydrodynamics of an Ideal Fluid with Free Surfaces," NASA-TT-F-12801, February, 1970.

Kopal, Z. (1969), "The Roche Coordinates and Their Use in Hydrodynamics of Celestial Mechanics," Astrophysics and Space Science, Vol. 5, pp. 360-384.

Koss, W.J. (1969), "Note on the Accumulated Error in the Numerical Integration of a Simple Forecast Model," Monthly Weather Review, Vol. 97, No. 12, p. 896.

Kosykh, A.P., and Minailos, A.N. (1970), "Explicit Schemes of the Method of Ascertainment in the Problem of Supersonic Flow Past a Blunted Body, Zh. Vychislitel'noi Matematiki i Matematicheskoi Fiziki, Vol. 10, pp. 514-520.

Kramer, H. (1969), "Some Analytical and Numerical Calculations for a Cylinder-Vortex Combination in Incompressible Flow," NLR-TR-69057-U, National Aerospace Laboratory, July, 1969.

Krause, E. (1969), "Comment on 'Solution of a Three-Dimensional Boundary-Layer Flow with Separation'," AIAA J., Vol. 7, No. 3, pp. 575-576.

Krause, E., and Hirschel, E.H. (1971), "Exact Numerical Solutions for Three-Dimensional Boundary Layers." See Holt (1971).

Kreiss, H.O. (1964), "On Difference Approximations of the Dissipative Type for Hyperbolic Differential Equations," Communications on Pure and Applied Mathematics, Vol. 17, pp. 335-353.

Kreiss, H.O., and Lundquist, E. (1968), "On Difference Approximations with Wrong Boundary Values," Mathematics of Computation, Vol. 22, pp. 1-12.

Kreiss, H.O. (1968), "Stability Theory for Difference Approximations of Mixed Initial Boundary Value Problems - I," Mathematics of Computation, Vol. 22, No. 104, pp. 703-714.

Krupp, J.A., and Murman, E.M. (1971), "The Numerical Calculation of Steady Transonic Flows Past Thin Lifting Airfoils and Slender Bodies," AIAA Paper No. 71-566, AIAA 4th Fluid and Plasma Dynamics Conference, Palo Alto, California, June 21-23, 1971.

Kurihara, Y. (1965), "On the Use of Implicit and Iterative Techniques for the Time Integration of the Wave Equation," U.S. Weather Bureau Monthly Weather Review, Vol. 93, pp. 33-46.

Kurzrock, J.W. (1966), "Exact Numerical Solutions of the Time-Dependent Compressible Navier-Stokes Equations," Ph.D. Thesis, CAL Rept. No. AG-2026-W-1, Cornell University, Ithaca, New York.

Kurzrock, J.W., and Mates, R.E. (9166), "Exact Numerical Solutions of the Time-Dependent Compressible Navier-Stokes Equations," AIAA Paper No. 66-30.

Kushner, H.J. (1969), "Probability Limit Theorems and the Convergence of Finite Difference Approximations of Partial Differential Equations," NASA CR-107124, October, 1969.

Kusic, G.L., and Lavi, A. (1968), "Stability of Difference Methods for Initial-Value Type Partial Differential Equations," J. of Computational Physics, Vol. 3, pp. 358-378.

Kusic, G. (1969), "On Stability of Numerical Methods for Systems of Initial-Value Partial Differential Equations," J. of Computational Physics, Vol. 4, pp. 272-275.

Kuskova, T.V. (1968), "Difference Method for Calculating the Flow of a Viscous Imcompressible Fluid," Sb. Rabot Vyshisl. Tsentra Most. Un-ta, Vol. 7, pp. 16-27.

Kutler, P. (1969), "Application of Selected Finite-Difference Techniques to the Solution of Conical Flow Problems," Ph.D. Thesis, Iowa State University, Iowa City, Iowa.

Kutler, P., and Lomax, H. (1971), "A Systematic Development of the Supersonic Flow Fields Over and Behind Wings and Wing-Body Configurations Using a Shock-Capturing Finite-Difference Approach," AIAA Paper No. 71-99, AIAA 9th Aerospace Sciences Meeting, New York, New York, January 25-27, 1971.

Kutler, P., and Lomax, H. (1971B), "The Computation of Supersonic Flow Fields about Wing-Body Combinations by 'Shock Capturing' Finite Difference Techniques." See Holt (1971).

Kuznetsov, B.G. (1970), "Numerical Methods for Solving Some Problems of Viscous Liquid," Fluid Dynamics Transactions, Vol. 4, pp. 85-89.

Kyriss, C.L. (1970), "A Time Dependent Solution for the Blunt Body Flow of a Chemically Reacting Gas Mixture," AIAA Paper No. 70-711, AIAA 3rd Fluid and Plasma Dynamics Conference, Los Angeles, California, June 29-July 1, 1970.

Laasonen, P. (1949), "Über eine Methode zur Lösung der Wärmeleitungsgleichung," Acta Mathematica, Vol. 81, p. 309.

Laasonen, P. (1958A), "On the Truncation Error of Discrete Approximations to the Solutions of Dirichlet Problems in a Domain with Corners," J. of the Association for Computing Machinery, Vol. 5, pp. 32-38.

Laasonen, P. (1958B), "On the Solution of Poisson's Difference Equation," J. of the Association for Computing Machinery, Vol. 5, pp. 370-382.

Ladyzhenskaya, O.A. (1963), The Mathematical Theory of Viscous Incompressible Flow, Gordon and Breach, New York, 1963.

Lamb, H. (1945), Hydrodynamics, Dover Publications, New York, 1945. 6th ed. Chapter IV.

Lancaster, P. (1970), "Explicit Solutions of Linear Matrix Equations," SIAM Review, Vol. 12, pp. 544-566.

Landshoff, R. (1955), "A Numerical Method for Treating Fluid Flow in the Presence of Shocks," LASL Rept. No. LA-1930, Los Alamos, New Mexico.

Lapidus, A. (1967), "A Detached Shock Calculation by Second-Order Finite Differences," J. of Computational Physics, Vol. 2, pp. 154-177.

Larkin, B.K. (1964), "Some Stable Explicit Difference Approximations to the Diffusion Equation," Mathematics of Computation, Vol. 18, pp. 196-202.

Larkin, B.K. (1966), "Numerical Solution of the Continuity Equation," American Institute of Chemical Engineering Journal, Vol. 12, No. 5, pp. 1027-1028.

Larsen, T. (1969), "Superelliptic and Related Coordinate Systems," Rept. No. 72, Lab. of Electromagnetic Theory, University of Denmark, Lyngby, Denmark, August, 1969.

Lathrop, K.D. (1969), "Spatial Differencing of the Transport Equation: Positivity vs. Accuracy," J. of Computational Physics, Vol. 4, pp. 475-598.

Laufer, J. (1969), "Thoughts on Compressible Turbulent Boundary Layers," Memo RM-5946-PR, Rand Corporation, Santa Monica, California, March, 1969.

Laval, P. (1969), "Pseudo-Viscosity Method and Starting Process in a Nozzle," Recherche Aerospatiale, No. 131, July/August, 1969, pp. 3-16.

Laval, P. (1969), "Pseudo-Viscosity Method and Starting Process in a Nozzle," NASA-TT-F-12,863; Aztec School of Languages, Inc., Maynard, Massachusetts, Research Translation Division.

Laval, P. (1971), "Time-Dependent Calculation Method for Transonic Nozzle Flows." See Holt (1971).

Lavan, Z., Nielsen, H., and Fejer, A.A. (1969), "Separation and Flow Reversal in Swirling Flows in Circular Ducts," Physics of Fluids, Vol. 12, No. 9, pp. 1747-1757.

Lax, P.D. (1954), "Weak Solutions of Nonlinear Hyperbolic Equations and Their Numerical Computation," Communication on Pure and Applied Mathematics, Vol. 7, pp. 159-193.

Lax, P.D., and Richtmyer, R.D. (1956), "Survey of the Stability of Linear Finite Difference Equations," Communications on Pure and Applied Mathematics, Vol. 9, pp. 267-293.

Lax, P.D. (1957), "Hyperbolic Systems of Conservation Laws II," Communications on Pure and Applied Mathematics, Vol. 10, pp. 537-566.

Lax, P.D. (1958), "Differential Equations, Difference Equations, and Matrix Theory," Communications on Pure and Applied Mathematics, Vol. 11, pp. 175-194.

Lax, P.D., and Wendroff, B. (1960), "Systems of Conservation Laws," Communications on Pure and Applied Mathematics, Vol. 13, pp. 217-237.

Lax, P.D.,(1961) "On the Stability of Difference Approximations to Solutions of Hyperbolic Equations with Variable Coefficients," Communications on Pure and Applied Mathematics, Vol. 14, pp. 497-520.

Lax, P.D., and Wendroff, B. (1964), "Difference Schemes with High Order of Accuracy for Solving Hyperbolic Equations," Communications on Pure and Applied Mathematics, Vol. 17, p. 381.

Lax, P.D., and Nirenberg, L. (1966), "On Stability for Difference Schemes, a Sharp Form of Gårding's Inequality," Communications on Pure and Applied Mathematics, Vol. 19, No. 4, pp. 473-492.

Lax, P.D. (1967), "Hyperbolic Difference Equations: A Review of the Courant-Friedrichs-Lewy Paper in the Light of Recent Developments," IBM J., March, 1967, pp. 235-238.

Lax, P.D. (1969), "Nonlinear Partial Differential Equations and Computing," SIAM Review, Vol. 11, No. 1, January, 1969, pp. 7-19.

Leal, L.G. (1969), "A Study of Steady, Closed Streamline Flows at Large Reynolds Number," Ph.D. Dissertation, Stanford University, Stanford, California.

LeBail, R. (1969), "Fast Reversal Mapping for the Solution of Two-Dimensional Operators," SUIPR Report No. 314, Institute for Plasma Research, Stanford University, Stanford, California, June, 1969.

LeBail, R. (1972), "Use of Fast Fourier Transforms for Solving Partial Differential Equations in Physics," J. Computational Physics, Vol. 9, No. 3, June, 1972, pp. 440-465.

Leblanc, L.L. (1967), "A Numerical Experiment in Predicting Stratus Clouds," Ph.D. Dissertation, Texas A & M University, College Station, Texas.

Lee, C.C., and Inman, S.J. (1964), "Numerical Analysis of Plug Nozzles by the Method of Characteristics," Brown Engineering Tech. Note R-101, May, 1964.

Lee, C.C. (1966), "Gasdynamic Structure of Jets from Plug Nozzle," AIAA J., Vol. 4, No. 6, pp. 1114-1115.

Lee, E.S., and Fan, L.T. (1968), "Quasilinearization Technique for Solution of Boundary Layer Equations," The Canadian Journal of Chemical Engineering, Vol. 46, pp. 200-203.

Lee, J.-S., and Fung, U.-C. (1970), "Flow in Locally Constricted Tubes at Low Reynolds Numbers," J. of Applied Mechanics, Vol. 37, Ser. E, No. 1, pp. 9-17.

van Leer, B. (1969), "Stabilization of Difference Schemes for the Equations of Inviscid Compressible Flow by Artificial Diffusion," J. of Computational Physics, Vol. 3, No. 4, pp. 473-485.

Leith, C. (1964), "Numerical Simulation of the Earth's Atmosphere," University of California, Livermore, California, UCRL 7986-T, 1964.

Leith, C.E. (1965), "Numerical Simulation of the Earth's Atmosphere," Methods in Computational Physics, Vol. 4, pp. 1-28.

Leith, C.E. (1969), "Diffusion Approximation to Spectral Transfer in Homogenous Turbulence." See Frenkiel and Stewartson (1969).

Leith, C.E. (1971), "Two-Dimensional Turbulence and Atmospheric Predictability." See Holt (1971).

Lemmon, E.C., and Heaton, H.S. (1969), "Accuracy, Stability, and Oscillation Characteristics of Finite Element Method for Solving Heat Conduction Equation," ASME Paper 69-WA/HT-35.

Leslie, L.M. (1971), "The Development of Concentrated Vortices: A Numerical Study," J. Fluid Mechanics, Vol. 48, Part 1, pp. 1-21.

Lester, W.G.S. (1961), Aerospace Research Committee Report Mem. (Gt. Bt.) No. 3240.

Levine, J.N. (1968), "Finite Difference Solution of the Laminar Boundary Layer Equations Including Second-Order Effects," AIAA Paper No. 68-739, AIAA Fluid and Plasma Dynamics Conference, Los Angeles, California, June 24-26, 1968.

Lewis, C.H. (1970A), "Numerical Methods for Nonreacting and Chemically Reacting Laminar Flows - Tests and Comparisons," AIAA Paper No. 70-808, AIAA 3rd Fluid and Plasma Dynamics Conference, Los Angeles, California, June 29-July 1, 1970.

Lewis, C.H. (1970B), "Nonreacting and Chemically Reacting Viscous Flows over a Hyperboloid at Hypersonic Condition," Advisory Group for Aerospace Research and Development, Paris, France.

Lewis, C.H. (1971), "Numerical Methods for Nonreacting and Chemically Reacting Laminar Flows - Tests and Comparisons," J. Spacecraft, Vol. 8, No. 2, pp. 117-122.

Lewis, C.H., Moore, F.G., and Black, R. (1971A), "Sharp and Blunt Cones at Angle of Attack in Supersonic Nonuniform Free Streams," AIAA Paper No. 71051, AIAA 9th Aerospace Sciences Meeting, New York, New York, January 25-27, 1971.

Lewis, C.H., Anderson, E.C., and Miner, E.W. (1971B), "Nonreacting and Equilibrium Chemically Reacting Turbulent Boundary-Layer Flows," AIAA Paper No. 71-597, AIAA 4th Fluid and Plasma Dynamics Conference, Palo Alto, California, June 21-23, 1971.

Li, C.P. (1971), "Time Dependent Solutions of Nonequilibrium Airflow Past a Blunt Body," AIAA Paper No. 71-595, AIAA 4th Fluid and Plasma Dynamics Conference, Palo Alto, California, June 21-23, 1971.

Lick, D.W., and Tunstall, J.N. (1968), "A Study of Non-Linear Dirichlet Problems," Rept. No. K-1737, Union Carbide Corporation, Nuclear Division, Oak Ridge Gaseous Diffusion Plant, Oak Ridge, Tennessee, February 27, 1968.

Lick, D.W. (1969), "A Divergence Theorem for a Nonlinear Dirichlet Problem," J. of Computational Physics, Vol. 4, pp. 142-143.

Liepman, H. (1918), "Die Angenaherte Ermittebung Harmonischer Functionen and konformer Abbildungen," Bayer. Akad. Wiss., Math.-phys. Klasse, Sitz.

Liepman, H.W., and Roshko, A. (1957), Elements of Gas Dynamics, Wiley.

Lighthill, M.J. (1954), "The Response of Laminar Skin Friction and Heat Transfer to Fluctuations in the Stream Velocity," Proceedings Royal Society of London, Series A., Vol. 224, pp. 1-23.

Lilly, D.K. (1965), "On the Computational Stability of Numerical Solutions of Time-Dependent Non-Linear Geophysical Fluid Dynamics Problems," U.S. Weather Bureau Monthly Weather Review, Vol. 93, No. 1, pp. 11-26.

Lilly, D.K. (1966), "Theoretical Models of Convection Elements and Ensembles," Advances in Numerical Weather Prediction, 1965-66, Seminar Series, Travelers Research Center, Inc., Hartford, Connecticut, pp. 24-33.

Lilly, D.K. (1969), "Numerical Simulation of Two-Dimensional Turbulence." See Frenkiel and Stewartson (1969).

Lilly, D.K. (1971), "Numerical Simulation of Developing and Decaying Two-Dimensional Turbulence," J. Fluid Mechanics, Vol. 45, Part 2, pp. 395-415.

Lipnitskii, Iu. M., and Lifshits, Tu. B. (1970), "On the Analysis of Transonic Flow Around Bodies of Revolution," Prikl. Matem. i Mekh., Vol. 34, No. 3, pp. 508-513.

Lipps, F.B., and Somerville, R.C.J. (1971), "Dynamics of Variable Wavelength in Finite-Amplitude Benard Convection," The Physics of Fluids, Vol. 14, No. 4, pp. 759-765.

Liu, C.Y., Hopper, A.T., and Hulbert, L.E. (1969), "Transient Heat Conduction with Time Dependent Boundary Conditions," Batelle Memorial Institute.

Lo, C.-F. (1969), "Numerical Solutions of the Unsteady Heat Equation," AIAA J., Vol. 7, No. 5, pp. 973-975.

Loc, T.P. (1970), "Transfert Thermique Instationnaire En Ecoulement Laminaire A L'Entree Des Tubes Circulaires," International J. Heat and Mass Transfer, Vol. 13, pp. 1767-1778.

Lock, R.C. (1970), "Test Cases for Numerical Methods in Two-Dimensional Transonic Flows," AGARD R-575-70, National Physical Laboratory, Teddington, England.

Loer, S. (1969), "Examination of the Stability of Disturbed Boundary-Layer-Flow by a Numerical Method." See Frenkiel and Stewartson (1969).

Lomax, H., Railev, H.E., and Ruller, F.B. (1969), "On Some Numerical Difficulties in Integrating the Equations for One-Dimensional Nonequilibrium Nozzle Flow," NASA TN D-5176, National Aeronautics and Space Administration, Ames Research Center, Moffett Field, California.

Lomax, H., Kutler, P., and Fuller, F.B. (1970), "The Numerical Solution of Partial Differential Equations Governing Convection," AGARD-AG-146-70, Advisory Group for Aerospace Research and Development, Paris, France.

Longley, H.J. (1960), "Methods of Differencing in Eulerian Hydrodynamics," LASL Rept. No. LAMS-2379, Los Alamos Scientific Lab, Los Alamos, New Mexico.

Lorenz, E.N. (1963), "Deterministic Nonperiodic Flow," J. of the Atmospheric Sciences, Vol. 20, No. 2, March, 1963, pp. 130-141.

Lu, Y.-P. (1967), "Finite Difference Solutions of the Acoustic Radiation Equation in the Near Field," Ph.D. Thesis, University of Houston, Houston, Texas.

Lucey, J.W., and Housen, K.F. (1964), "A Stable Method of Matrix Factorization," Trans. Amer. Nuclear Soc., Vol. 7, p. 259.

Luckinbill, D.L., and Childs, B. (1968), "Inverse Problems in Partial Differential Equations," Rept. RE 1-68, Houston University, Houston, Texas.

Ludford, G., Polachek, H., and Seeger, R.J. (1953), "On Unsteady Flow of Compressible Viscous Fluids," J. of Applied Physics, Vol. 24, pp. 490-495.

Ludloff, H.F., and Friedman, M.B. (1954), "Aerodynamics of Blasts - Diffraction of Blast Around Finite Corners," J. of the Aeronautical Sciences, January, 1955, pp. 27-34.

Lugt, H.J., and Rimon, Y. (1970), "Finite-Difference Approximations of the Vorticity of Laminar Flows at Solid Surfaces," Naval Ship Research and Development Center Rept. 3306, Washington, D.C., April, 1970.

Lugt, H.J., and Haussling, H.J. (1971), "Laminar Flows Past a Flat Plate at Various Angles of Attack." See Holt (1971).

Lynch, R.E., and Rice, J.R. (1968), "Convergence Rates of ADI Methods with Smooth Initial Error," Mathematics of Computation, Vol. 22, No. 102, pp. 311-335.

Lyness, J.N. (1970), "The Calculation of Fourier Coefficients by the Möbius Inversion of the Poisson Summation Formula, Part 1. Functions Whose Early Derivatives are Continuous," Mathematics of Computation, Vol. 24, No. 109, pp. 101-135.

Lysen, J.C. (1964), "Variable Mesh Difference Equation for the Stream Function in Axially Symmetric Flow," AIAA J., Vol. 2, No. 1, pp. 163-165.

Macagno, E.O. (1965), "Some New Aspects of Similarity in Hydraulics," La Houille Blanche, Vol. 20, No. 8, pp. 751-759.

Macagno, E.O., and Hung, T.K. (1967), "Pressure, Bernoulli Sum, and Momentum and Energy Relations in a Laminar Zone of Separation," Physics of Fluids, Vol. 10, No. 1, pp. 78-82.

MacCormack, R.W. (1969), "The Effect of Viscosity in Hypervelocity Impact Cratering," AIAA Paper No. 69-354.

MacCormack, R.W. (1971), "Numerical Solution of the Interaction of a Shock Wave with a Laminar Boundary Layer." See Holt (1971).

MacCormack, R.W., and Paullay, A.J. (1972), "Computational Efficiency Achieved by Time Splitting of Finite Difference Operators," AIAA Paper, January, 1972.

MacPherson, A.K. (1971), "The Formation of Shock Waves in a Dense Gas Using a Molecular-Dynamics Type Technique," J. Fluid Mechanics, Vol. 45, Part 3, pp. 601-621.

Mader, C.L. (1964), "The Two Dimensional Hydrodynamic Hot Spot," LASL Rept. LA-3077, Los Alamos Scientific Laboratory, Los Alamos, New Mexico.

Magnus, R., and Yoshihara, H. (1970), "Inviscid Transonic Flow over Airfoils," AIAA J., Vol. 8, No. 12, pp. 2157-2162.

Magomedov, K.M. (1966), "Method of Characteristics for Numerical Calculation of Spatial Flows of Gas," FTD-MT-66-131, Foreign Technology Division; also, Zhurnal Vychislitel'noy Matematiki i Matematicheskoy Fiziki, Vol. 6, No. 2, pp. 313-325.

Magomedov, K.M., and Kholodov, A.S. (1967), "Three-Dimensional Supersonic Flow Past a Delta Wing with Blunted Edges," An SSSR. Izvestiya, Mekhanika Zhidkosti i Gaza.

Magomedov, K.M., and Kholodov, A.S. (1967), "Supersonic Three-Dimensional Flow Around a Delta Wing with Blunted Leading Edges," Izv. AN SSSR. Mekhanika Zhidkosti i Gaza, Vol. 2, No. 4, pp. 159-163.

Makhin, N.A., and Syagaev, V.F. (1966), "Numerical Solution of the Problem of Supersonic Flow Past Conical Bodies at an Angle of Attack," Mekhanika Zhidkosti i Gaza, January-February, 1966, pp. 100-101.

Mancuso, R.L. (1967), "A Numerical Procedure for Computing Fields of Stream Function and Velocity Potential," J. of Applied Meteorology, Vol. 6, pp. 994-1001.

Mann, W.R., Bradshaw, C.L., and Cox, J.G. (1957), "Improved Approximations to Differential Equations by Difference Equations," J. Mathematics and Physics, Vol. 35, No. 4, pp. 408-415.

Marchuk, G.I. (1965), "A Theoretical Weather-Forecasting Model," Doklady of the Academy of Sciences of the U.S.S.R., Vol. 155, Nos. 1-6, pp. 10-12.

Marris, A.W., and Passman, S.L. (1968), "Generalized Circulation-Preserving Flows," Archive for Rational Mechanics and Analysis, Vol. 28, pp. 245-265.

Martelluci, A., Rie, H., and Sontowski, J.F. (1969), "Evaluation of Several Eddy Viscosity Models Through Comparison with Measurements in Hypersonic Flows," AIAA Paper No. 69-688, AIAA Fluid and Plasma Dynamics Conference, San Francisco, California, June 16-18, 1969.

Martin, D.W., and Tee, G.J. (1961), "Iterative Methods for Linear Equations with Symmetric Positive Definite Matrix," Computer J., Vol. 4, pp. 242-254.

Mascheck, H.-J. (1968), "Comments on the Numerical Integration of the Equations of Motion for Incompressible Flows at Large Reynolds Numbers," Wissenschaftliche Zeitschrift, Vol. 16, No. 4, pp. 1227-1228.

Masliyah, J.H., and Epstein, N. (1970), "Numerical Study of Steady Flow Past Spheroids," J. Fluid Mechanics, Vol. 44, Part 3, pp. 493-512.

Mason, D.S., and Thorne, B.J. (1970), "A Preliminary Report Describing the Rezoning Features of the WONDY IV Program," SC-DR-70-146, Sandia Laboratories, Albuquerque, New Mexico, March, 1970.

Masson, B.A., Taylor, T.D., and Foster, R.M. (1969), "Application of Godunov's Method to Blunt-Body Calculations," AIAA J., Vol. 7, No. 4, pp. 694-698.

Masson, B.S. (1968), "A Simplified Angle of Attack Program for Transient Blunt Body Flow Fields," AMC 68-36, Picatinny Arsenal, Dover, New Jersey.

Matin, S.A. (1968), "Numerical Iteration of an Elliptic Mixed Boundary-Value Problem in a Region with Curved Boundary," J. of Computational Physics, Vol. 3, pp. 327-330.

Matthes, W. (1970), "Simulation of Relaxation Processes by Monte Carlo," J. of Computational Physics, Vol. 6, pp. 157-167.

Matthews, C.W. (1969), "A Numerical Analysis of the Use of Perforated Walls to Control Shock Location and Movement in an Internal Compression Supersonic Inlet," NASA TR-R-317, NASA Langley Research Center, Langley Station, Virginia.

McCormick, W.T., Jr., and Hansen, K.F. (1969), "Numerical Solution of the Two-Dimensional Time-Dependent Multigroup Equations," MIT 3903-1, Dept. of Nuclear Engineering, Massachusetts Institute of Technology, Cambridge, Massachusetts.

McCreary, J. (1967), "The Mathematical Calculation of the Dynamic Behavior of a Three-Dimensional Atmospheric Convection," Ph.D. Thesis, University of Kansas, Lawrence, Kansas.

McDonald, P.W. (1971), "The Computation of Transonic Flow Through Two-Dimensional Gas Turbine Cascades," Paper 71-GT-89, American Society of Mechanical Engineers, Gas Turbine Conference and Products Show, Houston, Texas, March 28-April 1, 1971.

McDowell, L.K. (1967) "Variable Successive Over-Relaxation," Ph.D. Thesis, University of Illinois, Urbana, Illinois.

McGlaughlin, D.W., and Greber, I. (1967), "Experiments on the Separation of a Fluid Jet from a Curved Surface," in Advances in Fluidics; Fluidics Symposium, Chicago, Illinois, May 9-11, 1967, (ASME), pp. 14-30.

McGrath, F.J. (1967), "Nonstationary Plane Flow of Viscous and Ideal Fluids," Archive for Rational Mechanics and Analysis, Vol. 27, pp. 329-348.

McKee, S., and Mitchell, A.R. (1970), "Alternating Direction Methods for Parabolic Equations in Two Space Dimensions with a Mixed Derivative," The Computer J., Vol. 13, No. 1, pp. 81-86.

McNamera, W. (1966), "Axisymmetric Interaction of a Blast Wave with the Shock Layer of a High-Speed Blunt Body," MIT Aeroelastic and Structures Research Laboratory, ASRL TR 121-15.

McNamera, W. (1967), "FLAME Computer Code for the Axisymmetric Interaction of a Blast Wave with a Shock Layer on a Blunt Body," J. Spacecraft and Rockets, Vol. 4, pp. 790-795.

Mehta, U.B., and Lavan, Z. (1968), "Flow in a Two-Dimensional Channel with a Rectangular Cavity," NASA-CR-1245.

Mehta, U.B., and Lavan, Z. (1969), "Flow in a Two-Dimensional Channel with a Rectangular Cavity," J. of Applied Mechanics, Transactions of the ASME Series E, Vol. 36, Part 4, pp. 897-901.

Meladze, V. (1970), "Schemes of Higher Order Accuracy for Systems of Elliptic and Parabolic Equations," Zhurnal Vychislitel'noi Matematiki i Matematicheskoi Fiziki, Vol. 10, pp. 482-490.

Melnik, R.E., and Ives, D.C. (1971), "Subcritical Flows Over Two Dimensional Airfoils by a Multistrip Method of Integral Relations." See Holt (1971).

Meyer, K.A. (1969), "A Three-Dimensional Study of Flow between Concentric Rotating Cylinders," LASL Rept. LA 4202, Los Alamos Scientific Laboratory, Los Alamos, New Mexico.

Michael, P. (1966), "Steady Motion of a Disk in a Viscous Fluid," Physics of Fluids, Vol. 9, No. 3, pp. 466-471.

Migdal, D., Klein, K., and Moretti, G. (1969), "Time-Dependent Calculations for Transonic Nozzle Flow," AIAA J., Vol. 7, No. 2, pp. 372-374.

Miller, R.H. (1967), "An Experimental Method for Testing Numerical Stability in Initial-Value Problems," J. of Computational Physics, Vol. 2, pp. 1-7.

Mintz, Y. (1965), "Very Long-Term Global Integration of the Primitive Equations of Atmospheric Motion," WMO-IUGG Symposium on Research Development, Aspects of Long-Range Forecasting, Boulder, Colorado, 1964, WMO Tech. Note No. 66, p. 141.

Mitchell, A.R. (1969), Computational Methods in Partial Differential Equations, J. Wiley and Sons, Ltd.

Mitchell, T.M. (1970), "Numerical Studies of Asymmetric and Thermodynamic Effects on Cavitation Bubble Collapse," Rept. UMICH 03371-5-T, Mech. Eng. Dept., Michigan University, Ann Arbor, Michigan.

Miyakoda, K. (1962), "Contribution to the Numerical Weather Prediction - Computation with Finite Difference," Japanese J. of Geophysics, Vol. 3, pp. 75-190.

Moiseenko, B.D., and Rozhdestvenskii, B.L. (1970), "Numerical Solution of the Stationary Equations of Hydrodynamics in the Presence of Tangential Discontinuities," Zhurnal Vychislitel'noi Matematiki i Matematicheskoi Fiziki, Vol. 10, pp. 499-505.

Moiseenko, B.D., and Rozhdestvenskii, B.L. (1971), "The Calculation of Hydrodynamic Forces with Tangential Discontinuities." See Holt (1971).

Molenkamp, C.R. (1968), "Accuracy of Finite Difference Methods Applied to the Advection Equation," J. of Applied Meteorology, Vol. 7, pp. 160-167.

Moore, F.G., and DeJarnette, F.R. (1971), "Viscous Flow-Field Calculations on Pointed Bodies at Angle of Attack in Nonuniform Freestreams," AIAA Paper No. 71-624, AIAA 4th Fluid and Plasma Dynamics Conference, Palo Alto, California, June 21-23, 1971.

Moreno, J.B. (1967), "Inverse-Method and Method-of-Characteristics Flow-Field Programs," SC-DR-67-652, Sandia Laboratories, Albuquerque, New Mexico.

Moretti, G., and Abbett, M. (1966A), "Numerical Studies of Base Flow," GASL Tech. Rept. 584, March 7, 1966.

Moretti, G., and Abbett, M. (1966B), "A Time-Dependent Computational Method for Blunt Body Flows," AIAA J., Vol. 4, No. 12, pp. 2136-2141.

Moretti, G., and Bleich, G. (1967), "Three-Dimensional Flow Around Blunt Bodies," AIAA Reprint 67-222.

Moretti, G., and Bleich, G. (1968), "Three-Dimensional Inviscid Flow About Supersonic Blunt Cones at Angle of Attack," SC-RR-68-3728, Sandia Laboratories, Albuquerque, New Mexico, 87115, September, 1968.

Moretti, G. (1968A), "Inviscid Blunt Body Shock Layers - Two-Dimensional Symmetric and Axisymmetric Flows," Brooklyn Polytechnic Institute, PIBAL Rept. No. 68-15.

Moretti, G. (1968B), "The Importance of Boundary Conditions in the Numerical Treatment of Hyperbolic Equations," Polytechnic Institute of Brooklyn, PIBAL Rept. No. 68-34, November, 1968. See also Frenkiel and Stewartson (1969).

Moretti, G.,(1969A), "A Critical Analysis of Numerical Techniques: The Piston-Driven Inviscid Flow," Polytechnic Institute of Brooklyn, PIBAL Report No. 69-25, July, 1969.

Moretti, G. (1969B), "The Choice of a Time-Dependent Technique in Gas Dynamics," PIBAL Report No. 69-26, Polytechnic Institute of Brooklyn, Department of Aerospace Engineering and Applied Mechanics, Farmingdale, New York, July, 1969.

Moretti, G., and Salas, M.D. (1969), "The Blunt Body Problem for a Viscous Rarefied Gas Flow," AIAA Paper No. 69-139, AIAA 7th Aerospace Sciences Meeting, New York City, New York, January 20-22, 1969.

Moretti, G., and Salas, M.D. (1970), "Numerical Analysis of Viscous One-Dimensional Flows," J. of Computational Physics, Vol. 5, pp. 487-506.

Moretti, G. (1971), "Initial Conditions and Imbedded Shocks in the Numerical Analysis of Transonic Flows." See Holt (1971).

Morris, J.L. (1971), "On the Numerical Solution of a Heat Equation Associated with a Thermal Print Head. II," J. of Computational Physics, Vol. 7, pp. 102-119.

Morse, R.L., and Nielson, C.W. (1971), "Numerical Simulation of the Weibel Instability in One and Two Dimensions," The Physics of Fluids, Vol. 14, No. 1, pp. 830-840.

Motzkin, T.S., and Wasow, W. (1953), "On the Approximation of Linear Elliptic Differential Equations by Difference Equations with Positive Coefficients," J. Mathematics and Physics, Vol. 31, pp. 253-259.

Mouradoglou, A.J. (1967), "Numerical Studies on the Convergence of the Peaceman-Rachford Alternating Direction Implicit Method," Interim Tech. Rept. No. 14, Computation Center, The University of Texas, Austin, Texas, June, 1967.

Mueller, T.J., Hall, C.R., Jr., and Roache, P.J. (1970), "Influence of Initial Flow Direction on the Turbulent Base Pressure in Supersonic Axisymmetric Flow," J. Spacecraft, Vol. 7, No. 12, pp. 1484-1488.

Mueller, T.J., and O'Leary, R.A. (1970), "Physical and Numerical Experiments in Laminar Incompressible Separating and Reattaching Flows," Los Angeles, California, June 29-July 1, 1970.

Mufti, I.H. (1969), "Initial-Value Methods for Two-Point Boundary-Value Problems," National Research Council of Canada, NRC No. 11114, Ottawa, Canada, June, 1969.

Murman, E.M., and Cole, J.D. (1971), "Calculation of Plane Steady Transonic Flows," AIAA J., Vol. 9, No. 1, pp. 114-121.

Murman, E.M., and Krupp, J.A. (1971), "Solution of the Transonic Potential Equation Using a Mixed Finite Difference Scheme." See Holt (1971).

Murphey, W.D. (1963), "Numerical Analysis of Boundary Layer Problems," NYO 1480-63, Courant Institute of Mathematical Sciences, New York University, New York, New York.

Myshenkov, V.I. (1970), "Subsonic and Transonic Flows of a Viscous Gas in the Wake of a Flat Body," Akademiia Nauk SSSR, Izvestiia, Mekhanika Zhidkosti i Gaza, March-April, 1970, pp. 73-79.

Nagel, A.L. (1967), "Compressible Boundary Layer Stability by Time Integration of the Navier-Stokes Equations and An Extension of Emmons' Transition Theory to Hypersonic Flow," Boeing Scientific Research Laboratories Document D1-82-0655, Renton, Washington.

Nakayama, P.I. (1970), "Turbulence Transport Equations and Their Numerical Solution," AIAA Paper No. 70-3, AIAA 8th Aerospace Sciences Meeting, New York, New York, January 19-21, 1970.

Narang, B.S. (1967), "Numerical Treatment of Laminar Free Surface Flows," Ph.D. Thesis, University of Illinois, Urbana, Illinois.

Nash, J.F. (1969), "The Calculation of Three-Dimensional Turbulent Boundary Layers in Incompressible Flow," J. Fluid Mechanics, Vol. 37, Part 4, pp. 625-642.

Ndefo, D.E. (1969), "A Numerical Method for Calculating Steady Unsymmetrical Supersonic Flow Past Cones, Rept. No. AS-69-11, College of Engineering, University of California, Berkeley, California.

Nelson, P., Jr. (1970), "Application of Invariant Imbedding to the Solution of Partial Differential Equations by the Continuous-Space Discrete-Time Method," Americal Federation of Information Processing Societies, Spring Joint Computer Conference, Atlantic City, New Jersey, May 5-7, 1970.

NiCastro, J. (1968), "Similarity Studies on the Radiative Gas Dynamic Equations," Tech. Rept. A-47, Case Inst. of Technology, Cleveland, Ohio, NASA-CR-94409.

Nichols, B.D. (1971), "Recent Extensions to the Marker-and-Cell Method for Incompressible Fluid Flows." See Holt (1971).

Nickel, K.L.E. (1971), "Error Bounds in Boundary Layer Theory." See Holt (1971).

Nigro, B.J., Woodward, R.A., and Brucks, C.R. (1968), "A Digital Computer Program for Deriving Optimum Numerical Integration Techniques for Real-Time Flight Simulation," Aerospace Medical Research Laboratories Rept. AMRL-TR-68-4, Air Force Systems Command, Wright-Patterson Air Force Base, Ohio.

Niu, K., Higuchi, S., and Urashima, S. (1969), "Analysis of Time-Dependent Shock Wave," Japan Society for Aeronautical and Space Sciences, Transactions, Vol. 12, No. 21, pp. 50-54.

Noh, W.F., and Protter, M.H. (1963), "Difference Methods and the Equations of Hydrodynamics," J. of Mathematics and Mechanics, Vol. 12, No. 2, pp. 149-191.

Noh, W.F. (1964), "CEL: A Time-Dependent, Two-Space-Dimensional, Coupled Eulerian-Lagrange Code," Methods in Computational Physics, Vol. 3, pp. 117-180.

Nomura, R., and Deiters, R.M. (1968), "Improving the Analog Simulation of Partial Differential Equations by Hybrid Computation," Simulation, pp. 73-79.

Novack, B.B., and Cheng, H.K. (1971), "Numerical Analysis and Modeling of Slip Flows at Very High Mach Numbers." See Holt (1971).

O'Brien, G.D. (1970), "A Numerical Investigation of Finite-Amplitude Disturbances in a Plane Poiseuille Flow," Ph.D. Dissertation, U.S. Naval Postgraduate School, Monterey, California, June, 1970.

O'Brien, G.G., Hyman, M.A., and Kaplan, S. (1950), "A Study of the Numerical Solution of Partial Differential Equations," J. Mathematics and Physics, Vol. 29, pp. 223-251.

Ogura, M. (1969), "A Direct Solution of Poisson's Equation by Dimension Reduction Method," J. of the Meteorological Society of Japan, Ser. II, Vol. 47, No. 4, pp. 319-323.

Ohman, G.A. (1967), "Numerical Calculation of Steady Convective Mass Transfer Around a Circular Cylinder at Small Reynolds Numbers," Acta Academiae Aboensis, Vol. 27, No. 1.

Ohrenberger, J.T., and Baum, E. (1970), "A Theoretical Model of the Near Wake of a Slender Body in Supersonic Flow," AIAA Paper No. 70-792.

O'Leary, R.A., and Mueller, T.J. (1969), "Correlation of Physical and Numerical Experiments for Incompressible Laminar Separated Flows," Tech. Rept. No. THEMIS-UND-69-4, University of Notre Dame, College of Engineering, Notre Dame, Indiana, July, 1969.

Oleinik, O.A. (1967), "On the Solution of Prandtl Equations by the Method of Finite Differences," PMM, Vol. 31, No. 1, pp. 90-100.

Orszag, S.A. (1969), "Numerical Methods for the Simulation of Turbulence." See Frenkiel and Stewartson (1969).

Orszag, S.A. (1970), "Analytical Theories of Turbulence," J. of Fluid Mechanics, Vol. 41, No. 2, pp. 363-386.

Osher, S. (1969A), "On Systems of Difference Equations with Wrong Boundary Conditions," Mathematics of Computation, Vol. 23, No. 107, pp. 567-572.

Osher, S. (1969B), "Stability of Difference Approximations of Dissipative Type for Mixed Initial-Boundary Value Problems, I," Mathematics of Computation, Vol. 23, No. 106, pp. 335-340.

Owczarek, J.A. (1964), Fundamentals of Gas Dynamics, International Textbook Co., Scranton, Pennsylvania.

Pagnani, B.R. (1968), "An Explicit Finite-Difference Solution for Natural Convection in Air in Rectangular Enclosures," Ph.D. Thesis, Oregon State University, Corvallis, Oregon.

Paller, A.J., and Kaylor, R. (1968), "Numerical Studies of Penetrative Convective Instabilities," Maryland University, Tech. Note DN-554, College Park, Maryland.

Pal'tsev, B.V. (1970), "On the Convergence of Successive Approximations with Splitting of the Boundary Conditions for the Solution of a Boundary Value Problem for the Navier-Stokes Equations," Vychisl. Matem. i Matem. Fiz., Vol. 10, No. 3, pp. 785-788.

Pan, F., and Acrivos, A. (1967), "Steady Flows in Rectangular Cavities," J. Fluid Mechanics, Vol. 23, Part 4, pp. 643-655.

Pao, H.-P. (1970), "A Numerical Computation of a Confined Rotating Flow," J. of Applied Mechanics, July, 1970, pp. 480-487.

Pao, Y.-H. (1969A), "Origin and Structure of Turbulence in Stably Stratified Media," in Clear Air Turbulence and Its Detection, Plenum Press. Also available from Boeing Scientific Research Laboratories, P.O. Box 3981, Seattle Washington, 98124.

Pao, Y.-H., ed. (1969B), Symposium on Turbulence, Abstracts of Papers, Seattle, Washington, June 23-27, 1969.

Pao, Y.-H., and Daugherty, R.J. (1969), "Time-Dependent Viscous Incompressible Flow Past a Finite Flat Plate," Boeing Scientific Research Laboratories, D1-82-0822, January, 1969.

Paris, J., and Whitaker, S. (1965), "Confined Wakes: A Numerical Solution of the Navier-Stokes Equations," American Institute of Chemical Engineers J., Vol. 11, No. 6, pp. 1033-1041.

Parlett, B. (1966), "Accuracy and Dissipation in Difference Schemes," Communications in Pure and Applied Mathematics, Vol. 19, No. 1, pp. 111-123.

Parter, S.V. (1967), "Elliptic Equations," IBM J., March, pp. 244-247.

Passman, S.L. (1970), "Two Theorems in Classical Vorticity Theory," Zeitschrift fur Angewandte Mathematik und Physik, Vol. 21, No. 1, pp. 130-133.

Patankar, S.V., and Spalding, D.B. (1967A), "A Finite-Difference Procedure for Solving the Equations of the Two-Dimensional Boundary Layer," International J. Heat Mass Transfer, Vol. 10, pp. 1389-1411.

Patankar, S.V., and Spalding, D.B. (1967B), Heat and Mass Transfer in Boundary Layers, Morgan-Grampian, London.

Paul, R.J.A., and Ahmad, A.I.S. (1970), "Numerical Solution of 2nd-Order Hyperbolic Partial Differential Equations by the Method of Continuous Characteristics," Institution of Electrical Engineers, Proceedings, Vol. 117, June, 1970, pp. 1166-1174.

Pavlov, B.M. (1968), "Calculation of Supersonic Blunt Body Flow Using Complete Navier-Stokes Equations," Izvestiya Akademii Nauk SSSR, Mekhanika Zhidkosti i Gaza, No. 3, May/June, 1968, pp. 128-133.

Pavlov, B.M. (1969), "Numerical Solution of the Problem of Viscous Supersonic Gas Flow Around Blunt Bodies," Vychislitel'nye Metody if Programmirovanie, No. 11, pp. 32-41.

Payne, R.B. (1958), "Calculation of Unsteady Viscous Flow Past a Circular Cylinder," J. Fluid Mechanics, Vol. 4, p. 81.

Peaceman, D.W., and Rachford, H.H., Jr. (1955), "The Numerical Solution of Parabolic and Elliptic Differential Equations," J. Soc. Indust. Applied Mathematics, Vol. 3, No. 1, March, 1955, pp. 28-41.

Pearce, B.E., and Emery, A.F. (1970), "Heat Transfer by Thermal Radiation and Laminar Forced Convection to an Absorbing Fluid in the Entry Region of a Pipe," J. of Heat Transfer, Vol. 92, Series C, No. 2, pp. 221-230.

Pearson, C.E. (1964), "A Computational Method for Time-Dependent Two-Dimensional Incompressible Viscous Flow Problem," Sperry Rand Research Center Report SRRC-RR-64-17, February, 1964.

Pearson, C.E. (1965A), "A Computational Method for Viscous Flow Problems," J. Fluid Mechanics, Vol. 21, Part 4, pp. 611-622.

Pearson, C.E. (1965B), "Numerical Solutions for the Time-Dependent Viscous Flow Between Two Rotating Coaxial Disks," J. Fluid Mechanics, Vol. 21, Part 4, pp. 623-633.

Pearson, J.T., and Kaplan, B. (1970), "Computer Time Comparison of Point and Block Successive Over Relaxation," AFIT TR 70-6, Air Force Institute of Technology, Wright-Patterson Air Force Base, Ohio.

Pereyra, V. (1969), "Highly Accurate Numerical Solution of Casilinear Elliptic Boundary Value Problems in N Dimensions," MRC Technical Summary Report #1009, Mathematics Research Center, University of Wisconsin, Madison, Wisconsin.

Petschek, A.G., and Hanson, M.E. (1968), "Difference Equations for Two-Dimensional Elastic Flow," J. of Computational Physics, Vol. 3, pp. 307-321.

Phillips, H., and Wiener, N. (1923), "Nets and the Dirichlet Program," J. Mathematics and Physics, Vol. 2, pp. 105-124.

Phillips, N.A. (1959), "An Example of Non-Linear Computational Instability," Atmosphere and Sea in Motion, Rossby Memorial Volume, (B. Bolin, Ed.), Rockefeller Institute Press, New York.

Piacsek, S.A. (1966), "The Axisymmetric Theory of Flow in a Rotating Annulus: Numerical Experiments," Ph.D. Thesis, MIT (1966); also, Scientific Report No. SR6, Rotating Fluids Laboratory, MIT.

Piacsek, S.A., and Hide, R. (1966), "The Axisymmetric Regime of Thermal Convection in a Rotating Annulus of Liquids," I.U.T.A.M. Symposium of Rotating Fluid Systems, LaJolla, California, April, 1966. For proceedings see J. Fluid Mechanics, Vol. 26, 1966, p. 393.

Piacsek, S.A. (1967), "Numerical Experiments on Convective Flows in Geophysical Fluid Systems," DR 148, 7th Symposium Naval Hydrodynamics, Office of Naval Research, Department of the Navy.

Piacsek, S.A. (1968), "Numerical Experiments on Thermal Convection in Geophysical Fluid Systems," Proceedings of the 7th Naval Hydrodynamics Symposium, Rome, Italy, U.S. Government Printing Office, p. 753.

Piacsek, S.A. (1969A), "Penetrative Thermohaline Convection: Some Numerical Experiments," Midwest Geophysical Fluid Dynamics Colloquium, The University of Chicago, Chicago, Illinois, 1969.

Piacsek, S.A. (1969B), "Numerical Experiments on Penetrative Convection in a Rotating Fluid Cooled Uniformly from Above," Symposium on Fluid Mechanics in Natural Phenomena, University of Newcastle-Upon-Tyne, England, April, 1969.

Piacsek, S.A., and Williams, G. (1970), "Conservation Properties of Convection Difference Schemes," J. Computational Physics, Vol. 6, pp. 393-405.

Piacsek, S.A. (1970), "Three Dimensional Numerical Experiments on Penetrative Convection in a Rotating Fluid," Fourth Symposium on Numerical Simulation of Plasmas, Naval Research Laboratory, Washington, D.C., November, 1970.

Pierce, S. (1970), "Partially Discrete Numerical Solution of the Generalized Diffusion Equation," Ph.D. Thesis, University of California, Los Angeles, California.

Pipes, L.A., and Hovanessian, S.A. (1969), Matrix-Computer Methods in Engineering, John Wiley, New York, New York.

Piquet, J. (1970), "Resolution Numerique de Certaines Cauches Limites Incompressibles Instationnaires," La Resherche Aerospatiale, No. 1970-2.

Pletcher, R.H., and McManus, H.N., Jr. (1965), "The Fluid Dynamics of Three Dimensional Liquid Films with Free Surface Shear: A Finite Difference Approach," Developments in Mechanics, Vol. 3, Part 2, pp. 305-318.

Pletcher, R.H. (1969), "On a Finite-Difference Solution for the Constant-Property Turbulent Boundary Layer," AIAA J., Vol. 7, No. 2, pp. 305-311.

Plooster, M.N. (1970), "Shock Waves from Line Sources. Numerical Solutions and Experimental Measurements," The Physics of Fluids, Vol. 13, No. 11, pp. 2665-2675.

Plotkin, A. (1968), "A Numerical Solution for the Laminar Wake Behind a Finite Flat Plate," Ph.D. Thesis, Stanford University, Stanford, California.

Plotkin, A., and Flugge-Lotz, I. (1968), "A Numerical Solution for the Laminar Wake Behind a Finite Flat Plate," Tech. Rept. No. 179, Stanford University, Stanford, California.

Plows, W.H. (1968), "Some Numerical Results for Two-Dimensional Steady Laminar Benard Convection," Physics of Fluids, Vol. 11, No. 8, pp. 1593-1599.

Polezhaev, V.I. (1966), "Numerical Solution of the System of One-Dimensional Unsteady Navier-Stokes Equations for a Compressible Gas," Izv. AN SSSR. Mekhanika Zhidkosti i Gaza, Vol. 1, No. 6, pp. 34-44.

Polezhaev, V.I. (1967), "Numerical Solution of the System of Two-Dimensional Unsteady Navier-Stokes Equations for a Compressible Gas in a Closed Region," Izv. AN SSSR. Mekhanika Zhidkosti i Gaza, Vol. 2, No. 2, pp. 103-111.

Polger, P.D. (1971), "A Study of Non-linear Computational Instability for a Two-Dimensional Model," NOAA TM NWS NMC-49, U.S. Department of Commerce, National Oceanic and Atmospheric Administration, National Meteorological Center, Washington, D.C., February, 1971.

Polozhii, G.N. (1965), The Method of Summary Representation for Numerical Solution of Problems of Mathematical Physics, Pergammon Press, New York, New York.

Pomraning, G.C., Wilson, H., and Lindley, W. (1969), "Theoretical and Computational Radiation Hydrodynamics, Vol. I., Radiation-Hydrodynamics, Theoretical Considerations," GA 9530-Vol-1, Gulf General Atomic Inc., San Diego, California.

Popov, Iu. P., Samarskii, A.A. (1969), "Completely Conservative Difference Schemes," Vychisl. Matem. i Matem. Fiz., Vol. 9, No. 4, pp. 953-958.

Popper, L.A., Toong, T.Y., and Sutton, G.W. (1970), "Three-Dimensional Ablation Considering Shape Changes and Internal Heat Conduction," AIAA Paper No. 70-199, AIAA 8th Aerospace Sciences Meeting, New York, New York, January 19-21, 1970.

Powe, R.E., Carley, C.T., and Carruth, S.L. (1971), "A Numerical Solution for Natural Convection in Cylindrical Annuli," J. of Heat Transfer, May, 1971, pp. 210-220.

Powers, S.A., and O'Neill, J.B. (1963), "Determination of Hypersonic Flow Fields by the Method of Characteristics," AIAA J., Vol. 1, pp. 1693-1694.

Powers, S.A., Niemann, A.F., Jr., Der, J. Jr. (1967), "A Numerical Procedure for Determining the Combined Viscid-Inviscid Flow Fields over Generalized Three-Dimensional Bodies," Tech. Rept. AFFDL-TR-67-124, Vol. 1, Wright-Patterson Air Force Base, Ohio.

Pracht, W.E. (1971A), "A Numerical Method for Calculating Transient Creep Flows," J. of Computational Physics, Vol. 7, pp. 46-60.

Pracht, W.E. (1971B), "Implicit Solution of Creeping Flows, with Application to Continental Drift." See Holt (1971).

Prentice, J.J. (1971), "A Numerical Solution for the Flow Field Surrounding a Supersonic Cone at an Angle of Attack," Ph.D. Dissertation, Oklahoma State University, Stillwater, Oklahoma, May, 1971.

Presley, L.L., and Hanson, R.K. (1969), "Numerical Solutions of Reflected Shock-Wave Flow-Fields with Nonequilibrium Chemical Reactions," AIAA J., Vol. 7, No. 12, pp. 2267-2346.

Price, J.F. (1966), "Numerical Analysis and Related Literature for Scientific Computer Users," Boeing Scientific Research Laboratories, D1-82-0517, March, 1966.

Pritchett, J.W. (1971), "Incompressible Calculations of Underwater Explosion Phenomena." See Holt (1971).

Putre, H.A. (1970), "Computer Solution of Unsteady Navier-Stokes Equations for an Infinite Hydrodynamic Step Bearing, NASA-TN-D-5682, NASA Lewis Research Center, Cleveland, Ohio.

Quarmby, A. (1968), "A Finite Difference Analysis of Developing Slip Flow," Applied Scientific Research, Vol. 19, No. 1, pp. 18-33.

Quon, D., Dranchuk, P.M., Allada, S.R., and Leung, P.K. (1966), "Application of the Alternating Direction Explicit Procedure to Two-Dimensional Natural Gas Reservoirs," Society of Petroleum Engineers J., June, 1966, pp. 137-142.

Rachford, H.H., Jr. (1966), "Numerical Calculation of Immiscible Displacement by a Moving Reference Point Method," Society of Petroleum Engineers J., June, 1966, pp. 87-101.

Rakich, J.V. (1967), "Three-Dimensional Flow Calculation by the Method of Characteristics," AIAA J., Vol. 5, No. 10, pp. 1906-1908.

Rakich, J.V. (1969), "A Method of Characteristics For Steady Three-Dimensional Supersonic Flow with Application to Inclined Bodies of Revolution," NASA TN-D-5341, NASA, Ames Research Center, Moffett Field, California.

Rakich, J.V., and Cleary, J.W. (1970), "Theoretical and Experimental Study of Supersonic Steady Flow Around Inclined Bodies of Revolution," AIAA J., Vol. 8, No. 3, pp. 511-518.

Ransom, V.H., Thompson, H.D., and Hoffman, J.D. (1970), "Three-Dimensional Supersonic Nozzle Flowfield Calculations," J. Spacecraft and Rockets, Vol. 7, pp. 458-462.

Ransom, V.H., Thompson, H.D., and Hoffman, J.D. (1971), "Stability and Accuracy Studies on a Second-Order Method of Characteristics Scheme for Three-Dimensional, Steady, Supersonic Flow." See Holt (1971).

Reed, W.H., and Hansen, K.F. (1969), "Finite Difference Techniques for the Solution of the Reactor Kinetics Equations," MIT-3903-2, Department of Nuclear Engineering, Massachusetts Institute of Technology, May, 1969.

Reynolds, R.C. (1970), "Spshell, Oildrop, and Spclam: Three Programs to Solve Hydrodynamics Problems in Two Dimensions," AFWL-TR-70-96, Air Force Weapons Lab., Kirtland Air Force Base, New Mexico, December, 1970.

Rich, M. (1963), "A Method for Eulerian Fluid Dynamics," Rept. No. LAMS-2826, Los Alamos Scientific Laboratory, Los Alamos, New Mexico.

Richards, C.G. (1970), "A Numerical Study of the Flow in the Vortex Angular-Rate Sensor," Paper No. 70-WA/FE-5, American Society of Mechanical Engineers, Winter Annual Meeting, New York, New York, November 29-December 3, 1970.

Richardson, D. (1964), "The Solution of Two-Dimensional Hydrodynamic Equations by the Method of Characteristics," Methods in Computational Physics, Vol. 3, pp. 295-318.

Richardson, L.F. (1910), "The Approximate Arithemetical Solution by Finite Differences of Physical Problems Involving Differential Equations, with an Application to the Stresses in a Masonry Dam," Transactions of the Royal Society of London, Ser. A., Vol. 210, pp. 307-357.

Richardson, L.F. (1965), Weather Prediction by Numerical Process, Dover.

Richtmyer, R.D. (1957), Difference Methods for Initial-Value Problems, Interscience Publishers, Inc., New York, New York.

Richtmyer, R.D. (1963), "A Survey of Difference Methods for Nonsteady Fluid Dynamics," NCAR Technical Note 63-2, Boulder, Colorado.

Richtmyer, R.D., and Morton, K.W. (1967), Difference Methods for Initial-Value Problems, Second Edition, Interscience Publishers, J. Wiley and Sons, New York, New York.

Rigler, A.K. (1969), "A Choice of Starting Vectors in Relaxation Methods," J. of Computational Physics, Vol. 4, pp. 419-423.

Rimon, Y. (1968), "Numerical Solution of the Incompressible Time-Dependent Viscous Flow Past a Thin Oblate Spheroid," AML-24-68, Applied Mathematics Laboratory, July, 1968, Naval Ship Research and Development Center, Washington, D.C.

Rimon, Y., and Cheng, S.I. (1969), "Numerical Solution of a Uniform Flow over a Sphere at Intermediate Reynolds Numbers," Physics of Fluids, Vol. 12, No. 5, pp. 949-959.

Rimon, Y., and Lugt, H.J. (1969), "On Laminar Flows Past Oblate Spheroids of Various Thicknesses," Physics of Fluids, Vol. 12, No. 12, pp. 2465-2472.

Ringleb, F.O. (1963), "Geometrical Construction of Two-Dimensional and Axisymmetrical Flow Fields," AIAA J., Vol. 1, No. 10, pp. 2257-2263.

Roache, P.J., and Mueller, T.J. (1968), "Numerical Solutions of Compressible and Incompressible Laminar Separated Flows," AIAA Paper No. 68-741, AIAA Fluid and Plasma Dynamics Conference, Los Angeles, California, June 24-26, 1968.

Roache, P.J., and Mueller, T.J. (1970), "Numerical Solutions of Laminar Separated Flows," AIAA J., Vol. 8, No. 3, pp. 530-538.

Roache, P.J. (1970), "Sufficiency Conditions for a Commonly Used Downstream Boundary Condition on Stream Function," J. of Computational Physics, Vol. 6, No. 2, pp. 317-321.

Roache, P.J. (1971A), "A Direct Method for the Discretized Poisson Equation," SC-RR-70-579, Sandia Laboratories, Albuquerque, New Mexico, February, 1971.

Roache, P.J. (1971B), "A New Direct Method for the Discretized Poisson Equation." See Holt (1971).

Roache, P.J. (1971C), "On Artificial Viscosity," SC-RR-710301, Sandia Laboratories, Albuquerque, New Mexico, July, 1971. Also, J. of Computational Physics, Vol. 10, October 1972, pp. 169-184.

Roache, P.J. (1972), "Finite Difference Methods for the Steady-State Navier-Stokes Equations," SC-RR-72-0419, Sandia Laboratories, Albuquerque, New Mexico, December, 1972. Also, Proc. Third Int'l. Conference on Numerical Methods in Fluid Dynamics, Paris, France, July 3-7, 1972.

Robert, A.J., Shuman, F. G., and Gerrity, J.P., Jr. (1970), "On Partial Difference Equations in Mathematical Physics," U.S. Monthly Weather Review, Vol. 98, No. 1, pp. 1-6.

Roberts, G.O. (1971), "Computational Meshes for Boundary Layer Problems." See Holt (1971).

Roberts, K.V., and Weiss, N.O. (1966), "Convective Difference Schemes," Mathematics of Computation, Vol. 20, No. 94, pp. 272-299.

Robertson, S.J., and Willis, D.R. (1971), "Method-of-Characteristics Solution of Rarefied, Monatomic Gaseous Jet Expansion into a Vacuum," AIAA J., Vol. 9, No. 2, pp. 291-296.

Rodin, E.Y. (1970), "On Some Approximate and Exact Solutions of Boundary Value Problems for Burgers' Equation," J. of Mathematical Analysis and Applications, Vol. 30, No. 2, pp. 401-414.

Roesner, K. (1967), "On the Calculation of Three-Dimensional Unsteady Flow Fields in Compressible Media," Mitteilungen aus dem Max-Planck-Institute fur Stromungs Forchung und der Aerodynamischen Versuchsanstalt, No. 41.

Roesner, K. (1971), "Numerical Integration of the Euler-Equations for Three-Dimensional Unsteady Flows." See Holt (1971).

Rogers, E.H. (1967), "Stability and Convergence of Approximation Schemes," J. of Mathematical Analysis and Applications, Vol. 20, No. 3, pp. 442-453.

Rosenbaum, H. (1968), "Some Numerical Solutions of the Rotational Euler Equations," AIAA J., Vol. 6, No. 2, pp. 320-325.

von Rosenberg, D.U. (1969), Methods for the Numerical Solution of Partial Differential Equations, American Elsevier Publishing Company, New York.

Rosenhead, L. (ed.) (1963), Laminar Boundary Layers, Oxford University Press.

Roslyakov, G.S., and Chudov, L.A. (1963), Numerical Methods in Gas Dynamics, Israel Program for Scientific Translations, Jerusalem, 1966.

Ross, B.B., and Chung, S.I. (1971), "A Numerical Solution of the Planar Supersonic Near-Wake with its Error Analysis." See Holt (1971).

Royal, J.W. (1969), "Numerical Solution of the Rayleigh Cell Problem," Ph.D. Dissertation, Boston University, Boston, Massachusetts.

Rozhdestvenskii, B.L., and Lananko, N.N. (1969), "Homogeneous Schemes with Pseudo-Viscosity," Lockheed Missiles and Space Co., Trans., N69-33070, Palo Alto, California. From Sistemy Qrazilineinykh Urarnenil, Nanka Press, Moscow, 1968, pp. 393-407.

Rozsa, P., and Toth, I. (1970), "Eine Direkte Methode zur Numerischen Losung der Poissonschen Differentialgleichung mit Hilfe des 9-Punkte-Verfahrens," Zeitschrift fur Angewandte Mathematik und Mechanik, Vol. 50, No. 12, pp. 713-721.

Rubin, E.L., and Burstein, S.Z. (1967), "Difference Methods for the Inviscid and Viscous Equations of a Compressible Gas," J. of Computational Physics, Vol. 2, pp. 178-196.

Rubin, E.L., Gerstenbluth, C., and Khosla, P.K. (1967), "One-Dimensional Unsteady Flow of a Radiating Gas, PIBAL Rept. No. 990, Polytechnic Institute of Brooklyn, Department of Aerospace Engineering and Applied Mechanics, May, 1967.

Rubin, E.L., and Preiser, S. (1968), "On the Derivation of Three-Dimensional Second Order Accurate Hydrodynamic Difference Schemes," PIBAL Rept. No. 68-24, Polytechnic Institute of Brooklyn, Department of Aerospace Engineering and Applied Mechanics, July, 1968.

Rubin, E.L., and Preiser, S. (1970), "Three-Dimensional Second-Order Accurate Difference Schemes for Discontinuous Hydrodynamic Flows," Mathematics of Computation, Vol. 24, No. 109, pp. 57-63.

Rubin, E.L., and Khosla, P.K. (1970), "A Time-Dependent Method for the Solution of One-Dimensional Radiating Flow," Zeitschrift fur Angewandte Mathematik und Physik, Vol. 21, November, 1970, pp. 962-977.

Rubin, E.L. (1970), "Time-Dependent Techniques for the Solution of Viscous, Heat Conducting, Chemically Reacting, Radiating Discontinuous Flows," Lecture Notes in Mathematics, Conference on the Numerical Solution of Differential Equations, Springer-Verlag; Berlin, Heidelberg, New York.

Runchal, A.K., Spalding, D.B., and Wolfshtein, M. (1969), "Numerical Solution of the Elliptic Equations for Transport of Vorticity, Heat and Matter in Two-Dimensional Flow." See Frenkiel and Stewartson (1969).

Runchal, A.K., and Wolfshtein, M. (1969), "Numerical Integration Procedure for the Steady State Navier-Stokes Equations," J. Mechanical Engineering Science, Vol. 11, No. 5, pp. 445-453.

Ruo, S.-Y. (1967), "Evaluation of the Applicability of an Explicit Numerical Method to a Plane, Turbulent, Low Velocity, Partially Confined Jet," Ph.D. Thesis, Oklahoma State University, Stillwater, Oklahoma.

Ruptosh (1952), "Supersonic Wind Tunnels - Theory, Design and Performance," UTIA Review No. 5, University of Toronto, Ontario, Canada.

Rusanov, V.V. (1961), "Calculation of Interaction of Non-Steady Shock Waves with Obstacles," Zhur. Vychislitel'noi Mathematicheskoi Fiziki, Vol. 1, No. 2, pp. 267-279.

Rusanov, V.V., and Lyubimov, A.N. (1968), "Studies of Flows Around Blunt Bodies by Numerical Methods," Applied Mechanics, Proceedings of the 12th International Congress of Applied Mechanics, Stanford University, Stanford, California, August 26-31, 1968, pp. 356-363.

Rusanov, V.V. (1969), "Three-Dimensional Supersonic Flow Over a Blunt Body," JPRS 47180, Joint Publications Research Service, Department of Commerce, Washington, D.C.

Rusanov, V.V. (1970), "On Difference Schemes of Third Order Accuracy for Nonlinear Hyperbolic Systems," J. of Computational Physics, Vol. 5, pp. 507-516.

Rusanov, V.V. (1971), "Non-Linear Analysis of the Shock Profile in Difference Schemes." See Holt (1971).

Rushton, K.R., and Laing, L.M. (1968), "A Digital Computer Solution of the LaPlace Equation Using the Dynamic Relaxation Method," The Aeronautical Quarterly, November, 1968, pp. 375-387.

Russell, D.B. (1962), Aerospace Research Council, R. & M., No. 3331, United Kingdom.

Russell, D. (1963), "On Obtaining Solutions to the Navier-Stokes Equations with Automatic Digital Computers," A.R.C.R. & M. No. 3331, United Kingdom.

Rybicki, E.F., and Hopper, A.T. (1970), "Higher Order Finite Element Method for Transient Temperature Analysis of Inhomogeneous Materials," ASME WA 69-WA/HT-33, The American Society of Mechanical Engineers, United Engineering Center, New York, New York.

Sackett, S., and Healey, R. (1969), "JASON - A Digital Computer Program for the Numerical Solution of the Linear Poisson Equation $\underline{\nabla}\cdot(\eta\nabla\Phi) + \rho = 0$," UCRL-18721, Lawrence Radiation Laboratory, University of California, Berkeley, California, February, 1969.

Sadourny, R., Arakawa, A., and Mintz, Y. (1968), "Integration of the Non-Divergent Barotropic Vorticity Equation with an Icosahedral-Hexagonal Grid for the Sphere," U.S. Monthly Weather Review, Vol. 96, pp. 351-356.

Sadourny, R., and Morel, P. (1969), "A Finite-Difference Approximation of the Primitive Equations for a Hexagonal Grid on a Plane," U.S. Monthly Weather Review, Vol. 97, No. 6, pp. 439-445.

Saggendorf, F. (1971), Ph.D. Dissertation, Mechanical Engineering Department, University of Kentucky, Lexington, Kentucky.

Sakurai, A., and Iwasaki, M. (1970), "Finite Difference Calculation of One-Dimensional Navier-Stokes Shock Structure," Proceedings Japan National Committee for Theoretical and Applied Mechanics, 18th National Congress for Applied Mechanics, Tokyo Metropolitan University, Tokyo, Japan.

Sakurai, A. (1971), "Foundation of Approximate Solutions." See Holt (1971).

Salvadori, M., and Baron, M. (1961), Numerical Methods in Engineering, Prentice-Hall, Inc., Englewood Cliffs, New Jersey.

Samarski, A.A. (1962), "An Economical Difference Method for the Solution of a Multidimensional Parabolic Equation in an Arbitrary Region," USSR Computational Mathematics and Mathematical Physics, Vol. 2, pp. 894-926.

Samuels, M.R., and Churchill, S.W. (1967), "Stability of a Fluid in a Rectangular Region Heated from Below," American Institute of Chemical Engineers J., Vol. 13, No. 1, pp. 77-86.

Satofuka, Nobuyuki (1970), "A Numerical Study of Shock Formation in Cylindrical and Two-Dimensional Shock Tubes," Report No. 451, Institute of Space and Aeronautical Sciences, University of Tokyo, Tokyo, Japan, June, 1970.

Sauerwein, H., and Sussman, M. (1964), "Numerical Stability of the Three-Dimensional Method of Characteristics," AIAA J., Vol. 2, No. 2, pp. 387-389.

Sauerwein, H., and Diethelm, M. (1967), "Improved Calculations of Multi-Dimensional Fluid Flows by the Method of Characteristics," Aerospace Rept. No. TR-1001 (S2816-23)-1, Aerospace Corporation, San Bernardino, California.

Saul'yev, V.K. (1957), "On a Method of Numerical Integration of a Diffusion Equation," Dokl. Akad. Nauk SSSR, Vol. 115, No. 6, pp. 1077-1080. (In Russian)

Saul'yev, V.K. (1964), Integration of Equations of Parabolic Type by the Method of Nets, Translated from Russian by G.J. Tee, Pergamon Press, The MacMillan Company, New York.

Saunders, L.M. (1966), "Numerical Solution of the Flow Field in the Throat Region of a Nozzle," Brown Engineering Co., Tech. Note BSVD-P-66-TN-001, August, 1966.

Scala, S.M., and Gordon, P. (1966), "Reflection of a Shock Wave at a Surface," The Physics of Fluids, Vol. 9, June, 1966, pp. 1158-1166.

Scala, S.M., and Gordon, P. (1967), "Solution of the Time-Dependent Navier-Stokes Equations for the Flow Around a Circular Cylinder," AIAA Paper No. 67-221, AIAA 5th Aerospace Sciences Meeting, New York, New York, January 23-26, 1967. Also, AIAA J., Vol. 6, No. 5, May, 1968, p. 815.

Schechter, H. (1967), "Application of Pade Integration to the Partial Differential Equations of the Hypersonic Wake," Tech. Memo No. 165, General Applied Science Laboratories, Westbury, Long Island, New York.

Schelkunoff, S.A. (1965), Applied Mathematics for Engineers and Scientists, Bell Telephone Laboratories Series, Second Edition.

Schlichting, H. (1968), Boundary Layer Theory, 6th Edition, McGraw-Hill Book Co., Inc., New York.

Schoenauer, W. (1970), "Solution of the Three-Dimensional Time-Dependent Navier-Stokes Equations by the Difference Method," Ph.D. Dissertation, Karlsruhe University, Karlsruhe, West Germany, April, 1970.

Schoenherr, R.U., and Churchill, S.W. (1970), "The Use of Extrapolation for the Solution of Heat Transfer Problems by Finite-Difference Methods," J. of Heat Transfer, August 1970, pp. 564-565.

Schroeder, R.C., and Thomsen, J.M. (1969), "An Implicit Difference Solution of Multi-Dimensional Hydrodynamic Flow Problems," UCRL-71651, Lawrence Radiation Laboratory, University of California, Livermore, California.

Schulz, W.D., (1964), "Tensor Artificial Viscosity for Numerical Hydrodynamics," J. of Mathematical Physics, Vol. 5, No. 1, pp. 133-138.

Schwartz, R.A. (1967), "Computation of Relativistic Gravitational Collapse," Supernovae and Their Remnants, Gordon and Breach Science Publishers, New York.

Segal, B.M., and Ferziger, J.H. (1971), "Shock Wave Structure by Several New Modeled Boltzmann Equations." See Holt (1971).

Seinfeld, J.H., Lapidus, L., and Hwang, M. (1970), "Review of Numerical Integration Techniques for Stiff Ordinary Differential Equations," Industrial Engineering Chemical Fundamentals, Vol. 9, No. 2, 1970, pp. 266-275.

Shampine, L.F. (1968), "Boundary Value Problems for Ordinary Differential Equations," SIAM J., Numerical Analyses, Vol. 5, p. 219.

Shampine, L.F., and Thompson, R.J. (1970), "Difference Methods for Nonlinear First-Order Hyperbolic Systems of Equations," Mathematics of Computation, Vol. 24, No. 109, p. 45.

Shank, G.D. (1968), "Error Bounds for Difference Approximations to Initial-Boundary and Mixed Initial-Boundary Value Problems for Parabolic Equations of General Domains," Ph.D. Thesis, Maryland University, College Park, Maryland.

Shapiro, A.H. (1953), The Dynamics and Thermodynamics of Compressible Fluid Flow, Vols. I and II, Ronald Press, New York.

Shapiro, M.A., and O'Brien, J.J. (1970), "Boundary Conditions for Fine-Mesh Limited-Area Forecasts," J. of Applied Meteorology, Vol. 9, No. 3, pp. 345-349.

Shavit, G., and Lavan, Z. (1971), "Analytical and Experimental Investigation of Laminar Mixing of Confined Heterogeneous Jets," AIAA Paper No. 71-601, AIAA 4th Fluid and Plasma Dynamics Conference, Palo Alto, California, June 21-23, 1971.

Shchennikov, V.V. (1969), "A Class of Exact Solutions of the Navier-Stokes Equations for a Compressible Heat-Conducting Gas," PMM: J. of Applied Mathematics and Mechanics, Vol. 33, No. 3, pp. 570-573.

Sheldon, J.W. (1959), "On the Spectral Norms of Several Iterative Processes," J. Association for Computing Machinery, Vol. 6, p. 494.

Sheldon, J.W. (1962), "Iterative Methods for the Solution of Elliptic Partial Differential Equations," Mathematical Methods for Digital Computers, J. Wiley and Sons, New York.

Shelton, G.A., Jr. (1970), "Blunt-Body Electrostatic Probe Analysis," SC-RR-70-331, Sandia Laboratories, Albuquerque, New Mexico, August, 1970.

Shevelev, Yu.D. (1967), "Numerical Study of Three-Dimensional Boundary Layer in a Compressible Gas," Izv. AN SSSR. Mekhanika Zhidkosti i Gaza, Vol. 2, No. 4, pp. 171-177.

Shortley, G.H., and Weller, R. (1938), "The Numerical Solution of LaPlace's Equation," J. of Applied Physics, Vol. 9, pp. 334-348.

Sibulkin, M., and Dispaux, J.-C. (1968), "Numerical Solutions for Radiating Hypervelocity Boundary-Layer Flow on a Flat Plate," AIAA J., Vol. 6, No. 6, pp. 1098-1104.

Sielecki, A., and Wurtele, M.G. (1970), "The Numerical Integration of the Nonlinear Shallow-Water Equations with Sloping Boundaries," J. of Computational Physics, Vol. 6, pp. 219-236.

Sihna, R., Zakkay, V., and Erdos, J. (1970), "Flow Field Analysis of Plumes of Two-Dimensional Underexpanded Jets by a Time-Dependent Method," NYU-AA-70-04, New York University, School of Engineering and Science, University Heights, New York, New York, June, 1970.

Sills, J.A. (1969), "Transformations for Infinite Regions and Their Application to Flow Problems," AIAA J., Vol. 7, No. 1, pp. 117-123.

Simeonov, S.V. (1967), "A Method of Solving Systems of Nonlinear Equations of the Potential Type," PMM: J. of Applied Mathematics and Mechanics, Vol. 31, No. 3, pp. 527-532.

Sims, J.L. (1958), "Results of the Computations of Supersonic Flow Fields Aft of Circular Cylindrical Bodies of Revolution by the Method of Characteristics," Army Ballistic Missile Agency Report DA-R-49, Redstone Arsenal, Alabama.

Simuni, L.M. (1964), Inzhenernii Zhowinal, USSR, Vol. 4, p. 446.

Singleton, R.E. (1968), "Lax-Wendroff Difference Scheme Applied to the Transonic Airfoil Problem," AGARD Conference Proceedings No. 35, September, 1968.

Sinnott, D.H. (1960), "The Use of Interpolation in Improving Finite Difference Solutions of TEM Mode Structures," Department of Supply Tech. Note Pad 142, Australian Defense Scientific Service, Weapons Research Establishment, Salisbury, South Australia.

Skoglund, V.J., and Cole, J.K. (1966), "Numerical Analysis of the Interaction of an Oblique Shock Wave and a Laminar Boundary Layer," Bureau of Engineering Research Report No. ME-23, University of New Mexico, Albuquerque, New Mexico, June, 1966.

Skoglund, V.J., Cole, J.K., and Staiano, E.F. (1967), "Development and Verification of Two-Dimensional Numerical Techniques for Viscous Compressible Flows with Shock Waves," SC-CR-67-2679, Sandia Laboratories, Albuquerque, New Mexico, August, 1967.

Skoglund, V.J., and Gay, B.D. (1968), "Numerical Analysis of Gas Dynamics," Bureau of Engineering Research Progress Report PR-85(68)S-082, The University of New Mexico, Albuquerque, New Mexico.

Skoglund, V.J., and Gay, B.D. (1969), "Improved Numerical Techniques and Solution of a Separated Interaction of an Oblique Shock Wave and a Laminar Boundary Layer," Bureau of Engineering Research Report No. ME-41(69)S-068, The University of New Mexico, Albuquerque, New Mexico.

Skoglund, V.J., and Watkins, C.B., Jr. (1971), "Numerical Solutions of Non-Equilibrium Reflected Shock Waves in Air with Radiation," SC-RR-71-0203, Sandia Laboratories, Albuquerque, New Mexico, May, 1971.

Slotnick, D.L. (1971), "The Fastest Computer," Scientific American, Vol. 224, No. 2, February, 1971, pp. 76-87.

Slotta, L.S., Elwin, E.H., Mercier, H.T., and Terry, M.D. (1969), "Stratified Reservoir Currents. Parts I and II," Bulletin No. 44, Engineering Experiment Station, Oregon State University, Corvallis, Oregon.

Smagorinsky, J., Manabe, S., and Holloway, J.L. (1965), "Numerical Results from a Nine-Level General Circulation Model of the Atmosphere," U.S. Weather Bureau Monthly Weather Review, Vol. 93, p. 727.

Smith, A.M.O., and Clutter, D.W. (1965), "Machine Calculation of Compressible Laminar Boundary Layers," AIAA J., Vol. 3, No. 4, April, 1965, pp. 639-647.

Smith, J.H. (1970), "Survey of Three-Dimensional Finite Difference Forms of Heat Equation," SC-M-70-83, Sandia Laboratories, Albuquerque, New Mexico, March, 1970.

Smith, J. (1970), "The Coupled Equation Approach to the Numerical Solution of the Biharmonic Equation by Finite Differences," SIAM J. on Numerical Analysis, Vol. 7, pp. 104-111.

Smith, R.R., and McCall, D. (1970), "Systems of Hyperbolic P.D.E.," Communications of the ACM., Vol. 13, No. 9, September, 1970, pp. 567-570.

Smola, F.M., and McAvoy, T.J. (1969), "The Finite Difference-Delay Technique: A New Approach for Solving Certain First Order Partial Differential Equations," Massachusetts University, Amherst, Massachusetts.

Snedeker, R.S., and du P. Donaldson, C. (1966), "Observation of a Bistable Flow in a Hemispherical Cavity," AIAA J., Vol. 4, No. 4, pp. 735-736.

Sobey, R.J. (1970), "Finite Difference Schemes Compared for Wave-Deformation Characteristics in Mathematical Modeling of Two-Dimensional Long Wave Propagation," Army Coastal Engineering Research Center, Washington, D.C.

Son, J.S., and Hanratty, T.J. (1969), "Numerical Solution for the Flow Around a Cylinder at Reynolds Numbers of 40, 200 and 500," J. Fluid Mechanics, Vol. 35, Part 2, pp. 369-386.

Southwell, R.V. (1946), Relaxation Methods in Theoretical Physics, Oxford University Press, New York, New York.

Spalding, D.B. (1967), "A Theory of Turbulent Boundary Layer," Proceedings of the Canadian Congress of Applied Mechanics, University of Laval, Quebec, Canada, May 22-26, 1967.

Spanier, J. (1967), "Alternating Direction Methods Applied to Heat Conduction Problems," Mathematical Methods for Digital Computers, Vol. II, John Wiley and Sons, Inc., New York, New York.

Sparrow, E.M., Quack, H., and Boerner, C.J. (1970), "Local Nonsimilarity Boundary Layer Solutions," AIAA J., Vol. 8, No. 11, pp. 1936-1942.

Spikjer, M.N. (1966), "Convergence and Stability of Step-by-step Methods for the Numerical Solution of Initial-Value Problems," Numerische Mathematik, Vol. 8, pp. 161-177.

Spurk, J.H. (1970), "Experimental and Numerical Nonequilibrium Flow Studies," AIAA J., Vol. 8, No. 6, pp. 1039-1045.

Steger, J.L., and Lomax, H. (1971), "Generalized Relaxation Methods Applied to Problems in Transonic Flow." See Holt (1971).

Steiger, M.H., and Sepri, P. (1965), "On the Solution of Initial-Valued Boundary Layer Flows," PIBAL Rept. No. 872, Polytechnic Institute of Brooklyn, Dept. of Aerospace Engineering and Applied Mechanics, May, 1965.

Sternberg, H.M. (1970), "Similarity Solutions for Reactive Shock Waves," Quarterly J. of Mechanics and Applied Mathematics, Vol. 23, No. 1, February, 1970, pp. 77-99.

Stetter, H.J. (1970), "Discretization Methods for Differential Equations," AFOSR 70-2432TR, Institue fur Numerische Mathematik Teshnische Hochschule, Vienna, Austria, September, 1970.

Stewartson, K. (1964), The Theory of Laminar Boundary Layers in Compressible Fluids, Oxford Mathematical Monographs, Oxford at the Clarendon Press.

Stone, H.L., and Brian, P.L.T. (1963), "Numerical Solution of Convective Transport Problems," American Institute of Chemical Engineers J., Vol. 9, pp. 681-688.

Stone, H.L. (1968), "Iterative Solution of Implicit Approximations of Multi-Dimensional Partial Differential Equations," SIAM J. of Numerical Analysis, Vol. 5, No. 3, pp. 530-558.

Strang, W.G. (1963), "Accurate Partial Difference Methods. I: Linear Cauchy Problems," Arch. Rational Mech. Anal., Vol. 12, pp. 392-402.

Strang, W.G. (1964), "Accurate Partial Difference Methods. II: Non-Linear Problems," Numerische Mathematik, Vol. 6, pp. 37-46.

Strawbridge, D.R., and Hooper, G.T.J. (1968), "Numerical Solutions of the Navier-Stokes Equations for Axisymmetric Flows," J. of Mechanical Engineering Science, Vol. 10, No. 5, pp. 389-401.

Strelkoff, T. (1971), "An Exact Numerical Solution of the Solitary Wave." See Holt (1971).

Stubbe, P. (1970), "Simultaneous Solution of the Time Dependent Coupled Continuity Equations, Heat Conduction Equations, and Equations of Motion for a System Consisting of a Neutral Gas, An Electron Gas, and a Four Component Ion Gas," J. of Atmospheric and Terrestrial Physics, Vol. 32, pp. 865-903.

Sundberg, W.D. (1970), "Two Computerized Methods for Plotting Functions of Two Independent Variables," SC-DR-70-112, Sandia Laboratories, Albuquerque, New Mexico, February, 1970.

Sundquist, H. (1963), "A Numerical Forecast of Fluid Motion in a Rotating Tank and a Study of How Finite-Difference Approximations Affect Non-Linear Interactions," Tellus, Vol. 15, No. 1, pp. 44-55.

Swartz, B., and Wendroff, B. (1969), "Generalized Finite-Difference Schemes," Mathematics of Computation, Vol. 23, No. 105, pp. 37-49.

Swift, G.W. (1971), "The Solution of Simultaneous Second Order Coupled Differential Equations by the Finite Difference Method," VPI-E-71-3 Virginia Polytechnic Institute, Blacksburg, Virginia, February, 1971.

Takami, H., and Keller, H.B. (1969), "Steady Two-Dimensional Viscous Flow of an Incompressible Fluid Past a Circular Cylinder." See Frenkiel and Stewartson (1969).

Talley, W.K., and Whitaker, S. (1969), "Monte Carlo Analysis of Knudsen Flow," J. of Computational Physics, Vol. 4, pp. 389-410.

Taylor, D.B. (1968), "The Calculation of Steady Plane Supersonic Gas Flows Containing an Arbitrarily Large Number of Shocks," J. of Computational Physics, Vol. 3, pp. 273-290.

Taylor, P.J. (1968), "The Stability of Boundary Conditions in the Numerical Solution of the Time-Dependent Navier-Stokes Equations," ARC-30406, Aerospace Research Comm. (England).

Taylor, P.J. (1969), "A Technique for Treating Dirichlet Conditions at Infinity in Numerical Field Problems," J. of Computational Physics, Vol. 4, pp. 138-141.

Taylor, P.J. (1970), "The Stability of the DuFort-Frankel Method for the Diffusion Equation with Boundary Conditions Involving Space Derivatives," The Computer Journal, Vol. 13, No. 1, pp. 93-97.

Taylor, T.D. (1964), "Computing Transient Gas Flows with Shock Waves," Physics of Fluids, Vol. 7, No. 10, pp. 1713-1715.

Taylor, T.D., and Masson, B.S. (1970), "Application of the Unsteady Numerical Method of Godunov to Computation of Supersonic Flows Past Bell-Shaped Bodies," J. of Computational Physics, Vol. 5, pp. 443-454.

Taylor, T.D., and Ndefo, E. (1971), "Computation of Viscous Flow in a Channel by the Method of Splitting." See Holt (1971).

Taylor, T.D., Ndefo, E., and Masson, B.S. (1972), "A Study of Numerical Methods for Solving Viscous and Inviscid Flow Problems," J. Computational Physics, Vol. 9, pp. 99-119.

Tejeira, E.J. (1966), "Numerical and Experimental Investigation of a Two-Dimensional Laminar Flow with Non-Regular Boundaries," Rept. EM-66-8-1, Department of Engineering Mechanics, The University of Tennessee, Knoxville, Tennessee, August, 1966.

Temam, R. (1969), "On An Approximate Solution of the Navier-Stokes Equations by the Method of Fractional Steps: Part 1," Archive for Rational Mechanics and Analysis, Vol. 32, No. 2, pp. 135-153.

Tewarson, R.P. (1969), "Projection Methods for Solving Sparse Linear Systems," The Computer Journal, Vol. 12, No. 1, pp. 77-80.

Textor, R.E. (1968), "A Numerical Investigation of a Confined Vortex Problem," Rept. No. K-1732, Union Carbide Corporation, Computing Technology Center, Oak Ridge, Tennessee.

Textor, R.E., Lick, D.W., and Farris, G.J. (1969), "Solution of Confined Vortex Problems," J. of Computational Physics, Vol. 4, pp. 258-269.

Thom, A. (1928), "An Investigation of Fluid Flow in Two Dimensions," Aerospace Research Center, R. and M., No. 1194, United Kingdom.

Thom, A., and Orr, J. (1931), "The Solution of the Torsion Problem for Circular Shafts of Varying Radius," Proceedings of the Royal Society of London, A131, pp. 30-37.

Thom, A. (1933), "The Flow Past Circular Cylinders at Low Speeds," Proceedings of the Royal Society of London, A141, pp. 651-666.

Thom, A. (1953), "The Arithmetic of Field Equations," The Aeronautical Quarterly, Vol. 4, pp. 205-230.

Thom, A., and Apelt, C.J. (1961), Field Computations in Engineering and Physics, C. Van Nostrand Company, Ltd.

Thoman, D.C., and Szewczyk, A.A. (1966), "Numerical Solutions of Time Dependent Two Dimensional Flow of a Viscous, Incompressible Fluid Over Stationary and Rotating Cylinders," Tech. Rept. 66-14, Heat Transfer and Fluid Mechanics Lab., Dept. of Mechanical Engineering, University of Notre Dame, Notre Dame, Indiana, July, 1966.

Thoman, D., and Szewczyk, A.A. (1969), "Time Dependent Viscous Flow over a Circular Cylinder," The Physics of Fluids Supplement II, pp. 76-87.

Thomas, L.H. (1954), "Computation of One-Dimensional Flows Including Shocks," Communications in Pure and Applied Mathematics, Vol. 7, pp. 195-206.

Thomas, P.D., Vinokur, M., Bastianon, R., and Conti, R.J. (1971), "Numerical Solution for the Three Dimensional Hypersonic Flow Field of a Blunt Delta Body," AIAA Paper No. 71-596, AIAA 4th Fluid and Plasma Dynamics Conference, Palo Alto, California, June 21-23, 1971.

Thomee, V. (1969), "Stability Theory for Partial Difference Operators," SIAM Review, Vol. 11, No. 2, pp. 152-195.

Thommen, H.U., and D'Attorre, L. (1965), "Calculation of Steady, Three-Dimensional Supersonic Flow-Fields by a Finite Difference Method," AIAA Paper No. 65-26, AIAA 2nd Aerospace Sciences Meeting, New York, New York, January 25-27, 1965.

Thommen, H.U. (1966), "Numerical Integration of the Navier-Stokes Equations," Zeitschrift fur Angewandte Mathematik und Physik, Vol. 17, pp. 369-384.

Thommen, H.U. (1967), "Viscous Flow Near the Leading Edge of Wedges and Cones in Supersonic Freestream," AIAA Preprint No. 67-220.

Thompson, B.W. (1971), "Some Semi-Analytical Methods in Numerical Fluid Dynamics." See Holt (1971).

Thompson, J.F., Jr. (1968), "Numerical Solution of the Incompressible, Two-Dimensional, Time-Dependent Navier-Stokes Equations for a Body Oscillating in Pitch in a Moving Fluid," R.R. No. 86, Aerophysics and Aerospace Engineering Department, Mississippi State University, State College, Mississippi, October, 1968.

Thompson, J.F. (1969), "Optimized Acceleration of Convergence of an Implicit Numerical Solution of the Time-Dependent Navier-Stokes Equations," AIAA J., Vol. 7, No. 11, pp. 2186-2188.

Thompson, P.D. (1961) Numerical Weather Analysis and Prediction, MacMillan Company, New York, New York.

Thuraisamy, V. (1967), "Discrete Analogs for Mixed Boundary Value Problems of Elliptic Type," Ph.D. Thesis, University of Maryland, College Park, Maryland.

Thuraisamy, V. (1969A), "Approximate Solutions for Mixed Boundary Value Problems by Finite-Difference Methods," Mathematics of Computation, Vol. 23, No. 106, pp. 373-386.

Thuraisamy, V. (1969B), "Monotone Type Discrete Analogue for the Mixed Boundary Value Problem," Mathematics of Computation, Vol. 23, No. 106, pp. 387-394.

Tillman, C.C., Jr. (1969), "EPS: An Interactive System for Solving Elliptic Boundary-Value Problems with Facilities for Data Manipulation and General-Purpose Computation," MAC-TR-62, Massachusetts Institute of Technology, Cambridge, Massachusetts, June, 1969.

Tobey, R.G. (1969), Proceedings of the 1968 Summer Institute on Symbolic Mathematical Computation, NASA-CR-116915, June, 1969.

Tollmien, W. (1949), "Theory of Characteristics," NACA TM 1242, National Advisory Committee for Aeronautics, Washington, D.C., September, 1949.

Torrance, K.E. (1968), "Comparison of Finite-Difference Computations of Natural Convection," J. of Research of the National Bureau of Standards, Vol. 72B, No. 4, pp. 281-301.

Torrance, K.E., Orloff, L., and Rockett, J.A. (1969), "Experiments on Natural Convection in Enclosures with Localized Heating from Below," J. Fluid Mechanics, Vol. 36, Part 1, pp. 21-31.

Torrance, K.E., and Rockett, J.A. (1969), "Numerical Study of Natural Convection in an Enclosure with Localized Heating from Below -- Creaping Flow to the Onset of Laminar Instability," J. Fluid Mechanics, Vol. 36, Part 1, pp. 33-54.

Torrance, K., Davis, R., Eike, K., Gill, P., Gutman, D., Hsui, A., Lyons, S., and Zien, H. (1972), "Cavity Flows Driven by Buoyancy and Shear," J. Fluid Mechanics, Vol. 51, Part 2, pp. 221-231.

Trulio, J.G. (1964), "Studies of Finite Difference Techniques for Continuum Mechanics," Air Force Systems Command, Air Force Weapons Lab., Research and Technology Division, WL TDR 64-72, Kirtland Air Force Base, Albuquerque, New Mexico.

Trulio, J.G., Carr, W.E., Niles, W.J., and Rentfrow, R.E. (1966), "Calculation of Two-Dimensional Turbulent Flow Fields," NASA CR-530, May, 1966.

Trulio, J.G., and Walitt, L. (1969), "Numerical Calculation of Viscous Compressible Fluid Flow Around an Oscillating Rigid Cylinder," Applied Theory, Santa Monica, California.

Trulio, J.G., Walitt, L., and Niles, W.J. (1969), "Numerical Calculations of Viscous Compressible Fluid Flow over a Flat Plate and Step Geometry," NASA CR 1466, Applied Theory, Inc., Santa Monica, California.

Tsao, Nai-kuan (1970), "On Direct Solutions of Linear Algebraic Systems," Hawaii University, Department of Electrical Engineering, Honolulu, Hawaii.

Tsien, H.S. (1958), "The Equations of Gas Dynamics," High Speed Aerodynamics and Jet Propulsion, Vol. III, Princeton University Press, Princeton, New Jersey.

Tsuboi, A., and Ichikawa, M. (1969), "Error Analysis on Various Electrical Analogues Solutions for One, Two, and Three Dimensional Heat Conduction Partial Differential Equations," Memoirs of the Faculty of Engineering, Nagoye University, Vol. 19, No. 2, pp. 228-249.

Tyler, L.D. (1965), "Numerical Solutions of the Flow Field Produced by a Plane Shock Wave Emerging into a Crossflow," Ph.D. Dissertation, Dept. of Mechanical Engineering, Oklahoma State University, Stillwater, Oklahoma.

Tyler, L.D., and Zumwalt, G.W. (1965), "Numerical Solutions of the Flow Field Produced by a Plane Shock Wave Emerging into a Crossflow," SC-DC-65-1916, Sandia Laboratories, Albuquerque, New Mexico.

Tyler, L.D. (1968), "Numerical Results of Blast Wave Propagation in Tunnel Intersections," SC-RR-68-430, Sandia Laboratories, Albuquerque, New Mexico, September, 1968.

Tyler, L.D., and Ellis, M.A. (1970), "The TSHOK Code: Lax Version," SC-TM-70-153, Sandia Laboratories, Albuquerque, New Mexico, April, 1970.

Tyler, L.D. (1971), "Heuristic Analysis of Convective Finite Difference Techniques." See Holt (1971).

Underwood, R.L. (1969), "Calculation of Incompressible Flow Past a Circular Cylinder at Moderate Reynolds Numbers," J. Fluid Mechanics, Vol. 37, Part 1, pp. 95-114.

Ung, M.T., and Paul, J.F. (1968), "Solving a Partial Differential Equation by Serial Methods," LACC E. & T. Memo # 3, Electronic Associates, Inc., Los Angeles Computation Center, Education & Training Group, Los Angeles, California.

Vaglio Laurin, R., and Miller, G. (1971), "A Heuristic Approach to Three-Dimensional Boundary Layers." See Holt (1971).

van de Hulst, H.C. (1968), "Asymptotic Fitting, A Method for Solving Anisotropic Transfer Problems in Thick Layers," J. of Computational Physics, Vol. 3, pp. 291-306.

van de Vooren, A.I., and Dijkstra, D. (1970), "The Navier-Stokes Solution for Laminar Flow Past a Semi-Infinite Flat Plate," J. of Engineering Mathematics, Vol. 4, No. 1, pp. 9-27.

Van Dyke, M.D. (1958), "The Supersonic Blunt Body Problem -- Review and Extension," J. of Aeronautical Sciences, Vol. 25, p. 485.

Van Dyke, M. (1961), "Second Order Compressible Boundary Layer Theory with Application to Blunt Bodies in Hypersonic Flow," Hypersonic Flow Research, F.R. Ridell, ed., Academic Press, New York.

Van Dyke, M. (1962A), "Higher Approximations in Boundary-Layer Theory, Part I, General Analysis," J. Fluid Mechanics, Vol. 14, pp. 161-177.

Van Dyke, M. (1962B), "Higher Approximations in Boundary-Layer Theory, Part 2, Application to Leading Edges," J. Fluid Mechanics, Vol. 14, pp. 481-495.

Varapaev, V.N. (1968), "Numerical Study of Periodic Jet Flow of a Viscous Incompressible Fluid," Izv. AN SSSR. Mekhanika Zhidkosti i Gaza, Vol. 3, No. 3, pp. 170-176.

Varga, R.S. (1962), Matrix Iterative Analysis, Prentice-Hall, Englewood Cliffs, New Jersey.

Varshanskaia, T.S. (1969), "A Method of Formulating the Boundary Conditions in Problems of Viscous Fluid Flow," Akademiia Nauk SSSR, Izvestiia, Mekhanika Zhidkosti i Gaza, No. 3, pp. 142-144.

Varzhanskaya, T.S. (1970), "One Method for Establishing Boundary Conditions for Problems of Flow of a Viscous Fluid," NASA-TT-F-13027, Techtran Corp., Glen Burnie, Maryland.

Vasiliev, O.F. (1971), "Numerical Solution of the Non-Linear Problems of Unsteady Flows in Open Channels." See Holt (1971).

Vemuri, V. (1970), "An Initial Value Formulation of the CSDT Method of Solving Partial Differential Equations," AFIPS Conference Proceedings, Vol. 36, 1970, pp. 403-407.

Veronis, G. (1968), "Large-Amplitude Benard Convection in a Rotating Fluid," J. Fluid Mechanics, Vol. 31, Part 1, pp. 113-139.

Vichnevetsky, R. (1968), "Analog/Hybrid Solution of Partial Differential Equations in the Nuclear Industry," Simulation, pp. 269-281.

Victoria, K.J., and Steiger, M.H. (1970), "Exact Solutions of the 2-D Laminar Near Wake of a Slender Body in Supersonic Flow at High Reynolds Number," APP-0059 (S9990)-5, Aerospace Corporation, San Bernardino, California, October, 1970.

Viecelli, J.A. (1969), "A Method for Including Arbitrary External Boundaries in the MAC Incompressible Fluid Computing Technique," J. of Computational Physics, Vol. 4, pp. 543-551.

Vladimirova, N.H., Kuznetsov, B.G., and Yanenko, N.N. (1966), "Numerical Calculation of Symmetrical Viscous Incompressible Fluid Flow Past a Two-Dimensional Flat Plate," Certain Problems of Computational and Applied Mathematics, (Russian), Novosibirsk.

de Vogelaere, R. (1971), "The Reduction of the Stefan Problem to the Solution of an Ordinary Differential Equation." See Holt (1971).

Vogenitz, F.W., Bird, G.A., Broadwell, J.E., and Rungaldier, H. (1968), "Theoretical and Experimental Study of Low Density Supersonic Flows About Several Simple Shapes," Paper No. 68-6, AIAA 6th Aerospace Sciences Meeting, New York, New York, January 22-24, 1968.

Vogenitz, F.W., Broadwell, J.E., and Bird, G.A. (1970), "Leading Edge Flow by the Monte Carlo Direct Simulation Technique," AIAA J., Vol. 8, No. 3, pp. 504-510.

Vogenitz, F.W., and Takata, G.Y. (1970), "Monte Carlo Study of Blunt Body Hypersonic Viscous Shock Layers," Seventh International Rarefied Gas Dynamics Symposium.

Vogenitz, F.W., and Takata, G. (1971), "Rarefied Hypersonic Flow About Cones and Flat Plates by Monte Carlo Simulation," AIAA J., Vol. 9, No. 1, pp. 94-100.

Vonka, V. (1970), "CASES - An Algol Program for Solving Diffusion Type Elliptic Differential Equations in a Plane Involving Arbitrary Boundary Conditions and Shape," RCN 123, Reactor Centrum Nederland, Petten, August, 1970.

Von Neumann, J. (1944), "Proposal and Analysis of a Numerical Method for the Treatment of Hydrodynamical Shock Problems," Nat. Def. and Res. Com. Report AM-551, March, 1944.

Von Neumann, J., and Richtmyer, R.D. (1950), "A Method for the Numerical Calculation of Hydrodynamic Shocks," J. of Applied Physics, Vol. 21, pp. 232-257.

Vreugdenhil, C.B. (1969), "On the Effect of Artificial-Viscosity Methods in Calculating Shocks," J. of Engineering Mathematics, Vol. 3, No. 4, pp. 285-288.

Vulis, L.A., and Dzhaugashtin, Yu.E. (1968), "A Numerical Solution to the Problem of Laminar Flow in a Conduit," NASA TT F-12, 373, Zhurnal Prikladroy Mikhaniki i Tekhnichiskoy Fiziki, No. 6, pp. 120-123.

Wachpress, E.L. (1966), Iterative Solution of Elliptic Systems, Prentice-Hall, Englewood Cliffs, New Jersey.

Wachpress, E.L. (1968), "Solution of the ADI Minimax Problem," KAPL-3448, Knolls Atomic Power Laboratory, Schenectady, New York, June 7, 1968.

Wagstaff, R.A., and Lee, S.S. (1971), "Higher Order Effects in Laminar Boundary Layer Theory for Curved Surfaces." See Holt (1971).

Walden, D.C. (1967), "The Givens-Householder Method for Finding Eigenvalue and Eigenvectors of Real Symmetric Matrices," CFSTI, AD 66 1 277, MIT Lincoln Labs, TN 1967-51, Lexington, Massachusetts.

Walitt, L. (1969), "Numerical Studies of Supersonic Near-Wakes," Ph.D. Dissertation, Aeronautical Engineering Dept., University of California, Los Angeles, California.

Walker, W.F., and Zumwalt, G.W. (1966), "A Numerical Solution for the Interaction of a Moving Shock Wave with a Turbulent Mixing Region," SC-CR-67-2531, Sandia Laboratories, Albuquerque, New Mexico, May, 1966. See also ASME Paper No. 68-APM-M.

Waltman, P. (1969), "Existence and Uniqueness of Solutions of Boundary Value Problems for Second Order Systems of Nonlinear Differential Equations," Rept. No. 6, Math. Dept., Iowa University, Iowa City, Iowa.

Wang, C.-H. (1969), "A Method of Setting Up Finite-Difference Schemes for Prediction Equations," AFCRL-69-0099, Air Force Cambridge Research Laboratories, L.G. Hanscom Field, Bedford, Massachusetts, March, 1969.

Wang, Y.L., and Longwell, P.A. (1964), "Laminar Flow in the Inlet Section of Parallel Plates," American Institute of Chemical Engineers J., Vol. 10, No. 3, pp. 323-329.

Wantland, J.L. (1969), "A Numerical Evaluation of the Velocity and Temperature Structure in Laminar Cellular Convection Between Parallel Rigid Surfaces," Ph.D. Thesis, University of Tennessee, ORNL-4458, Oak Ridge National Laboratory, Oak Ridge, Tennessee.

Warlick, C.H., and Young, D.M. (1970), "A Priori Method for the Determination of the Optimum Relaxation Factor for the Successive Overrelaxation Method," Int. Tech. Report No. 25, The University of Texas at Austin, Computing Center, Austin, Texas, May, 1970.

Wasow, W. (1957), "Asymptotic Development of the Solution of Dirichlet's Problem at Analytic Corners," Duke Mathematics J., Vol. 24, pp. 47-56.

Watkins, C.B., Jr. (1970), "Numerical Solutions of Nonequilibrium Reflected Shock Waves in Air with Radiation," Ph.D. Thesis, University of New Mexico, Albuquerque, New Mexico, August, 1970.

Watson, J.D., and Godfrey, C.S. (1967), "An Investigation of Projectile Integrity Using Computer Techniques," Report No. PITR-67-10, Physics International Company, 2700 Merced Street, San Leandro, California, February, 1967.

Watson, J.D. (1969), "High-Velocity Explosively Driven Guns," Report No. PIFR-113, Physics International Company, 2700 Merced Street, San Leandro, California, July, 1969.

Webb, C. (1970), "Practical Use of the Fast Fourier Transform (FFT) Algorithm in Time-Series Analysis," ARL-TR-70-22, Applied Research Laboratories, The University of Texas at Austin, Austin, Texas, July, 1970.

Webb, H.G., Dresser, H.S., Adler, B.K., and Waiter, S.A. (1967), "Inverse Solution of Blunt-Body Flowfields at Large Angle of Attack," AIAA J., Vol. 5, June, 1967, pp. 1079-1085.

Weber, M.E. (1969), "Improving the Accuracy of Crank-Nicolson Numerical Solutions to the Heat-Conduction Equations," J. of Heat Transfer, Transactions of the ASME, Series C., Vol. 91, No. 1, February, 1969, pp. 189-191.

Weinbaum, S. (1966), "Rapid Expansion of a Supersonic Boundary Layer and its Application to the Near Wake," AIAA J., Vol. 4, No. 2, pp. 217-227.

Weinberger, H.F. (1965), Partial Differential Equations, Blaisdell Publishing Co., New York, New York.

Weinstein, H.G., Stone, H.L., and Kwan, T.V. (1969), "Iterative Procedure for Solution of Systems of Parabolic and Elliptic Equations in Three Dimensions," Industrial and Engineering Chemistry Fundamentals, Vol. 8, pp. 281-287.

Weiss, D. (1968), "On the Use of Similar Solutions for Testing Laminar Compressible Boundary Layer Methods," RM-433, Grumman Research Department, Grumman Aircraft Engineering Corporation, Bethpage, New York.

Weiss, R.F., Greenberg, R.A., and Biondo, P.P. (1966), "A New Theoretical Solution of the Laminar, Hypersonic Near Wake, Part I., Formation of the Method of Solution," BSD-TR-66-258, Avco Everett Research Laboratory, Research Rept. 256, Everett, Massachusetts, August, 1966.

Weiss, R.F., and Weinbaum, S. (1966), "Hypersonic Boundary-Layer Separation and the Base Flow Problem," AIAA J., Vol. 4, No. 8, August, 1966, pp. 1321-1330.

Welch, J.E., Harlow, F.H., Shannon, J.P., and Daly, B.J. (1966), "The MAC Method," LASL Report No. LA-3425, Los Alamos Scientific Laboratory, Los Alamos, New Mexico.

Wendroff, B. (1960), "On Centered Difference Equations for Hyperbolic Systems," J. SIAM, Vol. 8, pp. 549-555.

Wendroff, B. (1969), "A Difference Scheme for Radiative Transfer," J. of Computational Physics, Vol. 4, pp. 211-229.

Werle, M.J., and Wornom, S.F. (1970), "Longitudinal Curvature and Displacement Speed Effects on Incompressible Boundary Layers," VPI-E-70-24, Virginia Polytechnic Institute, Blacksburg, Virginia, December, 1970.

Werner, W. (1968), "Numerical Solution of Systems of Quasilinear Hyperbolic Differential Equations by Means of the Method of Nebencharacteristics in Combination with Extrapolation Methods," Numerische Mathematik, Vol. 11, pp. 151-169.

Westlake, J.R. (1968), A Handbook of Numerical Matrix Inversion and Solution of Linear Equations, J. Wiley and Sons, Inc., New York, New York.

Whitehead, R.E., and Davis, R.T. (1969), "Surface Conditions in Slip-Flow with Mass Transfer," VPI-E-69-11, College of Engineering, Virginia Polytechnic Institute, Blacksburg, Virginia.

Whiteman, J.R. (1967), "Singularities Due to Re-entrant Corners in Harmonic Boundary Value Problems," MRC Tech. Summary Rept. 829, Mathematics Research Center, United States Army, The University of Wisconsin, Madison, Wisconsin.

Whiting, K.B. (1968), "A Treatment for Singularities in Finite-Difference Solutions of LaPlace's Equation," Tech. Note WPD 77, Department of Supply, Australian Defense Scientific Service, Weapons Research Establishment, Salisbury, South Australia.

Whitney, A.K. (1971), "The Numerical Solution of Unsteady Free Surface Flows by Conformal Mapping." See Holt (1971).

Widlund, O.B. (1967), "On Difference Methods for Parabolic Equations and Alterating Direction Implicit Methods for Elliptic Equations," IBM J., March, 1967, pp. 239-243.

Wigley, N.M. (1969), "On a Method to Subtract off a Singularity at a Corner for the Dirichlet or Neuman Problem," Mathematics of Computation, Vol. 23, No. 106, pp. 395-401.

Wilkes, J.O., and Churchill, S.W. (1966), "The Finite-Difference Computation of Natural Convection in a Rectangular Enclosure," American Institute of Chemical Engineers J., Vol. 12, No. 7, pp. 161-166.

Wilkins, M.L. (1969), "Finite Difference Scheme for Calculating Problems in Two Space Dimensions and Time," UCRL-71724, Lawrence Radiation Laboratory, University of California, Livermore, California.

Wilkins, M.L. (1970), "Finite Difference Scheme for Calculating Problems in Two Space Dimensions and Times," J. of Computational Physics, Vol. 5, pp. 406-414.

Wilkins, M.L., French, S.J., and Sorem, M. (1971), "Finite Difference Schemes for Calculating Problems in Three Space Dimensions and Time." See Holt (1971).

Williams, G.P. (1969), "Numerical Integration of the Three-Dimensional Navier-Stokes Equations for Incompressible Flow," J. Fluid Mechanics, Vol. 37, Part 4, pp. 727-750.

Williamson, D. (1968), "Integration of the Barotropic Vorticity Equation on a Spherical Geodesic Grid," Tellus, Vol. 20, pp. 642-653.

Williamson, D. (1969), "Numerical Integration of Fluid Flow Over Triangular Grids," Monthly Weather Review, Vol. 97, No. 12, pp. 885-895.

Williamson, D.L. (1971), "A Comparison of First- and Second-Order Difference Approximations Over a Spherical Geodesic Grid," J. of Computational Physics, Vol. 7, pp. 301-309.

Wilson, L.N. (1967), "Inflections in Bow Shock Shape at Hypersonic Speeds," AIAA J., Vol. 5, No. 8, August, 1967, pp. 1532-1533.

Winograd, S. (1969), "Computational Complexity," IBM Watson Research Center, Yorktown Heights, New York, November, 1969.

Winslow, A.M. (1963), "'Equipotential' Zoning of Two-Dimensional Meshes," UCRL-7312, Lawrence Radiation Laboratory, University of California, Livermore, California, June 5, 1963.

Winslow, A.M. (1966), "Numerical Solution of the Quasilinear Poisson Equation in a Nonuniform Triangle Mesh," J. of Computational Physics, Vol. 1, No. 2, November, 1966, p. 149.

Wolfshtein, M. (1968), "Numerical Smearing in Onesided Difference Approximations to the Equations of Non-Viscous Flow," Imperial College of Science and Technology, London, England.

Wolfshtein, M. (1970), "Some Solutions of the Plane Turbulent Impinging Jet." Paper No. 70-FE-27, Transactions of the ASME.

Wood, W.L. (1971), "Note on Dynamic Relaxation," International J. for Numerical Methods in Engineering, Vol. 3, January-March, 1971, pp. 145-147.

Woods, L.C. (1953), "The Relaxation Treatment of Singular Points in Poisson's Equation," Quarterly J. Mechanics and Applied Mathematics, Vol. 6, pp. 163-183.

Woods, L.C. (1954), "A Note on the Numerical Solution of Fourth Order Differential Equations," Aeronautical Quarterly, Vol. 5, Part 3, p. 176.

Woods, W.A., and Daneshyar, H. (1970), "Boundary Conditions and Initial Value Lines for Unsteady Homentropic Flow Calculations," Aeronautical Quarterly, Vol. 21, pp. 145-162.

Xerikos, J. (1968), "A Time-Dependent Approach to the Numerical Solution of the Flow Field About An Axisymmetric Vehicle at Angle of Attack," Douglas Aircraft Company, Santa Monica, California, June, 1968. See also Xerikos, J., and Anderson, W.A., NASA-CR-61982, 1968.

Yanenko, N.N., and Shokin, Y.I. (1971), "On the Group Classification of Difference Schemes for Systems of Equations in Gas Dynamics." See Holt (1971).

Yanenko, N.N. (1971), The Method of Fractional Steps: The Solution of Problems of Mathematical Physics in Several Variables, English translation edited by M. Holt, Springer-Verlag, New York.

Yanowitch, M. (1969), "A Numerical Study of Vertically Propagating Waves in a Viscous Isothermal Atmosphere," J. of Computational Physics, Vol. 4, pp. 531-542.

Yee, S.Y.K. (1969), "Noniterative Solution of a Boundary Value Problem of the Helmholtz Type," AFCRL-69-0478, Air Force Cambridge Research Laboratories, L.G. Hanscom Field, Bedford, Massachusetts, November, 1969.

Yen, S.M. (1969), "Numerical Methods for Solving Rarefied Gas Flow Problems," Applied Mechanics Reviews, Vol. 22, No. 6, pp. 557-564.

Yoshizawa, A. (1970), "Laminar Viscous Flow Past a Semi-Infinite Flat Plate, J. of the Physical Society of Japan, Vol. 28, No. 3, pp. 776-779.

Young, D. (1954), "Iterative Methods for Solving Partial Difference Equations of Elliptic Type," Transactions of the American Mathematical Society, Vol. 76, pp. 92-111.

Young, D.M., and Kincaid, D.R. (1969), "Norms of the Successive Overrelaxation Method and Related Methods," TNN-94, The University of Texas at Austin Computation Center, Austin, Texas, September, 1969.

Young, D.M., and Eidson, H.M. (1970), "On the Determination of the Optimum Relaxation Factor for the SOR Method when the Eigenvalues of the Jacobi Method are Complex," Int. Techn. Rept. No. 26, University of Texas at Austin, Center for Numerical Analysis, Austin, Texas, September, 1970.

Zandbergen, P.J. (1971), "The Viscous Flow Around a Circular Cylinder." See Holt (1971).

Ziemniak, S.E. (1970), "The Calculation of Steady Turbulent Flows in Two Dimensions," Proceedings of the 1970 Heat Transfer and Fluid Mechanics Institute, Stanford University Press, Stanford, California, pp. 121-134.

Zienkiewicz, O.C. (1967), The Finite Element Method in Structural and Continuum Mechanics, McGraw-Hill, Inc., London, England.

Zienkiewicz, O.C. (1969), "The Finite Element Method: From Intuition to Generality," Applied Mechanics Reviews, pp. 249-256.

Zitko, J. (1970), "Generalization of the Minimax Method for Calculation of the Spectral Radius of a Matrix," Aplikace Matematiky, Vol. 15, No. 1, pp. 41-62.

Zuev, A.I. (1966), "Three-Layer Scheme for Numerical Integration of the Equations of Gas Dynamics and the Nonlinear Equation of Thermal Conductivity," Chislen. Metody Resheniya Zadach Matem. Fiz., Moscow, Nauka, pp. 230-236.

Zumwalt, G.W. (1967), "Numerical Methods for Computing the Diffusion of Disturbances on Sonic Boom Waves," Final Rept. on Contract NASA NGR-37-002-037, School of Mechanical Engineering, Oklahoma State University, Stillwater, Oklahoma.

Zwas, G., and Abarbanel, S. (1970), "Third and Fourth Order Accurate Schemes for Hyperbolic Equations of Conservation Law Form," AFOSR-TR-71-0038, Tel-Aviv University, Israel.

SUGGESTIONS FOR A COURSE USING THIS TEXT

The following are some observations of mine, based on teaching two courses in computational fluid dynamics, one as an in-hours continuing education course in industry, and one graduate course at a university.

1. The subject material is not of itself too difficult, but there is much unclear thinking and plenty of surprises because it is new.

2. I believe that it can be taught to selected seniors, and certainly to first-year graduate students.

3. The students (and the teacher, in my case) can learn a good deal about fluid dynamics from the course. It therefore need not be treated in isolation in curriculum planning, but possibly could be used to introduce (or at least , to reinforce) the concepts of vorticity transport and production, normalizing systems, control volumes, advection (or convection) and diffusion processes, sufficiency of boundary conditions, dissipation, stiff equations, ellipticity of incompressible flows, shock-wave phenomena, Mach lines, domain of influence of hyperbolic equations, mathematical aspects of the boundary-layer approximations and the Euler equations, existence and uniqueness of solutions, and singularities.

4. If the mesh size is kept coarse, qualitatively realistic solutions to two-dimensional incompressible flow problems may be obtained very inexpensively. For example, one of my students obtained an unequivocally converged solution to a driven cavity problem in a 5x5 mesh in about 20 seconds on an IBM 360-65. Similarly, transient one-dimensional shock propagation problems can be computed quite economically.

5. I have found it important to get the students onto the computer as early as possible. Consequently, in the classroom I do not rigidly follow the subject sequence of the text. The text covers all the methods for the vorticity transport equation, then all the methods for the elliptic stream-function equation, then all the methods for boundary conditions, and then all the initial conditions and convergence criteria; aspects of information processing are covered in the last chapter. But in the course, I assign the driven cavity problem as soon as we have covered a few basic methods and have done a little one-dimensional experimenting with the model advection-diffusion equation; I then give a single lecture on the most elementary methods for the elliptic stream function equation and boundary conditions on no-slip walls, and they plunge into the two-dimensional problem. The students work on this problem for several weeks, while I again follow the text.

6. Enthusiasm is high for the course, especially when they are computing.

7. For a two-semester-sequence, I would finish the first semester in Chapter III, probably after the temperature solutions. The second semester could start with the incompressible solutions in primitive variables.

8. There are many problems which can be assigned as class projects. Some of these are unworked research projects, making them excellent motivators. For two examples of publications that have resulted from class projects on the driven cavity problem, see Torrance et al., "Cavity Flows Driven by Bouyancy and Shear", Journal of Fluid Mechanics, Vol. 51, Part 2, 1972, pp.221-231, and "Numerical Studies of Incompressible Viscous Flow in a Driven Cavity", NASA SP-378, 1975.

9. The text may be supplementd with the method of characteristics and boundary-layer methods, if desired.

P.J.R.

SUBJECT INDEX